Titles in This Series

19 **Ingemar J. Cox, Pierre Hansen, and Bela Julesz, Editors,** Partitioning Data Sets

18 **Guy E. Blelloch, K. Mani Chandy, and Suresh Jagannathan, Editors,** Specification of Parallel Algorithms

17 **Eric Sven Ristad, Editor,** Language Computations

16 **Panos M. Pardalos and Henry Wolkowicz, Editors,** Quadratic Assignment and Related Problems

15 **Nathaniel Dean and Gregory E. Shannon, Editors,** Computational Support for Discrete Mathematics

14 **Robert Calderbank, G. David Forney, Jr., and Nader Moayeri, Editors,** Coding and Quantization: DIMACS/IEEE Workshop

13 **Jin-Yi Cai, Editor,** Advances in Computational Complexity Theory

12 **David S. Johnson and Catherine C. McGeoch, Editors,** Network Flows and Matching: First DIMACS Implementation Challenge

11 **Larry Finkelstein and William M. Kantor, Editors,** Groups and Computation

10 **Joel Friedman, Editor,** Expanding Graphs

9 **William T. Trotter, Editor,** Planar Graphs

8 **Simon Gindikin, Editor,** Mathematical Methods of Analysis of Biopolymer Sequences

7 **Lyle A. McGeoch and Daniel D. Sleator, Editors,** On-Line Algorithms

6 **Jacob E. Goodman, Richard Pollack, and William Steiger, Editors,** Discrete and Computational Geometry: Papers from the DIMACS Special Year

5 **Frank Hwang, Clyde Monma, and Fred Roberts, Editors,** Reliability of Computer and Communication Networks

4 **Peter Gritzmann and Bernd Sturmfels, Editors,** Applied Geometry and Discrete Mathematics, The Victor Klee Festschrift

3 **E. M. Clarke and R. P. Kurshan, Editors,** Computer-Aided Verification '90

2 **Joan Feigenbaum and Michael Merritt, Editors,** Distributed Computing and Cryptography

1 **William Cook and Paul D. Seymour, Editors,** Polyhedral Combinatorics

Partitioning Data Sets

DIMACS

Series in Discrete Mathematics
and Theoretical Computer Science

Volume 19

Partitioning Data Sets

DIMACS Workshop
April 19–21, 1993

Ingemar J. Cox
Pierre Hansen
Bela Julesz
Editors

NSF Science and Technology Center
in Discrete Mathematics and Theoretical Computer Science
A consortium of Rutgers University, Princeton University,
AT&T Bell Labs, Bellcore

American Mathematical Society

This DIMACS volume on Partitioning Data Sets: With Applications to Psychology, Vision and Target Tracking contains papers from talks at a workshop held at DIMACS, April 19–21, 1993.

1991 *Mathematics Subject Classification.* Primary 62H30, 68T10, 68U10, 92J30, 68Txx.

Library of Congress Cataloging-in-Publication Data

Partitioning data sets: DIMACS workshop, April 19–21, 1993 / Ingemar J. Cox, Pierre Hansen, Bela Julesz, editors.

p. cm. — (DIMACS series in discrete mathematics and theoretical computer science, ISSN 1052-1798; v. 19)

"NSF Science and Technology Center in Discrete Mathematics and Theoretical Computer Science, a consortium of Rutgers University, Princeton University, AT&T Bell Labs, Bellcore."

Includes bibliographical references.

ISBN 0-8218-6606-0

1. Cluster analysis. 2. Partitions (Mathematics) 3. Computer vision. I. Cox, I. J. (Ingemar J.) II. Hansen, P. (Pierre) III. Julesz, Bela. IV. NSF Science and Technology Center in Discrete Mathematics and Theoretical Computer Science. V. DIMACS (Group) VI. IEEE Information Theory Society. VII. Series.

QA278.P37 1995

621.36′7′01519535—dc20 94-25120

CIP

∞ The paper used in this book is acid-free and falls within the guidelines established to ensure permanence and durability.
♻ Printed on recycled paper.

This volume was prepared by the authors using AMS-TeX and AMS-LaTeX, the American Mathematical Society's TeX macro systems.

10 9 8 7 6 5 4 3 2 1   00 99 98 97 96 95

Contents

Foreword ix

Preface xi

Part 1. Cluster Analysis Methods

The Median Procedure for Partitions
JEAN-PIERRE BARTHÉLEMY AND BRUNO LECLERC 3

Structural Properties of Pyramidal Clustering
P. BERTRAND 35

Partitioning by Maximum Adjacency Search of Graphs
WEIQING CAI AND DAVID W. MATULA 55

From Data to Knowledge: Probabilist Objects for a Symbolic Data Analysis
EDWIN DIDAY 65

A Labeling Algorithm for Minimum Sum of Diameters Partitioning of Graphs
S. GÉLINAS, P. HANSEN, AND B. JAUMARD 89

Agreement Subtrees, Metric and Consensus for Labeled Binary Trees
WAYNE GODDARD, EWA KUBICKA, GRZEGORZ KUBICKI, AND F. R. MCMORRIS 97

How to Choose K Entities Among N
P. HANSEN, B. JAUMARD, AND N. MLADENOVIC 105

On the Classification of Monotone-Equivariant Cluster Methods
MELVIN F. JANOWITZ AND RUDOLF WILLE 117

Contiguity-Constrained Hierarchical Clustering
FIONN D. MURTAGH 143

Part 2. Target Tracking

Image Segmentation Based on Optimal Layering for Precision Tracking
ANIL KUMAR, YAAKOV BAR-SHALOM, AND ELIEZER ORON 155

Multidimensional Assignments and Multitarget Tracking
AUBREY B. POORE 169

Part 3. Computer Vision

Grouping Edges: An Efficient Bayesian Multiple Hypothesis Approach
INGEMAR J. COX, JAMES H. REHG, SUNITA L. HINGORANI, AND MATT L. MILLER 199

Finding Salient Convex Groups
DAVID W. JACOBS 237

Mixture Models for Optical Flow Computation
ALLAN JEPSON AND MICHAEL J. BLACK 271

Multilevel Detection of Stereo Disparity Surfaces
YIBING YANG AND ALAN L. YUILLE 287

Part 4. Human Vision

Some Problems of Visual Shape Recognition to which the Application of Clustering Mathematics Might Yield Some Potential Benefits
IRVING BIEDERMAN 313

Perceptual Models of Small Dot Clusters
JACOB FELDMAN 331

Subjective Contours in Early Vision and Beyond
BELA JULESZ 359

The Visual Perception of Surfaces, their Properties, and Relationships
DANIEL KERSTEN AND SUTHEP MADARASMI 373

Visual Computations and Dot Cluster
STEVEN W. ZUCKER 391

Foreword

This DIMACS volume on Partitioning Data Sets: With Applications to Psychology, Vision and Target Tracking contains refereed versions of papers from a workshop held at DIMACS at Rutgers University in April of 1993. The workshop was sponsored jointly by a grant from the Office of Naval Research (ONR), NEC of Princeton-New Jersey, and DIMACS, through grants from the National Science Foundation and the New Jersey Commission on Science and Technology.

We would especially like to thank the workshop organizers, Ingemar J. Cox, Pierre Hansen, and Bela Julesz, for organizing this workshop which brought together so many outstanding speakers in this field.

Andras Hajnal
Director, DIMACS

Andrew Yao
Co-Director, DIMACS

Stephen R. Mahaney
Associate Director, DIMACS

Preface

Partitioning data sets into disjoint groups is a problem arising in many domains. The mathematical theory of Cluster Analysis aims at finding such groups which are both homogeneous, i.e., such that entities in the same group are similar, and well separated, i.e., such that entities in different groups are dissimilar. Work on the axiomatic foundations and the computational complexity of such problems as well as on the design and analysis of exact or heuristic algorithms to solve them is rapidly expanding. Homogeneity and separation can be made precise in a variety of ways, leading to a variety of clustering problems. Applications in psychology, computer vision, target tracking and other domains are numerous. In each of these domains, specific partitioning problems arise as well as efficient solution methods that exploit particular characteristics of the domain. This book presents twenty essays and surveys on cluster analysis theory, methods and applications and related problems in psychology, computer vision and target tracking.

A first group of papers, devoted to cluster analysis methods and algorithms, covers the main paradigms of that field. Melvin F. Janowitz and Rudolf Wille consider hierarchical clustering and classify ordinal cluster methods according to the significance assigned to the input data. Patrice Bertrand studies structural properties of a recent extension of the hierarchical clustering scheme: pyramidal clustering. A detailed survey of the median procedure for partitions, both from the axiomatic and the algorithmic points of view, is provided by Jean-Pierre Barthelemy and Bruno Leclerc. Another consensus problem is addressed by Wayne Goddard, Ewa Kubicka, Grzegorz Kubicki and F. R. McMorris, who focus on agreement subtrees of labeled binary trees. Two papers are devoted to clustering on graphs: Weiqing Cai and David W. Matula give a linear heuristic for partitioning by maximum adjacency search of graphs; Sylvie Gélinas, Pierre Hansen and Brigitte Jaumard provide a low-order polynomial labelling algorithm for minimum sum-of-diameters bipartitioning, useful in divisive hierarchical clustering. Fionn Murtagh studies contiguity-constrained hierarchical clustering and explains when and why reversals occur, demonstrating the approach on quantized images from astronomy. Pierre Hansen, Brigitte Jaumard and Nenad Mladenovic outline a new sequential clustering paradigm, in which clearly apparent clusters are isolated and removed one at a time (a procedure reminiscent of sequential identification and removal of objects in image analysis). In a more speculative vein, Edwin Diday describes symbolic data analysis which applies to boolean or probabilistic objects well adapted to representing knowledge.

A second group of papers addresses problems of partitioning that arise in the context of multitarget tracking and surveillance. Aubrey B. Poore describes a fast, near optimal Lagrangian relaxation based algorithm to solve the NP-hard data association problem of partitioning real-time observations into tracks and false alarms. Anil Kumar, Yaakov Bar-Shalom and Eliezer Oron describe an algorithm for segmenting an image into target and background regions.

The third group of papers describes some partitioning problems in computer vision. Ingemar J. Cox, James H. Rehg, Sunita L. Hingorani and Matt L. Miller propose an algorithm for grouping edges and segmenting contours that is based on Reid's multiple hypothesis algorithm, originally developed in the context of multitarget tracking. David W. Jacobs presents a low-order polynomial algorithm that locates salient convex collections of lines in an image. Optical flow computation merges information over a local image patch to estimate the 2D image velocity at a point, assuming that only a single motion is present. Allan Jepson and Michael J. Black relax this single-motion assumption and apply mixture models and an extension of the EM algorithm to cluster the multiple motions present within a region. Yibing Yang and Allan L. Yuille present a multilevel algorithm for matching a pair of stereo images, in terms of surface detection in a three-dimensional spatio-disparity space.

Finally, a fourth group of papers focuses on human vision. Irving Biederman discusses problems of visual shape recognition for which clustering mathematics might yield some potential benefits. Daniel Kersten and Sutheg Madarasmi consider similarities and differences between human perception of surfaces, their properties and relationships and traditional studies of data clustering, focussing on three problems: occlusion, transparency and lightness. Two papers are devoted to problems of dot clusters: Jacob Feldman examines how human observers draw perceptually coherent clusters out of fields of dots, a process analogous to but different from statistical clustering. Special attention is given to molecular clusters and their classification. Steven W. Zucker proposes that visual perception of dots underlies clustering operations when more domain specific information is not available, and studies visual computations in this perspective. Bela Julesz reviews work on illusory contour perception, from early work on cyclopean contours using random-dot stereograms to recent results that show that illusory contours depend on global parameters such as closure of stacked lines as well as local interactions.

The multiplicity of approaches, methods, problems and algorithms made for very lively discussions at the DIMACS workshop. To a large extent, the language barrier, due to the multidisciplinary character of the workshop, was alleviated by the participants' goodwill and through many questions and answers. Any remaining difficulties were compensated for by new ideas and perspectives for theory and applications. As organizers, we thank the DIMACS center, NEC Research and the Office of Naval Research for financial support, as well as the AMS for its willingness to publish this volume.

Ingemar J. Cox
Pierre Hansen
Bela Julesz

PART I

Cluster Analysis Methods

DIMACS Series in Discrete Mathematics
and Theoretical Computer Science
Volume 19, 1995

The Median Procedure for Partitions

Jean-Pierre Barthélemy and Bruno Leclerc

ABSTRACT. Let X be an n-set; we are concerned with the problem of finding a consensus partition π on X that summarizes a k-tuple (*profile*) of partitions (several situations where such a problem may occur are briefly recalled). A classical approach is to consider a distance function d on the set Π of all the partitions of X and to search to minimize the *remoteness* $\sum_{1\leq i\leq k} d(\pi,\pi_i)$. We review some classical algorithmic features of this *median procedure*, study some of its properties, and compare it with some other consensus approaches. Especially, we address the consequences for the median problem of the existence of a well-known algebraic structure on the set Π. Besides the classical symmetric difference metric, various other distances are used; restricted sets of partitions related with contiguity constraints are also considered.

RÉSUMÉ. On considère un ensemble X à n éléments, et le problème de trouver une partition consensus π résumant un k-uplet (*profil*) de partitions de X (plusieurs situations où se rencontre ce problème sont brièvement évoquées). Une approche classique est de munir l'ensemble Π de toutes les partitions de X d'une métrique d, puis de chercher à minimiser l'*éloignement* $\sum_{1\leq i\leq k} d(\pi,\pi_i)$. Nous faisons le point sur les aspects algorithmiques classiques de cette *procédure médiane*, nous en étudions certaines propriétés, et nous la confrontons à d'autres approches du problème. Nous abordons particulièrement les apports de la prise en considération de la structure de treillis de l'ensemble Π. Des métriques de types variés sont envisagées, à côté de la classique distance de la différence symétrique ; on considère aussi le problème pour des ensembles restreints de partitions avec contraintes de contiguité.

1. Introduction

Let $X = \{x_1, ..., x_n\}$ be a finite set of *n objects*. A *partition* of X is a set π of non intersecting subsets of X, called the *classes* of π, that covers X. Partitions are basic in cluster analysis. They correspond, in particular, to descriptions of X with categorical variables V: two objects x and y are in the same class of the partition $\pi(V)$ associated with V if and only if they take the same value for the variable V. In this case x and y are said to be *linked* in $\pi(V)$. In fields like image processing,

1991 *Mathematics Subject Classification.* Primary 62H30; Secondary 06Bxx.
This paper is in final form and no version of it will be submitted for publication elsewhere.

computer vision, geography, robotics,..., partitions correspond to cuts of a real space (discretized as the finite set X) into regions. In the case of objects described by features, a partition π will link pairs of objects that tend to share enough common features.

A *consensus partition* of a family, or *profile* $\pi^* = (\pi_1,\ldots,\pi_k)$ of k partitions is a partition of X that summarizes π^* in some useful sense. The so-called *consensus problem* consists of making clear what is meant by "useful sense" and determining explicitly the related consensus partitions. It arises when considering:

- k methods (for instance, algorithms) providing classifications (partitions) π_i, $1 \leq i \leq k$, that are supposed to be "approximations" of a true classification that we search to recover.
- k categorical variables $V_1,\ldots,V_k$ describe the population X. Following the statistical idea of a "central value", we search for a partition π that is close to the profile $\pi^* = (\pi(V_1),\ldots,\pi(V_k))$.
- The case where the profiles π^* result from measurements at times $t, t+1,\ldots, t+k-1$. We may then consider the notion of a *moving consensus* smoothing the series $\pi_1,\ldots, \pi_k, \pi_{k+1},\ldots, \pi_{k+t},\ldots$ of partitions (or categorical variables).

Three overlapping approaches have been used to tackle the consensus problem:

1°) The *axiomatic approach* (Mirkin [33], Leclerc [23], Neumann and Norton [39], Monjardet [37],...) is related to the Arrowian approach in the framework of social choice theory (Arrow [4]). In such an approach one retains only, as a consensus partition, a partition satisfying some conditions that arise from experimental evidence or from ethical considerations. This leads to the problem of the existence and uniqueness of partitions satisfying these conditions and leads to possibility/impossibility theorems.

2°) The *constructive approach*, where a way to construct a consensus is explicitely given. The most obvious constructive approach is surely the unanimity (or Pareto) rule, called also the *strict consensus rule*. This rule states that two objects are linked in a consensus partition if and only if they are linked in all the partitions of the profile. Such a consensus is generally not adequate, since a prohibitively large number of (very small) classes is obtained in most cases. Thus more sophisticated constructive rules are needed.

3°) The *combinatorial optimization approach*. Here we have at our disposal some criterion measuring the *remoteness* $R(\pi,\pi^*)$ of any partition π to the given profile π^*; we search for a partition solution of:

$$\text{Min } R(\pi,\pi^*)$$

in the set Π of all the partitions of X.

This approach goes back to Régnier [40] and is related to the notion of central value in statistics (the term *partitions centrales* was used by Régnier). Indeed, we can consider a distance function d on Π and define the remoteness of π to $\pi^* = (\pi_1,\ldots,\pi_k)$ as the sum of the distances $d(\pi,\pi_i)$ (median case), or as the sum of the squares of these distances (center case), and so on. As far as we know, this approach was first proposed by Kemeny [21] in the domain of social choice (see Arabie and Boorman [3] and Day [15] for the existence of various actually computable metrics on partitions, Barthélemy and Monjardet [11, 12] for a survey on the median procedure).

Throughout the paper, the following definitions and notations will be used: the set of all the partitions of X is denoted by Π and we shall usually say "partition" instead of "partition of X". Partitions are in a one-to-one correspondence with the equivalence relations on X: the equivalence relation $E[\pi]$ associated with the partition π is defined by $xE[\pi]y$ iff x and y are in the same class of π and the partition $\pi[E]$ associated with the equivalence relation E admits as classes the equivalence classes of E. A partition into two classes is called a *bipartition*. Bipartitions are associated with binary categorical variables.

The notation of a subset Y of X is abbreviated as xyz.... A partition π the classes of which are $A, B,..., N$, is denoted by $/A/B/.../N/$, and $|\pi|$ is the number of classes of π ($1 \leq |\pi| \leq n$). The partition $/X/$ is also denoted by ω; the partition π, with $|\pi| = n$-1, where only two distinct elements x, y of X are linked is denoted π_{xy}; the partition with n classes, each with a unique element, is denoted by α. Let π be a partition into at least two classes, and set A and B be two classes of π. The partition with the same classes as π, except that the two classes A, B have been replaced by the single class $A \cup B$ is denote $\pi(A+B)$; thus $\pi(A+B) = (\pi-\{A,B\})\cup\{A\cup B\}$.

The set Π is naturally ordered by the refinement order : given $\pi, \pi' \in \Pi$, $\pi \leq \pi'$ means that each class of π is included in a class of π' (or equivalently that the classes of π' are obtained as the union of classes of π); it is then said that π is *finer* than π', or that π' is *coarser* than π. In other terms, the equivalence relation $E[\pi]$ is included in the equivalence relation $E[\pi']$. Especially, π' *covers* π if and only if $\pi' = \pi(A+B)$, for some classes A and B of π. We write then $\pi \prec \pi'$; in such a case we have $\pi < \pi'$ and $\pi \leq \pi'' < \pi'$ implies $\pi = \pi''$.

A lattice structure is associated with the refinement order, with the meet and the join binary operations defined as follows: the *meet* $\pi\wedge\pi'$ of two partitions π and π' is their greatest lower bound, that is the coarsest of all the partitions finer than both π and π'; their *join* $\pi\vee\pi'$ is their least upper bound, that is the finest of all the partitions coarser than both π and π'. Thus, a subset C of X is a class of $\pi\wedge\pi'$ if and only if $C = A\cap B$ for some classes A of π and B of π'. In other terms, two distinct elements x, y of X are linked in $\pi\wedge\pi'$ if and only if they are linked in both π and π' and the meet of π and π' corresponds to the intersection of the equivalence relations $E[\pi]$ and $E[\pi']$. Two objects x and y are linked in $\pi\vee\pi'$ if and only if there is a sequence $x = x_1$, $x_2, x_3,..., x_p = y$ such that x_i and x_{i+1} are linked in π or in π', for $i = 1,..., p-1$. The join of π and π' corresponds to the transitive closure of the union of $E[\pi]$ and $E[\pi']$. The lattice Π is isomorphic to the lattice of all the equivalence relations of X, ordered by inclusion.

We call a *profile* any k-tuple $\pi^* = (\pi_1,\ldots,\pi_k)$ of elements of Π, with k a positive integer, called the *length* of π^*. The partitions π_i are called the *components* of π^*. We denote by Π^* the set $\cup_k \Pi^k$ of all profiles. The *concatenation* $\pi\pi'$ of two profiles π and π' is defined as usual.

From the three approaches we considered for consensus partitions (axioms, construction and optimization) arise three ways to tackle the consensus problem for partitions. We shall refer to them as k-procedures, complete procedures, or complete multiprocedures:

A *k-procedure* is a map $f: \Pi^k \to \Pi$.
A *complete procedure* is a map $\Pi^* \to \Pi$.
A *complete multiprocedure* is a map $\Pi^* \to 2^{\Pi}\text{-}\{\emptyset\}$.

Axiomatic characterizations and constructions correspond in general to procedures (complete or not), while consensus partitions obtained from optimization correspond to complete multiprocedures.

The paper is organized as follows: in the next section, we illustrate the three approaches and the three kinds of procedures with three examples. One of them introduces the median procedure which is the focus of this paper. In Section 3, we investigate this median procedure for a class of distance functions (the weight metrics); we briefly describe various computational approaches and state a possibility of dualizing the weight metrics. Sections 4, 5 and 6 especially address the consequences for the median problem of the lattice structure of Π. In Section 4, results concerning the consensus problem in abstract lattices are reviewed. They are applied to the partition lattices in Section 5. Other partitions lattices accounting for contiguity constraints are introduced in Section 6 and a theory of median consensus under contiguity constraints is sketched.

2. Examples of Consensus Partitions

2.1. Examples of procedures : axiomatic approach and oligarchies. We say that a k-procedure f is *Paretian* if and only if for each profile $\pi^* = (\pi_1,\ldots,\pi_k)$, we have:

$$\wedge_{1\leq i\leq k}\, \pi_i \leq f(\pi^*).$$

We say that f is *independent* when, for each π^*, π'^*, and $x, y \in X$:

$$\boldsymbol{k}_{xy}(\pi^*) = \boldsymbol{k}_{xy}(\pi'^*) \text{ implies: } xE[f(\pi^*)]y \text{ iff } xE[f(\pi'^*)]y,$$

where $\boldsymbol{k} = \{1,\ldots,k\}$ and $\boldsymbol{k}_{xy}(\pi^*) = \{i \in \boldsymbol{k}: xE[\pi_i]y\}$.

"Paretian" means that if x and y are linked in all the partitions of π^*, then they are linked in the consensus partition of π^*. "Independent" means that the fact that x and y are (resp. are not) linked in the consensus partition depends only upon the fact that they are (resp. are not) linked in the various partitions of the profile, independently of what happens for the other elements of X.

We say that a procedure $f\colon \Pi^k \to \Pi$ is *oligarchic* whenever there exists a subset I of k such that for each profile π^*, $f(\pi^*) = \wedge_{i\in I}\pi_i$. Oligarchic procedures may be, on some way, compared with practices in factor analysis. In this last case, objects described by numerical variables are geometrically structured by a limited number of factors, which are linear combinations of the original variables. Here objects described by categorical variables are clustered by a new categorical variable, which is a conjunction of some of the original variables. Notice that the *unanimity rule*, $u(\pi^*) = \wedge_{1\le i\le k}\pi_i$, is an oligarchic procedure and that an oligarchic procedure f reduces to the unanimity rule if and only if it is *symmetrical*, *i.e.* $f(\pi_1,\dots,\pi_k) = f(\pi_{\sigma(1)},\dots,\pi_{\sigma(k)})$, for each permutation σ of $\{1,\dots,k\}$.

The following result is stated under Leclerc [23] formulation:

THEOREM 1 (Mirkin [33]). *A k-procedure $\Pi^k \to \Pi$ is independent and Paretian if and only if it is oligarchic.*

Thus, we can state as an immediate consequence that: *a k-procedure reduces to the unanimity rule if and only if it is independent, Paretian and symmetrical.*

2.2. Examples of complete procedures : constructive approaches and quota rules. In the introduction, we have pointed out that the unanimity rule is generally not adequate ($u(\pi^*)$ is generally equal, or very close, to α). The quota rules may be viewed as a way to remedy to this drawback. The idea is that two objects are linked in a consensus partition as soon as they are linked in sufficiently many partitions from the profile.

Let $\pi^* = (\pi_1,\dots,\pi_k)$ be a profile with k components, and x, y a pair of objects. Denote by $\sigma(xy,\pi^*) = |k_{xy}(\pi^*)|$ the number of components of the profile π^* in which x and y are linked; the *index* $\gamma(xy,\pi^*)$ is defined as the ratio $\gamma(xy,\pi^*) = \frac{1}{k}\sigma(xy,\pi^*)$.

Let q, $0 \le q \le 1$, a real number. Consider the two symmetric relations $C_q[\pi^*]$ and $B_q[\pi^*]$ defined by :

$$xC_q[\pi^*]y \text{ if and only if } \gamma(xy,\pi^*) > q;$$

$$xB_q[\pi^*]y \text{ if and only if } \gamma(xy,\pi^*) \ge q.$$

These relations are not transitive in general (except for $B_1[\pi^*]$ which is the equivalence relation associated to the unanimity partition), and we have $C_q[\pi^*] \subseteq B_q[\pi^*]$. Notice also that if π^* is reduced to a single partition π, then $C_0[\pi^*] = E[\pi] = B_1[\pi^*]$. We denote by $c_q(\pi^*)$ and $b_q(\pi^*)$ the partitions associated with the transitive closure of respectively $C_q[\pi^*]$ and $B_q[\pi^*]$. Thus the classes of $c_q(\pi^*)$ and $b_q(\pi^*)$ are nothing else that the connected classes of the graphs associated with $C_q[\pi^*]$ and $B_q[\pi^*]$ respectively (we shall use, from now, the same notations: $C_q[\pi^*]$, $B_q[\pi^*]$ to represent the graph and the symmetrical relation). We have obtained two complete procedures c_q and b_q, called respectively the *q-quota rule* and the *q-weak quota rule*; obviously, for any profile π^*, $c_q(\pi^*) \le b_q(\pi^*)$.

The partitions $c_q(\pi^*)$ and $b_q(\pi^*)$ may be expressed using the join operation in the lattice Π.

$$c_q(\pi^*) = \vee\{\pi_{xy}: \gamma(xy,\pi^*) > q\},$$
$$b_q(\pi^*) = \vee\{\pi_{xy}: \gamma(xy,\pi^*) \geq q\}.$$

Other expressions and dualizations involving the lattice structure of Π will be discussed in Section 4. Notice that the mechanism of transitive closure has the effect to link some pairs x, y of objects such that $\gamma(xy,\pi^*)<q$. Examples of quota rules are :
- the *unanimity rule* $u(\pi^*) = b_1(\pi^*) = \vee\{\pi_{xy}: \gamma(xy,\pi^*) = 1\} = \wedge_{1\leq i\leq k} \pi_i$;
- the *dual unanimity rule* $u'(\pi^*) = c_0(\pi^*) = \vee\{\pi_{xy}: \gamma(xy,\pi^*) > 0\} = \vee_{1\leq i\leq k} \pi_i$;
- the *majority rule* $c = c_{1/2}$ and the *weak majority rule* $b = b_{1/2}$.

EXAMPLE 1. With $n \geq 3$, let π^* be the profile of all the bipartitions of X. Thus, $k = 2^{n-1}-1$ is an odd number, and $c(\pi^*) = b(\pi^*)$. It is not difficult to see that $c_q(\pi^*) = \vee\{\pi_{xy}\} = \omega$, for $q \leq \frac{1}{2} - \frac{2n-5}{2^{n-1}-1}$ and $c_q(\pi^*) = c(\pi^*) = \alpha$, otherwise.

EXAMPLE 2. Let π^* be the following profile of six partititions, with $X = \{a,b,c,d,e,f,g\}$:

$$\pi_1 = \pi_2 = /adg/b/cef/,\ \pi_3 = /abcd/eg/f/,\ \pi_4 = /abf/c/dg/e/,$$
$$\pi_5 = /abc/defg/,\ \pi_6 = /abd/ce/fg/.$$

The values of the index $\gamma(xy,\pi^*)$ are given in Table 1.

Table 1

	b	c	d	e	f	g
a	2/3	1/3	2/3	0	1/6	1/3
b		1/3	1/3	0	1/6	0
c			1/6	1/2	1/3	0
d				1/6	1/6	2/3
e					1/3	1/3
f						1/2

From Table 1, we remark that $b_1(\pi^*) = c_{5/6}(\pi^*) = b_{5/6}(\pi^*) = c_{2/3}(\pi^*)$; $b_{2/3}(\pi^*) = c_{1/2}(\pi^*)$; $b_{1/2}(\pi^*) = c_{1/3}(\pi^*)$; $b_{1/3}(\pi^*) = c_{1/6}(\pi^*)$ and $b_{1/6}(\pi^*) = c_0(\pi^*)$. The graphs $C_q[\pi^*]$, for q = 2/3, 1/2, 1/3, 1/6 and 0 are given in Figure 1. Finally, we have: $b_1(\pi^*) = c_{5/6}(\pi^*) = b_{5/6}(\pi^*) = c_{2/3}(\pi^*) = \alpha$; $b_{2/3}(\pi^*) = c_{1/2}(\pi^*)$ =/abdg/c/e/f/; $b_{1/2}(\pi^*) = c_{1/3}(\pi^*)$ = /abdgf/ce/; $b_{1/3}(\pi^*) = c_{1/6}(\pi^*) = b_{1/6}(\pi^*) = c_0(\pi^*) = \omega$.

The quota rules are Paretian and symmetrical. Theorem 1 admits Proposition 2 as a straighforward consequence.

PROPOSITION 2. *A quota rule is independent if and only if it is the unanimity rule* b_1.

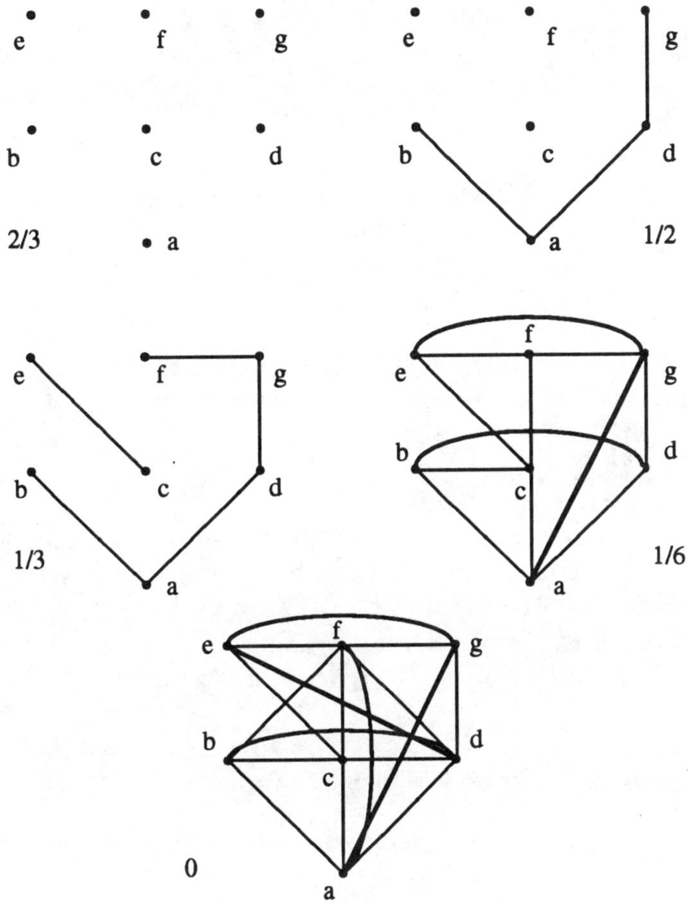

Figure 1.

2.3. Examples of complete multiprocedures: optimization and medians. Let $\pi^* = (\pi_1, \ldots, \pi_k)$ be the profile associated with k categorical variables $V_1, \ldots, V_k$ on X. Régnier [40], then Mirkin [32], proposed, as a clustering induced by the V_i, a partition, called the *median* of π^*, which minimizes the quantity:

$$\Sigma_{1 \leq i \leq k}\, \delta(\pi, \pi_i),$$

with δ the *symmetric difference distance* on Π. This distance is defined by (each unordered pair being counted twice in $E[\pi]$):

$$2\delta(\pi,\pi') = |E[\pi]\Delta E[\pi']| = |E[\pi]\cup E[\pi']| - |E[\pi]\cap E[\pi']| = |E[\pi]-E[\pi']| + |E[\pi']-E[\pi]|.$$

Thus, $\delta(\pi,\pi')$ is the number of disagreements between π and π'. It can be expressed directly from the classes of π and π' by the formula:

$$\delta(\pi,\pi') = \nu(\pi) + \nu(\pi') - 2\nu(\pi\wedge\pi'),$$

with $2v(\pi) = \sum_{A \in \pi} |A|^2$ (this formula corresponds to the fact that δ is a lattice metric, *cf.* Sections 4 and 5). If we associate, with each profile π^*, the set $\mathcal{M}(\pi^*)$ of all its medians, we get a complete multiprocedure $\mathcal{M}$ called the *median procedure*.

Following the work of Kemeny [21] on median linear orders, Mirkin and Chernyi [34], justified the use of δ by an axiomatic characterization in a "geometric" framework (cf. also Barthélemy [9] for an order theoretic framework).

This median procedure has some nice properties (Régnier [40], Barthélemy [7], Barthélemy and Monjardet [11, 12]):

Young consistency (introduced by Young [48], in the field of social choice): if the profiles π^* and π'^* are such that $\mathcal{M}(\pi^*) \cap \mathcal{M}(\pi'^*) \neq \emptyset$, then $\mathcal{M}(\pi^*\pi'^*) = \mathcal{M}(\pi^*) \cap \mathcal{M}(\pi'^*)$.

Stability for objects:
(i) Let μ be a median of the profile π^*, let A be the union of some classes of μ and $\mu_{|A}$ be the restriction of μ to A. Let $\pi^*_{|A}$ be the profile on A whose components are the restrictions of the partitions π_i to A. Then $\mu_{|A}$ is a median of $\pi^*_{|A}$.
(ii) Moreover, if π^* is a profile and π a partition of X, then π is a median of π^* if and only if, for each union A of classes of π, the partition $\pi_{|A}$ is a median of $\pi^*_{|A}$.

Pareto principle: If μ is a median of $\pi^* = (\pi_1,\ldots,\pi_k)$, then:

$$\wedge_{1 \le i \le k} \pi_i \le \mu.$$

Co-Pareto Principle: If μ is a median of $\pi^* = (\pi_1,\ldots,\pi_k)$, then:

$$\mu \le \vee_{1 \le i \le k} \pi_i.$$

Notice that the median procedure is also symmetrical. Leclerc [25, 26] established properties much stronger than the Pareto and the co-Pareto principles. They are reported in Sections 4 and 5.

The medians of a profile correspond to the solutions of a 0/1-integer programming, as first noticed by Régnier [40].

Let $\pi^* = (\pi_1,\ldots,\pi_k)$ be a profile and denote by $r_{ij}(p)$ the characteristic vector of $E[\pi_p]$, $1 \le i, j \le n$, $1 \le p \le k$ (we have $r_{ij}(p) = 1$ if x_i an x_j are linked in the partition π_p and $r_{ij}(p) = 0$ otherwise). The *Régnier problem*:

$$\begin{array}{ll} & \text{Minimize } \sum_{1 \le p \le k} \delta(\pi,\pi_p) \\ \text{subject to:} & \pi \in \Pi; \end{array}$$

becomes:

$$\text{Minimize } \Sigma_{1\leq p\leq k}\, \Sigma_{1\leq i,j\leq n}\, (r_{ij}(p)-r_{ij})^2,$$

subject to:

(1) $r_{ij} = r_{ji}$, for $1 \leq i,j \leq n$,

(2) $r_{ii} = 1$, for $1 \leq i \leq n$,

(3) $r_{ij} + r_{jh} - r_{ih} \leq 1$, for $1 \leq i,j,h \leq n$,

(4) $r_{ij} \in \{0,1\}$, for $1 \leq i,j \leq n$.

In this formulation r_{ij} is the characteristic vector of $E[\pi]$ (constraint (4)); constraints (1), (2) and (3) are the statements of, respectively, the symmetry, the reflexivity and the transitivity of $E[\pi]$.

We remark that: $\Sigma_{1\leq p\leq k}\, \Sigma_{1\leq i,j\leq n}\, (r_{ij}(p)-r_{ij})^2$

$= \Sigma_{1\leq p\leq k}\, \Sigma_{1\leq i,j\leq n}\, r_{ij}(p) + \Sigma_{1\leq p\leq k}\, \Sigma_{1\leq i,j\leq n}\, (r_{ij}-2r_{ij}(p)r_{ij})$

$= K + \Sigma_{1\leq p\leq k}\, \Sigma_{1\leq i,j\leq n}\, r_{ij}(1-2r_{ij}(p))$, and:

$\Sigma_{1\leq p\leq k}\, \Sigma_{1\leq i,j\leq n}\, r_{ij}(1-2r_{ij}(p)) = \Sigma_{1\leq i,j\leq n}\, \Sigma_{1\leq p\leq k}\, r_{ij}(1-2r_{ij}(p))$

$= \Sigma_{1\leq i,j\leq n}\, (r_{ij}-2\Sigma_{1\leq p\leq k}\, r_{ij}(p))$.

So, setting $c_{ij} = k - 2\Sigma_{1\leq p\leq k}\, r_{ij}(p)$, for $i \neq j$ and $c_{ii} = 0$, we get finally Problem (P):

$$\begin{array}{ll} & \text{Minimize } \Sigma_{1\leq i,j\leq n}\, c_{ij}r_{ij} \\ \text{subject to:} & (1), (2), (3) \text{ and } (4). \end{array} \qquad \text{(P)}$$

As a special case of this last problem, we obtain the *Zahn Problem* (Zahn [49]), which is characterized by $c_{ij} \in \{-1,+1\}$. It corresponds to the approximation of a symmetric relation by an equivalence relation: consider a symmetric relation S with s_{ij} as characteristic vector; the characteristic vectors r_{ij} of equivalence relations E minimizing $\delta(E,S)$ are the solutions of (P) with $c_{ij} = 1-2s_{ij}$.

The polytope defined by continuous relaxation of problem (P) has been studied by Régnier [41]. Grötschel and Wakabayashi [18, 19]) made a systematic study of the facets of the polytopes associated with the constraints of (P).

Surprisingly, the problem (P) is not more general that the Régnier Problem. In fact:

LEMMA 3 (Debord [16]). *If the integer coefficients* c_{ij} *(positive or negative)* $1\leq i,j\leq n$ *define a symmetric 0-diagonal matrix with the same nondiagonal entries's parity, then there exists a profile* $\pi^* = (\pi_1,\ldots,\pi_k)$ *such that* $c_{ij} = k-2\Sigma_{1\leq p\leq k}\, r_{ij}(p)$. *Moreover, such a profile may be constructed in a polynomial time from the* c_{ij}*'s.*

Krivanek and Moravek [22] proved the NP-hardness of the Zahn problem:

PROPOSITION 4. *The following problem is NP-complete:*

Instance: A set X, a symmetric relation S on X, an integer h.

Question: does there exist an equivalence relation E on X such that $\delta(R,S) \leq h$?

From this result, we deduce that (P), stated as a decision problem, is NP-complete. Hence, from Lemma 3, the decision version of the Régnier Problem is NP-complete as well. It is worth noticing that, in this last case, the number k is not fixed (the NP-status of the Régnier problem for a fixed number k is unknown; for more details, see Wakabayashi [47]).

More generally, we can consider any distance function d on Π and define the d-medians of a profile π^* as the solutions of:

$$\mathrm{Min}_{\pi\in\Pi} \Sigma_{1\leq p\leq k}\, d(\pi,\pi_p).$$

We shall denote by $\mathcal{M}_d$, and call the *d-median procedure*, the corresponding complete multiprocedure. Notice that, for any metric d, the d-median procedure is consistent: if the profiles π^* and π'^* are such that $\mathcal{M}_d(\pi^*)\cap\mathcal{M}_d(\pi'^*) \neq \varnothing$; then $\mathcal{M}_d(\pi^*\pi'^*) = \mathcal{M}_d(\pi^*)\cap\mathcal{M}_d(\pi'^*)$ (Barthélemy and Janowitz [10]).

3. Weight and co-Weight Metrics

3.1. Weight metrics. We consider a positive, symmetric, real valued function w defined on $(X\times X)$-Δ_X, with $\Delta_X = \{(x,x): x\in X\}$. So, we have:

W1: $w(x,x)$ is undefined for each $x \in X$;
W2: $w(x,y) > 0$, for all $x, y \in X$ with $x \neq y$;
W3: $w(x,y) = w(y,x)$, for all $x, y \in X$ with $x \neq y$.

We shall say that $w(x,y)$ (with $x \neq y$) is the *weight* of the pair x, y. If we consider this pair as a two element set, we state $w(xy) = w(x,y)$. In fact, we are concerned here by relativized consensus partitions: such weights may be regarded as resulting from some previous knowledge about X. For instance, assume that two objects x and y are linked in all the components of the profile π^* ($\gamma(xy,\pi^*) = 1$); we know (by the Pareto principle) that they are linked in any median partition, and it is then tempting to reduce the size of the problem by identifying y and x. If we want to address exactly the same problem as the initial one, we must then give a weight 2 to all the pairs xz, $z \neq x, y$.

Let S be a symmetric relation on X; we define the *weight* $w(S)$ of S as:

$$\frac{1}{2}\Sigma_{(x,y)\in S}\, w(x,y).$$

The *weight* of the partition π is the weight of the equivalence relation $E[\pi]$. In particular, the weight of the partition π_{xy} is $w(xy)$. By convention, we set $w(\alpha) = 0$.

The weight metric d_w on Π is defined by:

$$d_w(\pi,\pi') = w(E[\pi]\Delta E[\pi']) = w(E[\pi]\cup E[\pi'])\text{-}w(E[\pi]\cap E[\pi'])$$
$$= w(E[\pi]\text{-}E[\pi']) + w(E[\pi']\text{-}E[\pi]).$$

It can be expressed (*cf.* Sections 4 and 5) as :

$$d_w(\pi,\pi') = w(\pi)+w(\pi')\text{-}2w(\pi\wedge\pi').$$

The symmetric difference distance corresponds to the constant weight 1 ($w(xy) = 1$ for every pair xy): with such weights, one has $\delta = 2d_w$.

The *weight median procedure* is the complete multiprocedure $\mathcal{M}_{(w)}$ which associates with each profile $\pi^* = (\pi_1,\ldots,\pi_k)$ the solutions (the *d_w-medians*, called also the *w-weight medians*) of:

$$\mathrm{Min}_{\pi\in\Pi} \sum_{1\le p\le k} d_w(\pi,\pi_p).$$

Weight medians are solutions of the problem (WP) below, with $v_{ij} = w_{ij}c_{ij}$:

$$\begin{array}{ll} & \text{Minimize } \sum_{1\le i,j\le n} v_{ij}r_{ij} \\ \text{subject to:} & (1), (2), (3) \text{ and } (4). \end{array} \qquad \text{(WP)}$$

If a weighted pair is separated (linked) in a majority of components of a profile, then it tends to be separated (linked) in a consensus partition, even if, by transitive closure, the usual median procedure produces the converse.

(WP) can be seen as a *clique partitioning problem*: Consider the complete graph K_n. Define the *length* of the edge ij, $1 \le i, j \le n$, by v_{ij}, the *length* of a clique C of G as the sum of lengths of the edges in C and the *length* of a partition π into cliques as the sum of the lengths of the cliques of π. Clearly, to solve (WP) is equivalent to search for a clique partitioning with a minimal length.

This interpretation in term of clique partitioning shows that $\mathcal{M}_{(w)}$ satisfies the property of stability for objects: if π^* is a profile and π a partition of X, π is a d_w-median of π^* iff for each union A of classes of π, the partition $\pi_{|A}$ is a d_w-median of $\pi^*_{|A}$.

However $\mathcal{M}_{(w)}$ does not fulfil the Pareto Principle. A counter-example is provided by the profile $\pi^* = (\pi_1,\pi_2,\pi_3)$ of the set $X = \{a,b,c,d\}$ defined by π_1 = /abc/d/, π_2 = /a/bcd/ and π_3 = /abcd/. Set $w(ab) = w(ad) = w(cd) = 10$ and $w(ac) = w(bc) = w(bd) = 1$. The unique median is μ = /ab/cd/ which is not coarser than $u(\pi^*)$ = /a/bc/d/.

3.2. Computational results. We shall review the many algorithmic attempts to solve (WP), with a classification into four classes: exact methods, relaxation methods, hill-climbing heuristics, metaheuristics. It is fortunate that many of these methods have been used, for the Régnier problem, on the same data sets, for instance: classification of *cetacea* (Vescia [46]; $n = 36$, $k = 15$), classification of wild cats (Marcotorchino [28]; $n = 30$, $k = 14$), votes of all member states of the U.N. ($n = 158$, various values for k). Except for branch and bound, these methods compute just one solution (the dispersion of all the solutions of (WP) remains an open problem).

3.2.1. *Exact methods.* Branch and bound (Tüshaus [45]): it performs only for very small n.

Cutting planes (Wakabayashi [47], Grötschel and Wakabayashi [18, 19]): This approach is based on a careful study of the clique partitioning polytope (it is defined by the constraints 1, 2, 3 and 4). It performs efficiently for large n (U.N. votes).

3.2.2. *Relaxation.* Linear programming: the idea of relaxing the constraint (4) into (4'): $0 \le r_{ij} \le 1$, goes back to Régnier [41] who began to study the corresponding polytope. However the so-obtained linear program admits a large number of constraints (in $O(n^3)$) and needs intensive use of the dual problem together with an efficient linear programming code. This approach was followed by Michaud and Marcotorchino [31]. Optimality is not guaranteed since the associated polytope may have fractional vertices. However this method works for large n and, in practice, often provides an integral solution.

Lagrangean relaxation (Schader and Tüshaus [42]): this approach is based on the dual integer programming of (WP) and the construction of a sub-optimal feasible solution from the optimal Lagrange multipliers. Here also the performance is not guaranteed.

3.2.3. *Hill climbing heuristics.* These heuristics transform the current solution into another, while the lengths of the corresponding clique partitionings decrease (local improvement strategy). So, we get only a sub-optimal solution that is locally optimal relatively to the used transformation. In other terms, we consider a connected graph Γ with Π as vertex set and we walk in this graph from one vertex to one neighbor, until the lengths of the corresponding clique partitioning stop to decrease.

Exchange heuristics (Régnier [40]): the edges of Γ are the exchanges of an object from one class to another. The number of classes may decrease (exchange from a singleton), it may also increase if we admit exchanges into the empty set. In this last case, the graph Γ is connected. It can be shown that if a partition π admits p classes, p_1 one-element classes and p_2 two-elements classes, then the degree of π in Γ is: $\frac{1}{2}(2np-p_1-2p_2-p_1^2)$. So, the complexity of one iteration is reasonable. The shortest path metric in Γ has been considered by Arabie and Boorman [3]; Day [15] has shown that its computation is an NP-hard problem.

The exchange heuristic has been generalized by Marcotorchino and Michaud [29], by considering the exchange of two objects from different classes. Both heuristics may perform arbitrarily bad (de Amorim et al. [2]).

The neighborhood graph (undirected Hasse diagram) of the lattice Π does not seem to be used for heuristics, probably because the degrees of its vertices are very large.

3.2.4. *Meta-heuristics.* Simulated annealing (SA) and tabu search (TS) are studied and compared by de Amorim et al. [2], with the Régnier heuristics as hill climbing. They both perform almost efficiently on problems for which an optimal solution is known (from cutting planes), despite the fact that SA did not perform very well on other problems of combinatorial data analysis (De Soete et al. [17]). Notice however that TS seems to be somewhat more efficient than SA.

General clique partitioning problems have been approached by genetic algorithms (Jones and Beltramo [20], Talbi and Bessiere [44]). The main problem is to code partitions as chromosomes, in order to make meaningful operations as crossover or mutations. To avoid these difficulties, Snyers [43] uses dynamic programming and, mainly, a decision tree coding. He obtains performances comparable with thoses of SA and TS (with solutions interpretable from a "conceptual" point of view).

3.3. Co-weight metrics. We now introduce a family of metrics that appears to be dual of the family of weight metrics. This duality is established in Sections 4 and 5; here we shall introduce it from a "naive" point of view.

We consider a positive, co-symmetric real function w' defined on 2^X-$\{\emptyset,X\}$. By that we mean:

co-W1: $w'(\emptyset)$ and $w'(X)$ are undefined;
co-W2: $w'(A) > 0$, for each $A \subseteq X$ with $A \neq X, \emptyset$;
co-W3: $w'(A) = w'(X\text{-}A)$, for each $A \subseteq X$ with $A \neq X, \emptyset$ (co-symmetry).

We shall say that $w'(A)$ (with $A \neq X, \emptyset$) is the *co-weight* of the subset A of X.

Let $\mathcal{S}$ be a set of subsets of X such that $X \notin \mathcal{S}$ and $A \in \mathcal{S}$ implies $X\text{-}A \in \mathcal{S}$. We define the *co-weight* $w'(\mathcal{S})$ of $\mathcal{S}$ as $\frac{1}{2}\Sigma_{A\in \mathcal{S}}\, w'(A)$.

With each partition $\pi = \{A_1,\ldots,A_p\}$ is associated the set $M(\pi)$ of all unions of classes of π, other than X:

$$M(\pi) = \{\cup_{i\in P'} A_i \colon P' \subset \{1,\ldots,p\},\, P' \neq \emptyset\}.$$

The *co-weight* of π is defined as the co-weight of the set $M(\pi)$ of subsets of X:

$$w'(\pi) = \frac{1}{2}\Sigma_{P'\subseteq\{1,\ldots,p\},\, P'\neq\emptyset}\, w'(\cup_{i\in P'} A_i).$$

In particular, the co-weight of the bipartition $/A/X\text{-}A/$ is $w'(A)$. The co-weights define an antitone function on Π ($\pi \leq \pi'$ implies $w'(\pi) \geq w'(\pi')$). By convention, we set $w'(\omega) = 0$. As in the case of weight metrics, we can consider the weight of the symmetric difference between $M(\pi)$ and $M(\pi')$ and we get the co-weight metric $d^{w'}$ on Π, defined by:

$$d^{w'}(\pi,\pi') = w'(M(\pi)\Delta M(\pi')).$$

Obviously $d^{w'}$ is a distance function; it will be seen in Section 5.2 that it may be stated as:

$$d^{w'}(\pi,\pi') = w'(\pi) + w'(\pi')\text{-}2w'(\pi\vee\pi').$$

Each weight w defines a co-weight w' by:

$$w'(A) = \Sigma_{x\in A,\, y\in X\text{-}A}\, w(xy).$$

This formula accounts for the two main points in clustering: internally well linked (w) *vs.* externally well separated (w').

In this case, we get, for a partition $\pi = 2^{p-2}\Sigma_{1\leq i\leq p}\, w'(A_i)$. Notice that not every co-weight is defined on this way, as shown by the following counter-example.
Set $X = \{1,2,3,4\}$; a co-weight w' is defined from a weight w if and only if:

$w'(1) = w'(2,3,4) = w(1,2)+w(1,3)+w(1,4);$
$w'(2) = w'(1,3,4) = w(1,2)+w(2,3)+w(2,4);$
$w'(3) = w'(1,2,4) = w(1,3)+w(2,3)+w(3,4);$
$w'(4) = w'(1,2,3) = w(1,4)+w(2,4)+w(3,4);$
$w'(1,2) = w'(3,4) = w(1,3)+w(1,4)+w(2,3)+w(2,4);$
$w'(1,3) = w'(2,4) = w(1,2)+w(1,4)+w(2,3)+w(3,4);$
$w'(1,4) = w'(2,3) = w(1,2)+w(1,3)+w(2,4)+w(3,4).$

This system admits a solution if and only if:

$$w'(1)+w'(2)+w'(3)+w'(4) = w'(1,2)+w'(1,3)+w'(1,4).$$

Despite the fact that the definition of a co-weight metric looks somewhat satisfactory (essentially in the spirit of the obtention of a co-weight from a weight), we shall see that its use in median procedures leads to unexpectedly poor results.

4. Consensus Results in Lattices

4.1. Some definitions about finite lattices. General information on lattice theory and, especially, on the partition lattices, may be found in the book of Birkhoff [13]. The following definitions and properties will be particularly useful in the sequel.

Let L be a finite ordered set; a *lower bound* of a subset S of L is an element $t \in L$ such that $t \leq s$ for all $s \in S$. Dually, t is an *upper bound* of S if $s \leq t$ for all $s \in S$. When any pair s, t of elements of L has both a least upper bound (l.u.b), denoted by $s \vee t$ (the *join* of s and t) and an upper lower bound (u.l.b), denoted by $s \wedge t$ (the *meet* of s and t), L is endowed with a *lattice* algebraic structure. The binary operations $\vee$ and $\wedge$ are associative, commutative and idempotent ($s \vee s = s \wedge s = s$). Moreover, they are linked by the *absorption laws*: $s \vee (t \wedge s) = s \wedge (t \vee s) = s$, for all $s, t \in L$; the order on L is recovered from the algebraic structure by the equivalences $s \leq t \Leftrightarrow s \wedge t = s \Leftrightarrow s \vee t = t$.

A finite lattice L is a *complete lattice*: any subset S of L has also a join $\vee S$ and a meet $\wedge S$. The minimum and maximum elements are, as in Section 2, $\alpha = \wedge L$ and $\omega = \vee L$. The obvious property: $T \subseteq S$ implies $\vee T \leq \vee S$ and $\wedge T \geq \wedge S$, leads to set $\vee \varnothing = \alpha$ and $\wedge \varnothing = \omega$.

An element j of L is *join irreducible* if, for $S \subseteq L, j = \vee S$ implies $j \in S$; an equivalent property is that j covers exactly one element of L; recall s covers t (denoted $t \prec s$) means that $t \leq s' < s$ implies $t = s'$. In concrete lattices, join irreducible elements may be often thought of as elementary objects. The set of all these join irreducibles is denoted J (with $\alpha \notin J$, since $\vee \varnothing = \alpha$ and $\alpha \notin \varnothing$); we set $J(s) = \{j \in J: j \leq s\}$ for any $s \in L$. The set J is the unique minimal subset generating the entire lattice by the join operation, with $s = \vee J(s)$ for all $s \in L$.

Dually, $m \in L$ is a *meet irreducible* if $m = \wedge S$ implies $m \in S$; equivalently, m is covered by exactly one element of L. The set of all the meet irreducible elements of L is denoted by M and, for $s \in L$, $M(s) = \{m \in M: s \leq m\}$, with $s = \wedge M(s)$.

A set K of join irreducibles is said to be *closed* if it exists an element s of L such that $J(s) = K$. The representation of s by its closed set $J(s)$ (or, dually, by $M(s)$) is one-to-one, with the properties: $J(s \wedge t) = J(s) \cap J(t)$ and $M(s \vee t) = M(s) \cap M(t)$ (but only $J(s) \cup J(t) \subseteq J(s \vee t)$ and $M(s) \cup M(t) \subseteq M(s \wedge t)$). These representations are not minimal: generally, one may find, for instance, a proper subset K of $J(s)$ with $s = \vee K$.

We also briefly describe some classes of lattices. A lattice L is *distributive* if it satisfies the *distributivity laws*: for all $s, s', t \in L$, $(s \vee s') \wedge t = (s \wedge t) \vee (s' \wedge t)$, or, equivalently, $(s \wedge s') \vee t = (s \vee t) \wedge (s' \vee t)$. If L is finite, this property has the following useful characterization (D) bearing on the join irreducible elements:

(D) For $j \in J$ and $S \subseteq L$, the inequality $j \leq \vee S$ implies that there exists $s \in S$ such that $j \leq s$.

The finite lattice L is *upper semimodular* if, for every $s, t \in L$, $s \wedge t \prec s$ and $s \wedge t \prec t$ implies $s \prec s \vee t$ and $s \prec s \vee t$. The lower semimodularity is dually defined. If it is both upper and lower semimodular, then L is *modular*. Distributive lattices are modular.

The lattice L is *ranked* if it admits a real-valued *rank function* r (i.e. such that $r(\alpha) = 0$ and $s \prec t$ implies $r(t) = r(s)+1$). Upper and lower semimodular lattices are ranked.

Several recent works give evidence of the fact that lattice theory is particularly appropriate for the abstract study of consensus problems. Almost all the obtained results apply in fact to semilattices - where only one of the meet and join operations is always defined. Since we have only to deal with lattices in this paper, we summarize these results in the (easier) lattice case, with no claim to exhaustivity.

4.2. A class of algebraic procedures: the *quota rules*. As in Section 3, we denote by L^* the set $\cup_k L^k$ of all profiles of L. A first way of aggregating a profile $s^* = (s_1,\ldots,s_k)$ into a unique element s is provided by the algebraic structure of L. Given a real number $q \in [0,1]$, the *q-quota rule* consensus function c_q may be defined on two ways: first, generalizing the definition of Section 2.2, by:

$$c_q(s^*) = \vee\{j \in J: \gamma(j) > q\},$$

where the *index* $\gamma(j) = \gamma(j,s^*)$ is the number $\frac{|\mathbf{k}_j(s^*)|}{k}$, with $\mathbf{k} = \{1,\ldots,k\}$ and $\mathbf{k}_j(s^*) = \{i \in \mathbf{k}: j \leq s_i\}$.

An alternative expression for the q-quota rule is the lattice polynomial:

$$c_q(s^*) = \vee_{A \in \mathcal{P}(\mathbf{k}),\, |A| > qk} (\wedge_{i \in A}\, s_i).$$

A proof of the equality of the two expressions in any lattice may be found, for instance, in Leclerc [24]. Similarly, the *weak q-quota rule* consensus function b_q is defined, with, again, an alternative polynomial expression, by:

$$b_q(s^*) = \vee\{j \in J: \gamma(j) \geq q\},$$

The *dual q-quota rule* consensus function c'_q is defined by the following two equivalent expressions:

$$c'_q(s^*) = \wedge_{A \in \mathcal{P}(\mathbf{k}),\, |A| > qk} (\vee_{i \in A}\, s_i) = \wedge\{m \in M: \kappa(m) > q\},$$

where, for each meet irreducible m, the *dual index* $\kappa(m) = \kappa(m,s^*)$ is defined by $\boldsymbol{k}'_m(s^*) = \{i \in \boldsymbol{k}: s_i \leq m\}$ and $\kappa(m) = \frac{|\boldsymbol{k}'_m(s^*)|}{k}$.

Finally, the *weak dual q-quota rule* consensus function b'_q is given by:

$$b'_q(s^*) = \wedge\{m \in M: \kappa(m) \geq q\},$$

The following inequalities result from these definitions: $q \leq q'$ imply $c_q(s^*) \geq c_{q'}(s^*)$ and $c'_q(s^*) \leq c'_{q'}(s^*)$. Moreover, assume $qk = p$, an integer between 0 and k; let I and I' be two subsets of $\boldsymbol{k}$ such that $|I| \geq p$ and $|I'| \geq k\text{-}p\text{+}1$; then, $I \cap I' \neq \emptyset$, and one has $\wedge_{i \in I}\, s_i \leq \vee_{i \in I'}\, s_i$; finally, $c_{p/k}(s^*) \leq c'_{(k-p+1)/k}(s^*)$. The equalities $c_{p/k}(s^*) = c'_{(k-p+1)/k}(s^*)$ (any dual quota rules is also a quota rule) hold for all profiles of length k and integers p, $0 \leq p \leq k$, if and only L is distributive.

Certain values of q correspond to the algebraic formalizations of classical consensus rules; the following notations and terminology will be used in the sequel:

$q = 0.5$	$b(s^*) = \vee\{j \in J: \gamma(j) \geq 0.5\}$	(weak majority rule)
	$b'(s^*) = \wedge\{m \in M: \kappa(m) \geq 0.5\}$	(dual weak majority rule)
	$c(s^*) = \vee\{j \in J: \gamma(j) > 0.5\}$	(majority rule)
	$c'(s^*) = \wedge\{m \in M: \kappa(m) > 0.5\}$	(dual majority rule)
$q = 1$	$u(s^*) = \vee\{j \in J: \gamma(j) = 1\}$	(unanimity rule)
	$u'(s^*) = \wedge\{m \in M: \kappa(m) = 1\}$	(dual unanimity rule)

In all cases, $c(s^*) \leq b(s^*) \leq c'(s^*)$ and $c(s^*) \leq b'(s^*) \leq c'(s^*)$; if k is odd, one has $c(s^*) = b(s^*) \leq c'(s^*) = b'(s^*)$; if L is distributive, $c(s^*) = b'(s^*) \leq c'(s^*) = b(s^*)$.

4.3. An axiomatic characterization of some classes of k-procedures. We consider consensus functions f which associate a unique element s of L to each k-profile s^* of L:

f is said to be *decisive* if for all $j \in J$, $s^*, t^* \in L^k$,

$$[\boldsymbol{k}_j(s^*) = \boldsymbol{k}_j(t^*)] \text{ implies } [j \leq f(s^*) \Leftrightarrow j \leq f(t^*)];$$

f is *neutral monotonic* if for all $j, j' \in J$, $s^*, t^* \in L^k$,

$$[\boldsymbol{k}_j(s^*) \subseteq \boldsymbol{k}_{j'}(t^*)] \text{ implies } [j \leq f(s^*) \Rightarrow j' \leq f(t^*)];$$

f is *Paretian* if for every $s^* \in L^k$, $\wedge_{1 \leq i \leq k}\, s_i \leq f(x^*)$.

A *federation* on $\boldsymbol{k} = \{1,\ldots,k\}$ is a family $\mathcal{F}$ of subsets of $\boldsymbol{k}$ satisfying the monotonicity property: $[A \in \mathcal{F}, B \supseteq A] \Rightarrow [B \in \mathcal{F}]$. A federation consensus function $f_{\mathcal{F}}$ on L is associated with any federation $\mathcal{F}$ by $f_{\mathcal{F}}(s^*) = \vee_{A \in \mathcal{F}} (\wedge_{i \in A}\, s_i)$. Especially, if $\mathcal{F} = \{I' \subseteq \boldsymbol{k}: I' \supseteq I\}$, for a given subset I of $\boldsymbol{k}$, then $f_{\mathcal{F}}(s^*) =$

$\wedge_{i \in I} s_i$ is an *oligarchic* consensus function. The quota rules defined above constitute a special class of federation consensus functions, the unanimity being the oligarchic function with $I = k$.

In the lattice L, a dependence relation β is defined on the set J of the join irreducible elements by: $j \beta j'$ if $j \neq j'$ and there exists $s \in L$ such that $j, j' \nleq s$ and $j < j' \vee s$. Then, since $j < j' \vee \alpha$, $j < j'$ implies $j \beta j'$. It may be shown that the relation β is just the restriction to J of the lattice order of L if and only if L is distributive. In the opposite case, where the graph (J, β) is strongly connected, the lattice L is said β-*strong*. The main result in Monjardet [37] (see also Leclerc and Monjardet [27]) is, in the lattice case:

THEOREM 5. *If L is distributive, then f is a federation consensus function if and only if it is neutral monotonic. Otherwise, f is oligarchic if and only if it is neutral monotonic and does not always provide α or always provide ω as consensus element.*

Moreover, if L is β-strong, then f is oligarchic if and only if it is decisive and Paretian.

4.4. Some results on medians for metrics related with the lattice structure. Let v be a strictly isotone real function on L. Such a function is said to be a *lower valuation* if it satisfies one the following two equivalent properties:

(LV1) For all $s, t \in L$, $v(s) + v(t) \leq v(s \vee t) + v(s \wedge t)$

(LV2) The real function d_v defined on L^2 by the following formula (5) is a metric on L:

$$d_v(s,t) = v(s) + v(t) - 2v(s \wedge t). \tag{5}$$

The equivalence of conditions (LV1) and (LV2) in lattices goes back to Birkhoff [13]; see Barthélemy [8] and Monjardet [36] for the more general cases of semilattices and other ordered sets. A characteristic property of lower semimodular lattices is that their rank function is a lower valuation. *Upper valuations* are dually defined, with their metrics:

(UV1) $v(s) + v(t) \geq v(s \vee t) + v(s \wedge t)$, for all $s, t \in L$.

(UV2) The real function $d_v(s,t) = 2v(s \vee t) - v(s) - v(t)$ (6) on L^2 is a metric on L.

In both cases (lower or upper valuations), the metric d_v is a minimum path length metric in the neighborhood graph $\mathcal{N}(L)$ of the lattice L. The vertices of $\mathcal{N}(L)$ are the elements of L, and an unordered pair st of elements of L is an edge if $s \prec t$ or $t \prec s$; the length of this edge is $|v(s) - v(t)|$. The metric d_r associated in this way to the rank function r of a graded lattice will be denoted ∂ and, sometimes, called the *lattice metric* on L. For this metric, all the edges of $\mathcal{N}(L)$ have a unit length.

A function v on L which is both a lower and an upper valuation is a *valuation*. One has then $v(s) + v(t) = v(s \vee t) + v(s \wedge t)$ and $d_v(s,t) = v(s \vee t) - v(s \wedge t)$. It is known that only modular lattices admit valuations, and that a lattice is modular if and only if its rank function is a valuation.

We are particularly concerned with special types of lower and upper valuations, which exist on any lattice and include the weight and co-weight metrics of Section 3. A real function v on L is said to be a *weight valuation* if there exists a real strictly positive mapping w on J such that, for any $s \in L$, $v(s) = \sum_{j \in J(s)} w(j)$ and $v(\alpha) = 0$. Any weight valuation is a lower valuation and the metric d_v, now also called a *weight metric*, is given by (5), and also by the alternative formula (7), where Δ is the symmetric difference of subsets:

(7) $$d_v(s,t) = \sum_{j \in J(s)\Delta J(t)} w(j).$$

Especially, with constant, unit, weights, one gets the *symmetric difference metric* δ: $\delta(s,t) = |J(s)\Delta J(s)|$.

Similarly, let w' be a real strictly positive mapping defined on the set M. A *co-weight valuation* v is defined on L by: for any $s \in L$, $v(s) = \Theta - \sum_{m \in M(x)} w'(m)$, where $\Theta = \sum_{m \in M} w'(m)$. Then, v is an upper valuation, with $v(\alpha) = 0$ and $v(\omega) = \Theta$. The corresponding minimum path length metric (co-weight metric) is given by (6) or by:

(8) $$d_v(s,t) = \sum_{m \in M(s)\Delta M(t)} w'(m),$$

with the dual *symmetric difference metric* δ': $\delta'(s,t) = |M(s)\Delta M(s)|$.

Let $s^* = (s_1,\ldots,s_k)$ be a profile of L, and d a lower valuation metric on L. We consider, as in Section 2, the multiprocedure $\mathcal{M}_d$ which consists of searching for the *medians* of s^* for the metric d, that is the elements μ of L such that the *remoteness* $R(\mu,s^*) = \sum_{1 \le i \le k} d(\mu,s_i)$ is minimum. These medians are related with the majority rule consensus elements as defined above. First, there is a median characterization (Leclerc [25]) of the lower valuations by the following property (LV3):

(LV3) For any profile s^* and for any median μ for the metric d_v, the inequality $\mu \le c'(s^*)$ holds.

THEOREM 6. *A strictly isotone real function* v *on* L *is a lower valuation if and only if the property* (LV3) *is satisfied by the associated metric* d_v.

Moreover, if v *is a lower valuation, there is always at least one median* μ_0 *such that the inequality* $\mu_0 \le b'(s^*)$ *holds.*

In fact, one has always the inequality $R(s \wedge c'(s^*),s^*) \le R(s,s^*)$ in the case of a lower valuation. Dually, v is an upper valuation if and only if it satisfies the property (UV3). In addition, there exists in this case at least a median superior to $b(s^*)$.

(UV3) For any profile s^* and for any median μ with regard to the metric d_v, the inequality $\mu \ge c(s^*)$ holds.

Of course, if L is modular, a strictly isotone real function v is a valuation if and only if the inequalities $c(s^*) \leq \mu \leq c'(s^*)$ are satisfied by any median of any profile. The procedure $\mathcal{M}_d$, with $d = d_v$, will be also denoted $\mathcal{M}_{(v)}$.

Stronger results are available in the case where v is a weight valuation (Leclerc [25, 26]). In this case, the remoteness function $R(s,s^*)$ may be written as:

(9) $$R(s,s^*) = k(\Sigma - \Sigma_{j\in J(s)} (2\gamma(j)-1)w(j)),$$

where the constant $\Sigma = \Sigma_{j\in J}\, \gamma(j)w(j)$ depends only on the profile s^*. It is then easy to see that $c(s^*)$ (resp. $b(s^*)$) is a median whenever the set $C(s^*) = \{j\in J: \gamma(j) > 0.5\}$ (resp. $B(s^*) = \{j\in J: \gamma(j) \geq 0.5\}$ is closed. Such a property occurs very rarely in practice, except, as it is recalled below, when L is distributive; but one has also the following result:

PROPOSITION 7. *The following results hold for any weight (resp. co-weight) metric* d_v*:*

(WM) *For any profile* s^* *and for any median* μ *for the metric* d_v*, the inequality* $\mu \leq b(s^*)$ *holds. Moreover, there is at least one median* μ_0 *with* $\mu_0 \leq c(s^*)$.

(WM') *For any profile* s^* *and for any median* μ *for the metric* d_v*, the inequality* $\mu \geq b'(s^*)$ *holds. Moreover, there is at least one median* μ_0 *with* $\mu_0 \geq c'(s^*)$.

The weight valuations on a distributive lattice L are exactly the valuations (Barbut and Monjardet [6], ch.IV), and also, by duality, the co-weight valuations. In this case, it follows from the property (D) that the set $\{j \in J: \gamma(j) > q\}$ is closed for any number $q \in [0,1]$. From the expression (4) of the remoteness function, the set of all the medians is the lattice interval $[c(s^*),b(s^*)] = \{s \in L: c(s^*) \leq s \leq b(s^*)\}$ (Monjardet [35]; see also Leclerc [26]). As an immediate consequence, $c(s^*)$ is the unique median of any profile s^* with odd length (Barbut [5]). Note that these medians do not depend on the particular weight valuation v, that is on the particular weight function w on J, which is considered.

The convergence between the majority rule and the median procedure is, to some extent, characteristic of the distributivity property: if v is a strictly isotone real function on a lattice L such that, for any profile s^* with odd length, the majority rule element $c(s^*)$ is a median of s^* for the metric d_v, then v is a valuation and L is distributive (Leclerc [26]).

In a distributive lattice, the rank function is a weight valuation, and, so, a valuation; in fact, it corresponds with unit weights on all the join irreducibles, and the symmetric difference metric δ is the same as the lattice metric ∂. Barthélemy and Janowitz [10] have given an axiomatic characterization of the median procedure for the metric ∂ (and, so, for any weight metric) by five properties. The main two are the Young consistency and the following *quasi-Condorcet* (QC) property:

(QC) For any profile s^*, for any $j \in J$ such that $\gamma(j) = 0.5$, and for any $s \in L$, s is a consensus element if and only if $s \vee j$ is a consensus element.

We may conclude this section by emphasizing two points on which the three approaches converge:
- the importance of the role of the irreducible elements: they allow simple expressions of quota rules, of axioms (in Monjardet's work as well as in Barthélemy and Janowitz's), and of weight and co-weight metrics; these expressions are very powerful tools for generating consensus results;
- the privileged status of distributive lattices in the domain of consensus: in these lattices, quota and dual quota rules coincide; they all satisfy the strong (NM) axiom; the majority rules directly provide medians.

5. Quota Rules and Medians in the Partition Lattice

5.1. Quota rules. It was recalled in the introduction that the set Π of all the partitions of a finite set X, endowed with the *refinement order*, is a lattice, whose meet and join operations have been described. The lattice Π has a rank function $r(\pi) = n-|\pi|$, with $r(\alpha) = 0$ and $r(\omega) = n-1$. The partitions π_{xy}, of rank 1, which cover α, are the *atoms* of Π, while the bipartitions $/A/X-A/$ of X are the *coatoms*, covered by ω. More precisely, the following important properties will be of use in the sequel:

1) The lattice Π is *upper semimodular.*
2) The lattice Π is *atomic* and *coatomic*, that is the join irreducible are exactly the atoms, and the meet irreducibles are the coatoms. So, every partition π is the join of the elementary partitions π_{xy}, for all the pairs xy of elements of X linked in π, and the meet of all the bipartitions coarser than π. The sets of all the atoms and all the coatoms are denoted, as in Section 4, by J and M. A set K of atoms is closed if and only if $\pi_{xy} \in K$ as soon as there is a series $x = x_1, x_2, x_3, \ldots, x_p = y$ with $\pi_{x_ix_{i+1}} \in K$, for $i = 1, \ldots, p$.

Once recognized the sets of join or meet irreducibles, we are able to particularize the results of Section 4 about consensus in lattices to the case of the lattice of partitions. Several methods or results of Sections 2 and 3 will be recovered on this way. For instance, it may be shown that the partition lattice is β-strong: Theorem 1 is in fact as a special case of the last part of Theorem 5 (see Neumann and Norton [39] and Monjardet [37] for a dual result).

New procedures will be also considered and we begin by investigating their relevance.

Given a profile $\pi^* = (\pi_1,\ldots,\pi_k)$ of k partitions, the quota rules c_q and b_q of Section 2.2 immediately appear to be a particularization of those of Section 4. We may also particularize the dual quota rules c'_q and b'_q. The dual index of a bipartition $/A/X-A/$ is the number $\kappa(A,\pi^*) = \frac{1}{k}|\{i \in k: /A/X-A/ \geq \pi_i\}|$. The lattice formalizations of classical consensus rules of the previous section become:

weak majority rule:	$b(\pi^*) = \vee\{\pi_{xy}: \gamma(xy,\pi^*) \geq 0.5\}$
dual weak majority rule:	$b'(\pi^*) = \wedge\{/A/X\text{-}A/: \kappa(A,\pi^*) \geq 0.5\}$
majority rule:	$c(\pi^*) = \vee\{\pi_{xy}: \gamma(xy,\pi^*) > 0.5\}$
dual majority rule:	$c'(\pi^*) = \wedge\{/A/X\text{-}A/: \kappa(A,\pi^*) > 0.5\}$
unanimity rule:	$u(\pi^*) = \vee\{\pi_{xy}: \gamma(xy,\pi^*) = 1\}$
dual unanimity rule:	$u'(\pi^*) = \wedge\{/A/X\text{-}A/: \kappa(A,\pi^*) = 1\}$

So, in Example 1 of Section 2.2, one has $c(\pi^*) = b(\pi^*) = \alpha$ and, since $\kappa(A,\pi^*) = {}^1/_k$ for any subset $A \neq \varnothing, X$, $c'(\pi^*) = b'(\pi^*) = \omega$; in Example 2, $c(\pi^*) =$ /abdg/c/e/f/, $b(\pi^*) =$ /abdgf/ce/, and $c'(\pi^*) = b'(\pi^*) = \omega$. These examples suggest two general remarks:

1. The majority rule consensus partition $c(\pi^*)$ is $\vee$-generated by the majority atoms, i.e. the atom partitions the scores of which are superior to 0.5. For $|X| > 2$, Π is not a distributive lattice and the set $\{j \in J: \gamma(j) > 0.5\}$ is generally not closed; then, $J(c(\pi^*))$ may contain atoms with low score. In Example 2, $J(c(\pi^*)) = \{\pi_{ab},\pi_{ad},\pi_{ag},\pi_{bd},\pi_{bg},\pi_{dg}\}$, while only π_{ab}, π_{ad} and π_{dg} are majority atoms; though $\gamma(\pi_{bg}) = 0$, b and g are linked in $c(\pi^*)$.
2. In the partition lattice, the duality between the c_q and the c'_q rules is essentially formal. This lattice has 2^{n-1}-1 coatoms, an exponential function of n, and $2^{|\pi|-1}$-1 coatoms are coarser than a given partition π. If the numbers $|\pi_i|$ of classes of the partitions of the profile are small compared with n, few coatoms have a non-null score and it is unlikely that some of them reach a significant score: in most cases, one has $c'_q(\pi^*) = \omega$, except for very low values of q, and, so, $c'(\pi^*) = b'(\pi^*) = \omega$: such a situation may be observed in Examples 1 and 2. If not, the involved sets of coatoms become quickly difficult to handle when n grows. The situation is not the same for the quota rules, like the majority one, since the number of atoms, n(n-1)/2, is of the order of n^2.

5.2. Valuations and medians. Consider a lower or an upper valuation, and the corresponding metric, defined on the lattice Π. In the second case, since this lattice is upper semimodular, the *quadrilateron condition* (UV1') (Barthélemy [8]) is equivalent to the conditions (UV1) and (UV2); this local condition is frequently easier to examine than (UV1). A quadrilateron $(\pi_1,\pi_2,\pi_3,\pi_4)$ of Π is a quadruple of distinct partitions such that $\pi_1, \pi_2 \prec \pi_3$ and $\pi_4 \prec \pi_1, \pi_2$:

(UV1') For any quadrilateron $(\pi_1, \pi_2,\pi_3,\pi_4)$ of Π,

$$v(\pi_1) + v(\pi_2) \geq v(\pi_3) + v(\pi_4).$$

PROPOSITION 8. *A strictly isotone real function* v *on* Π *is an upper valuation if and only if it satisfies the property* (UV1').

The following Table 2 gives an example of each of the four types of real functions defined in Section 4, with the corresponding metric (since, for $n > 3$, Π is not modular, valuations do not exist). Here, n_i is the cardinality of the i-th class of a partition π, for $i = 1,\ldots, |\pi|$.

The first three metrics of the table are studied in papers of Arabie and Boorman [3], Boorman and Olivier [14] and Day [15].

- The function $|J(\pi)|$ is the simplest weight valuation: it is obtained when all the atom partitions have a unit weight; the corresponding distance is the symmetric difference metric of Section 2.

- Since the lattice Π is upper semimodular, its rank function is an upper valuation; recall that the corresponding lattice metric ∂ is the minimum path length metric in the neighborhood graph $\mathcal{N}(\Pi)$ when all its edges have a unit length.

- The entropy $I(\pi)$ is one of the lower valuations recognized in the literature; if a partition π' covers π, with $\pi' = \pi(A+B)$, then, with $n_A = |A|$ and $n_B = |B|$, $d_I(\pi,\pi') = \frac{n_A+n_B}{n}\left(\frac{n_A}{n_A+n_B}\log\frac{n_A+n_B}{n_A} + \frac{n_B}{n_A+n_B}\log\frac{n_A+n_B}{n_B}\right)$. This quantity may be interpreted as a conditional information. Note that many other lower or upper valuations might be considered on the partition lattice, several of them being related with some types of information (for instance, $I'(\pi) = \Sigma_{1\leq i\leq|\pi|} \log n_i$ is another example of a lower valuation).

For sake of completeness, the simplest co-weight valuation, obtained with $w'(A) = 1$ for any bipartition $/A/S\text{-}A/$ has been added in Table 1. It is associated with the co-weight valuation $v(\pi) = |M|-|M(\pi)| = (2^{n-1}-1)-(2^{|\pi|-1}-1)$, the number of bipartitions which are not coarser than π. The dual symmetric difference metric δ' is then a particularization of the formula for the co-weight metrics $d^{w'}$ of Section 3.3.

Table 2 : Examples of upper or lower valuations on the partition lattice.

Function v	Type of v	$d_v(\pi,\pi')$
$\lvert J(\pi)\rvert = \frac{1}{2}\Sigma_{1\leq i\leq\lvert\pi\rvert} n_i(n_i-1)$	weight valuation	$\delta(\pi,\pi') = \lvert J(\pi)\Delta J(\pi')\rvert$ $= \lvert J(\pi)\rvert+\lvert J(\pi')\rvert-2\lvert J(\pi\wedge\pi')\rvert$ (symmetric difference metric)
$I(\pi) = \Sigma_{1\leq i\leq\lvert\pi\rvert} \frac{n_i}{n}\log\frac{n_i}{n}$ (Shannon's Entropy)	lower valuation	$d_I(\pi,\pi')=I(\pi)+I(\pi)-2I(\pi\wedge\pi')$ (conditional information sum)
$r(\pi)$ (rank function)	upper valuation	$\partial(\pi,\pi') = \lvert\pi\rvert+\lvert\pi'\rvert-2\lvert\pi\vee\pi'\rvert$ (lattice metric ∂)
$\lvert M\rvert-\lvert M(\pi)\rvert = \frac{1}{2}(2^n-2^{\lvert\pi\rvert})$	co-weight valuation	$\delta'(\pi,\pi') = \lvert M(\pi)\Delta M(\pi)\rvert$ $= \lvert M(\pi)\rvert+\lvert M(\pi')\rvert-2\lvert M(\pi\vee\pi')\rvert$ (dual symmetric difference metric)

The consequences of the results of Section 4 for the medians in the examples of Section 2.2 are summarized below in Table 3.

In the first example, it results from this table that α is the unique median for the metric δ and ω for δ'; no information is provided on the medians for d_I and ∂, though

ω is obviously the unique median for ∂ (all the elements of the profile are adjacent to ω in the graph $\mathcal{N}(\Pi)$). In the second example, ω is again the unique median for δ' and no information is provided on the medians for d_I; finally, for metrics δ and ∂, the set of partitions where the medians may be found is markedly reduced: for instance, we know that at least one of the two partitions /abdgf/ce/ and ω is a median for the metric ∂ (the reader may verify that the median is ω).

Table 3. Relations between medians and majorities:

Metric	Example 1 (k odd)	Example 2 (k even)
δ	$\mu \leq c(\pi^*) = \alpha$	$\mu \leq b(\pi^*) =$ /abdgf/ce/ There exists $\mu_0 \leq c(\pi^*) =$ /abdg/c/e/f/
d_I	$\mu \leq c'(\pi^*) = \omega$	$\mu \leq b'(\pi^*) = \omega$ There exists $\mu_0 \leq c'(\pi^*) = \omega$
∂	$\mu \geq c(\pi^*) = \alpha$	$\mu \geq c(\pi^*) =$ /abdg/c/e/f/ There exists $\mu_0 \geq b(\pi^*) =$ /abdgf/ce/
δ'	$\mu \geq c'(\pi^*) = \omega$	$\mu \geq b'(\pi^*) = \omega$ There exists $\mu_0 \geq c'(\pi^*) = \omega$

More generally, the following observations can be made:

1. Due to the particularities of the lattice Π, one has in most cases $c'(\pi^*) = b'(\pi^*) = \omega$. With the above results, that means, on the one hand, that ω is often the only median for the dual weight valuation metrics and, on the other hand, that (LV3) provides poor information about medians for lower valuations metrics.
2. The medians for upper valuations and weight valuations are positioned in relation to the majority rule element. For instance, assume k is an odd number; if μ is a median for an upper valuation metric and μ' a median for a weight metric, the inequalities $\mu \geq c(\pi^*) \geq \mu'$ hold. Especially, any median for the symmetric difference metric δ is at least as fine as any median for the lattice metric ∂.
   The Property (WM) of Proposition 7 has been partly established by Michaud [30] and completed by Neumann and Boukaabar [38], both in the special case of the symmetric difference metric δ on the partition lattice Π.
3. In Section 2.3, it was recalled that the median procedure $\mathcal{M}$ for the symmetric difference metric has the property of "stability for objects". The extension of this property to the median procedures $\mathcal{M}_{(w)}$ for medians for weight metrics has been noticed in Section 3.1. With this property and the relation $\mu_0 \leq c(\pi^*)$, the problem of the search of medians may be divided into subproblems, one for each class of the majority rule partition, this reduction being significant whenever $c(\pi^*)$ has more than one class. For instance, in Example 2, we have just to find the medians in the sublattice of the partitions of $Y =$ abdg. For the metric δ, one founds the unique solution /ab/dg/ and, so, $\mu_0 =$ /ab/c/dg/e/f/ in the entire lattice.

If k is an even number, it is straightforward to obtain all the medians from those finer than $c(\pi^*)$: for any median μ, there is a median $\mu' \leq c(\pi^*)$ such that $\gamma(xy) = 0.5$ for any atom partition $\pi_{xy} \in J(\mu)\text{-}J(\mu')$. In Example 2, there is a second median μ_1 = /ab/ce/dg/f/, finer than $b(\pi^*)$, but not than $c(\pi^*)$.

6. Lattices of Constrained Partitions

6.1. Partitions with contiguity constraints. In this section, we begin to extend the preceding results to more general lattices of partitions, corresponding to restricted subsets of Π. Let $G = (S, \mathcal{C})$ be a connected undirected graph on S, where $\mathcal{C}$ is a subset of the set $X^{(2)}$ of all the unordered pairs of distinct elements of X. This graph is assumed to represent a *contiguity* relation: an edge xy of G means that the elements x and y are contiguous. As for the weights of Section 3.1, but on a different way, such a relation may be issued from previous knowledge on the considered partitioning problem.

A partition $\pi(\mathcal{E})$ is associated to each subset $\mathcal{E}$ of $\mathcal{C}$, the classes of $\pi(\mathcal{E})$ being the connected classes of the subgraph $(X,\mathcal{E})$ of G. The set $\Pi(\mathcal{C}) = \{\pi(\mathcal{E}): \mathcal{E} \subseteq \mathcal{C}\}$ of all the partitions obtained on this way is exactly the set of all the partitions compatible with the contiguity relation represented by the graph G.

EXAMPLE 3. Consider the ten planar cells of Figure 2. The corresponding contiguity graph G is given in Figure 3: an edge between two vertices of G represents two cells with a common border. Partitions like /aj/b/c/d/e/f/g/h/i/ or /adfhj/bcegi/ are not in $\Pi(\mathcal{C})$, since they have classes constituted of not neighboring cells.

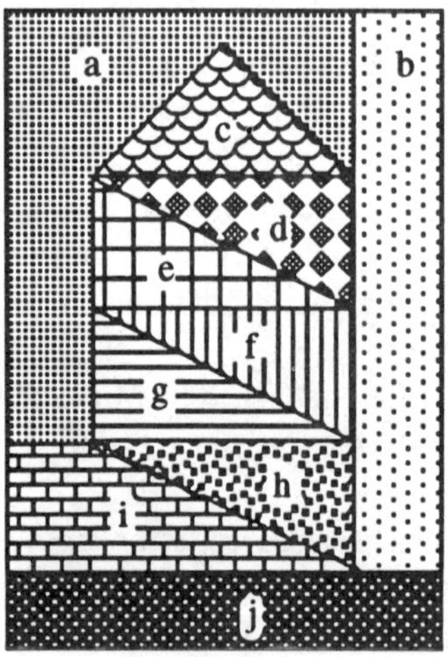

Figure 2

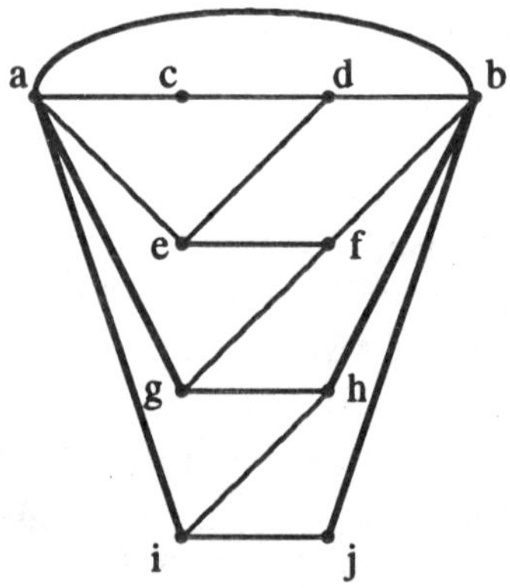

Figure 3

It may be first pointed out that the set $\Pi(\mathcal{C})$ is, again with the refinement order, endowed with a lattice structure of the same type as Π: atomic and upper semimodular, that is *geometric*.

Recall, following Aigner [1, ch. II and VI] and his formulation of the fundamental Whitney Theorem (p.274), that a subset $\mathcal{E}$ of $\mathcal{C}$ is a *flat* of the polygon matroid of the graph G if, for all $x, y \in X$ with $xy \in \mathcal{C}$, if x and y are in the same connected component of the subgraph $(X,\mathcal{E})$, then $xy \in \mathcal{E}$. In other terms, if x and y are linked in the partition $\pi(\mathcal{E})$ and $xy \in \mathcal{C}$, then $xy \in \mathcal{E}$. Then, it becomes straightforward to observe that the correspondence $\mathcal{E} \to \Pi(\mathcal{E})$ is one-to-one between, on one hand, the set Φ of all the flats, and, on the other hand, the partitions in $\Pi(\mathcal{C})$. Moreover, the refinement order on $\Pi(\mathcal{C})$ and the inclusion order on Φ are isomorphic. Since it is well-known that Φ is a geometric lattice, so is $\Pi(\mathcal{C})$. The entire partition lattice corresponds with the case $\mathcal{C} = X^{(2)}$.

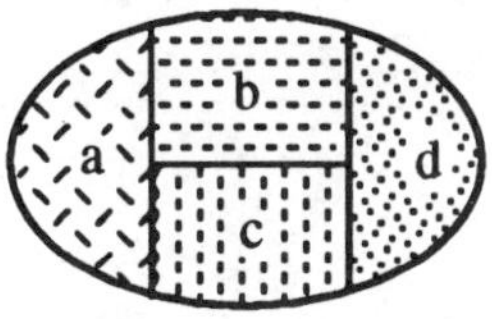

Figure 4

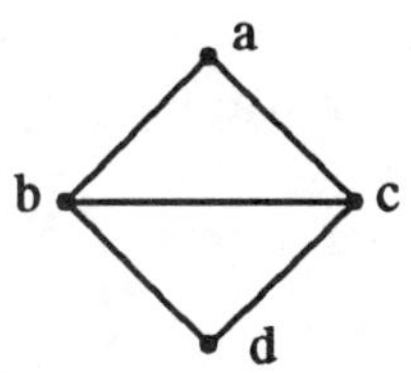

Figure 5

EXAMPLE 4. In Figure 4, an oval is divided into four cells a, b, c and d; the corresponding contiguity graph is shown in Figure 5, with the edge set $\mathcal{C}$ = {ab,ac,bc,bd,cd}. The neighborhood graph of the lattice $\Pi(\mathcal{C})$ is given in Figure 6. The partitions removed from Π in order to obtain $\Pi(\mathcal{C})$ are those with the class ad, that is /ad/b/c/ and /ad/bc/.

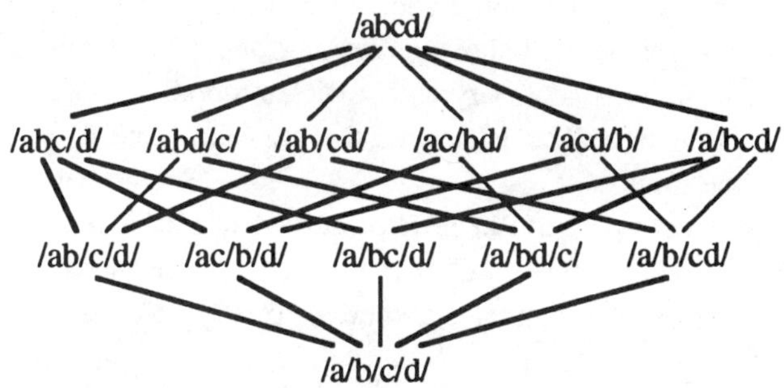

Figure 6

EXAMPLE 5. Assume X = {a,b,c,d,e,f} and G is the 6-cycle on this set (Figure 7). A subset $\mathcal{E}$ of $\mathcal{C}$ is a flat if $\mathcal{E} = \mathcal{C}$ or $|\mathcal{E}| \leq 4$; so, the lattice of flats, isomorphic to $\Pi(\mathcal{C})$, is obtained by the deletion of all the sets of five edges from the boolean lattice $\mathcal{P}(\mathcal{C})$.

The extreme partitions α and ω belong to $\Pi(\mathcal{C})$. Its set of atoms is $J_{\mathcal{C}} = \{\pi_{xy}: xy \in \mathcal{C}\}$ and its set of coatoms is $M_{\mathcal{C}} = \{/A/X\text{-}A/$: both A and X-A generate connected subgraphs of $G\}$. This set $M_{\mathcal{C}}$ is not empty, since G is connected. The rank function and the join operation of Π are preserved in $\Pi(\mathcal{C})$, but, generally, the meet operation $\wedge_{\mathcal{C}}$ of $\Pi(\mathcal{C})$ differs from the meet of Π, with, only, $\pi \wedge_{\mathcal{C}} \pi' \leq \pi \wedge \pi'$. For instance, in Example 5 above, the meet of the partitions /abcd/e/f/ and /adef/b/c/ is α, instead of π_{ad} in the lattice Π.

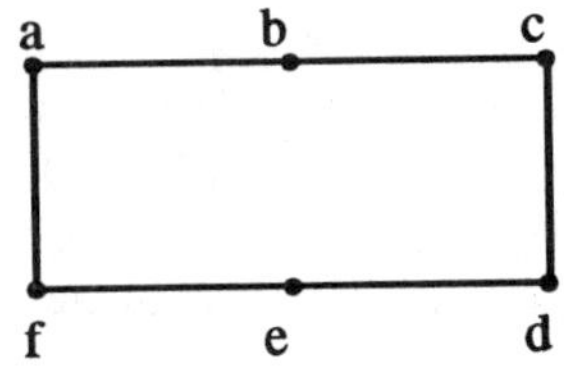

Figure 7

Let $\pi^* = (\pi_1,\ldots,\pi_k)$ be a profile of $\Pi(\mathcal{C})$; we must distinguish the quota rule in the lattice Π, denoted again by c_q, and the quota rule $c_q^{\mathcal{C}}$ in $\Pi(\mathcal{C})$:

$$c_q^{\mathcal{C}}(\pi^*) = \vee_{A \in \mathcal{P}(k),\, |A| > qk} (\wedge_{\mathcal{C}} \{\pi_i : i \in A\}) = \vee\{j \in J_{\mathcal{C}} : \gamma(j,\pi^*) > q\}.$$

Similarly, the dual quota rule $c'^{\mathcal{C}}_q$ in $\Pi(\mathcal{C})$ is:

$$c'^{\mathcal{C}}_q(\pi^*) = \wedge_{\mathcal{C}}\{(\vee_{i \in A} \pi_i) : A \in \mathcal{P}(k),\, |A| > qk\} = \wedge_{\mathcal{C}}\{m \in M_{\mathcal{C}} : \kappa(m,\pi^*) \geq q\}.$$

The weak quota rules and the majority rule elements $c^{\mathcal{C}}$, $b^{\mathcal{C}}$, $c'^{\mathcal{C}}$ and $b'^{\mathcal{C}}$ are defined on the same way. Notice that one has $c_q^{\mathcal{C}}(\pi^*) \leq c_q(\pi^*)$, $c'^{\mathcal{C}}_q(\pi^*) \leq c'_q(\pi^*)$, and so on...

From the results of Section 4.4, the situations where $\Pi(\mathcal{C})$ is distributive, or, even, where it is modular are particularly interesting.

Distributivity. It is well-known that a geometric lattice is distributive if and only if it is boolean. With $J_{\mathcal{C}}$ as defined above as set of atoms, this implies that every subset of $\mathcal{C}$ is a flat. An obvious necessary and sufficient condition is that G has no cycle, that is, that G is a tree. The lattice $\Pi(\mathcal{C})$ has then exactly n-1 coatoms, the minimum possible number.

In this case, the meet operation in Π is preserved in $\Pi(\mathcal{C})$: a class A of a partition $\pi \in \Pi(\mathcal{C})$ generates a subtree of G; if B is a class of another partition π' in $\Pi(\mathcal{C})$, the intersection $A \cap B$ also generates a subtree; finally, $\pi \wedge \pi' \in \Pi(\mathcal{C})$ and $\Pi(\mathcal{C})$ is a sublattice of Π.

Modularity. The following result may be found in the book of Aigner [1, p.59]: a geometric lattice L is modular if and only if $s \wedge m \neq \alpha$, for any coatom m and element s of rank 2. It is then an elementary exercise to see that $\Pi(\mathcal{C})$ is modular if and only if the graph G has no cycles of length superior to three. Of course, such a condition is a very strong one.

Again, it may be shown in this case that the intersection of two classes of elements of $\Pi(\mathcal{C})$ generates a connected subgraph of Π, and, so, that $\Pi(\mathcal{C})$ is a sublattice of Π.

6.2. Metrics and medians for contiguity constrained partitions. Let us examine the valuations and metrics of Section 5.2, when restricted to a geometric lattice $\Pi(\mathcal{C})$.

Upper valuations. An upper valuation v on Π is still an upper valuation on $\Pi(C)$: since $v(\pi) + v(\pi') \geq v(\pi \vee \pi') + v(\pi \wedge \pi') \geq v(\pi \vee \pi') + v(\pi \wedge_C \pi')$, the property (UV1) of Section 4.4 is satisfied. More, since the join is preserved, the corresponding metric, given by the expression (2'), is just the restriction of d_v to $\Pi(C)$. Especially, the rank function r is still an upper valuation (a consequence also of the upper semimodularity of $\Pi(C)$) and the lattice metric on $\Pi(C)$ is just the restriction of ∂.

Especially, if G has no cycle of length superior to 3, the rank r is a valuation and the medians for the lattice metric ∂ belong to the lattice interval $[c^C(\pi^*),c'^C(\pi^*)] = [c(\pi^*),c'(\pi^*)]$ of $\Pi(C)$. If G is a tree, this interval, equal to $[c(\pi^*),b(\pi^*)]$, is the set of all the medians, reduced to the unique element $c(\pi^*)$ when the profile π^* has an odd length.

Co-weight valuations. A co-weight valuation on Π is generally not a co-weight valuation on $\Pi(C)$; see for instance the unit co-weight valuation of Table 2 when restricted to the lattice of Figure 6. A co-weight valuation on $\Pi(C)$ must only take on account the coatoms belonging to M_C. The associated co-weight metric has the form $d_v(\pi,\pi') = \sum_{j \in M_C(\pi) \Delta M_C(\pi')} w(j)$, where $M_C(\pi) = \{m \in M_C : \pi \leq m\}$ (and, similarly, $J_C(\pi) = \{j \in J_C : j \leq \pi\}$).

Lower valuations. A lower valuation v on Π may no longer satisfy the condition (LV) on $\Pi(C)$. For instance, consider the Shannon entropy, and the lattice of partitions associated with the contiguity graph of Example 5: for the partitions π = /abcd/e/f/ and π' = /adef/b/c/, one has $\pi \wedge_C \pi' = \alpha$ and $\pi \vee \pi' = \omega$. Then, $I(\pi) + I(\pi') = \frac{2}{3}\log 2 - 2\log 3 > I(\alpha) + I(\omega) = -\log 2 - \log 3$. Of course, this cannot occur when $\Pi(C)$ is a sublattice of Π.

Weight valuations. The weight valuations on $\Pi(C)$ differ from those defined on Π: only the atoms $\pi_{xy} \in J_C$ are now considered. So, we have, for instance, a new symmetric difference metric δ_C defined by $\delta_C(\pi,\pi') = |J_C(\pi) \Delta J_C(\pi')|$. According to the property (WM) of Proposition 7, any median μ for the metric δ_C satisfies the inequality $\mu \leq b^C(\pi^*)$, and there is one median μ_0 with $\mu_0 \leq c^C(\pi^*)$.

In particular, if G is a tree, the lattice metric ∂ and the symmetric difference metric δ_C are the same.

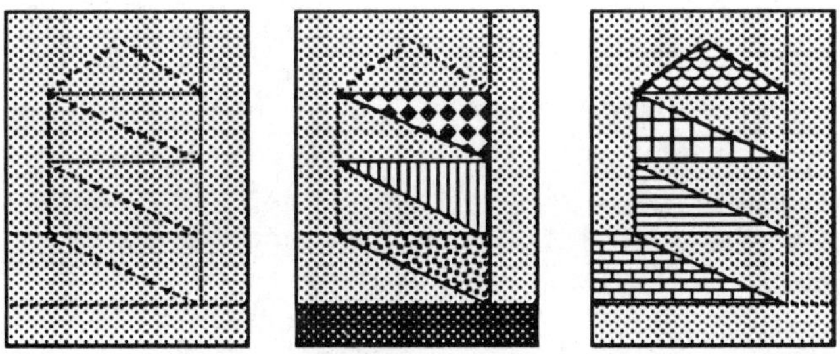

Figure 8

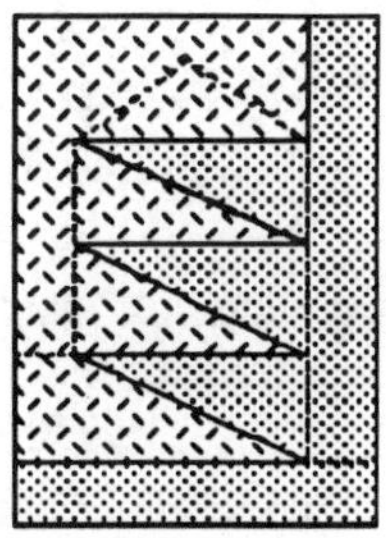

Figure 9

The following counter-example shows that the median procedure in $\Pi(C)$ with the metric δ_C does not always satisfy the Pareto principle: consider the set of cells of

Example 3 above, and the profiles of three partitions of this set into, respectively, one, five and five classes given in Figure 8 below. With the symmetric difference metric δ_C, the unique median partition under contiguity constraints is the partition of Figure 9, where the unanimity pair ab is not linked.

7. Conclusion

The three approaches of the introduction have been used throughout the paper, with some emphasis on the particularization of the median procedure to the case of partitions, and on the contributions of lattice theory.

Table 4

	Pareto	Co-Par	Independence	Symmetry	Consistency	Stability
Unanimity rule:						
	YES	*YES*	*YES*	*YES*	*YES*	*YES*
Oligarchic procedures:						
	YES	*YES*	*YES*	*NO*	not relevant	*YES*
Quota rules:						
	YES	*YES*	*NO*	*YES*	*NO*	*YES*
Dual quota rules:						
	YES	*YES*	*NO*	*YES*	*NO*	*YES*
Median procedure for the symmetric difference metric						
- in the partition lattice Π:						
	YES	*YES*	*NO*	*YES*	*YES*	*YES*
- in a lattice $\Pi(C)$ of constrained partitions (general case):						
	NO	*YES*	*NO*	*YES*	*YES*	*YES*
- in a distributive lattice $\Pi(C)$ of constrained partitions:						
	YES	*YES*	*YES*	*YES*	*YES*	*YES*
Median procedure for weight metrics						
- in Π or in not distributive lattice $\Pi(C)$ of constrained partitions:						
	NO	*YES*	*NO*	*YES*	*YES*	*YES*
- in a distributive lattice $\Pi(C)$ of constrained partitions:						
	YES	*YES*	*YES*	*YES*	*YES*	*YES*
Median procedure for the lattice metrics ∂_C						
- in the partition lattice Π or in a lattice $\Pi(C)$ of constrained partitions (general case):						
	YES	open problem	*NO*	*YES*	*YES*	*NO*
- in a modular lattice $\Pi(C)$ of constrained partitions:						
	YES	*YES*	*NO*	*YES*	*YES*	open problem
- in a distributive lattice $\Pi(C)$ of constrained partitions ($\partial_C = \delta_C$):						
	YES	*YES*	*YES*	*YES*	*YES*	*YES*

In fact, various types of procedures have been considered, and in various instances. Some relations between these procedures have been established, and also some properties that they satisfy or not. Table 4 summarizes our knowledge on the properties fulfilled by a selection of the procedures we met. To get it, we used the facts that independence always implies the stability for objects, and that the quota rules are not consistent, but stable.

Stability of, say, c_q, comes from the fact that if $c_q(\pi^*) = \pi$, then the union A of some classes of π is an union of connected classes of $C_q(\pi^*)$ and that $\pi_{|A}$ is the set of these classes, while $C_q(\pi^*_{|A})$ admits as connected classes the connected classes of $C_q(\pi^*)$ that are included in A. In other words: $C_q(\pi^*)_{|A} = C_q(\pi^*_{|A})$, and also: $B_q(\pi^*)_{|A} = B_q(\pi^*_{|A})$.

As a counter-example to consistency of c_q, we can consider the following: $X = \{a,b,c\}$; $\pi^* = (\pi_1,\pi_2,\pi_3)$ and $\pi'^* = (\pi'_1,\pi'_2,\pi'_3)$ are defined by: π_1 = /ab/c/, $\pi_2 = \pi'_2$ = /abc/, $\pi_3 = \pi'_3$ = /a/bc/, π'_1 = /ac/b/; we have $c_{1/2}(\pi^*)$ = /abc/, $c_{1/2}(\pi'^*)$ = /abc/. But $c_{1/2}(\pi^*\pi'^*)$ = /a/bc/. Counter-examples involving c'_q, b_q or b'_q are let to the reader. Discussion about consistency and quota rules in abstract lattices may be found in Barthélemy and Janowitz [10].

A counter-example to stability of medians for the metric ∂ is: $X = \{a,b,c,d,e\}$; $\pi^* = (\pi_1,\pi_2,\pi_3,\pi_4,\pi_5)$ where π_1 = /ac/b/de/, π_2 = /abcd/e/, π_3 = /ab/c/d/e/, π_4 = /abd/ce/ and π_5 = /abc/de/. The unique median for ∂ is /abc/de/; of course, $\pi^*_{|de}$ has /d/e/ as median for ∂.

For weight metrics, the highly desirable Pareto property is not satisfied in several cases. Then, the problem of the search of medians may be changed into: "minimize $R(\pi,\pi^*)$, subject to $\gamma(xy,\pi^*) = 1$ implies $\pi_{xy} \leq \pi$". It may be shown that Property (WM) still holds for the solutions of this modified problem. In fact, it is equivalent to identify the elements x and y of each unanimity pair xy into one element, with an adequate change on the weights of the pairs xz, $z \neq x, y$. On the same way, mixed consensus procedures may be considered, like "minimize the remoteness, subject to $\gamma(xy,\pi^*) \geq q$ implies $\pi_{xy} \leq \pi$", where q is a threshold percentage, as .9 or .95.

References

1. M. Aigner *Combinatorial Theory*, Springer-Verlag, Berlin, 1979.

2. S.G. de Amorim, J.-P. Barthélemy, C.C. Ribeiro, Clustering and clique partitioning: simulated annealing and tabu search approaches, *J. of Classification* **9** (1992), 17-41.

3. P. Arabie, S.A. Boorman, Multidimensional scaling of measures of distances between partitions, *J. Math. Psychol.* **17** (1973), 31-63.

4. K.J. Arrow, *Social Choice and Individual Values*, Wiley, New York, 1951.

5. M. Barbut, Médiane, distributivité, éloignements, repr. (1980), *Math. Sci. hum.* **70** (1961), 5-31.

6. M. Barbut, B. Monjardet, *Ordre et Classification, Algèbre et Combinatoire*, Hachette, Paris, 1970.

7. J.-P. Barthélemy, Sur les éloignements symétriques et le principe de Pareto, *Math. Sci. hum.* **56** (1976), 97-125.

8. J.P. Barthélemy, Remarques sur les propriétés métriques des ensembles ordonnés, *Math. Sci. hum.* **61** (1978), 39-60.

9. J.P. Barthélemy, Caractérisations axiomatiques de la distance de la différence symétrique entre des relations binaires, *Math. Sci. hum.* **67** (1979), 85-113.

10. J.P. Barthélemy, M.F. Janowitz, A formal theory of consensus, *SIAM J. on Discrete Math.* **4** (1991), 305-322.

11. J.P. Barthélemy, B. Monjardet, The Median Procedure in Cluster Analysis and Social Choice Theory, *Math. Soc. Sci.* **1** (1981), 235-268.

12. J.P. Barthélemy, B. Monjardet, The Median Procedure in Data Analysis: New Results and Open Problems, in *Classification and related methods of data analysis* (ed. H.H. Bock), North-Holland, Amsterdam, 1988, pp. 309-316.

13. G. Birkhoff, *Lattice Theory* (3rd ed.), Amer. Math. Soc., Providence, 1967.

14. S.A. Boorman, D.C. Olivier, Metrics on spaces of finite trees, *J. Math. Psychol.* **10** (1973), 26-59.

15. W.H.E. Day, The complexity of computing metric distances between partitions, *Math. Soc. Sci.* **1** (1981), 269-287.

16. B. Debord, *Axiomatisation des procédures d'agrégation de préférences*, Doctorate Thesis, University of Grenoble, 1987.

17. G. De Soete, L. Hubert and P. Arabie, The comparative performance of simulated annealing on two problems of combinatorial data analysis, in *Data Analysis and Informatics* (eds. E. Diday et al.), North Holland, Amsterdam, 1988, pp. 489-496.

18. M. Grötschel and Y. Wakabayashi, A cutting plane algorithm for a clustering problem, *Math. Programming* **45** (1989), 59-96.

19. M. Grötschel and Y. Wakabayashi, Facets of the Clique partitioning Polytope, *Math. Programming* **47** (1990), 367-387.

20. D. Jones and M. Beltramo, Solving partitioning problems with GA, *4-th Conf. on Genetic Algorithms*, 1991.

21. J. Kemeny, Mathematics without numbers, *Daedalus* **88** (1959), 577-591.

22. M. Krivanek and J. Moravek, NP-hard problems in hierarchical-tree clustering, *Acta Informatica* **23** (1986), 311-323.

23. B. Leclerc, Efficient and binary consensus functions on transitively valued relations, *Math. Soc. Sci.* **8** (1984), 45-61.

24. B. Leclerc, Medians and majorities in semimodular lattices, *SIAM J. on Discrete Math.* **3** (1990), 266-276.

25. B. Leclerc, Lattice valuations, medians and majorities, *Discrete Math.* **111** (1993), 345-356.

26. B. Leclerc, Medians for weight metrics in the covering graphs of semilattices, *Discrete Applied Math.* **49** (1994), 281-297.

27. B. Leclerc and B. Monjardet, Latticial Theory of Consensus, 1992, *in* W. Barnett, H. Moulin, M. Salles, N. Schofield, eds., *Social Choice, Welfare, and Ethics*, Cambridge University Press, to appear (1995).

28. J.-F. Marcotorchino, *Agrégation des similarités en classification automatique*, Doctorate Thesis, University of Paris 6, 1981.

29. J.-F. Marcotorchino and P. Michaud, Heuristic approach to the similarity aggregation problem, *Methods of Operation Research* **43** (1981), 395-404.

30. P. Michaud, *Hommage à Condorcet*, Étude F.094, R.R. IBM France, Centre Scientifique de Paris, 1985.

31. P. Michaud and J.-F. Marcotorchino, Optimisation en analyse des données relationnelles, in *Data Analysis and Informatics* (eds. E. Diday et al.),North Holland, Amsterdam, 1980, pp. 655-670.

32. B.G. Mirkin, The problems of approximation in Spaces of Relations and Qualitative Data Analysis, *Automatika y Telemecanika*, engl. trans. *Information and Remote Control* **35** (1974), 1424-1431.

33. B.G. Mirkin, On the problem of reconciling partitions, in *Quantitative Sociology, Intern. Persp. on Math. and Statist. Modelling*, Academic Press, New York, 1975, pp. 441-449.

34. B.G. Mirkin, L.B. Chernyi, Measurement of distance between distinct partitions of a finite set of units, *Automatika i Telemecanika*; engl. trans. *Information and Remote Control* **31**, n°5 (1970), 786-792.

35. B. Monjardet, Théorie et application de la médiane dans les treillis distributifs finis, *Annals of Discrete Math.* **9** (1980), 87-91.

36. B. Monjardet, Metrics on partially ordered sets - A survey, *Discrete Math.* **35** (1981), 173-184.

37. B. Monjardet, Arrowian characterization of latticial federation consensus functions, *Math. Soc. Sci.* **20** (1990), 51-71.

38. D.A. Neumann, K. Boukaabar, *The median problem with respect to the L^1-metric on the partition lattice*, preprint, Bowling Green State University, 1991.

39. D.A. Neumann, V. Norton, Clustering and Isolation in the Consensus Problem for Partitions", *J. of Classification* **3**, n°2 (1986), 281-297.

40. S. Régnier, Sur quelques aspects mathématiques des problèmes de classification automatique, *ICC Bull.* **4** (1965), 175-191, repr. *Math. Sci. hum.* **82** (1983), 13-29.

41. S. Régnier, Etudes sur le polyèdre des partitions, *Math. Sci. Hum* **82** (1983), 85-111.

42. M. Schader and U. Tüshaus, Ein Subgradienten-Verfahren zur Klassification qualitativer Daten, *O. R. Spektrum* **7** (1985), 1-5.

43. D. Snyers, Clique partitioning problem and genetic algorithms, *Int. Conf. on Artificial Neural Nets and Genetic Algorithms*, Innsbrück, Austria, 1993.

44. E.-G. Talbi and P. Bessière, A parallel genetic algorithm for the graph partitioning problem, *ACM Conf. on Supercomputing*, Köln, Germany, 1991.

45. U. Tüshaus (1983), *Aggregation binärer Relationen in des qualitativen Datenanalyse*, Berlin: Athenäum, Mathematical Systems in Economics 82.

46. G. Vescia, Automatic classification of cetaceans by similarity aggregation, in *Data Analysis in Real Life Environment : Ins and Outs of Solving Problems* (eds. J.-F. Marcotorchino, J.F. Proth, J. Jansed), Elsevier, Amsterdam, 1985, pp. 7-24.

47. Y. Wakabayashi, *Aggregation of binary relations: algorithmic and polyhedral investigations*, Doctoral Thesis, Universität Augsburg, 1986.

48. H.P. Young, An axiomatization of Borda's rule, *J. Econ. Theory* **9** (1974), 43-52.

49. C.T. Zahn, Approximating symmetric relations by equivalence relations, *SIAM J. on Appl. Math.* **12** (1964), 840-847.

LABORATOIRE D'INTELLIGENCE ARTIFICIELLE ET SYSTEMES COGNITIFS, TÉLÉCOM BRETAGNE, BREST, FRANCE.
Current Address: Laboratoire d'Intelligence Artificielle et Systèmes Cognitifs, Télécom Bretagne, B.P. 832, F-29285 BREST CEDEX, FRANCE.
E-mail Address: jp.barthelemy@enst-bretagne.fr

CENTRE D'ANALYSE ET DE MATHÉMATIQUE SOCIALES, ÉCOLE DES HAUTES ÉTUDES EN SCIENCES SOCIALES, 54 Boulevard Raspail, F-75270 PARIS CEDEX 06, FRANCE.
E-mail Address: leclerc@msh-paris.fr

DIMACS Series in Discrete Mathematics
and Theoretical Computer Science
Volume 19, 1995

STRUCTURAL PROPERTIES OF PYRAMIDAL CLUSTERING

P. BERTRAND

ABSTRACT. Pyramidal clustering is an extension of the hierarchical clustering method. In this paper, our aim is to describe the mathematical properties of the pyramidal clustering model. Most of them generalize the properties of the indexed hierarchies. We first extend the well-known one-to-one correspondence between ultrametrics and indexed hierarchies to the case of pyramidal structures. Then, we turn our attention to the order features of the pyramidal clustering model. It leads to the enumeration of the orders such that for a given binary pyramidal structure, each cluster is a set of consecutive elements.

1. INTRODUCTION

In spite of well-known drawbacks, Hierarchical Clustering is one of the most widely used methodology in classification. Two attractive features of this model are worth being noticed in order to perceive why hierarchical trees are so frequently used. First, if E is a set of n objects to be clustered, then a hierarchical tree provides no more than $n-1$ non single object clusters among the 2^n subsets of the power set $P(E)$. Second, a dendrogram provides an intuitive graphic representation of the data set.

Pyramidal Clustering, proposed by Diday [11], is one extension scheme of Hierarchical Clustering. The indexed structures obtained from pyramidal clusterings are called *weakly indexed pyramids.*

Weakly indexed pyramids are numerically stratified systems of clusters. They define a cophenetic proximity measure which is the level where two objects first meet in the same cluster (see for example Janowitz [18]). In the pyramidal case, the clusters may overlap, provided that two requirements are fulfilled. First, any intersection of clusters remains a cluster, and second, the objects can be ranked according to some total order (which is said to be *compatible*) so that each cluster is an interval of this order. In comparison with hierarchies, weakly indexed pyramids induce Robinsonian proximities that include ultrametrics as a very particular case (Diday [11], Bertrand & Diday [5]). Then, Pyramidal Clustering provides a better data fit than Hierarchical Clustering. Algorithms for Pyramidal Clustering have been proposed by Diday [11], Bertrand [3] and Durand [12].

1991 *Mathematics Subject Classification.* 62H30; Secondary 06A05.
Key words and phrases. Classification, Partitioning, Pyramidal clustering, Seriation.

Although Pyramidal Clustering leads to a structure of clusters which is slightly more complicated than Hierarchical Clustering does, it should be further pointed out that the two mentioned attractive characteristics of hierarchies are not greatly altered in the pyramidal case. First, if n objects are to be clustered, then the number of non single object clusters in each weakly indexed pyramid is no more than $n(n-1)/2$. Moreover, it is possible to prune any pyramidal structure in order to decrease the number of clusters while keeping the loss of fit below a given threshold. Second, each cluster is still graphically represented by an horizontal line and joined to its successors by vertical or slanted lines. The picture obtained called hereafter *pyramidal representation*, is highly similar to a dendrogram, except that it is no more a tree but a planar graph.

Let us consider the simple example provided by Figure 1. In this figure a dendrogram (Figure (1a)) and a pyramidal representation (Figure (1b)) are placed side by side. The set of objects to be clustered is $E = \{a,b,c,d\}$. Figure (1a) shows

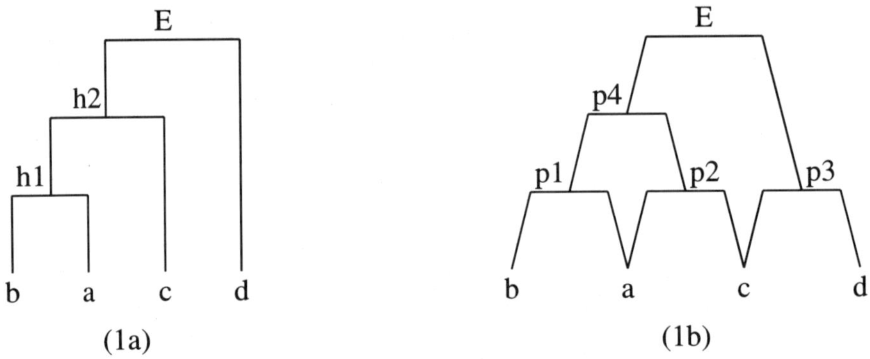

FIGURE 1. a dendrogram (1a) and a pyramidal representation (1b).

a hierarchy for which the non trivial clusters are $h_1 = \{a,b\}$ and $h_2 = \{a,b,c\}$, whereas Figure (1b) shows a pyramidal representation for which the non trivial clusters are $p_1 = \{a,b\}$, $p_2 = \{a,c\}$, $p_3 = \{c,d\}$ and $p_4 = \{a,b,c\}$.

As is suggested by this particular case, pyramidal clustering generally provides expected clusters that are overlapping, or non exclusive, and such that all clusters can be embedded in a subspace of dimensionality reduced to one, for example the clusters may be described as convex parts of a reasonably smooth curve.

Nevertheless, one might object that in many cases the actual data structure is more accurately described by a representation where the data units are partially ranked, instead of being totally ranked. Such a partial order occurs when not all objects, but only subsets of objects are ranked according to a total order, or in other words, the objects are clustered according to an ordered partition. Pyramidal representations provide such partitions. They can be identified by selecting the smallest level value at which the representation may be cut so that the associated clusters can be ranked according to the initial proximity between the objects.

A necessary condition for using the pyramidal model is the ease of interpretation of its outputs. However, we will not turn our attention to the different ways of discovering useful interpretations, but it should be clear that the main results of our research yield useful aids for interpreting pyramidal representations. The next section of this paper is devoted to definitions and basic properties and leads to the extension of the one-to-one correspondence between indexed hierarchies and ultrametrics. The third section describes the structure of the orders that are compatible with a given weakly indexed pyramid. This section also provides a formula which indicates the total number of orders that are compatible with a binary pyramidal structure. We close the text by presenting several possible extensions of this work.

2. Definitions and basic properties

The results of this section are not new: they have been already presented by Diday [11], and then by Bertrand and Diday [4], Bertrand [3], Durand and Fichet [13], Durand [12], Batbedat [1] and, Bertrand and Diday [5]. Here, we give an English version of all these basic results and propose relevant proofs by using a threshold graph proximity approach.

In the following, the symbol $\subseteq$ will be used to denote set inclusion, whereas the symbol $\subset$ will denote strict set inclusion, that is $A \subset B$ if A is a subset of B and $A \neq B$.

We will consider a finite set E of n objects to be clustered. The symbol θ will denote a total order on E. A subset A of E is called *convex* for a total order θ if for all x, $y \in A$, the closed interval $[x,y]$ of θ is included in A.

It is clear that since E is finite and the order θ is total, then a subset A is convex for θ, if and only if A is an interval of θ.

An element x of E will be called an *extremum* of the order θ if it is a minimal or a maximal element of E according to the order θ.

Recall that a dissimilarity d is an application from E^2 into $\mathbf{R}^+$ satisfying the following properties:

$$\begin{array}{lll} (D1) & \text{for all } x \in E, & d(x,x) = 0 \\ (D2) & \text{for all } x,y \in E, & d(x,y) = d(y,x) \end{array}$$

The dissimilarity d is said to be *definite* if d satisfies the following condition:

$$(DD) \quad \text{for all } x,y \in E, \quad d(x,y) = 0 \Rightarrow x = y$$

The following inequalities are restrictive conditions that are frequently considered:

$$\begin{array}{lll} (DM) & \text{for all } x,y,z \in E, & d(x,y) + d(y,z) \geq d(x,z) \\ (DU) & \text{for all } x,y,z \in E, & max\{d(x,y), d(y,z)\} \geq d(x,z) \end{array}$$

Condition (DM), which is also called the triangle inequality, characterizes *metric* dissimilarities, whereas the dissimilarities that satisfy condition (DU), are said to be *ultrametric* dissimilarities. It has been shown that indexed hierarchies are in one-to-one correspondence with ultrametrics (Jardine and Sibson [19], or Benzecri [2]). This correspondence associates each indexed hierarchy $(\mathbf{H}, f)$ to the cophenetic proximity that is induced by $(\mathbf{H}, f)$.

In this paper, we will focus our attention on dissimilarities d that satisfy the following restrictive condition :

$$\begin{array}{ll} (DR) & \text{there exists at least one total order } \theta \text{ defined on } E \text{ such that:} \\ & x\ \theta\ y\ \theta\ z \Rightarrow max\{d(x,y), d(y,z)\} \leq d(x,z) \end{array}$$

Such dissimilarities are called *Robinsonian*: they have been introduced by Robinson [24] in order to propose a model of seriation of archeological artefacts. The set of definite and Robinsonian dissimilarities will be denoted by $\mathcal{D}_R$. The order θ that satisfies the condition (DR) is generally not unique, and is said to be *d-compatible*, or *compatible with d*.

It has been pointed out that a given dendrogram has no crossing if and only if the objects are displayed on the base line of the dendrogram according to an order which is compatible with the cophenetic proximity induced by the dendrogram (Diday [10]). This condition ensures that no crossing occur and is of course an essential property to obtain an easily readable picture of any hierarchical clustering.

It was proposed by Diday [11] to extend the set of indexed hierarchies to the set of clusterings that induce cophenetic proximities that are Robinsonian. Such clusterings, which are built so that they may be represented by a planar graph, are called *pyramids* (Diday [11]) or *pseudo-hierarchies* (Durand and Fichet [13]). If an appropriate index measure is associated to each cluster, they are said to be *weakly indexed pyramids* (Diday [11]). More precisely, a *pyramid* on E is a set $\mathbf{P}$ of non empty subsets of E such that :

$(p1)$ $E \in \mathbf{P}$ and for all $x \in E$, $\{x\} \in \mathbf{P}$,
$(p2)$ for all $A, B \in \mathbf{P}$, $A \cap B \in \mathbf{P} \cup \{\emptyset\}$,
$(p3)$ there is a total order on E such that $\mathbf{P}$ is a set of intervals of this order.

An order that satisfies condition $(p3)$, is generally not unique and is said to be $\mathbf{P}$-*compatible*, or *compatible with* $\mathbf{P}$. The elements of $\mathbf{P}$ will be called *clusters*.

The finite set $\mathbf{P}$ is naturally ordered by set inclusion. Then, a cluster A is said to be a *successor* (resp. *predecessor*) of a cluster B if A is a maximum (resp. minimum) element of the set of clusters which are included in A (resp. that contain A). We will show that any cluster has no more than two predecessors (see Proposition 2.1).

Let f be an application on a pyramid $\mathbf{P}$ taking nonnegative real values, then $(\mathbf{P}, f)$ is *a weakly indexed pyramid* if for all A, B in $\mathbf{P}$, we have:

$(p4)$ $f(A) = 0 \quad \Leftrightarrow \quad |A| = 1$
$(p5)$ $A \subseteq B \quad \Rightarrow \quad f(A) \leq f(B)$
$(p6)$ $A \subset B$ and $f(A) = f(B) \quad \Rightarrow \quad$ A has two distinct predecessors

For each A in $\mathbf{P}$, $f(A)$ is called the *level* of A. In the following, the set of weakly indexed pyramid on E will be denoted by $\mathbf{\Pi}_E$ or $\mathbf{\Pi}$, if there is no ambiguity.

Each pyramid is ordered by set inclusion, and is entirely defined by the graph associated with its predecessor relation. An element A of a pyramid $\mathbf{P}$ is represented by an horizontal segment. If A is a predecessor of B, B lies below A and the two clusters A and B are joined by a segment. Moreover, the level of each cluster A is represented by the height of the segment associated with A, which is proportional to $f(A)$. The picture obtained represents the weakly indexed pyramid $(\mathbf{P}, f)$, and is called a *pyramidal representation* (see Figure 2).

Each weakly indexed pyramid $(\mathbf{P}, f)$, and more generally, each couple $(\mathbf{Q}, g)$ where $\mathbf{Q}$ is a family of subsets that contains E and g maps $\mathbf{Q}$ into $\mathbf{R}^+$, induces a cophenetic dissimilarity ρ defined by:

$$\rho(x, y) = min\{g(A) \mid A \in \mathbf{Q}, x \in A, y \in A\}$$

In other words, for all x, y in E, the dissimilarity $\rho(x, y)$ is the level at which two objects x and y are first met in the same cluster.

Example.
Let us consider the following weakly indexed pyramid $(\mathbf{P}, f)$ described by:

$$E = \{a,b,c,d,e\}; \mathbf{P} = \{E, \{a\}, \{b\}, \{c\}, \{d\}, \{e\}, \{a,b\}, \{d,c\}, \{d,c,e\}, \{a,b,c,d\}\}$$

$$\text{for all } x \in \mathbf{P}, f(\{x\}) = 0 \;\; ; \;\; f(E) = 3 \; ;$$

$$f(\{a,b,c,d\}) = 2.5 \;\; ; \;\; f(\{a,b\}) = f(\{c,d\}) = f(\{d,c,e\}) = 1$$

In this case, it is easy to see that there are more than two $\mathbf{P}$-compatible orders. Examples of compatible orders are (*a,b,d,c,e*), (*a,b,c,d,e*), (*d,c,e,a,b*), (*e,c,d,a,b*), etc. Figure 2 shows the pyramidal representation of $(\mathbf{P}, f)$ with its induced cophenetic dissimilarity.

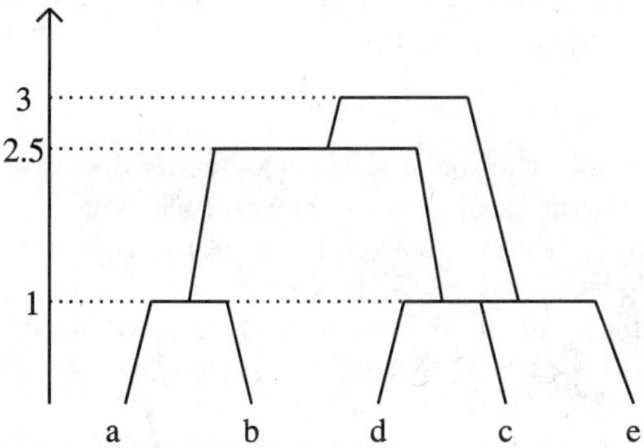

(2a)

a	b	d	c	e	
0	1	2.5	2.5	3	a
	0	2.5	2.5	3	b
		0	1	1	d
			0	1	c
				0	e

(2b)

FIGURE 2. A pyramidal representation and its induced cophenetic dissimilarity

Proposition 2.1. *Each cluster in a pyramid has no more than two predecessors.*

Proof. If the result were not true, there would exist an element A of a pyramid $\mathbf{P}$ which has at least three distinct predecessors B_1, B_2, B_3. We then deduce that:

$$A = B_1 \cap B_2 = B_1 \cap B_3 = B_2 \cap B_3 \tag{1}$$

Let us consider an order θ on E such that θ is $\mathbf{P}$-compatible. Since each element of $\mathbf{P}$ is an interval of the order θ, the clusters B_i can be expressed as follows:

$$B_i = [x_i, y_i],$$

where x_i and y_i belong to E and the index i takes values 1, 2 or 3. We can assume without any loss of generality that :

$$x_1 \; \theta \; x_2 \; \theta \; x_3$$

Therefore, the second equality of (1) can be rewritten as follows:

$$[x_2, min(y_1, y_2)] = [x_3, min(y_2, y_3)]$$

It follows that $x_2 = x_3$ and consequently $B_2 \subseteq B_3$ or $B_3 \subseteq B_2$ holds. These inclusions contradict the hypothesis that B_2 and B_3 were predecessors of the cluster A. $\square$

Proposition 2.2. *Each ultrametric is Robinsonian.*

Proof. Let u be an ultrametric defined on E, and $\mathbf{T}$ be a minimum spanning tree (MST) for u. Recall that for each pair x, y of elements of E, there is a unique path, denoted $\mathbf{T}(xy)$ between x and y. If d is a dissimilarity, a well-known characteristic property of MST is (see, e.g., Hu [15] or Leclerc [22]):

$$\text{for each } x, y, x', y' \in E, \quad x'y' \in \mathbf{T}(xy) \Rightarrow d(x', y') \leq d(x, y) \tag{2}$$

Furthermore, there are at least $2n-2$ chains among these MSTs (see Brossier [7]). By using (2), we deduce that there are at least $2n-1$ u-compatible orders, and u is Robinsonian. $\square$

DEFINITIONS. The diameter $\delta_d(A)$ of a subset A according to a dissimilarity d is the largest dissimilarity between two elements of A, that is:

$$\delta_d(A) = max\{d(x, y) \mid x \in A, y \in A\}$$

If α is a positive real value, let us consider the set of all subsets whose diameter is α. This set is ordered by inclusion and the family of its maximal elements will be denoted by $\mathbf{G}_{d,\alpha}$. It is clear that $\mathbf{G}_{d,\alpha}$ is also the set of cliques of the threshold graph associated with the dissimilarity level α, that is, $\mathbf{G}_{d,\alpha}$ is the graph where each node represents an object and the nodes x and y are linked if their dissimilarity is less than α. Let us consider the set $\bigcup_{\alpha \geq 0} \mathbf{G}_{d,\alpha}$, which is called the *filtering set* associated with d. We denote by $\mathbf{G}_d$ the intersection closure of the filtering set, defined as:

$$\mathbf{G}_d = \{A \mid A \text{ is a finite intersection of elements in } \bigcup_{\alpha \geq 0} \mathbf{G}_{d,\alpha}\}.$$

Properties of the filtering set have been described by Degenne, Flament and Verges [9]. The next proposition gives some properties of the set $\mathbf{G}_d$. A slightly different version of the following property *(iii)* may be found in Durand and Fichet [13].

Proposition 2.3. *If d is a dissimilarity on E, then:*

(i) The cophenetic dissimilarity induced by $(\mathbf{G}_d, \delta_d)$ is equal to d,
(ii) Each d-compatible order is $\mathbf{G}_d$-compatible,
(iii) If d is definite and Robinsonian then $(\mathbf{G}_d, \delta_d)$ is a weakly indexed pyramid.

Proof. *(i)* is an immediate consequence of the definition of $\mathbf{G}_d$.

(ii) Let us consider an order θ which is d-compatible and α a positive real value. According to the definition of $\mathbf{G}_d$, we have to show that each element A in $\mathbf{G}_{d,\alpha}$ is convex w.r.t. θ. If this result were not true, there would exist three elements x, y, z such that :

$$x \ \theta \ y \ \theta \ z \text{ with } x \in A, z \in A \text{ and } y \notin A.$$

Since $A \in \mathbf{G}_{d,\alpha}$, there exists a in A such that $d(y, a) > \alpha$.

If $a\ \theta\ y$, then $a\ \theta\ y\ \theta\ z$ and so $\alpha < d(a,y) \leq d(a,z)$.

Similarly, if $y\ \theta\ a$, then $x\ \theta\ y\ \theta\ a$ and it follows that $\alpha < d(a,y) \leq d(a,x)$.

Hence, in either cases we obtain a contradiction to the assumption that the diameter of A is α. Therefore, A is convex for the order θ and the result holds.

(*iii*) If $\alpha = \delta_d(E)$, then E belongs to $\mathbf{G}_{d,\alpha}$. Since d is definite, then:

$$\{x\} \in \mathbf{G}_{d,0} \subseteq \mathbf{G}_d, \text{ for each } x \text{ in } E.$$

Therefore, $\mathbf{G}_d$ satisfies $(p1)$ and from the construction of $\mathbf{G}_d$, it follows that condition $(p2)$ is also satisfied.

Property (*ii*) shows that $\mathbf{G}_d$ is a family of convex subsets according to any d-compatible order. Hence $\mathbf{G}_d$ is a pyramid.

Since d is definite, δ_d satisfies $(p4)$. The condition $(p5)$ is obviously verified. Let us prove $(p6)$.

If A and B in $\mathbf{G}_d$ are such that $\delta_d(A) = \delta_d(B)$ and $A \subset B$, then A is not a maximal element among all the subsets whose diameter is equal to $\delta_d(A)$. From the definition of $\mathbf{G}_d$, we deduce that there exists X, Y in $\mathbf{G}_d$ such that $A = X \cap Y$ with X and Y distinct from A. Consequently, $(\mathbf{G}_d, \delta_d)$ verifies $(p6)$ and the result follows. □

The next propositions of this section leads to the extension of the well-known one-to-one correspondence between indexed hierarchies and ultrametrics.

Recall that this one-to-one correspondence is the mapping which associates each indexed hierarchy to its cophenetic proximity. We now consider the extension Φ of this mapping to the set $\mathbf{\Pi}$. So, Φ is a mapping from $\mathbf{\Pi}$ into the set of dissimilarities, and $\Phi((\mathbf{P}, f))$ denotes the cophenetic dissimilarity induced by the weakly indexed pyramid $(\mathbf{P}, f)$. In order to simplify the notation, we will write $\Phi(\mathbf{P}, f)$ instead of $\Phi((\mathbf{P}, f))$.

The following two lemmas are immediate:

Lemma 2.1. *If $(\mathbf{P}, f)$ is a weakly indexed pyramid, then each $\mathbf{P}$-compatible order is $\Phi(\mathbf{P}, f)$-compatible.*

Lemma 2.2. *Φ maps $\mathbf{\Pi}$ into $\mathcal{D}_R$.*

Lemma 2.3. *Consider two weakly indexed pyramids $(\mathbf{P}, f)$ and $(\mathbf{Q}, g)$ such that $\mathbf{P} \subseteq \mathbf{Q}$. If $g_{|\mathbf{P}} = f$ and $\Phi(\mathbf{P}, f) = \Phi(\mathbf{Q}, g)$, then $(\mathbf{P}, f) = (\mathbf{Q}, g)$.*

Proof. If the result were not true, then $\mathbf{P} \subset \mathbf{Q}$, that is, there would exist an element A_1 of $\mathbf{Q}$ which is not in $\mathbf{P}$. Let us consider a $\mathbf{Q}$-compatible order θ, and let x, y be the extreme points of A_1 according to θ, i.e. $A_1 = [x,y]$. Each element of $\mathbf{Q}$ that contains x and y, contains also A_1. Then, if ρ denotes the Robinsonian dissimilarity $\Phi(\mathbf{Q}, g)$, we can write:

$$\rho(x,y) = g(A_1)$$

Let us consider the element B of $\mathbf{P}$ such that $f(B)$ is the level at which x and y are first in the same cluster of $(\mathbf{P}, f)$. From $\Phi(\mathbf{P}, f) = \Phi(\mathbf{Q}, g) = \rho$, we deduce that if an element B of $\mathbf{P}$ is such that $f(B)$ is the level at which x and y are first in the same cluster of $(\mathbf{P}, f)$, then:

$$f(B) = \rho(x, y)$$

Therefore, $g(A_1) = f(B)$, and since $g_{|\mathbf{P}} = f$, we get $g(A_1) = g(B)$. By construction of B, x and y are in B which must be an interval of θ, since $B \in \mathbf{P} \subseteq \mathbf{Q}$. Then, we deduce:

$$A_1 = [x, y] \subseteq B$$

But, from our working hypothesis, $A_1 \notin \mathbf{P}$ and then $A_1 \neq B$. From $(p6)$ there are two elements A_{11} and A_{12} in $\mathbf{Q}$ such that:

$$A_1 = A_{11} \cap A_{12} \text{ with } A_{11} \neq A_1 \text{ and } A_{12} \neq A_1$$

Note that at least one of the two clusters A_{11} and A_{12} is not in $\mathbf{P}$, for $A_1 \notin \mathbf{P}$.

Let us denote by A_2 one of these clusters that does not belong to $\mathbf{P}$. Applying the previous arguments to A_2 shows that there exists a cluster A_3 in $\mathbf{Q}$ but not in $\mathbf{P}$, such that A_2 is strictly included into A_3. By induction, we construct in this way an infinite sequence of clusters of $\mathbf{Q}$ that are strictly ordered by set inclusion. This contradicts the fact that E is finite. Consequently, $\mathbf{P} = \mathbf{Q}$ and $f = g$. □

Proposition 2.4. *If* $(\mathbf{P}, f)$ *is a weakly indexed pyramid, then* $(\mathbf{P}, f) = (\mathbf{G}_d, \delta_d)$, *where* d *denotes the cophenetic dissimilarity induced by* $(\mathbf{P}, f)$.

Proof. Since the dissimilarity d is definite and Robinsonian (Lemma 2.2), $(\mathbf{G}_d, \delta_d)$ lies in the set $\mathbf{\Pi}$ (Proposition 2.3). The property (i) of Proposition 2.3 proves that $\Phi(\mathbf{G}_d, \delta_d) = d$, hence :

$$\Phi(\mathbf{G}_d, \delta_d) = \Phi(\mathbf{P}, f)$$

According to Lemma 2.3, the result will be true if we show that $\mathbf{G}_d \subseteq \mathbf{P}$ and $f_{|\mathbf{G}_d} = \delta_d$. Let us first note that an order which is $\mathbf{P}$-compatible, is both d-compatible (Lemma 2.1), and $\mathbf{G}_d$-compatible (Proposition 2.3 $-(ii)$). In this proof, we will denote by θ such an order.

Since $\mathbf{P}$ is closed under intersection, we will set $\mathbf{G}_d \subseteq \mathbf{P}$ by only proving that for all $\alpha \geq 0$, $\mathbf{G}_{d,\alpha} \subseteq \mathbf{P}$. The order θ is $\mathbf{G}_d$-compatible. Therefore, if $\alpha \geq 0$ and $A \in \mathbf{G}_{d,\alpha}$, there exists u, v in E such that $A = [u, v]$, i.e. A is an interval of the order θ and u, v are the extreme points of A. But θ is also d-compatible, thus:

$$d(u, v) = \delta_d(A) = \alpha$$

Now, the weakly indexed pyramid $(\mathbf{P}, f)$ induces d, so there is a cluster B in $\mathbf{P}$ such that:

$$u \in B,\ v \in B \text{ and } f(B) = \alpha \tag{3}$$

The cluster B is convex for θ and $A = [u, v] \subseteq B$. Assume that this inclusion is strict, then $\delta_d(B) > \alpha$, for A is in $\mathbf{G}_{d,\alpha}$. Therefore, there are x and y in B such that $d(x, y) > \alpha$.

But, by definition, $d(x, y) = min\{f(C) \mid x \in C,\ y \in C \text{ and } C \in \mathbf{P}\}$ and $f(B) > \alpha$, which contradicts (3). Consequently, $A = B$, and $\mathbf{G}_d \subseteq \mathbf{P}$.

Now, let us show that $f_{|\mathbf{G}_d} = \delta_d$. Let A be an element of $\mathbf{G}_d$ and u, v be the extreme points of A according to an order θ which is $\mathbf{P}$-compatible. Since we know that θ is also d-compatible, we have:

$$d(u,v) = \delta_d(A) \ . \tag{4}$$

From the definition of f, it follows that:

$$f(A) \geq d(u,v) \ . \tag{5}$$

Let us consider the element B of $\mathbf{P}$ such that $f(B)$ is the level at which u and v first meet in the same cluster of $(\mathbf{P}, f)$. We have $d(u,v) = f(B)$. From $A = [u,v] \subseteq B$, we deduce:

$$f(A) \leq f(B) = d(u,v) \ . \tag{6}$$

Finally by (4), (5) and (6), we get $f(A) = \delta_d(A)$, i.e. $f_{|\mathbf{G}_d} = \delta_d$. □

Proposition 2.5. *Let θ be a total order on E and let $(\mathbf{P}, f)$ be a weakly indexed pyramid, then the two following assertions are equivalent:*

(*i*) *θ is compatible with $\Phi(\mathbf{P}, f)$,*
(*ii*) *θ is compatible with $\mathbf{P}$.*

Proof. (*ii*) $\Rightarrow$ (*i*) is exactly Lemma 2.1.
(*i*) $\Rightarrow$ (*ii*). Let θ be a $\Phi(\mathbf{P}, f)$-compatible order. From Proposition 2.4, $\mathbf{P} = \mathbf{G}_d$ and by Proposition 2.3 – (*ii*), the order θ is $\mathbf{G}_d$-compatible, and so is also $\mathbf{P}$-compatible. □

As a consequence of Proposition 2.2 and Proposition 2.5, there follows:

Corollary 2.1. *Each indexed hierarchy is a weakly indexed pyramid.*

The following result is due to Diday [11] and, in a different form to Durand and Fichet [13].

Theorem 2.1. *The mapping Φ is a one-to-one correspondence from $\mathbf{\Pi}$ onto $\mathcal{D}_R$.*

Proof. The injectivity of Φ is a consequence of Proposition 2.4. By properties (*i*) and (*iii*) of Proposition 2.3, Φ is also surjective. □

3. Properties of the compatible orders.

Let $\mathcal{F}$ be a family of subsets of E, and consider an order $\ll$ on $\mathcal{F}$ such that:

(C_1) Two distinct elements A and B in $\mathcal{F}$ are comparable for $\ll$ if and only if they are not comparable for set inclusion.

We will denote by $O_{\mathcal{F}}$ the set of orders on $\mathcal{F}$ satisfying condition (C_1).

The purpose of the next paragraph is to show that if $\mathcal{F}$ is a family of subsets of E such that all singletons of E are in $\mathcal{F}$ then there is a bijection between $O_{\mathcal{F}}$ and $C(\mathcal{F})$, where $C(\mathcal{F})$ denotes the set of $\mathcal{F}$-compatible orders.

3.1. A one-to-one correspondence between $O_{\mathcal{F}}$ and $C(\mathcal{F})$. In this paragraph, $\mathcal{F}$ will be a family of subsets of E which contains all the singletons of E.

The idea is as follows: if $\mathcal{F}$ contains all singletons and $\ll$ is an order in $O_{\mathcal{F}}$, then $\ll$ induces a total order on E, since two singletons are always comparable according to the order $\ll$. So each order in $O_{\mathcal{F}}$ will be mapped to its unique induced order defined on E.

According to such an induced order, the minimum (resp. maximum) of an arbitrary set A in $\mathcal{F}$ will be denoted by $minA$ (resp. $maxA$). Hence, for all a in A, we get:

$$\{minA\} \ll \{a\} \ll \{maxA\} .$$

The next result is an immediate consequence of the definition of $O_{\mathcal{F}}$.

Lemma 3.1. *Assume that $\ll$ is an element of $O_{\mathcal{F}}$. If A, B, C are in $\mathcal{F}$ and x is in E, then :*

$$\begin{array}{llcl}
(a1) & A \subseteq B \ll C & \Rightarrow & A \subset C \text{ or } A \ll C \\
(a2) & C \ll B \supseteq A & \Rightarrow & A \subset C \text{ or } C \ll A \\
(a3) & A \supseteq B \ll C & \Rightarrow & C \subset A \text{ or } A \ll C \\
(a4) & C \ll B \subseteq A & \Rightarrow & C \subset A \text{ or } C \ll A \\
(b1) & \{x\} \ll A & \Rightarrow & \{x\} \ll \{minA\} \\
(b2) & A \ll \{x\} & \Rightarrow & \{maxA\} \ll \{x\} .
\end{array}$$

Lemma 3.2. *Each order in $O_{\mathcal{F}}$ induces an $\mathcal{F}$-compatible order on E.*

Proof. let $\ll$ in $O_{\mathcal{F}}$ and A in $\mathcal{F}$. By proving the implication:

$$\text{for all } u,v \text{ in } A, \quad \{u\} \ll \{x\} \ll \{v\} \quad \Rightarrow \quad x \in A$$

we will show that A is an interval of the order induced by $\ll$.

Assume that u, v in A and x in E are such that $\{u\} \ll \{x\} \ll \{v\}$, then:

$$A \supseteq \{u\} \ll \{x\} \quad \text{and} \quad \{x\} \ll \{v\} \subseteq A$$

By applying $(a3)$ and $(a4)$, we get:

$$(x \in A \text{ or } A \ll \{x\}) \quad \text{and} \quad (x \in A \text{ or } \{x\} \ll A) .$$

Therefore, if $x \notin A$ then $A \ll \{x\} \ll A$, hence $\{x\} = A$, which contradicts $x \notin A$. Consequently, $x \in A$ and this proves that A is an interval of the order induced by $\ll$. □

Lemma 3.3. *Let $\ll$ be in $O_{\mathcal{F}}$ and let $\leq$ denotes the order induced by $\ll$ on E. If A and B are two distinct elements of $\mathcal{F}$, then the following two assertions are equivalent:*

$$\begin{array}{ll}
(i) & A \ll B \\
(ii) & minA < minB \text{ and } maxA < maxB .
\end{array}$$

Proof. $(i) \Rightarrow (ii)$. If (i) holds, then $A \ll B \supseteq \{minB\}$ and it follows by using $(a2)$:

$$\{minB\} \subset A \ \text{ or } \ A \ll \{minB\}$$

From $(b2)$ we deduce that in both cases $minA \leq minB$. For the same reasons, $maxA \leq maxB$. If one of these inequalities becomes an equality, then $A \subseteq B$ or $B \subseteq A$, which is inconsistent with the fact that A and B are distinct and satisfy $A \ll B$. Therefore, (ii) holds.

$(ii) \Rightarrow (i)$. Assume (ii), that is:

$$minA < minB \quad \text{and} \quad maxA < maxB \; .$$

Let us note that:

$$A \subseteq B \quad \Leftrightarrow \quad minB \leq minA \quad \text{and} \quad maxA \leq maxB$$

$$B \subseteq A \quad \Leftrightarrow \quad minA \leq minB \quad \text{and} \quad maxB \leq maxA \; .$$

It follows that neither $A \subseteq B$ nor $B \subseteq A$ are verified, and consequently, $A \ll B$ or $B \ll A$ holds. If $B \ll A$ then by use of $(i) \Rightarrow (ii)$, $minB < minA$ and this contradicts $minA < minB$. Hence $A \ll B$, and (ii) implies (i). □

Proposition 3.1. *Let $f : O_{\mathcal{F}} \to C(\mathcal{F})$, where $f(\ll)$ denotes the order induced on E by $\ll$ for an order $\ll$ in $O_{\mathcal{F}}$. The application f is a one-to-one correspondence between $O_{\mathcal{F}}$ and $C(\mathcal{F})$.*

Proof. Note that f takes its values into $C(\mathcal{F})$ according to Lemma 3.2. We first prove that f is injective. Let $\ll_1$ and $\ll_2$ be two orders in $O_{\mathcal{F}}$ such that:

$$f(\ll_1) = f(\ll_2),$$

and let denote by $\leq$ this common induced order on E.

From Lemma 3.3, if A and B are two distinct elements of $\mathcal{F}$, then for all $i \in \{1, 2\}$:

$$A \ll_i B \quad \Leftrightarrow \quad (minA < minB \ \text{ and } \ maxA < maxB)$$

Therefore, $\ll_1$ and $\ll_2$ are identical, and f is injective.

We now prove that f is surjective. If $\leq$ is an arbitrary order of $C(\mathcal{F})$, we consider the reflexive binary relation $\mathcal{R}$ which is defined for all distinct A, B in $\mathcal{F}$, by:

$$A \, \mathcal{R} \, B \quad \Leftrightarrow \quad (minA < minB \ \text{ and } \ maxA < maxB)$$

where $minA$ (resp. $maxA$) denotes the smallest (resp. greatest) element of the set A according to the order $\leq$.

It is clear that $\mathcal{R}$ is an order on $\mathcal{F}$. Let us prove that $\mathcal{R}$ is in $O_{\mathcal{F}}$, i.e. that condition (C_1) is verified. If A and B are two distinct elements of $\mathcal{F}$, there are five possible cases:

$$\begin{array}{lllll}
(i) & minA < minB & \text{and} & maxA < maxB \\
(ii) & minA < minB & \text{and} & maxA \geq maxB \\
(iii) & minA > minB & \text{and} & maxA \leq maxB \\
(iv) & minA > minB & \text{and} & maxA > maxB \\
(v) & minA = minB & \text{and} & maxA \neq maxB \; .
\end{array}$$

In cases (ii), (iii) and (v), $A \subseteq B$ or $B \subseteq A$. Case (i) implies $A \, \mathcal{R} \, B$ and, case (iv) implies $B \, \mathcal{R} \, A$. Hence, condition (C_1) necessarily holds. In other words,

$\mathcal{R} \in O_{\mathcal{F}}$. Moreover, it is clear that $f(\mathcal{R})$ and $\leq$ are identical. Consequently, f is surjective and the result holds. □

3.2. Number of orders which are compatible with a binary pyramid. Let $\mathcal{A}$ and $\mathcal{F}$ be two families of subsets of E such that $\mathcal{A} \subseteq \mathcal{F}$. Consider an order $\ll$ defined on $\mathcal{F}$, and the restriction $\ll_{\mathcal{A}}$ of $\ll$ on $\mathcal{A}$. It is clear that if $\ll$ is in $O_{\mathcal{F}}$ then $\ll_{\mathcal{A}}$ is also in $O_{\mathcal{A}}$.

Let us consider the intersection graph $\Omega(\mathcal{F})$ of $\mathcal{F}$ and its maximal connected subgraphs which naturally induces a partition $\Gamma = \{\mathcal{C}_1, \dots, \mathcal{C}_r\}$ on the family $\mathcal{F}$. The r families $\mathcal{C}_i$ of subsets will be called *connected components* of $\mathcal{F}$.

Lemma 3.4. *Consider an order $\ll$ in $O_{\mathcal{F}}$ and let A and B be in arbitrary distinct connected components $\mathcal{C}_i$ and $\mathcal{C}_j$. If $A \ll B$, then:*

$$\textit{for all } (X,Y) \textit{ in } \mathcal{C}_i \times \mathcal{C}_j, \quad X \ll Y\ .$$

Proof. For reasons of symmetry, we only have to handle the following three cases:

$$\begin{array}{llll} (case1) & X \subseteq A & \text{and} & Y \subseteq B, \\ (case2) & X \subseteq A & \text{and} & B \subseteq Y, \\ (case3) & A \subseteq X & \text{and} & B \subseteq Y. \end{array}$$

By Lemma 3.1 each of these cases leads to $X \ll Y$. □

From this lemma, we deduce that each order $\ll$ in $O_{\mathcal{F}}$ induces a unique total order on the set Γ of all the connected components. Let us consider the equivalence relation $\sim$ for which two orders of $O_{\mathcal{F}}$ are $\sim$-equivalent whenever they induce the same order on Γ.

Lemma 3.5. *The number of equivalence classes of the relation $\sim$ is equal to r!*

Proof. Recall that each total order on Γ can be identified to a permutation on Γ. Assume that the order θ is identified with the identity permutation and let τ be a transposition on Γ. We will denote by θ_τ the order identified to τ, that is, θ_τ is deduced from θ by reversing all the order comparisons between the two elements exchanged by the transposition τ. It is known that, given two permutations ρ_1 and ρ_2, permutation ρ_2 may be deduced from ρ_1 by successive transpositions of two elements.

Consider an order θ induced on Γ by an order $\ll$ in $O_{\mathcal{F}}$ and let us denote by τ an arbitrary transposition on Γ which exchanges the components $\mathcal{C}_u$ and $\mathcal{C}_v$. From the previous remark, the result will be true if we can prove that θ_τ is induced by at least one order in $O_{\mathcal{F}}$.

The order θ_τ gives the same comparisons as the order θ does, except for the pairs of components that lie between $\mathcal{C}_u$ and $\mathcal{C}_v$ and for which the comparisons are reversed.

In a similar way, we define the relation $\ll_\tau$ on $\mathcal{F}$ which gives the same comparisons as the order $\ll$ does, except for the pairs of clusters $\{X, Y\}$ such that $X \in \mathcal{C}_i$ and $Y \in \mathcal{C}_j$ with:

$$\mathcal{C}_u\ \theta\ \mathcal{C}_i\ \theta\ \mathcal{C}_j\ \theta\ \mathcal{C}_v$$

for which the comparisons are reversed. The verification that $\ll_\tau$ is an order on $\mathcal{F}$ is left to the reader and it is clear that $\ll_\tau$ is in $O_\mathcal{F}$. By construction, $\ll_\tau$ induces the order θ_τ on Γ, and the result follows. $\square$

Lemma 3.6. *The number of elements of each equivalence class of the relation* $\sim$ *is equal to* $\prod_{i=1}^{r} \mid O_{\mathcal{C}_i} \mid$.

Proof. Consider an order $\ll_0$ in $O_\mathcal{F}$ and let θ be the order induced on Γ by $\ll_0$. Assume that there are m orders in the equivalence class of $\ll_0$.

If we number the equivalence classes of the relation $\sim$ according to the order θ, then each order $\ll$ which is equivalent to $\ll_0$, satisfies:

$$\text{for all } (X_1, \dots, X_r) \text{ in } \mathcal{C}_1 \times \dots \times \mathcal{C}_r, \quad X_1 \ll \dots \ll X_r \,. \tag{7}$$

Moreover, it is clear that the restriction of the order $\ll$ on each class $\mathcal{C}_i$ is in $O_{\mathcal{C}_i}$. Consequently:

$$m \leq \prod_{i=1}^{r} \mid O_{\mathcal{C}_i} \mid \tag{8}$$

Conversely, let us consider a binary relation $\mathcal{R}$ on $\mathcal{F}$ which satisfies (7) and whose restriction on each $\mathcal{C}_i$ is an order of $O_{\mathcal{C}_i}$. It is then immediate to show that $\mathcal{R}$ is an order which is in $O_\mathcal{F}$. Moreover, $\mathcal{R}$ is clearly equivalent to $\ll_0$. Therefore, $m \geq \prod_{i=1}^{r} \mid O_{\mathcal{C}_i} \mid$ and by (8) the result holds. $\square$

By applying Lemma 3.5 and Lemma 3.6, we obtain:

Corollary 3.1. *If $\mathcal{F}$ has r connected components, then:*

$$\mid O_\mathcal{F} \mid = r! \prod_{i=1}^{r} \mid O_{\mathcal{C}_i} \mid$$

DEFINITIONS. Let $\mathcal{F}$ be a family of subsets of E and $\mathcal{A}$ be a family of subsets of E such that $\mathcal{A} \subseteq \mathcal{F}$.

- If X is a subset of E, the complement of X in E will be denoted by X_c, that is $X_c = E - X$.
- The *ideal* $\mathcal{A}^-$ generated by the family $\mathcal{A}$ is:

$$\mathcal{A}^- = \{X \in \mathcal{F} \,:\, \exists A \in \mathcal{A} \text{ such that } X \subseteq A\}$$

- The *principal ideal* X^- generated by an element X of $\mathcal{F}$ is:

$$X^- = \{Y \in \mathcal{F} \,:\, Y \subseteq X\}$$

- The *filter* $\mathcal{A}^+$ generated by the family $\mathcal{A}$ is:

$$\mathcal{A}^+ = \{X \in \mathcal{F} \,:\, \exists A \in \mathcal{A} \text{ such that } A \subseteq X\}$$

- The *principal filter* X^+ generated by an element X of $\mathcal{F}$ is:

$$X^+ = \{Y \in \mathcal{F} \,:\, X \subseteq Y\}$$

- An element X of $\mathcal{F}$ is said to be *hierarchical* if:

$$\forall Y \in \mathcal{F}, \quad X \cap Y \in \{X, Y, \emptyset\}$$

- The set of successors of an element X in $\mathcal{F}$ will be denoted by $S(X)$.

The following lemma is immediate.

Lemma 3.7. *Let $\mathcal{F}$ be a family of subsets of E. If an element X of $\mathcal{F}$ is hierarchical then $\{X_c^- \cup X^+ - \{X\},\ X^-\}$ is a partition of $\mathcal{F}$.*

Lemma 3.8. *Let $\mathcal{F}$ be a binary pyramid on E and let v be an element of E which is extremal according to an $\mathcal{F}$-compatible order. The family $\mathcal{M}$ of subsets of E defined by $\mathcal{M} = \{X \in \mathcal{F} - \{E\} : X$ is hierarchical and contains $v\}$ satisfies the two conditions:*

(*i*) *$\mathcal{M}$ is a nonempty family which is totally ordered by set inclusion,*
(*ii*) *The maximum of $\mathcal{M}$ ranked by set inclusion is also an extremum of $\mathcal{F}$ ranked by any arbitrary order in $O_{\mathcal{F}}$.*

Proof. (*i*). The family $\mathcal{M}$ is nonempty since it contains $\{v\}$. All clusters in $\mathcal{M}$ admit v as a common element and v is an extremity of an $\mathcal{F}$-compatible order. Therefore $\mathcal{M}$ is totally ordered by set inclusion.

(*ii*). We proceed by induction on the number of elements of E. The result is trivial if E has 2 or 3 elements.

Let us now consider the induction step. Assuming that the result holds for any binary pyramid defined on a set of size $k < n$, we consider the case of a binary pyramid $\mathcal{F}$ defined on a set E of size n.

First of all, note that since E has only two successors then, if one of them is hierarchical then the other is also hierarchical.

Suppose that the two successors of E are hierarchical. Then M is equal to one of these successors and so M is clearly an extremum of $\mathcal{F}$ ranked by any arbitrary order in $O_{\mathcal{F}}$.

Suppose now that the two successors of E are not hierarchical. Give the label S_2 to the successor that contains M and the label S_1 to the other successor. Since S_1 and S_2 are not hierarchical, it follows that:

$$S_1 \cap S_2 \quad \notin \quad \{\emptyset, S_1, S_2\}.$$

There exists an $\mathcal{F}$-compatible order such v is an extremal element of E, therefore M^+ is totally ordered by set inclusion. So M is not included in $S_1 \cap S_2$, otherwise M would be included in S_1 and in S_2, and these are not comparable under set inclusion. Moreover M is hierarchical, then we deduce $M \cap S_1 = \emptyset$ and so $M \subseteq S_2 - (S_1 \cap S_2)$.

Let us denote by $\ll$ an arbitrary order in $0_{\mathcal{F}}$ and consider the successor S_3 of S_1 that does not contain $S_1 \cap S_2$.

Note that a cluster X is comparable with M if and only if $X \subseteq M'$. Therefore the clusters S_3 and $S_1 \cap S_2$ are mutually comparable.

Suppose that $S_3 \ll S_1 \cap S_2$, then $S_3 \ll S_1 \cap S_2 \subseteq M$ and by Lemma 3.1, $S_3 \ll M$. Another use of Lemma 3.1 proves that $S_1 \ll M$. Therefore, each cluster X in S_1^- satisfies $X \ll M$.

If X is not in S_1^-, then X is included in S_2. Using the induction hypothesis for the binary pyramid S_2^- and the fact that $S_1 \cap S_2 \ll M$, we deduce that $X \ll M$.

Consequently we have proved that if $S_3 \ll S_1 \cap S_2$ then each cluster X which is comparable with M satisfies $X \ll M$. By using the argument of symmetry, it is clear that if $S_1 \cap S_2 \ll S_3$ then each cluster X which is comparable with M satisfies $M \ll X$.

Therefore by induction, M is an extremum of $\mathcal{F}$ ranked by any arbitrary order in $O_{\mathcal{F}}$. It follows that the proof of assertion (*ii*) is complete. □

Proposition 3.2. *Let $\mathcal{F}$ be a binary pyramid on E and suppose that $S(E)$ has only one connected component. If an element M of $\mathcal{F}$ is hierarchical and is an extremum of each order in $O_{\mathcal{F}}$, then :*

$$| O_{\mathcal{F}} | = | O_{M^-} || O_{M_c^- \cup M^+ - \{M\}} | .$$

Proof. Consider the application ϕ from $O_{\mathcal{F}}$ into $O_{M_c^- \cup M^+ - \{M\}} \times O_{M^-}$ which is defined by $\phi(\ll) = (\ll_{M_c^- \cup M^+ - \{M\}}, \ll_{M^-})$.

The proposition will be proved if it can be proved that ϕ is a one-to-one correspondence.

Consider an arbitrary order $\ll$ in $O_{\mathcal{F}}$. From Lemma 3.7 and since each element of M^- is included in each element of M^+, the set of pairs of elements that are $\ll$-comparable is associated with the partition $\{ P_1, P_2, P_3 \}$ where the sets P_i are defined as:

$$\begin{aligned} P_1 &= \{ \{X,Y\} : \{X,Y\} \subseteq M^- \text{ and } X \cap Y \notin \{X,Y\} \} \\ P_2 &= \{ \{X,Y\} : \{X,Y\} \subseteq M_c^- \cup M^+ - \{M\} \text{ and } X \cap Y \notin \{X,Y\} \} \\ P_3 &= \{ \{X,Y\} : X \in M^-, \ Y \in M_c^- \text{ and } X \cap Y \notin \{X,Y\} \} \end{aligned}$$

Then the injectivity of ϕ can be proved by setting the following condition (C):

(C) For each order in $O_{\mathcal{F}}$, the comparison between two arbitrary elements X and Y such that $\{X,Y\} \in P_3$ can be deduced from comparisons between two elements Z and T such that $\{Z,T\} \in P_1 \cup P_2$.

Let $\{X,Y\}$ be an arbitrary pair in P_3. Recall that M is an extremum of any order in $O_{\mathcal{F}}$. By using Lemma 3.3 and Proposition 3.1, M contains an extremum of any $\mathcal{F}$-compatible order. Therefore M^+ is totally ordered by set inclusion and the predecessor of M is unique. We denote by L this predecessor and since $\mathcal{F}$ is a binary pyramid, L has two successors that are M and $L - M = L \cap M_c$.

Now $L \neq E$ for $S(E)$ has only one connected component. Hence there is an element u in $E - L = M_c - L$. Since they are disjoint, the clusters $\{u\}$ and $L \cap M_c$ are $\ll$-comparable. Suppose that:

$$\{u\} \ll L \cap M_c$$

By using lemma 3.1 twice, $\{u\} \ll L$ and $\{u\} \ll M$. But, since M is an extremum of the order $\ll$:

$$\forall Y \in M_c^-, \quad Y \ll M$$

Again Lemma 3.1 shows that:

$$\forall X \in M^-, \forall Y \in M_c^-, \quad Y \ll X$$

We have then shown the following implication:

$$\{u\} \ll L \cap M_c \quad \Rightarrow \quad \forall \{X, Y\} \in P_3, \quad Y \ll X$$

By symmetry, we obtain:

$$L \cap M_c \ll \{u\} \quad \Rightarrow \quad \forall \{X, Y\} \in P_3, \quad X \ll Y$$

Consequently, condition (C) is satisfied and it follows that ϕ is injective.

Let us prove that ϕ is surjective. Consider an element $(\ll_1, \ll_2)$ in $O_{M^-} \times O_{M_c^- \cup M^+ - \{M\}}$. As it has already been mentioned, $E - L = M_c - L$ is not empty. Let us denote by u an element of $M_c - L$ and consider the binary relation $\mathcal{R}$ such that according to $\mathcal{R}$, the only comparisons between two distinct elements of $\mathcal{F}$ are defined as follows:

- $\mathcal{R}_{M^-}$ is identical to $\ll_1$,
- $\mathcal{R}_{M_c^- \cup M^+ - \{M\}}$ is identical to $\ll_2$,
- if $\{u\} \ll L \cap M_c$, then for each $(X, Y) \in M^- \times M_c^-$ set $Y \mathcal{R} X$
  if $L \cap M_c \ll \{u\}$, then for each $(X, Y) \in M^- \times M_c^-$ set $X \mathcal{R} Y$

It is clear that $\mathcal{R}$ is an order. Since the inclusion comparisons restricted to M^- are not modified by adding the family $M_c^- \cup M^+ - \{M\}$, and vice versa, then it is easy to see that $\mathcal{R}$ is in $O_{\mathcal{F}}$. Moreover, we have $\phi(\mathcal{R}) = (\ll_1, \ll_2)$. Then ϕ is surjective and the required result follows. □

Theorem 3.1. *If $\mathcal{F}$ is a binary pyramid on E, then the number of $\mathcal{F}$-compatible orders is equal to $2^{h-|E|}$, where h denotes the number of hierarchical elements of $\mathcal{F}$.*

Proof. The theorem is clearly true if E has 2 or 3 elements.

We now consider the induction step. Assuming that the theorem is true for any set E of size $k \leq n-1$, we consider a set E of size n.

Suppose that $S(E)$ has more than one connected component. Since $\mathcal{F}$ is a binary pyramid, $S(E)$ has two connected components which are the two hierarchical successors of E and which we denote by S_1 and S_2. Therefore, S_1^- and S_2^- are the two connected components of $\mathcal{F} - \{E\}$ and so by Corollary 3.1, we obtain:

$$| O_{\mathcal{F}-\{E\}} | \quad = \quad 2! \, | O_{S_1^-} | \, | O_{S_2^-} |$$

By applying the induction hypothesis and since $O_{\mathcal{F}-\{E\}} = O_{\mathcal{F}}$, we get:

$$| O_{\mathcal{F}} | \quad = \quad 2^{h_1 + h_2 - n_1 - n_2 + 1}$$

where h_i denotes, for $i \in \{1, 2\}$ the number of hierarchical elements of S_i^- and where $n_1 + n_2 = | S_1 | + | S_2 | = n$. Then since $\mathcal{F} = S_1^- \cup S_2^- \cup \{E\}$, we get:

$$| O_{\mathcal{F}} | \quad = \quad 2^{h - |E|}$$

where h denotes the number of hierarchical elements of $\mathcal{F}$.

Suppose now that $S(E)$ has only one connected component. From Lemma 3.8 we know that there exists a cluster M in $\mathcal{F}$ which is hierarchical and extremal for any order in $O_{\mathcal{F}}$. Then by Proposition 3.2:

$$(9) \qquad \mid O_{\mathcal{F}} \mid \;=\; \mid O_{M^-} \mid \mid O_{M_c^- \,\cup\, M^+ \,-\, \{M\}} \mid$$

M^- is a binary pyramid on the set M, so the induction hypothesis gives:

$$(10) \qquad \mid O_{M^-} \mid \;=\; 2^{\,h_1 \,-\, |M|}$$

where h_1 designates the number of hierarchical elements of M^-.

In order to evaluate $\mid O_{M_c^- \,\cup\, M^+ \,-\, \{M\}} \mid$, let us first note that if $\mathcal{A}$ and $\mathcal{B}$ are two families of subsets such that the two graphs $(\mathcal{A}, \subseteq)$ and $(\mathcal{B}, \subseteq)$ are isomorphic, then $O_{\mathcal{A}}$ and $O_{\mathcal{B}}$ are also isomorphic.

We deduce that $O_{M_c^- \,\cup\, M^+ \,-\, \{M\}}$ where M is a set of arbitrary size, is isomorphic to $O_{M_c^- \,\cup\, \{w\}^+ - \{\{w\}\}}$ which is the same expression except that we have replaced M (resp. M^+) by $\{w\}$ (resp. $\{w\}^+$) where w is an arbitrary element of M. Consequently:

$$(11) \qquad \mid O_{M_c^- \,\cup\, M^+ \,-\, \{M\}} \mid \;=\; \mid O_{M_c^- \,\cup\, \{w\}^+ \,-\, \{\{w\}\}} \mid .$$

But now $M_c^- \cup \{w\}^+$ is a binary pyramid on $M_c \cup \{w\}$, so taking into account that $\mid O_{\{w\}} \mid = 1$, Proposition 3.2 shows:

$$(12) \qquad \mid O_{M_c^- \,\cup\, \{w\}^+} \mid \;=\; \mid O_{M_c^- \,\cup\, \{w\}^+ - \{\{w\}\}} \mid .$$

We can also apply the induction hypothesis to the binary pyramid $M_c^- \cup \{w\}^+$ and, since $\{w\}$ and E are the only hierarchical elements of $\{w\}^+$, we deduce:

$$(13) \qquad \mid O_{M_c^- \,\cup\, \{w\}^+} \mid \;=\; 2^{\,h_2 \,+\, 2 \,-\, |M_c \,\cup\, \{w\}|}$$

where h_2 designates the number of hierarchical elements of M_c^-.

From (11), (12) and (13), it follows that:

$$(14) \qquad \mid O_{M_c^- \,\cup\, M^+ \,-\, \{M\}} \mid \;=\; 2^{\,h_2 \,+\, 1 \,-\, |M_c|}$$

Thus, taking into account (9), (10) and (14), we get:

$$\mid O_{\mathcal{F}} \mid \;=\; 2^{\,h_1 \,-\, |M| \,+\, h_2 \,-\, |M_c| \,+\, 1}$$

where h_1 (resp. h_2) denotes the number of hierarchical elements of M^- (resp. M_c^-).

Therefore, since E is the only hierarchical element of $\mathcal{F}$ which is not in $M^- \cup M_c^-$, we have $h = h_1 + h_2 + 1$ and then we get:

$$\mid O_{\mathcal{F}} \mid \;=\; 2^{\,h \,-\, |E|}$$

where h denotes the number of hierarchical elements of $\mathcal{F}$. Then, by using Proposition 3.1, the proof of the theorem is complete. □

4. Conclusion

Several extensions of this work are possible:

1 - First, we have set the one-to-one correspondence between Robinsonian dissimilarities and weakly indexed pyramids without any reference to the order theoretic model for cluster analysis proposed by Janowitz [18]. We could also have studied how to introduce new conditions in order to include the pyramidal structure in this model.

2 - In this paper, the orders that are compatible with a given pyramid have been enumerated only in the particular case where the pyramid is binary. In the general case, we conjecture that the number of compatible orders with a pyramid **P** is:

$$\prod_{X \in \mathbf{P}} n_h(X)!\, 2^{m_h(X)}$$

where $n_h(X)$ is the number of connected components of $S(X)$ that are hierarchical and $m_h(X)$ is the number of connected components of $S(X)$ that are hierarchical and which have size greater than one. This suppose that we extend the definition of a hierarchical subset to the case of a hierarchical family of subsets.
This result would also provide a way to test if a graph is an interval graph and it should be compared to the approach proposed by Booth and Lueker [6].

3 - Now from a more practical point of view, since we know the total number of compatible orders, it could be interesting to verify whether this number provides valuable information about the possibility to represent data sets by totally ordered sets.

References

1. Batbedat A., *Les approches pyramidales dans la classification arborée*, Masson, Paris, 1990.
2. Benzecri J.P., *L'Analyse des Données*, Dunod, Paris, 1973.
3. Bertrand P., *Etude de la représentation pyramidale*, Thèse de troisième cycle, Université Paris IX-Dauphine, Paris, 1986.
4. Bertrand P., Diday E., *A visual representation of the compatibility between an order and a dissimilarity index : the pyramids*, Computational Statistics Quaterly, **2**, (1985), 31–44.
5. ———, *Les pyramides classifiantes: une extension de la structure hiérarchique*, C.R. Acad. Sci. Paris, t. 313, Série I, (1991), pp. 693–696.
6. Booth K.S., Lueker G.S., *Testing for the Consecutive Ones Property, Interval Graphs, and Graph Planarity Using PQ - Tree Algorithms*, J. of Computer and System Sciences, **13**, (1976), 335–379.
7. Brossier G., *Représentation ordonnée des classifications hiérarchiques*, Statistique et Analyse des Données, **2**, (1980), 31–44.
8. Critchley F., Van Cutsem B., *Predissimilarities, prefilters and ultrametrics on an arbitrary set*, Rapport de Recherche, IMAG, Université Grenoble I, 1989.
9. Degenne A. , Flament C. , Verges P. , *L'analyse de similitude fondée sur un indice de ressemblance*, Séminaire IRIA Classification automatique et perception par ordinateur, (1976), 231–248.
10. Diday E., *Croisements, ordres et ultramétriques*, Math. Sci. humaines, **21**, no. 83, (1983), pp.31–54.
11. ———, *Une représentation visuelle des classes empiétantes : les pyramides*, Rapport I.N.R.I.A. no. 291, Rocquencourt, 1984, also published in R.A.I.R.O. APII, **20** (1986), 5, 475–526.

12. Durand C., *Ordres et graphes pseudo-hiérarchiques : théorie et optimisation algorithmique*, Thèse, Université de Provence, 1989.
13. Durand C., Fichet B., *One-to-one correspondences in pyramidal representation : a unified approach*, Classification and Related Methods of Data Analysis, H.H. Bock (ed.), Amsterdam, North- Holland, (1988), 85–90.
14. Guénoche A., *Méthodes combinatoires de sériation à partir d'une dissimilarité*, Data Analysis and Informatics 5, E. Diday et al. (eds.), Amsterdam, North-Holland, (1987).
15. Hu T.C., *The maximum capacity route problem*, Operations Research, **8**, (1960), 733–736.
16. Hubert L., *Some applications of graph theory and related non- metric techniques to problems of approximate seriation : the case of symmetric proximity measures*, British J. of Math. and Stat. Psychology, **27**, 2, (1974), 133–153.
17. ———, *Some applications of graph theory to clustering*, Psychometrika, **39** (1974), 283–309.
18. Janowitz M.F., *An order theoretic model for cluster analysis*, SIAM J. APPL. MATH., **34**, no. 1, (1978).
19. Jardine N., Sibson R., *A model of Taxonomy*, Mathematical Biosciences, **2**, (1968), 465–482.
20. Johnson S.C., *Hierarchical clustering schemes*, Psychometrika, 12, 3, (1967), 241–254.
21. Kendall D.G., *Incidence Matrices, Interval Graphs and Seriation in Archaeology*, Pacific Journal of Mathematics, 28, (1969), 3, 565–570.
22. Leclerc B., *Description combinatoire des ultramétriques*, Math. Sci. humaines, 73, (1981), 5–37.
23. ———, *Arbres minimums communs et compatibilité de données de types variés*, Math. Sci. humaines, **98**, (1987), 41–67.
24. Robinson W.S., *A method for chronological ordering of archeological deposits*, American Antiquity, **16** (1951), 4, 293–301.

UNIVERSITÉ PARIS IX - DAUPHINE, LISE-CEREMADE, PL. DU MAR. DE LATTRE DE TASSIGNY, 75 775 PARIS CEDEX 16, FRANCE

Current address: I.N.R.I.A., projet CLOREC, Domaine de Voluceau, BP 152, 78 156 Le Chesnay Cedex, FRANCE

E-mail address: bertrand@icare.inria.fr

DIMACS Series in Discrete Mathematics
and Theoretical Computer Science
Volume 19, 1995

Partitioning by Maximum Adjacency Search of Graphs

WEIQING CAI AND DAVID W. MATULA

April 12, 1993

ABSTRACT. We describe a generalized *maximum adjacency search* (MAS) procedure for graphs and show it can be implemented in time linear in the number of edges of the graph. MAS is shown to provide a nontrivial lower bound on the number of edge disjoint paths between certain pairs of vertices. We describe many properties of consecutive sequences of vertices identified by the search order of the MAS. We illustrate how MAS may be employed for partitioning the vertex set based on this "number of disjoint paths" bound.

1. Introduction

The edge connectivity of an undirected graph is a fundamental parameter with applications in various areas. Fast algorithms are pursued by many authors [EH84],[ET75],[Ma87],[NI92a],[NI92b],[Ga91]. Applications in cluster analysis are described in [Li72],[Ma77]. For many applications it would be sufficient to have a fast algorithm to find a good lower bound on the edge connectivity. In this paper we discuss some linear time algorithms which can find good connectivity bounds for a large range of graphs. We further show that employing "forward" and "reverse" passes of our basic algorithm yields a useful partitioning of the vertices based on their pairwise connectivities.

In §2 we describe a graph search procedure herein termed maximum adjacency search and list some of its important properties. We define a level function associated with the order in which vertices are scanned during a certain maximum adjacency search on a given graph. These enable us to bound, in linear time, the maximum number of edge-disjoint paths between consecutive pairs of vertices in search order.

1991 *Mathematics Subject Classification.* Primary 05C40, 92B10; Secondary 05C85.
The expanded version of this paper will be submitted for publication elsewhere.

In §3 we define a lower bound on edge connectivity derived from our MAS.

In §4 we analyze the time complexity of MAS.

In §5 we describe an arc labeling version of the MAS procedure. It gives more information on the connectivities between consecutive sequences of vertices in search order and leads to a useful vertex partitioning.

2. Maximum Adjacency Search

Given an undirected graph $G = (V, E)$, a search on graph G is a procedure to give each edge of G a label and to give the vertex set an order. A *Maximum Adjacency Search* (MAS) gives each edge a label of an integer in the range of $(1, |E|)$, and gives each vertex x integers $order(x)$, $pick\_level(x)$ and $final\_level(x)$ in the range of $(0, |V|)$. The term Maximum Adjacency Search comes from the fact that during the procedure, a vertex with maximum adjacency with the vertices already visited is picked up next. Maximum adjacency search can be started at any vertex.

ALGORITHM MAS:

Initialize:
For each vertex x let the current degree $d(x) = 0$, and
$pick\_level(x)$=$final\_level(x) = 0$
Let $order = 1$. Let the maximum current degree equal 0.
Mark all vertices as unvisited.
Step 1:
Pick up an unvisited vertex x with $d(x)$ equal to the maximum current degree.
If more than one vertex has this property, pick up any one of them.
Let $pick\_level(x) = d(x)$. Let $order(x) = order$ and update $order$ to be $order + 1$.
Step 2:
Consider each edge (x, y) incident to the picked vertex x, with vertex y unvisited.
If $d(x) = d(y)$, then $d(x)$ is set to $d(x) + 1$ and $d(y)$ is set to $d(y) + 1$, and give edge (x, y) a label of $d(y)$;
if $d(x) > d(y)$, then $d(y)$ is set to $d(y) + 1$, and give edge (x, y) a label of $d(y)$.
Step 3:
If all the edges incident to vertex x are labeled, let $final\_level(x) = d(x)$. Mark vertex x as visited.
Repeat Step 1, 2 3 until all the vertices are visited.

Figure 1 shows a graph with 9 vertices $A, B, \ldots, I$ labeled with the result of a maximum adjacency search, started at vertex A. The number in parentheses close to vertex x is *order*(x). The number close to an edge is the label of the edge, for example the label of (A, B) is 1.

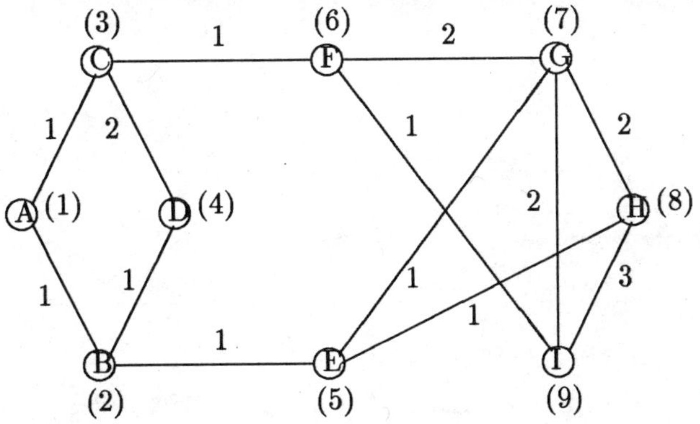

FIGURE 1. Vertex order and edge labels corresponding to an MAS

Let E_i be the subset of E with all edges labeled by i. Let $F_i = (V, E_i)$ be the subgraph of G induced by E_i. Let $G_i = (V, E_1 \cup E_2 \cup \ldots \cup E_i)$.

The proofs of the following four lemmas can be derived from [NI92a].

LEMMA 2.1. *$F_i = (V, E_i)$ is a forest for each i.*

LEMMA 2.2. *If an edge (x, y) is labeled k, then vertices x and y are in the same component of F_i for any $i \leq k$.*

LEMMA 2.3. *If vertices x and y are in the same component of F_k, then x and y are in the same component of F_i for any $i \leq k$.*

LEMMA 2.4. *If vertices x and y are in the same component of F_k, there exist at least k edge-disjoint paths connecting x and y in G. (Note: one path in each F_i for any $i \leq k$.)*

Note that the labeling identifies a particular path in each forest F_i using only edges labeled i for $1 \leq i \leq k$. Also for any cut $(A, \bar{A})$ of the graph G, if there exists an edge $(x, y) \in (A, \bar{A})$ labeled k, then there exist $k - 1$ edges in $(A, \bar{A})$ labeled $1, 2, \ldots, k - 1$ respectively. The following useful conclusion is obtained by Nagamochi and Ibaraki [NI92a].

THEOREM 2.1. *If the edge connectivity of graph $G = (V, E)$ is λ, then $G_i = (V, E_1 \cup E_2 \cup \ldots \cup E_i)$ is an i-connected subgraph of G for any $i \leq \lambda$.*

In the balance of this paper we assume a particular MAS is chosen and the vertices are then correspondingly indexed $x_1, x_2, \ldots, x_n$, so that $order(x_i) = i$ for all indices $1 \leq i \leq n$. Associated with levels and order, the level function $\ell(z)$ is defined on the real interval $1 \leq z \leq n$ such that on the interval $i < z < i + 1$ for any $1 \leq i \leq n - 1$ the value of $\ell(z)$ equals $pick\_level(x_{i+1})$. The value of $\ell(z)$

at each of the integral points $z = 1, 2, \ldots, n$ is $final\_level(x_z)$. Figure 2 gives the level function for the MAS illustrated in Figure 1.

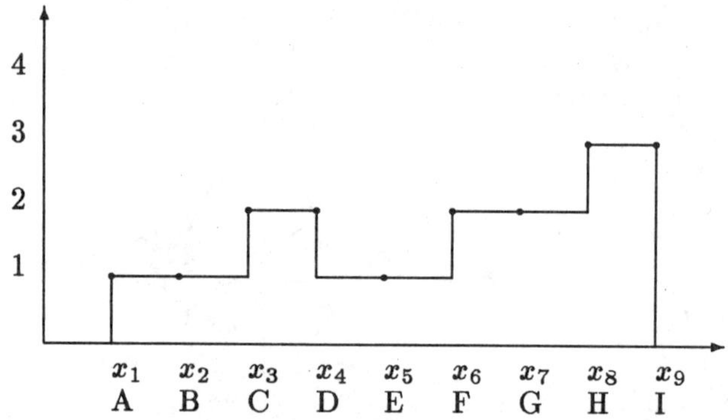

FIGURE 2. The level function associated with the MAS illustrated in Figure 1.

From the level function we can see graphically the connections between two successive vertices in the MAS visitation order. This pairwise connection can be extended to more vertices in MAS visitation order.

The following results are readily derived from the MAS construction and the previous lemmas.

Each component of F_i for every i has vertices that are consecutive in the MAS search order.

Note that $\{x_3, x_4\}$ and $\{x_6, x_7, x_8, x_9\}$ are the vertex sets of the two nontrivial components of F_2 in Figure 1, and $\{x_8, x_9\}$ is the vertex set of the only nontrivial component of F_3.

LEMMA 2.5. *A consecutive sequence $x_p, x_{p+1}, \ldots, x_{p+j}$ that is maximal in search order with $\ell(z) \geq k$ for $p \leq i \leq p+j$ forms the vertex set of a component of F_k.*

LEMMA 2.6. *Given x_i, x_j, let $k = \min\{\ell(z) | i \leq z \leq j, z : real\}$. Then there is a path from x_i to x_j using only edges labeled m and only vertices of $x_1, x_2, \ldots, x_j$ for each $m \leq k$.*

Thus from Lemma 2.6 there must be two paths from $x_6 = F$ to $x_8 = H$ using only vertices preceding x_8 in search order, where one path (F, G, H) uses edges labeled 2 and another path (F, C, A, B, E, H) uses edges labeled 1. Paths identified by the MAS edge labeling between x_i and x_j for $i < j$ guarantee a lower bound on the number of edge disjoint paths between x_i and x_j. The labeling leaves open the possibility that more edge disjoint paths exist employing intermediate vertices from $\{x_{j+1}, x_{j+2}, \ldots, x_n\}$. Note in Figure 1 that (F, I, H) is another edge-disjoint path from x_6 to x_8 of this type. We consider a methodical approach to find such additional paths by an extension of the MAS to an "arc" labeling procedure in section 5.

The level function may be directly related to edge labels of a cut rather than the pick level as follows:

LEMMA 2.7. *The maximum value of the labels on the edges of the cut* $(A, \bar{A})$, *where* $A = \{x_1, x_2, \ldots, x_i\}$ *and* $\bar{A} = \{x_{i+1}, x_{i+2}, \ldots, x_n\}$ *is the value* $\ell(z)$ *for* $i < z < i+1$ *for any* $1 \leq i \leq n-1$.

LEMMA 2.8. *If* $\ell(z) = k$ *for* $i < z < i+1$, *then the induced subgraph* $< \{ x_{i+1}, x_{i+2}, \ldots, x_n\} >$ *has minimum degree less by at most* k *than the minimum degree of the whole graph.*

By the last lemma we infer that after a low value of $\ell(z)$, such as $\ell(z) = 1$ over the interval $5 < z < 6$ in Figure 1, the remaining vertices x_6, x_7, x_8, x_9 constitute a relatively highly linked subset in that each vertex has all but at most one of its adjacencies to other vertices of this subset. Here $\{x_6, x_7, x_8, x_9\}$ itself has minimum degree 2.

3. Lower Bound

Let $V' \subset V$ be a subset of V such that for any two vertices x and y in V', there are at least k edge-disjoint paths in G connecting x and y. Call such a subset a *k-connected subset*. If there exist at least k edge-disjoint paths in G connecting x and $V' \subset V$, where $x \notin V'$, we say that x is k-connected to V'.

LEMMA 3.1. *Let* $V' \subset V$ *be a k-connected subset. Let* $v \notin V'$ *be a vertex k-connected to* V'. *Then,* $V' \cup \{v\}$ *is also a k-connected subset.*

Let x_{n-1} and x_n be the last two vertices visited by an MAS. Then $V' = \{x_{n-1}, x_n\}$ is degree(x_n)-connected because vertex x_n has the property that each edge incident to it has a different label. Obviously, degree(x_n) here is the maximum number of edge-disjoint paths connecting x_{n-1} and x_n .

Let λ_i be the maximum number of edge-disjoint paths connecting x_i to $V_{i+1} = \{x_{i+1}, \ldots, x_n\}$. Then it is readily shown that for the edge connectivity λ:

LEMMA 3.2. $\lambda = \min_{1 \leq i \leq n-1}\{\lambda_i\}$.

Consider a sequence $\bar{\lambda}_i$ of bower bounds to the values of λ_i, that is $\bar{\lambda}_i \leq \lambda_i$ for $1 \leq i \leq n-1$. From these we obtain that

$$\lambda^* = \min_{1 \leq i \leq n-1} \{\bar{\lambda}_i\} \leq \min_{1 \leq i \leq n-1} \{\lambda_i\} = \lambda$$

gives a lower bound to the edge connectivity.

A readily obtained sequence of lower bounds is given by Algorithm MAS with the associated vertex order by

$$\bar{\lambda}_i = [pick\_level(x_{i+1}) - pick\_level(x_i)] + [deg(x_i) - 1].$$

A better lower bound is derived as follows.

The edge-disjoint paths connecting x_i to V_{i+1} based on an MAS can be classified as two kinds. One is of edges between x_i and some vertices in V_{i+1}. They are paths of length 1. We call them *forward paths.* The number of these path is denoted by $fwd(x_i)$, which equals to $deg(x_i) - pick\_level(x_i)$.

The number of paths of the other kind is at most k, where $k = pick\_level(x_i)$. These are paths containing vertices $x_1, x_2, \ldots, x_i$, plus one edge between a visited vertex and an unvisited vertex, respectively. The number of these path is denoted by $bwd(x_i)$.

Let us extend the computations associated with an MAS as follows. At any point in the search, $out(x)$ is the total number of unvisited neighbors of the visited vertex x. When x is visited, $out(x)$ is set to $fwd(x)$ and any neighbor y previously visited has $out(y)$ decreased by 1. When the search is finished, $out(x) = 0$ for any vertex s. Then we can bound the number of edge-disjoint paths between two vertices x_i and x_j by the following lemma.

LEMMA 3.3. *Given two vertices x_j, x_i, $j < i$, define $P(x_i, x_j)$ to be the number of edge-disjoint paths connecting x_i to the preceding vertex x_j based on the MAS, and $P(x_i, x_i) = \infty$. Then,*

$$P(x_i, x_j) = \min\{P(x_i, x_{j+1}), pick\_level(x_{j+1}) + [fwd(x_j) - out(x_j)] - 1\}$$

Start from x_i, chose any one path from these $P(x_i, x_j)$ paths, plus one edge (a forward path) from the $out(x_j)$ of them, we obtain one back ward path. So, the number of backward paths (they are edge-disjoint) connecting x_i to V_{i+1} through x_j is

$$P_j(x_i, V_{i+1}) = \min\{P(x_i, x_j), out(x_j)\}.$$

To find the total number of backward path $bwd(x_i)$, set $bwd(x_i)$ to be 0 to start. Then we compute $P_{i-1}(x_i, V_{i+1})$. If $P_{i-1}(x_i, V_{i+1}) = k$, stop. Because k is the maximum of the number of backward paths. If not, compute $P_{i-2}(x_i, V_{i+1})$ and set $bwd(x_i)$ to be $bwd(x_i) + \min\{k - bwd(x_i), P_{i-2}(x_i, V_{i+1})\}$. Continue to do so until either we have $bwd(x_i) = k$, or we finished with $P_j(x_i, V_{i+1})$ where $j = 1$.

The sequence of lower bounds

$$\lambda'_i = fwd(x_i) + bwd(x_i) \tag{3.1}$$

with $bwd(x_i)$ so computed as described above gives a better lower bound on the edge connectivity. By this bounding procedure it can be confirmed that the MAS of Figure 1 is sufficient to confirm the graph has $\lambda = 2$.

4. Time Complexity

When vertex x is visited in Algorithm MAS, all its neighbors will be scanned. For each of its visited neighbor y, $out(y)$ will be updated in constant time. It is obvious then that the computing of $fwd(x_i)$ and updating of $out(y)$ for all the neighbors y of vertex x can be done in time proportional to the degree of x.

In computing $bwd(x_i)$, we need to consider $x_{i-1}, x_{i-2}, \ldots$ following the opposite visitation order. At each vertex x_{i-j}, the computing of $P(x_i, x_j)$ is in constant time because $pick\_level(x_{j+1})$, $fwd(x_j)$ and $out(x_j)$ are all being recorded and $P(x_i, x_{j+1})$ is computed recursively. Therefore the total time taken at each vertex is constant. So if $out(x_j) > 0$ for any s, the time to compute $bwd(x_i)$ is clearly proportional to the degree of x_i. Because each step back can add at least 1 to λ'_i. Since the minimum degree δ is an upper bound on the edge connectivity and $\delta \leq degree(x_i)$, at most δ steps back are needed to compute $bwd(x_i)$.

But if some $out(x_j) = 0$ then considering x_j will not add anything to λ'_i and we could have to step back up to x_1. This is in worst case $O(n)$.

To solve this problem, we keep a list of visited vertices x with $out(x) > 0$ in the visitation order. Whenever $out(x_j)$ becomes 0 during MAS, we delete x_j from the list. When a x_j is deleted, we need to save some useful information to compute $P(x_i, x_{j-1})$, which requires the knowledge of $P(x_i, x_j)$.

Notice that by our definition, in computing $P(x_i, x_j)$, the $pick\_level(x_{j+1}) + [fwd(x_j) - out(x_j)] - adj(x_j, x_{j+1})$ becomes a constant since $out(x_j) = 0$, therefore is independent of i for any $i > j$. Compute and save it when x_j is deleted from the list. Follow the new list when stepping back. By the argument above, $bwd(x_i)$ can be computed in time proportional to the degree of x_i.

Therefor our lower bound λ'_i can be computed in time proportional to degree of x_i, and

THEOREM 4.1. *The lower bound* $\min_i\{\lambda'\}$ *on edge connectivity given by formula (3.1) can be computed in linear time.*

5. MAS Arc Labeling

An undirected graph can be viewed as a directed graph with each edge replaced by two arcs, one in each direction. Given a certain MAS, we view each edge (x, y) as two arcs with arc (x, y) labeled where $order(x) < order(y)$. We now describe a procedure to label certain other arcs.

After an MAS, start from the vertex visited last by the MAS, x_n. We shall now obtain a "reverse" MAS by deleting arcs as necessary to obtain that $x_n, x_{n-1}, x_{n-2}, \ldots, x_2, x_1$ is a legal search order for this "reverse" MAS. Here we label each arc exactly as we described in the original MAS procedure. Instead of choosing a vertex with maximum current degree, we choose x_{n-1}. If x_{n-1} is not the vertex with maximum current degree, we delete arcs from those vertices with current degree higher than that of x_{n-1} until x_{n-1} has maximum current degree. This way we follow the reverse order of the MAS and keep the properties stated in the lemmas of section 2 for this reverse MAS.

For the graph and MAS in Figure 1, we delete arcs (F, C) and (E, B). Figure 3 shows the level function of the corresponding reverse MAS (in reverse order).

Combining the original MAS and this reverse arc labeling, we can find more edge disjoint paths connecting x_i and x_{i+1}. Because the pick level of x_{i+1} in

MAS is a guaranteed number of edge disjoint paths connecting x_i and x_{i+1} only containing vertices among $x_1, x_2, \dots, x_{i+1}$; the reverse pick level of x_i in arc labeling is a guaranteed number of edge disjoint paths connecting x_i and x_{i+1} only containing vertices among $x_i, x_{i+1}, x_{i+2}, \dots, x_n$. Therefore the sum of these two numbers minus the one for any edge between x_i and x_{i+1}, (that edge was counted twice), is a guaranteed number of edge disjoint paths connecting x_i and x_{i+1} in the graph. For example pick level of D is 2, reverse pick level of C is 1 and C and D are adjacent. So C and D are connected by at least 2+1-1=2 disjoint paths.

By Figure 2 and 3, we can see that vertices A, B, C, D are pairwise connected by two edge disjoint paths, and E, F, G, H, I are pairwise connected by at least three edge disjoint paths. This is summarized by Figure 4. Therefore it is reasonable here to identify $\{A, B, C, D\}$ as one cluster and $\{E, F, G, H, I\}$ as another.

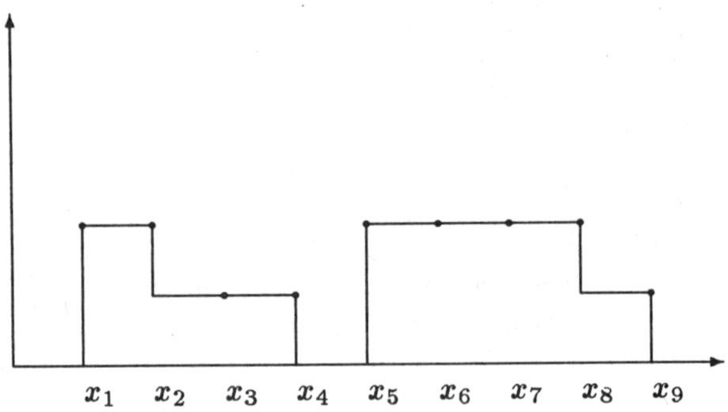

FIGURE 3. The level function associated with the arc labeling illustrated in Figure 1.

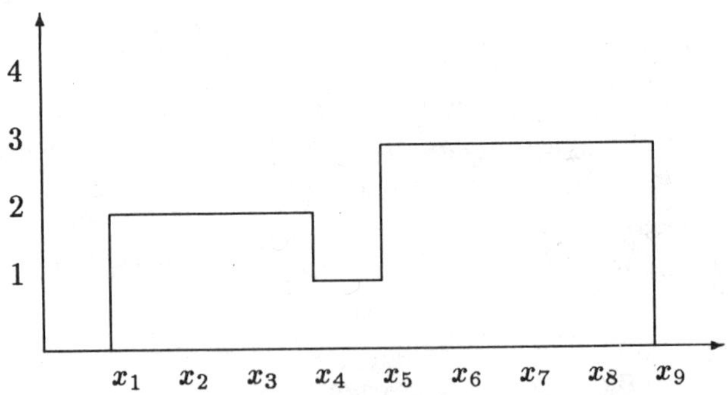

FIGURE 4. The connections between vertices in the graph of Figure 1

References

[BarMa] Bartholomew, P. and Matula, David, Test Graph: a generation and visualization package for creating random graphs with embedded structure, Research report for the Stockholm conference on Random graphs and applications, April 25-27, 1989.

[Eh84] Esfahanian, A.H. and Hakimi, S. L., On computing the connectivities of graphs and digraphs, Networks 14 (1984)355-366.

[ET75] Even, A. and Tarjan, R.E., Network flow and testing graph connectivity, SIAM J. Computing 4 (1975) 507-581.

[Ga91] Gabow, H.N., A matroid approach to finding edge connectivity and packing arborescences, Proc. 23rd Annual ACM Symp. on Theory of Comp., 1991, pp. 112-122.

[Li72] Ling, R. R., On the theory and construction of k-clusters, Computer J. 15, 1972, pp326-332.

[Ma77] Matula, D.W., Graph theoretic techniques for cluster analysis algorithms, Classification and Clustering, Edited by J. Van Ryzin, Academic Press, Inc. 1977, pp.93-129.

[Ma87] Matula, D.W., Determining edge connectivity in $O(nm)$, Proc. 28th IEEE Symp. on Foundations of Computer Science, 1987, pp.249-251.

[Ma93] Matula, D.W., A linear time $2 + \epsilon$ approximation algorithm for edge connectivity. Proceedings of the fourth annual ACM-SIAM Symposium on Discrete Algorithms.

[NI92a] Nagamochi, H. and Ibaraki, T., A linear time algorithm for finding a sparse k-connected spanning subgraph of a k-connected graph, Algorithmica, 7, 1992, pp.583-596.

[NI92b] Nagamochi, H. and Ibaraki, T., Computing edge connectivity in multigraphs and capacitated graphs, SIAM J. Discrete Math., Vol.5 No.1, 1992, pp54-66.

[Ha69] Harary, F, Graph Theory, Addison-Wesley, 1969.

Department of Computer Science and Engineering Southern Methodist University, Dallas, Texas 75275

Current address: DEL NORTE Technology, Inc. 1100 Pamela Drive, P.O. Box 696, Euless, TX 76039

E-mail address: cai@seas.smu.edu

Department of Computer Science and Engineering Southern Methodist University, Dallas, Texas 75275

E-mail address: matula@seas.smu.edu

DIMACS Series in Discrete Mathematics
and Theoretical Computer Science
Volume 19, 1995

From Data to Knowledge : Probabilist Objects for a Symbolic Data Analysis

Edwin DIDAY

ABSTRACT. The main aim of the symbolic approach in data analysis is to extend problems, methods and algorithms used on classical data to more complex data called "symbolic objects"which are well adapted to representing knowledge and which can "unify" unlike usual observations which characterize "individual things". We introduce two kinds of symbolic objects : boolean and probabilist. We briefly present some of their qualities and properties; a theorem shows how Probability, theory may be extended on these objects. Finally four kinds of data analysis problems including the symbolic extension are illustrated by several algorithms which induce knowledge from classical data or from a set of symbolic objects.

Key-words : Knowledge analysis, symbolic data analysis, Probability, Possibility, Evidence theory, uncertainty logic.

Introduction

If we wish to describe the fruits produced by a village, by the fact that "the weight is between 300 and 400 grammes and the color is white or red and if the color is white then the weight is lower than 350 grammes", it is not possible to put this kind of information in a classical data table where rows represent villages and columns descriptors of the fruits. This is because there will not be a single value in each cell of the table (for instance, for the weight) and also because it will not be easy to represent rules (if..., then...) in this table. It is much easier to represent this kind of information by a logical expression such as :

a_i = [weight = [300,400]] $\wedge$ [color = {red, white}] $\wedge$ [if [color = white] then [weight $\leq$ 350]], where a_i, associated to represents the ith village, is a mapping defined on the set of fruits such that for a given fruit w, $a_i(w)$ = true if the weight of w belongs to the interval [300,400], its color is red or white and if it is white then its weight is less than 350 gr. Following the terminology of this paper a_i is a kind of symbolic object. If we have a set of 1000 villages represented by a set of 1000 symbolic objects $a_1, \ldots, a_{1000}$, an important problem is to know how to apply statistical methods to statistics on it. For instance, what is a histogram or a probability law for such a set of

1991 *Mathematics Subject Classification.* Primary 68P05.

objects ? The aim of symbolic data analysis (Diday 1990,1991) is to provide tools for answering this problem.

In some fields a boolean representation of the knowledge (a_i (w) = true or false) is sufficient to get the main information, but in many cases we need to include uncertainty to represent the real world with more accuracy. For instance, if we say that in the ith village "the color of the fruits is often red and seldom white" we may represent this information by a_i = [color = often red, seldom white]. More generally, in the case of boolean objects or objects where frequency appears, we may write a_i = [color = q_i] where q_i is a characteristic function in the boolean case, and a probability measure in the second case. More precisely, in the boolean case, if a_i = [color = red, white] we have q_i (red) = q_i (white) = 1 and q_i = 0, for the other colors ; in the probabilist case, if a_i = [color = 0.9 red, 0.1 white] we have q_i (red)=0.9, q_i (white)=0.1. If an expert says that the fruits are red we may represent this information by a symbolic object a_i = [color = q_i] where q_i is a "possibilist" function in the sense of Dubois and Prade (1986) ; we will have for instance q_i (white) = 0, q_i (pink) =0.5 and q_i (red) = 1. If an expert who has to study a representative sample of fruits from the ith village, says that 60% are red, 30% are white and the color is unknown for 10% which were too rotten, we may represent this information by a_i = [color = q_i] where q_i is a belief function such that q_i (red) = 0.6, q_i (white) = 0.3 and q_i (O) = 1,where O is the set of possible colors. Depending on the kind of the mapping q_i used, a_i has been called a boolean, probabilist, possibilist or belief object. In all these cases a_i is a mapping from Ω in [0,1]. Now, the problem is to know how to compute a_i (w) ; if there is doubt about the color of a given fruit w, for instance, if the expert says that "the color of w, is red or pink" then, w may be described by a charateristic function r and represented by a symbolic object w^s = [color = r] such that r (red) = r (pink) = 1 and r = 0 for the other colors. Depending on the kind of knowledge that the user wishes to represent, r may be a probability, possibility or belief function. Having a_i =[color = q_i] and w^s = [color = r] to compute a_i (w) we introduce a comparison function g such that a_i (w)= g (q_i,r) measures the fit between q_i and r. What is the meaning of a_i (w) ? May we say that a_i (w) measures a kind of probability, possibility or belief that w belongs to the class of fruits described by a_i when q_i and r are respectively charateristic, probability, possibility or belief functions ? To answer this question we have extended a_i to a "dual" mapping a_i^* (such that a_i (w) = a_i^* (w^s)) defined on the set of symbolic objects of the a_i kind denoted $\mathfrak{a}_x$ and an extension of the union, intersection and complementary operators of classical sets denoted $OP_x = \{ \cup_x, \cap_x, c_x \}$ where x depends upon the kind of knowledge used; then, we have shown that when x represents probability , then a_i^* satisfies the axioms of probability measures by using OP_{pr}(x= **pr**obability) and in the case of possibilist objects that a_i^* satisfies the axioms of possibility functions by using some given operators denoted OP_{pos} (see Diday (1992) for more details).

In probability theory , very little is said about events which are generally identified as parts of the sample set Ω. In computer science, object oriented languages consider more general events called objects or "frames" defined by intention. In data analysis (multidimensional scaling, clustering, exploratory data analysis etc.) more importance is given to the elementary objects which belong to the sample Ω than in classical statistics where attention is focused on the probability laws of Ω ; however, objects of data analysis are generally identified to points of $\mathbb{R}^p$ and hence are unable to treat complex objects coming for instance from large data bases, and knowledge bases. Our aim is to define complex objects called "symbolic objects" inspired by those of oriented object languages in such a way that data analysis becomes generalized in knowledge analysis. Objects will be defined by intention by the properties of their extension. More precisely, we distinguish objects which "unify" rather than elementary observed objects which characterize "individual things" (their extension) : for instance "the customers of my shop" instead of "a customer of my shop", "a species of mushroom" instead "the mushroom that I have in my hand".

The aim of this paper is to reduce the gap between statistics or data analysis (where limited interest has, as yet, been shown in treating this kind of objects) and artificial intelligence (where the emphasis is more on knowledge representation, reasoning and learning than on knowledge analysis, by extending data analysis methods to symbolic objects).

We have not used the notion of "predicates" from classical logic, firstly, because by using only functions, things seem more understandable, especially to statisticians ; secondly, because they cannot be used simply in the case of probabilist, possibilist and belief objects (see Diday 92) where uncertainty is present.

1. Boolean Symbolic Objects

We consider Ω a set of individual things called "elementary objects" and a set of descriptor functions $y_i : \Omega \rightarrow O_i$.

A basic kind of symbolic object are "*events* ". An event denoted $e_i = [y_i=V_i]$ where $V_i \subseteq O_i$ is a function $\Omega \rightarrow \{\text{true, false}\}$ such that $e_i(w) = \text{true}$ iff $y_i(w) \in V_i$. For instance, if $e_i = [\text{color=red,white}]$, then $e_i(w) = \text{true}$ iff the color of w is red or white. When $y_i(w)$ is meaningless (the kind of computer used by a company without computer) $V_i = \phi$ and when it has a meaning but this is not known $V_i = O_i$. The extension of e_i in

Ω denoted by ext (e_i/Ω) is the set of elements $w \in \Omega$ such that $e_i(w)$ = true.

An assertion is a conjunction of events $a = \underset{i}{\wedge} [y_i=V_i]$; the extension of a denoted $ext(a/\Omega)$ is the set of elements of Ω such that $\forall i \ y_i(w) \in V_i$.

A "*horde* " is a symbolic object which appears, for instance, when we need to express relations between parts of a picture that we wish to describe. More generally a horde is a function h from Ω^p in {true, false} such that $h(u) = \underset{i}{\wedge} [y_i(u_i) = V_i]$ if $u = (u_1,....,u_p)$. For example : $h = [y_1(u_1) = 1] \wedge [y_2(u_2) = \{3,5\}] \wedge [y_3(u_1) = [30,35]] \wedge$ [neighbour (u_1,u_2) = yes].

A synthesis object is a conjunction or a semantic link between hordes denoted in the case of conjunction by $s = \underset{i}{\wedge} h_i$ where each horde may be defined on a different set Ω_i by different descriptors. For instance Ω_1 may be individuals, Ω_2 location, Ω_3 kind of job etc. All these objects are detailed in Diday (1991).

2. Modal Objects

Suppose that we wish to use a symbolic object to represent individuals of a set satisfying the following sentence : "It is possible that their weight be between 300 and 500 grammes and their color is often red or seldom white" ; this sentence contains two events e_1 = [color = {red, white}] which lack the modes *possible* , *often* and *seldom* , a new kind of event, denoted f_1 and f_2, is needed if we wish to introduce them f_1 = *possible* [height = 300,500]] and f_2 = [color = {*often* red , *seldom* white}] ; we can see that f_1 contains an *external* mode *possible* affecting e_1 whereas f_2 contains *internal* modes affecting the values contained in e_2. Hence, it is possible to describe informally the sentence by a modal assertion object denoted $a = f_1 \wedge_x f_2$ where $\wedge_x$ represents a kind of conjunction related to the background knowledge of the domain. The case of modal assertions of the kind $a = \underset{i}{\wedge} f_i$ where all the f_i are events with external modes has been studied, for instance, in Diday (1990). This paper is concerned with the case where all the f_i contain only internal modes.

3. Internal Modal Objects

3.1. A formal definition of internal modal objects. Let x be the background knowledge and

. M^x a set of modes, for instance M^x = {often, sometimes, seldom, never} or $M^x = [0,1]$.

. $Q_i = \{q_i^j\}_j$ a set of mappings q_i^j from O_i in M^x, for instance O_i = {red, yellow, green}, $M^x = [0,1]$ and $q_i^j(\text{red}) = 0.1$; $q_i^j(\text{yellow}) = 0.3$; $q_i^j(\text{green}) = 1$, where the meaning of the values 0.1, 0.3, 1 depends on the background knowledge (for instance q_i^j may express a possibility, see §.4.1).

. y_i is a descriptor (the *color* for instance) ; it is a mapping from Ω in Q_i. Notice that in the case of boolean objects y_i was a mapping from Ω in O_i, and not Q_i.

Example : if O_i and M^x are chosen as in the previous example and the color of w is red then $y_i(w) = r$ means that $r \in Q_i$ be defined by $r(\text{red}) = 1$, $r(\text{yellow}) = 0$, $r(\text{green}) = 0$.

. $OP_x = \{\cup_x, \cap_x, c_x\}$ where $\cup_x, \cap_x$ expresses a kind of union and intersection between subsets of Q_i and $c_x(q_i)$ (sometimes denoted $\bar{q}_i$, the complementary of $q_i \in Q_i$). We denote by Q_i^x the smallest stable set for OP_x (i.e Q_i^x is the set of any $*_x$ or c_x combination of elements $q_i^j \in Q_i\}$.

If $q_i^j \in Q_i$ and $Q_i^j \subseteq Q_i$, we settle : $b(Q_i^j) = \{\cup_x q_i / q_i \in Q_i^j\}, Q_i^1 *_x Q_i^2 = b(Q_i^1) *_x b(Q_i^2)$ where $*_x \in \{\cup_x, \cap_x\}$ and $c_x(Q_i^j) = 1 - b(Q_i^j)$.

Example : if $q_i^j \in Q_i$ and $Q_i^j \subseteq Q_i$ $q_i^1 \cup_x q_i^2 = q_i^1 + q_i^2 - q_i^1 q_i^2 q_i^1 \cap_x q_i^2 = q_i^1 q_i^2$ where $q_i^1 q_i^2 (v) = q_i^1(v)\, q_i^2(v)$; $c_x(q_i) = 1 - q_i$

This choice of OP_x is "archimedian" because it satisfies a family of properties studied by Shweizer and Sklar (1960) and recalled by Dubois and Prade (1988).

. g_x is a "comparison" mapping from $Q_i^x \times Q_i^x$ in an ordered space L^x.

Example : $L^x = M^x = [0,1]$ and $g_x(q_i^1, q_i^2) = \langle q_i^1, q_i^2 \rangle$ the scalar product

. f_x is an "aggregation" mapping from $P(L^x)$ the power set of L^x in L^x. For instance, $f_x(\{L_1,...,L_n\}) = \text{Max } L_i$.

Let $Y = \{y_i\}$ be a set of descriptors and $V = \{V_i\}$ a set of subsets of Q_i^x such that $V_i = \{q_i^j\}_j \subseteq Q_i^x$ Now we are able to give the formal definition of an internal modal object (called "im" object).

Definition of an im assertion - Given OP_x, g_x and f_x, an im assertion is a mapping a_{YV} from Ω in an ordered space L^x denoted $a = \wedge_i [y_i = \{q_i^j\}_j]$ such that $w \in \Omega$ is described for any i by $y_i(w) = \{r_i^j\}_j$ then $a_{YV}(w) = f_x(\{g_x(\cup x_j\, q_i^j, \cup x_j\, r_i^j)\}_i)$.

We denote by $\mathfrak{a}_x$ the set of im objects associated to background knowledge x and φ the mapping from Ω in $\mathfrak{a}_x$ such that $(w) = w^s = \wedge x_i\, [y_i = y_i\ (w)]$.

Notice that more complex objects may occur when instead of only one, as in the preceding definition, several events concern the same variable ; for instance if we have $a = \wedge_i a_i$ with $a_i = \wedge x_\ell\, [y_i = q_i^\ell]$; in this case, it is necessary to introduce a third mapping h from $P(L^x)$ in L^x such that $a_i(w) = h(\{g(q_i^\ell, r_i)\}_\ell)$; hence, more generally, if $a = \wedge x_i\, a_i = \wedge x_i \wedge x_\ell\, [y_i = q_i^\ell]$ then $a(w) = f_x\ (\{a_i\ (w)\}_i) = f_x\ (\{h_x\ (\{g_x\ (q_i^\ell, r_i)\}_\ell)\}_i)$.

Example : Let $M_i^x = [0,1]$, $O_i = \{v_1, v_2\}$, and Q_i be the set of probability measures $P(O_i) \rightarrow [0,1]$; y is a mapping from a set Ω in Q_i and $w^s = [y_i = r]$ is such that $r(v_1) = r(v_2) = \frac{1}{2}$; the set of im assertions $e_i = [y = q_i]$ such that $a_i(w) \geq \frac{1}{2}$ is defined by the set of probability measures q_i which satisfy the inequality $e_i\ (w) = f_x\ (g_x\ (q_i, r)) \geq \frac{1}{2}$; if f_x is the mean and g_x is the scalar product we get $e_i\ (w) = \text{Mean}\ (\{<q_i, r>\}) = <q_i, r>$ as there is only one variable. Hence q_i has to satisfy the following inequality : $e_i\ (w) = <q_i, r> = q_i\ (v_1)\ r(v_1) + q_i(v_2)\ r(v_2) \geq \frac{1}{2}$ which is equivalent to $\frac{1}{2}\ q_i(v_1) + \frac{1}{2}\ q_i(v_2) \geq$

$\frac{1}{2}$ which is satisfied by any assertion a, as $q(v_1)+q(v_2) = 1$ for any measure of probability defined on O_i. Let be $a_i = \wedge x_{\ell} \{e_i^{\ell} / \{e_i^{\ell}(w) \geq \frac{1}{2}\}\}$ then $a_i(w) = h_x(\{e_i^{\ell}(w)\}_{\ell})$; if $h_x = \min$ then $a_i(w) = \text{Min}(\{e_i^{\ell}(w)\}_{\ell}) = \frac{1}{2}$.

3.2. Extension of im objects. There are at least two ways to define the extension of an im object a. The first consists in considering that each element $w \in \Omega$ is more or less in the extension of a according to its weight given by a(w); in this case the extension of a denoted Ext (a/Ω) will be the set of couples $\{(w,a(w))/w \in \Omega\}$. The second requires a given threshold α and then, the extension of a will be Ext $(a/\Omega, \alpha) = \{(w, a(w)) / w \in \Omega, a(w) \geq \alpha\}$.

3.3. Semantic of im objects. In addition to the modes, several other notions may be expressed by an im object a :

a) Certitude : a(w) is not true or false as for boolean objects but expresses a degree of certitude.

b) Variation : this appears at two levels in an im object denoted $a = \wedge x_i [y_i = \{q_i^j\}_j]$ first in each q_i^j, for instance if y_i is the color and $q_i^1(\text{red}) = 0.5$, $q_i^1(\text{green}) = 0.3$ it means that a variation exists between the individual objects which belong to the extension of a (for instance a species of mushrooms) where some are red and others are green; second, for given description y_i between the q_i^j (each q_i^j expresses for instance the variation in a different kind of species).

c) Doubt : if we say that the color of a species of mushroom is red "or" green, it is an "or" of variation, but if we say that the color of the mushroom which is in my hand is red "or" green, it is an "or" of doubt. Hence, if we describe $w \in \Omega$ by $\varphi(w) = w^s = \wedge_i [y_i = y_i(w)]$ where $y_i(w) = \{r_i^j\}_j$ we express a doubt in each r_i^j and among the r_i^j provided, for instance, by several experts.

4. Probabilist Objects

4.1. The probabilist approach. First we recall the well known axioms of Kolmogorov :

If $C(\Omega)$ is a σ-algebra on Ω (i.e. a set of subsets stable for numerable intersection or union and for complementary).We say that p is a measure of probability on $(\Omega, C(\Omega))$ if

i) $p(\Omega) = 1$
ii) $p(\cup_i A_i) = \Sigma\, p(A_i)$ if $A_i \in C(\Omega)$ and $A_i \cap A_j = \phi$.

There are several semantics which follow these axioms : for instance luck in games, frequencies, some kind of uncertainty by subjective probability. Let Q_i be a set of measures of probabilities defined on $(O_i, C(O_i))$. We suppose that the $w^s = [y_i=y_i(w)]$ are such that $y_i(w) \in Q_i$.

Definition - A probabilist assertion is an im assertion which takes its values in $L^{pr} = [0,1]$ OP_{pr} : $\forall\, q_i^1, q_i^2 \in Q_i$ $q_i^1 \cup_{pr} q_i^2 = q_i^1 + q_i^2 - q_i^1 q_i^2$; $q_i^1 \cap_{pr} q_i^2 = q_i^1 q_i^2$ which is the mapping which associate to $v \in O_i$, $q_i^1(v)\, q_i^2(v)$; $c_{pr}(q) = \bar{q} = 1 - q$. $g_{pr} : \forall\, \{q_i^1, q_i^2\} \in Q_i^x \times Q_i$ $g_{pr}(q_i^1, q_i^2) = < q_i^1, q_i^2 > = \Sigma\, \{q_i^1(v)\, q_i^2(v) \,/\, v \in O_i\}$. f_{pr} : $f_{pr}(\{L_i\})$ = mean of the L_i.

Notice that if there are some characteristic dependancies between variables, then, an event of the form $[y_i = q_i]$ may represent them;for instance,if the expert wishes to describe the dependencies between y_1,y_3,y_7, then,this information may be represented by the event denoted $[y_{137} = pr(y_1,y_3,y_7)]$ where $pr(y_1,y_3,y_7)$ represents the conjoint probability of y_1,y_3,y_7 ; this event is of the form $[y_i = q_i]$ where $y_i = y_{137}$ and $q_i = pr(y_1,y_3,y_7)$.

In the case where "causalities" or "influences" among set of variables are given by the expert to describe a symbolic object, propagation technics (see Pearl (1988)), Lauritzen & Spiegelhalter (1988)) may be used which induce other mappings g_{pr} and f_{pr}.

To give an intuitive idea of the notion of union of measures of probabilities it is easy to see that if q_i^1 and q_i^2 are the measure of probabilities associated to two dice, $q_i^1 \cup_{pr} q_i^2\,(V)$ is the probability that the event V occurs, for one dice **or** (not exclusive) for the other, when the two dices are thrown independently. Notice that $q_i^1 \cup_{pr} q_i^2$ is not a measure of probability because even if $q_i^1 \cup_{pr} q_i^2\,(v) \in [0,1]$ the sum of the $q_i^1 \cup_{pr} q_i^2\,(v)$ on O_i is larger then 1. Also, $q_i^1 \cap_{pr} q_i^2$ is not necessarily a measure of probability because the sum of the $q_i^1 \cap_{pr} q_i^2$ (v) on O_i may be lower than 1.

4.2. Example. An object w is described by its color y_1 (w) which may be red or blue and its roundness y_2(w) which may be round or flat.

Let $a = [y_1 = q_1^1, q_1^2] \wedge_{pr} [y_2 = q_2]$ and $w^s = [y_1 = r_1] \wedge_{pr} [y_2 = r_2]$ where q_1^1(red)=0.9; q_1^1(blue)=0.1; q_1^2(red)=0.5; q_1^2(blue)=0.5; q2(round)=0.2; q2(flat) = 0.8.

It results that a is described by two kind of objects : either often red and rarely blue, or red or blue with equal probability.

By using $q_1^3 = q_1^1 \cup_{pr} q_1^2 = q_1^1 + q_1^2 - q_1^1 q_1^2$ we obtain

$$q_1^3 \text{ (red)} = 0.9 + 0.5 - 0.9 \text{ x } 0.5 = 0.95$$

$$q_1^3 \text{ (blue)} = 0.1 + 0.5 - 0.1 \text{ x } 0.5 = 0.55$$

If r_1 and r_2 are defined as follows :

r_1 (red) = 1, r_1 (blue) = 0; r_2 (round) = 1, r_2 (flat) = 0, it results that $a(w) = g_{pr}(q_1^3, r1) \wedge_{pr} g_{pr}(q_2, r_2) = (0.95 \text{ x } 1 + 0.55 \text{ x } 0) \wedge_{pr} (0.2 \text{ x } 1 + 0.8 \text{ x } 0) = 0.95 \wedge_{pr} 0.20 = \frac{1}{2}(0.95 + 0.20) = 0.57$, which represents a membership degree for w to the im object defined by a.

5. The Particular Case of Boolean Objects

A boolean object $a = \hat{\imath}\, [y_i = V_i]$ is an im object $a_b = \hat{\imath}\, [y_i = q_i]$ where q_i is the characteristic mapping of V_i in O_i, $OP_b = \{\cup_b, \cap_b, c_b\}$ is such that $q_1 \cup_b q_2 = \text{Max}(q_1, q_2)$, $q_1 \cap_b q_2 = \min(q_1, q_2)$ and $c_b(q) = 1 - q$; if $w = \hat{\imath}\, [y_i = r_i]$ where r_i is the characteristic mapping of $y_i(w)$ in O_i, $g_b(q_i, r_i) = \langle q_i, r_i \rangle$ and f_b = min ; it results that if there exists only a single $v \in O_i$ such that $r_i(v) \neq 0$ then $a_b(w) = 1$ (thus $r_i \leq q_i$) $\Leftrightarrow$ a(w) = true and then $a_b(w) = 0 \Leftrightarrow$ a(w) = false.

If we denote $|a|_\Omega$ the set of elements of Ω such that a(w) = true, we have $|a|_\Omega = \text{Ext}(a_b / \Omega, \alpha)\ \forall \alpha \in\ ]0,1]$.

6. Some Qualities and Properties of Symbolic Objects

6.1. Order, union and intersection between im objects. It is possible to define a partial preorder $\leq_\alpha$ on the im objects by setting that : $a_1 \leq_\alpha a_2$ iff $\forall w \in \Omega \;\; \alpha \leq a_1(w) \leq a_2(w)$.

We deduce from this preorder an equivalence relation R by $a_1 R a_2$ iff $\mathrm{Ext}(a_1 / \Omega, \alpha) = \mathrm{Ext}(a_2 / \Omega, \alpha)$ and a partial order denoted $\leq_\alpha$ and called "symbolic order" on the equivalence classes induced from R.

We say that a_1 inherits from a_2 or that a_2 is more general than a_1, at the level α, iff $a_1 \leq_\alpha a_2$ (which implies $\mathrm{Ext}_\alpha(a_1 / \Omega, \alpha) \subseteq \mathrm{Ext}_\alpha(a_2 / \Omega, \alpha)$).

We call intention at the level α of a subset $\Omega_1 \subset \Omega$ the symbolic object b defined by the conjunction of events whose extension at the level α contains Ω_1.

The symbolic union $a_1 \cup_{x,\alpha} a_2$ (resp. intersection $a_1 \cap_{x,\alpha} a_2$) at the level α is the intention of $\mathrm{Ext}(a_1 / \Omega, \alpha) \cup \mathrm{Ext}(a_2 / \Omega, \alpha)$ (resp. $\mathrm{Ext}(a_2 / \Omega, \alpha) \cap \mathrm{Ext}(b / \Omega, \alpha)$).

6.2. Some qualities of symbolic objects. As in the boolean case, see Brito, Diday (1990), it is possible to define different kinds of qualities of symbolic objects (refinement, simplicity, completeness etc.).

For instance, we say that a symbolic object s is "complete" iff the properties which characterize its extension are exactly those whose conjunction defines the object; in other words s is a complete symbolic object if it is the intention of its extension. More intuitively, if I can see some white dogs and I state "I can see some dogs", my statement doesn't describe the dogs in a complete way, since I am not saying that they are white.

On the other hand,the simplicity at level α of an im object is the smallest number of elementary events whose extension at level α coincides with the extension of s at the same level.

6.3. Some properties of im objects : lattice and completeness. It may be shown, see Diday (1992) for instance, that given a level α the set of im objects is a lattice for the symbolic order and that the symbolic union and intersection define the supremum and infimum of any couple. To do so, f_x, g_x and h_x (see § 3.1) have to be well chosen and we introduce a "full" and an "empty" symbolic object denoted Ω^s and ϕ^s such that $\forall w \in \Omega$, $\Omega^s(w) = 1$ and $\phi^s(w) = 0$.

It may also be shown that the symbolic union and intersection of complete im objects are complete im objects and hence that the set of complete im objects is also a lattice.

7. An Extension of Probabilities Assertions on Symbolic Objects

Our aim is to extend an im assertion $a = \hat{\wedge}_i\, [y_i = q_i]$ (where q_i depends on the choice of x and may be for instance a possibility, a probability or a belief function), to a dual im assertion denoted a* defined on subsets of $\mathfrak{a}_x$ (the set of im assertions associated to x), and "$*_x$ combinations" of such subsets of the kind $A *_x B$ where $*_x \in \{\cup_x, \cap_x\}$ and to show that a* is itself a kind of probability function. More precisely :

Given $A, B \subseteq \mathfrak{a}_x$ we denote a^*_ℓ a "dual" measure of $a_\ell = \hat{\wedge}_{ix}\, [y_i = q_i^\ell]$ and Q_i^A the set of q_i^j such that $a_j = \hat{\wedge}_{ix}\, [y_i = q_i^j] \in A$; then, we settle $a^*_\ell(A) = f_x(\{g_x(q_i^\ell, \{\cup_{x\,j}\; q_i^j / q_i^j \in Q_i^A\})\}_i)$ and if $*_x \in \{\cup_x, \cap_x\}$ then $A *_x B = \hat{\wedge}_{ix}\, [y_i = \cup_{x\,j}\; \{q_i^j / q_i^j \in Q_i^A\} *_x \cup_{x\,j}\; \{q_i^j / q_i^j \in Q_i^B\}]$. Then in the case of probabilist objects we have the following result (Diday 1992) :

THEOREM.

i) $a^*(\mathfrak{a}_{pr}) = 1 \quad a^*(\phi) = 0$

ii) $\forall A_1, A_2 \subseteq \mathfrak{a}_{pr}\; a^*(A_1 \cup_{pr} A_2) = a^*(A_1) + a^*(A_2) - a^*(A_1 \cap_{pr} A_2)$.

8. Data Analysis of Symbolic Objects

Several studies have recently been carried out in this field : for histograms of symbolic objects, see De Carvalho & al (1990) and (1991) ; for generating rules by decision graph on im objects in the case of possibilist objects with typicalities as modes see Lebbe and Vignes (1991) ; for generating overlapping clusters by pyramids on symbolic objects see Brito, Diday (1990).

More generally, four kinds of data analysis may roughly be defined depending on the input and output : **a)** numerical analysis of classical data tables **b)** symbolic analysis of classical data tables, (for instance obtaining a factor analysis or a clustering automatically interpreted by symbolic objects) **c)** numerical analysis of symbolic objects (for instance by defining distances between objects) **d)** symbolic analysis of symbolic objects where the input and output of the methods are symbolic objects.

To illustrate these four approaches, on a simple example, a similarity between symbolic objects defined as follows will be used :

Let $a_\ell = \hat{i}\,[y_i = q_i^\ell] \in \mathfrak{a}_x$ be the set of im assertions. We denote a_ℓ^* a mapping $\mathfrak{a}_x \to [0,1]$ such that $a_\ell^*(a_k) = f_x(\{g_x(\{q_i^\ell, q_i^k\}_i)$; then, we settle :

(1) $$s(a_\ell,a_k) = \frac{1}{2}(a_\ell^*(a_k) + (a_k^*(a_\ell))/\sqrt{a_\ell^*(a_\ell)\,a_k^*(a_k)}\ ;$$

in the case where g_x is symmetric (which happens when we have probabilist, possibilist and plausibilist assertions) s may be written :

$$s(a_\ell,a_k) = a_\ell^*(a_k)/\sqrt{a_\ell^*(a_\ell)\,a_k^*(a_k)} = a_k^*(a_\ell)/\sqrt{a_\ell^*(a_\ell)\,a_k^*(a_k)}\ .$$

We illustrate these four approaches by applying three data analysis methods : principal components, hierarchical and pyramidal clustering.

Let T be the following data table where the set of individual objects is $\Omega = \{w_1,...,w_5\}$ which are five companies described by two variables y_1 : the employment rate and y_2 : the profit.

	w_1	w_2	w_3	w_4	w_5
y_1	-1/2	1/2	2	1	2
y_2	-1/2	1/2	1	2	2

Table T

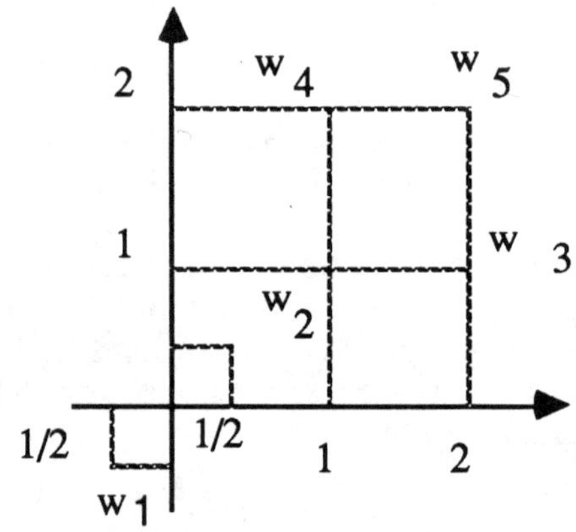

Graphical representation of table T

a) *Numerical analysis of classical data table.*

• Principal component analysis of Table T : From the covariance matrix $V = \begin{pmatrix} 0.9 & 0.7 \\ 0.7 & 09 \end{pmatrix}$ we deduce the eigen values $\lambda_1 = 1.6$ and $\lambda_2 = 0.2$ and the eigen vectors $u_1^T = \frac{1}{\sqrt{2}}(1\ 1)$, $u_2^T = \frac{1}{\sqrt{2}}(1\ -\ 1)$. Finally we get the principal component representation given in figure 1, where the projection of w_j on the axis i is given by $F_i\,(w_j) = u_i^T\ x_j$ where $x_j^T = (y_1(w_1) - Y_1,\ y_2\,(w_2) - Y_2)$ where $Y_i = 1$, is the mean of y_i ;for instance, $F_1(w_1) = \frac{1}{\sqrt{2}}(1\ 1)\begin{pmatrix} -3/2 \\ -3/2 \end{pmatrix}$.

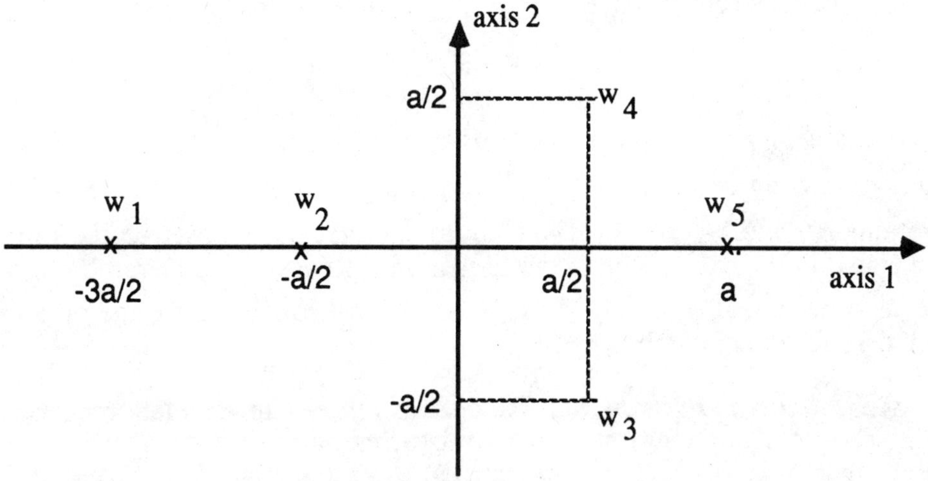

Figure 1. Principal component analysis of table T whith $a = \sqrt{2}$.

• Hierarchical and Pyramidal clustering of table T :

We make the single link hierarchy based on the city-block distance defined by $d(w_\ell, w_k) = \sum_{j=1}^{2} |y_j\,(w_\ell\,) - y_j\,(w_k)|$.

The algorithm is the following : starting from 5 classes $C_i = \{w_i\}$ where $w_i \in \Omega$ we merge at each step the two classes which minimize $\delta(C_i, C_j)$:

$\delta\ (C_i, C_j) = \text{Max}\ \{d(w_i, w_j)\ /\ w_i \in C_i, w_j \in C_j\}$. When two classes are merged their elements are supressed from the set to be classified and the process continues until only one class remains.

To obtain a pyramid, we may use a similar algorithm where classes may be merged twice (instead only once in the case of hierarchies) if they respect a common order (for more details see for instance Brito, Diday (1990)).

By using these algorithms we get the hierarchy and the pyramid given in figures 2a and 2b.

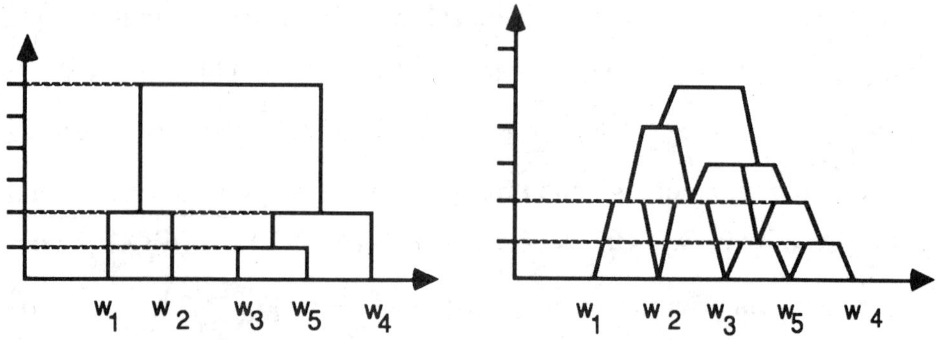

(a) - The hierarchy of table T (b) - The pyramid of table T

Figure 2

Remark : if we associate a dissimilarity σ induced by the hierarchy and the pyramid by settling : $\sigma(w_i, w_j)$ = {height of the lower level which contains w_i and w_j}, then, it is easy to see that σ is closer to the initial distance d in the case of the pyramid than in the case of the hierarchy; more precisely we get $|d-\sigma| = \sum |d(w_i, w_j) - \sigma(w_i, w_j)|$ equal to 3 for the pyramid and to 11 for the hierarchy.

b) *Symbolic analysis of a classical data table.* The correlation between $(w_1,\ldots, w_5)$ with the first axis of the principal component analysis is respectively (-1, -0.707, 0.707, 0.707, 1) ; if we associate to each side of the first axis the objects whose correlation is higher than 0.707 or lower than -0.707, we obtain two classes of objects; the first class, $C_1 = \{w_1, w_2\}$, explains the left side of the axis and the second one $C_2 = \{w_3, w_4, w_5\}$ explains the right side. By using these classes, we get two kinds of symbolic interpretation of the first axis, by using assertion we may say that the left side is explained by : $a_1=[y_1=-1/2,1/2] \wedge [y_2=-1/2,1/2]$; the right side is explained by $a_2=[y_1=1,2] \wedge [y_2=1,2]$. If at the input we have a taxonomy saying that the rate of employment and the profit are low when they are lower than $\frac{1}{2}$ and high when they are higher than 1, we may use the assertions a_1 and a_2 to get the following explanation of the first axis : it is explained by two opposite assertions which characterize two classes of companies :

$$a_1=[\text{Rate of employment} = \text{low}] \wedge [\text{Profit} = \text{low}]$$

$$a_2=[\text{Rate of employment} = \text{high}] \wedge [\text{Profit} = \text{high}]$$

Of course, in real examples things become much more complicated; for instance, to get more accuracy when the two classes contain numerous objects, each side of the axis may be explained by a disjunction of assertions obtained by a symbolic interpretation of a clustering done on each class. We may also enrich the interpretation by adding certain properties; for instance, we may add to a_1 the following rules : [if $y_1 = \frac{1}{2}$ then $y_2 = -\frac{1}{2}$] $\wedge$ [if $y_1 = \frac{1}{2}$ then $y_2 = \frac{1}{2}$] and to a_2 the rule [if $y_1 = 1$ then $y_2 = 2$].

We may also give an interpretation of the first axis by a horde object $h : h = a_1(u_1) \wedge a_2(u_2)$ = [Rate of employment(u_1) = low]$\wedge$[Profit(u_1) = low] $\wedge$ [Rate of employment (u_2) = high] $\wedge$ [Profit (u_2) = high] whose extension is composed of couples of companies (w_i, w_j) the first element of the couple w_i,being of low rate of employment and profit and the second one w_j, of high rate of employment and profit. If an external variable gives the age of the companies the horde object h may become : $h = a_1(u_1) \wedge a_2(u_2) \wedge$ [age(u_1) <age(u_2)].

A symbolic analysis of a classical data table may also be done by an automatic interpretation of a clustering by symbolic objects : for instance, it is possible to associate to each level of the hierarchy a complete symbolic object (see §. 8.2); more precisely, if we denote $h_1 = \{w_3, w_5\}$ then, we may associate to h_1, the assertion $a_1 = [y_1 = 2] \wedge [y_2 = 1,2]$; a_1 is complete, because : i) it is defined by the intention of h_1, in other words, by the conjunction of all the events $e_i=[y_i=V_i]$ whose extension contains h_1 and ii) its extension is h_1; in the same way $h_2 = \{w_1,w_2\}$, $h_3 = \{w_3,w_4,w_5\}$ and $h_4 = \Omega$ may be respectively associated to the complete assertions $a_2 = [y_1=-1/2,1/2] \wedge [y_2=-1/2,1/2]$, $a_3 = [y_1=1,2] \wedge [y_2=1,2]$, $a_4 = [y_1=0_1] \wedge [y_2 = 0_2]$ where 0_1 and 0_2 are the set of all the values taken by y_1 and y_2 in the data table T. In using the fact that each level is represented by a complete assertion we deduce from any level $h_\ell = h_i \cup h_k$ the rule $a_\ell \rightarrow a_i \vee a_k$. Hence, from the hierarchy we get the two following rules : $R_1 : a_4 \rightarrow a_2 \vee a_3$ and $R_2 : a_3 \rightarrow a_1 \vee w_4^s$ where $w_4^s = [y_1=1] \wedge [y_2 = 2]$ is the symbolic object associated to w_4. All the rules from the bottom-up such as $a_1 \rightarrow a_3$ are true because the a_i and b_i are complete objects. Finally we have induced from the hierarchy given in a) a graph (see figure 3(a)) whose nodes are assertions and rules are expressed between them by directions.

The same kind of symbolic interpretation may be done by starting from the pyramid given in figure 3 ; hence, we obtain the graph given in figure 3(b) ; in this way, we obtain more assertions and more rules between them. If we denote $h_1 = \{w_1,w_2\}$, $h_2 = \{w_2,w_3\}$, $h_3 = \{w_3,w_5\}$, $h_4 = \{w_4,w_5\}$, $h_5 = \{w_3,w_4,w_5\}$, $h_6 = \{w_2,w_3,w_5\}$, $h_7 = \Omega | w_1$, $h_8 = \{w_1,w_2,w_3\}$, $h_9 = \Omega$, the associated complete assertions are :

$b_1 = [y_1 = -1/2, 1/2] \wedge [y_2 = -1/2, 1/2]$, $b_2 = [y_1 = 1/2, 2] \wedge [y_2 = 1/2, 1]$

$b_3 = [y_1 = 2] \wedge [y_2 = 1,2]$, $b_4 = [y_1 = 1,2] \wedge [y_2 = 2]$

$b_5 = [y_1 = 1,2] \wedge [y_2 = 1,2]$, $b_6 = [y_1 = 1/2,2] \wedge [y_2 = 1/2,1,2]$

$b_7 = [y_1 = 1/2,1,2] \wedge [y_2 = 1/2,1,2]$

$b_8 = [y_1 = -1/2, 1/2, 2] \wedge [y_2 = -1/2, 1/2, 1]$, $b_9 = [y_1 = 0_1] \wedge [y_2 = 0_2]$.

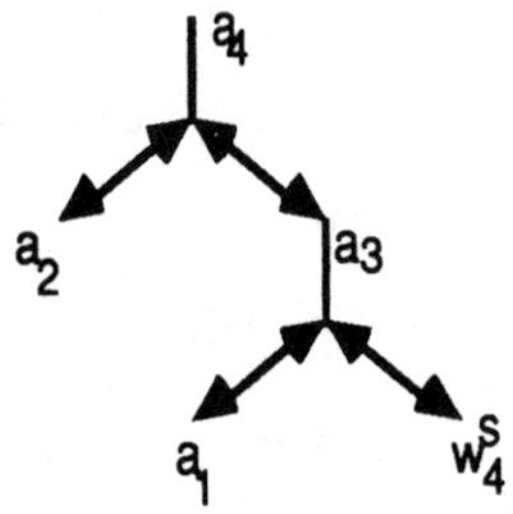

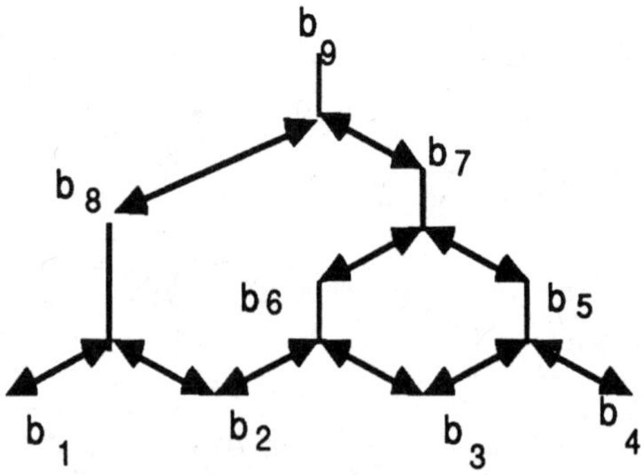

Figure 3 - Induced graph of rules between assertions
(a) from the hierarchy, (b) from the pyramid.

Hence we can induce the following rules :

$r_1 : b_9 \rightarrow b_8 \vee b_7$ $r_2 : b_7 \rightarrow b_6 \vee b_5$ $r_3 : b_8 \rightarrow b_1 \vee b_2$

$r_4 : b_6 \rightarrow b_2 \vee b_3$ $r_5 : b_5 \rightarrow b_3 \vee b_4$.

We have $b_1 = a_2$, $b_3 = a_1$, $b_5 = a_3$ and $b_9 = a_4$; hence, it is possible to deduce from the rules r_i given by the pyramid, the rules given by the hierarchy; to do so, we need to use the following property :
if $r : b_i \rightarrow b_j \vee b_k \vee b_\ell$ and $\text{Ext}(b_j / \Omega) = \text{Ext}(b_\ell / \Omega)$, then r may be simplified in $b_i \rightarrow b_j \vee b_\ell$. Hence, for instance, from r_1, r_2 and r_3 we get $b_9 \rightarrow b_1 \vee (b_2 \vee b_6) \vee b_5$ and then $b_9 \rightarrow b_1 \vee b_5$ which is $R_1 : a_4 \rightarrow a_2 \vee a_3$, obtained from the hierarchy (see figure 3a).

c) *Numerical analysis of symbolic objects* . The given set of symbolic objects is supposed to be the set of the five first symbolic objects defined by the pyramid : $\{b_1, b_2, b_3, b_4, b_5\} = B$.

A simple way to make a bridge with classical data analysis methods is to compute a measure of similarity between the objects of B; having this measure it is then possible to use multidimensional scaling, clustering etc. To do so, we may compute the similarity s which has been defined by (1); as B is a set of symbolic objects, we have to use the mapping f_b and g_b defined in §. 7. We have, for instance, $s_b(b_1,b_2) = b_1^*(b_2) / \sqrt{b_1^*(b_1)\, b_2^*(b_2)}$ with $b_1 = [y_1 = q_1^1] \wedge_b [y_2 = q_2^1]$ where q_1^1 and q_2^1 are characteristic mappings such that : $q_1^1(-\frac{1}{2}) = q_1^1(\frac{1}{2}) = 1$ and $q_2^1(-\frac{1}{2}) = q_2^1(\frac{1}{2}) = 1$ and $q_1^1(v) = q_2^1(v) = 0$ elsewhere.

	b1	b2	b3	b4	b5
b1	1	1/2	0	0	0
b2		1	$\frac{\sqrt{2}}{2}$	0	1/2
b3			1	1	$\frac{\sqrt{2}}{2}$
b4				1	$\frac{\sqrt{2}}{2}$
b5					1

(a)

	b1	b2	b3	b4	b5
b1	0	0.5	1	1	1
b2		0	0.3	1	0.5
b3			0	0	0.3
b4				0	0.3
b5					0

(b)

	b1	b2	b3	b4	b5
b1	0	1.2	1.7	1.7	1.7
b2		0	1	1.7	1.2
b3			0	0.7	1
b4				0	1
b5					0

(c)

Figure 4

We have $b_2 = [y_1 = q_1^2] \wedge_b [y_2 = q_2^2]$ and $q_1^2(v) = 1$ if $v \in \{\frac{1}{2}\ \frac{1}{2}\}$, $q_1^2(v) = 0$ elsewhere, $q_2^2(v) = 1$ if $v \in \{\frac{1}{2}, 1\}$ and $q_2^2(v) = 0$ elsewhere. As we have (see §7), $b^*{}_1(b_2) = f_b(\{g_b(q_i^1, q_i^2)\}_i) = \text{Min}(<q_1^1, q_1^2>, <q_2^1, q_2^2>) = \text{Min}(\Sigma\{q_1^1(v)\, q_1^2(v) / v \in O_1\}, \Sigma\{q_2^1(v)\, q_2^2(v) / v \in O_2\}) = \text{Min}(q_1^1(\frac{1}{2})\, q_1^2(\frac{1}{2}), q_2^1(\frac{1}{2})\, q_2^2(\frac{1}{2})) = \text{Min}(1,1) = 1$.

We have $b_1^*(b_1) = \text{Min}\ (<q_1^1,q_1^1>,\ <q_2^1,q_2^1>) = \text{Min}\ (2,2) = 2$ and also $b_2^*(b_2) = 2$; hence, $s_b\ (b_1,b_2) = 1 / \sqrt{2 \times 2} = 1/2$.

By computing in the same way all the similarities $s_b\ (b^{(i)}, b^{(j)})$ we finally get the symmetric table of similarities given in figure 4 (a).The similarity s_b is transformed in a dissimilarity $d = 1\text{-}s_b$ given in 4 (b); If we choose $c = \text{Max}\ d(b_i,b_j) - M$ where M is the sum of the two couples (b_i,b_j) of smallest dissimilarity $d(b_i,b_j)$, then $c \geq \text{Max}\ (d(b_i,b_j) - d(b_i,b_k) - d(b_k,b_j))$ and D such that $D\ (b_i,b_j) = d(b_i,b_j)$, $D\ (b_i,b_i) = 0$, is a distance, because $\forall\, i,j,k$ $d(b_i,b_j) + c \leq d(b_i,b_k) + c + d(b_k,b_j) + c$. It is easy to see that M=0+0.3 and c = 1-0.3 = 0.7; it is then possible to change d in a distance D such that $D(b_i,b_j) = d(b_i,b_j) + 0.7$ which is given in 4 (c).It is then possible to apply many existing methods of classical data analysis by using s, d or D as input.

d) *Symbolic analysis of symbolic objects.* At input we have the following set of probabilist objects : $B = \{b_1,...,b_5\}$ such that $b_j = [y_1 = q_1^j] \wedge_{pr} [y_2 = q_2^j]$ where q_1^j is a measure of probability from $P(O_j) \to [0,1]$ where $O_j = \{-\frac{1}{2}, \frac{1}{2}, 1, 2\}$ and $P(O_j)$ is the power set of O_j. If we settle $b_j = \hat{\wedge}_i [y_i = (q_i^j(v_1))\ v_1, (\ (q_i^j(v_2))\ v_2, ...]$ the b_j are defined as follows, where the value v_ℓ associated to $q_i^j\ (v_\ell) = 0$ does not appear :

$$b_1 = [y_1 = (\tfrac{1}{2})\ -\tfrac{1}{2}, (\tfrac{1}{2})\ \tfrac{1}{2}\ ] \wedge_{pr} [y_2 = (\tfrac{1}{2})\ -\tfrac{1}{2}\ , (\tfrac{1}{2})\ \tfrac{1}{2}\ ]$$

$$b_2 = [y_1 = (\tfrac{1}{2})\ \tfrac{1}{2}, (\tfrac{1}{2})\ 2\ ] \wedge_{pr} [y_2 = (\tfrac{1}{2})\ \tfrac{1}{2}\ , (\tfrac{1}{2})\ 1\ ]$$

$$b_3 = [y_1 = (1)\ 2] \wedge_{pr} [y_2 = (\tfrac{1}{2})\ 1, (\tfrac{1}{2})\ 2\ ]$$

$$b_4 = [y_1 = (\tfrac{1}{2})\ 1, (\tfrac{1}{2})\ 2\ ] \wedge_{pr} [y_2 = (1)\ 2\ ]$$

$$b_5 = [y_1 = (\tfrac{1}{2})\ 1, (\tfrac{1}{2})\ 2\ ] \wedge_{pr} [y_2 = (\tfrac{1}{2})\ 1, (\tfrac{1}{2})\ 2\ ]$$

To treat this set of probabilist objects the first way may be to compute the similarity $s_{pr}\ (b_i,b_j) = b_i^*\ (b_j) / \sqrt{b_i^*\ (b_i)\ b_j^*\ (b_j)}$ and then, to use multidimensional scaling or clustering methods interpreted by symbolic objects as has already been done in b).

For instance, for the couple (b_1,b_2), $b_1^*\ (b_2) = f_{pr}\ (\{g_{pr}\ (q_i^1,q_i^2\ )\}_i)$ is computed as follows :

$b_1^*(b_2) = \text{Mean}\,(<q_1^1, q_1^2>, <q_2^1, q_2^2>)$; therefore :

$b_1^*(b_2) = \text{Mean}\,(\Sigma\,\{q_1^1(v).\,q_1^2(v)\,/\,v \in O_1\},\ \Sigma\,\{q_2^1(v).\,q_2^2(v)\,/\,v \in O_2\})$.

Hence $b_1^*(b_2) = \text{Mean}\,(\frac{1}{2} \times 0 + \frac{1}{2} \times \frac{1}{2} + 0 \times 0 + 0 \times \frac{1}{2}, \frac{1}{2} \times 0 + \frac{1}{2} \times \frac{1}{2} + 0 \times \frac{1}{2} + 0 \times 0) = \text{Mean}\,(\frac{1}{4}, \frac{1}{4}) = (\frac{1}{4} + \frac{1}{4})\,\frac{1}{2} = \frac{1}{4}$.

$b_1^*(b_1) = \text{Mean}\,(\frac{1}{2}, \frac{1}{2}) = \frac{1}{2}$; $b_2^*(b_2) = \text{Mean}\,(\frac{1}{2}, \frac{1}{2}) = \frac{1}{2}$.

Finally by setting $\alpha = \sqrt{\frac{3}{2}}$ we obtain the following similarity :

$$\{s_{pr}(b_i,b_j)\} = \begin{matrix} 1 & 1/2 & 0 & 0 & 0 \\ & 1 & \alpha/2 & \alpha/6 & 1/2 \\ & & 1 & 2/3 & 2\alpha/3 \\ & & & 1 & 2\alpha/3 \\ & & & & 1 \end{matrix} = \begin{matrix} 1 & 0.5 & 0 & 0 & 0 \\ & 1 & 0.6 & 0.2 & 0.5 \\ & & 1 & 0.7 & 0.8 \\ & & & 1 & 0.8 \\ & & & & 1 \end{matrix}$$

For treating B, another way is to obtain directly from B, clusters of symbolic objects represented by an "inheritance" hierarchy, where each node is expressed by a complete probabilist assertion a_{jk}, or an approximation of it such that if $a_{jk} = a_j \cup_{x,\alpha} a_k$ then $a_{jk} \geq_\alpha \text{Max}\,(a_j,a_k)$ where $\cup_{pr,\alpha}$ and $\geq_\alpha$ have been defined in §8.1. To do so, we may use the following algorithm on a set of symbolic objects A :

First step : $a_{jk} = a_j \cup_{x,\alpha} a_k$ is computed $\forall\ a_j, a_k \in A$.

Second step : the a_{jk} of smaller extension constitute the first levels of the hierarchy, their height is the cardinality of their extension.

Third step : the retained a_{jk} at step 2 are added to A and a_j, a_k are suppressed from A; then, we go back to the first step until the cardinality of A becomes equal to 1.

In practice, how can we compute $a_{jk} = a_j \cup_{pr,\alpha} a_k$? By definition a_{jk} is the conjunction of the elementary events $a_{jk}^i = [y_i{=}q_i]$ such that $\text{Ext}(a_{jk}^i/\,\Omega,\alpha)$ contains $\Omega_1 = \text{Ext}(a_j/\Omega,\alpha) \cup \text{Ext}(a_k\,/\,\Omega,\alpha)$. Hence, for any $w \in \Omega_1$ such that $w^s = \hat{}_i[y_i{=}\,r_i]$ we have $a_{jk}(w) \geq \alpha$; this condition is satisfied if we have $\forall$i, $g(q_i,r_i) \geq \alpha$ because $a_{jk}(w) = f(\{g(q_i,r_i)\}_i)$ and, by definition of f it, is the mean; hence, if we denote $x_j^i = q_i\,(v_j)$, we have the inequality :

(1) $$g(q_i,r_i) = \Sigma\{x_j^\ell.r_i(v_j)/v_j \in O_j\} \geq \alpha;$$

hence, we have to solve a system of card(Ω_1) inequations where the unknown are the x_j^ℓ. If this system has several solutions, for each i we denote them $[y_i=q_i^\ell]$; hence, we obtain $a_{jk} = \hat{\wedge}_i{}_{pr}\,(\hat{\vee}_\ell{}_{pr}[y_i=q_i^\ell])$; by choosing h_{pr}=min (see §.3.1) the extension of a_{jk} at level α is $\Omega_2=\{w/a_{jk}(w) = f(\{Min\{g(q_i^\ell,r_i)\}_\ell) \geq \alpha\}$.

To obtain the inheritance hierarchy on B given by the algorithm, the first step consists in computing the $a_{jk} = b_j \cup_{pr,\alpha} b_k$ whose extension is of minimum cardinality; we choose $\alpha = \frac{1}{2}$ and to compute for instance $a_{12} = b_1 \cup_{pr,\alpha} b_2$ we do as follows : first we settle $a_{jk}=a_{jk}^1 \wedge a_{jk}^2$ where $a_{jk}^\ell = [y_\ell=q_\ell]$ is such that $a_{12}^\ell(b_1) \geq \frac{1}{2}$ and $a_{12}^\ell(b_2) \geq \frac{1}{2}$. Hence, $x_j^\ell=q_\ell(v_j)$ where $\{v_1,...,v_4\} = \mathbf{O}_1 = \mathbf{O}_2 = \{-\frac{1}{2}, \frac{1}{2}, 1, 2\}$, we have to solve the following inequations, where the x_j^ℓ are the unknowns,with the constraint

$\sum_j q_\ell(v_j)= \sum_j x_j^\ell =1$: $a_{12}^\ell(b_1) = g_{pr}(q_\ell,r_1^\ell) = \Sigma\ \{q_\ell(v_i)\ r_\ell^1(v_i)\ /\ v_i\in \mathbf{O}_\ell\}$; hence, we obtain : $a_{12}^1(b_1) = \frac{1}{2}x_1^1 + \frac{1}{2}x_2^1 \geq \frac{1}{2}$; $a_{12}^1(b_2) = \frac{1}{2}x_2^1 + \frac{1}{2}x_4^1 \geq \frac{1}{2}$ from which we deduce that $x_1^1 + x_2^1 = 1$, therefore (as $\Sigma\{x_i^1\ /\ i=1, 4\} =1$) we get $x_4^1 = 0$ and $x_2^1 = 1$, $x_i^1 = 0$ if i≠2. $a_{12}^2(b_1) = \frac{1}{2}x_1^2 + \frac{1}{2}x_2^2 \geq \frac{1}{2}$; $a_{12}^1(b_2) = \frac{1}{2}x_2^2 + \frac{1}{2}x_3^2 \geq \frac{1}{2}$, from which it results that x_2^2=1 and x_i^2=0 if i≠2.Finally we obtain : $a_{12} = a_{12}^1 \wedge_{pr} a_{12}^2 = [y_1=(1)\frac{1}{2}] \wedge_{pr} [y_2=(1)\frac{1}{2}]$ (which is equivalent to the boolean object $[y_1 = \frac{1}{2}] \wedge_b [y_2 = \frac{1}{2}]$).

In the same way, we get : $a_{13}^1(b_1) = \frac{1}{2}x_1^1 + \frac{1}{2}x_2^1 \geq \frac{1}{2}$ and $a_{13}^1(b_3) = x_4^1 \geq \frac{1}{2}$ this is contradictory because the first equation implies $x_4^1 = 0$. Hence, the only symbolic object whose extension contains b_1 and b_2 is the full object Ω^s whose extension is Ω ; $\Omega^s = \hat{\wedge}_i[y_i=q_i]$ is defined in the case of probabilist objects by functions $q_i : P(\mathbf{O}_i) \to \{1\}$ (which are not, of course, probabilities !), then it is easy to see that $\Omega^s(w) = 1\ \forall\ w\in\Omega$. In the same way we get : $a_{14}=a_{15}=a_{24}=\Omega^s$; $a_{23}=[y_1=(1)2] \wedge_{pr}[y_2=(1)1]$; $a_{25}=a_{23}$; $a_{34}=[y_1=(1)2] \wedge_{pr}[y_2=1(2)]$; a_{35} is computed as follows : $a^1_{35}(b_3) = x^1_4 \geq \frac{1}{2}$ and $a^1_{35}(b_5) = \frac{1}{2}\, x^1_3 + \frac{1}{2}\, x^1_4 \geq \frac{1}{2}$ implies $x^1_4 = 1$ and $x^1_i = 0$ if $i \neq 4$; $a^2_{35}(b_3) = \frac{1}{2}\, x^2_3 + \frac{1}{2} x^2_4 \geq \frac{1}{2}$ and $a^2_{35}(b_5) = \frac{1}{2} x^2_3 + \frac{1}{2} x^2_4 \geq \frac{1}{2}$; we have three solutions

i) $x^2_3 = x^2_4 = \frac{1}{2}$;

ii) $x^2_3 = 1,\ x^2_i = 0$ for $i \neq 3$;

iii) $x^2_4 = 1,\ x^2_i = 0$ for $i \neq 4$; therefore $a_{35} = [y_1=(1)2] \wedge_{pr} [y_2 = (\frac{1}{2})1, (\frac{1}{2})2] \wedge_{pr} [y_2=(1)1] \wedge_{pr}[y_2=(1)2]$.

In a similar way we get :

$a_{45} = [y_1 = (\frac{1}{2})1,(\frac{1}{2})2] \wedge_{pr} [y_1=(1)1] \wedge_{pr} [y_1=(1)2] \wedge_{pr} [y_2=(1)2]$

In the following table we give in the cell of the ith row and jth column the extension of $a_{ij} = b_i \cup_{pr,1/2} b_j$:

$\text{Ext}(a_{ij}/B,\frac{1}{2}) =$

	1	2	3	4	5
1		b_1b_2	Ω	Ω	Ω
2			$b_2b_3b_5$	Ω	$b_2b_3b_5$
3				$b_3b_4b_5$	b_3b_5
4					b_4b_5
5					

In using this table it is easy to do the inheritance hierarchy, in merging at each step the couple of least extension. Hence, the first couples are $(b_1,b_2),(b_3,b_5),(b_4,b_5)$; to get a hierarchy it is not possible to retain simultaneously (b_3,b_5) and (b_4,b_5) therefore if there are no external constraints on the clusters (for instance, constraint of geographical proximity) we have to choose randomly one of them; if we retain, for instance (b_3,b_5) the first couples to be merged are finally (b_1,b_2) and (b_3,b_5); therefore, we obtain the two first levels of the hierarchy characterized by $a_{12} = b_1 \cup_{pr,1/2} b_2$ and $a_{35} = b_3 \cup_{pr,1/2} b_2$. Hence, it remains b_4 to be merged with (b_1,b_2) or (b_3,b_5).

It is then easy to see that $a^2_{124}(b_1) = \frac{1}{2}\, x^2_1 + \frac{1}{2}\, x^2_2 \geq \frac{1}{2}$ and $a^2_{124}(b_4) = \frac{1}{2}\, x^4_2 \geq \frac{1}{2}$ which give no solution such that $\sum_{i=1,4} x^2_i = 1$;

therefore, $a_{124} = \Omega^s$ whose extension is B. We have already seen that $Ext(a_{34}/B,\frac{1}{2}) = \{b_3,b_4,b_5\}$, therefore $a_{345} = a_{34}$; hence, the next couple to be merged will be $(b_4,(b_3,b_5))$ which gives a third level represented by $a_{345} = a_{34}$; the last level merges (b_1,b_2) with (b_3,b_4,b_5) and is represented by the full object Ω^s. To summarize, we have finally obtained four levels whose representation and extension are given in table 2. 0

Level	Representation	Extension
1	$a_{12} = [y_1=(1)\frac{1}{2}] \wedge_{pr} [y_2=(1)\frac{1}{2}]$	$\{b_1,b_2\}$
2	$a_{35}=[y_1=(1)2] \wedge_{pr} [y_2=((\frac{1}{2})1,(\frac{1}{2})\ 2)]$ $\wedge_{pr} [y_2=(1)1] \wedge_{pr} [y_2=(1)2]$	$\{b_3,b_4\}$
3	$a_{345} = [y_1=(1)2] \wedge_{pr} [y_2=(1)2]$	$\{b_3,b_4,b_5\}$
4	$a_{12345} = \sum_{i=1,2} [y_i=(1) - \frac{1}{2}, (1)\frac{1}{2}, (1)1, (1)2]= \Omega^s$	B

Table 2

In using the fact that the height of each level is the cardinality of the extension of it associated probabilistic assertion, it is then easy to make the inheritance hierarchy associated to the set of probabilist objects B represented in figure 5.

We say that we have a rule between two probabilistic assertions a_1 and a_2 at level (α_1,α_2) denoted $R : a_1 \overset{(\alpha_1,\alpha_2)}{\rightarrow} a_2$ when $Ext(a_1/B, \alpha_1) \subseteq Ext(a_2/B,\alpha_2)$; in other words, the rule R is true if, when b is in the extension of a_1 at level α_1, then, it is in the extension of a_2 at level α_2; when $\alpha_1=\alpha_2=\alpha$ this rule is denoted $a_1 \overset{\alpha}{\rightarrow} a_2$. By using this denotation it is easy to induce from the inheritance hierarchy of figure 5, by going bottom-up, the rule : $a_{35} \overset{1/2}{\rightarrow} a_{345}$; it is possible to induce top-down the following rule : $\Omega^s \overset{(1,1/2)}{\rightarrow} a_{12} \vee a_{345}$ which means that if b is in the extension of Ω^s at level 1, it is in the extension of a_{12} or a_{345} ,at the level 1/2; in the same way we get also $a_{345} \overset{1/2}{\rightarrow} b_4 \vee a_{35}$.

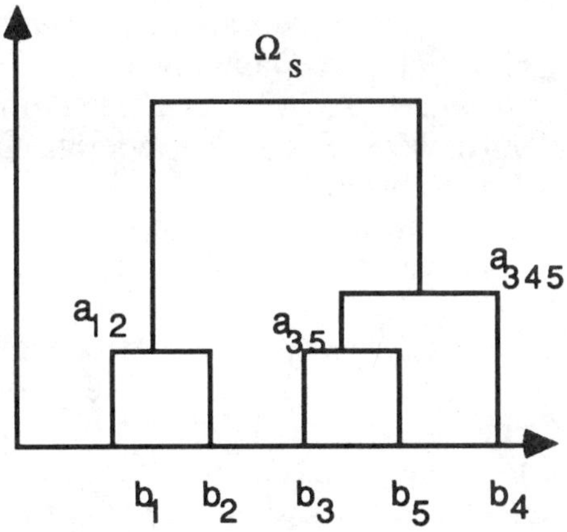

Figure 5 : Inheritance hierarchy on probabilist objects

Conclusion

Starting from classes of individual objects defined by intention (in contrast to classes only defined by the set of their members) we have given several ways to define them in a mapping denoted a_x on Ω (the set of individual objects) depending on the background knowledge x. A question naturally arose : is it possible to say that a_x (w) measures a "probability", "possibility" or a "belief" that w belongs to the class represented by a_x ? To answer this question we needed to extend a_x to a^*_x defined on $\mathfrak{a}_x$ a set of symbolic objects and to define set operators $OP_x = \{\cup_x, \cap_x, c_x\}$ in $\mathfrak{a}_x$ adapted to x. If we say that classical sets represent a knowledge level of order 0; probability, possibility and belief, a knowledge level of order 1, the question was now to know if a^*_x represents a knowledge level of order 2. In other words, if it is a probability of probability, a possibility of possibility a belief of belief respectively associated with the corresponding operators OP_x; the theorems 1,2,3 show that this is the case, if OP_x and the functions g and f are well chosen. Unlike most work carried out in Artificial Intelligence, symbolic data analysis constitutes a "critique of pure reasoning" by giving less importance to the reasoning and more importance to the statistical study of knowledge bases, considered as a set of "symbolic objects". A wide field of research is opened by extending classical statistics to statistics of intentions and more specially by extending problems, methods and algorithms of data analysis to symbolic objects.

References

[1] Brito P., Diday E., *Pyramidal representation of symbolic objects*, in NATO ASI Series, (1990), Vol. F 61, Knowledge Data and computer-assisted Decisions edited by Schader and W. Gaul. Springer Verlag.

[2] De Carvalho F.A.T., *Histogramme en Analyse des Données Symboliques*, (1992), Thèse Paris IX-Dauphine et INRIA.
[3] Diday E., *Knowledge representation and symbolic data analysis*, in NATO ASI Series, (1990), Vol. F 61, Knowledge Data and computer-assisted Decisions edited by Schader and W. Gaul. Springer Verlag.
[4] Diday E., *From data to Knowledge : new objects for a statistical analysis*, (1992), in New Techniques and Technologies for statistics conference, GMD, Bonn.
[5] Dubois D., Prade H., *Possibility theory*, (1988), Plenum New York.
[6] Dubois D., Prade H., *A set-theoretic view of belief functions*, International Journal General Systems, (1986), Vol. 12, pp 193-226.
[7] Lauritzen S.L. and Spiegelhalter D.J., *Local computation with probabilities on graphical structures and their application to expert system,* In Readings in uncertain reasoning, (1990) edited by G. Shafer and J.Pearl.Morgan Kaufman Publishers.
[8] Lebbe J., Vignes R., Darmoni S., *Symbolic numeric approach for biological knowledge representation : a medical example with creation of identification graphs*, in : Proc. of Conf. on Data Analysis, Learning Symbolic and Numerical Knowledge, (1990), Antibes ed. E. Diday, Nova Science Publishers, Inc., New York.
[9] Pearl J., *Probabilist reasoning in intelligent systems,* (1988), Morgan Kaufman, San Mateo.

Edwin DIDAY, INRIA, DOMAINE DE VOLUCEAU, 78153 LE CHESNAY CEDEX, FRANCE.
E-mail : Edwin.Diday@inria.fr

DIMACS Series in Discrete Mathematics
and Theoretical Computer Science
Volume 19, 1995

A Labeling Algorithm for Minimum Sum of Diameters Partitioning of Graphs

S. Gélinas, P. Hansen, and B. Jaumard

ABSTRACT. An $O(mn \log n)$ labeling algorithm is proposed for minimum sum of diameters bipartitioning of the vertex set of a weighted graph (with n vertices and m edges). The diameter of a set of vertices is defined as the largest weight of an edge joining two vertices in that set. This algorithm has, in fact, a lower complexity than a previous one due to Monma and Suri.

1. Introduction

Let $G = (V, E)$ be an undirected edge-weighted graph. We assume the weights to be non-negative. Let $d_{k\ell}$ be the weight of the edge $\{v_k, v_\ell\} \in E$. Let $V_j \subset V$ be a subset of vertices of G. The *diameter* $d(V_j)$ of V_j is the largest weight of an edge with both endpoints in V_j:

$$d(V_j) = \max_{k,\ell: v_k \in V_j, v_\ell \in V_j} d_{k\ell}.$$

The *minimum sum-of-diameters graph partitioning* problem is to find a partition $V_1, V_2, \ldots, V_M$ of the vertices of G into M subsets, i.e., $V_i \neq \emptyset, V_i \cap V_j = \emptyset$ for $i, j = 1, 2, \ldots, M$, $i \neq j$ and $\bigcup_{i=1}^{M} V_i = V$, such that the sum of the diameters $d(V_1), d(V_2), \ldots, d(V_M)$ is minimum.

A related problem is, given positive numbers $d_1, d_2, \ldots, d_M$, to determine if there exists a partition of the vertices of G such that $d(V_1) \leq d_1$, $d(V_2) \leq d_2$, $\ldots$, $d(V_M) \leq d_M$. We call this the *given-diameters problem.*

1991 Mathematics subject Classification. Primary 05C78; Secondary 62H30.
The first author has been supported by a CRSNG doctoral fellowship. The second and third authors have been supported by the Office of Naval Research grant #N00014-92-J-1194 and FCAR (Fonds pour la Formation de Chercheurs et l'Aide à la Recherche) grant 92EQ1048. The second author has also been supported by NSERC (Natural Sciences and Engineering Council of Canada) grant to HEC and NSERC grant GP0105574. The third author has also been supported by NSERC grant GP0036426 and a FRSQ (Fonds pour la Recherche en Santé du Québec) fellowship.
This paper is in final form and will not be submitted for publication elsewhere.

Minimum sum-of-diameters graph partitioning has applications in cluster analysis, location theory and computational geometry. In cluster analysis (e.g. Hartigan, 1975, Gordon, 1981, and Kaufman and Rousseeuw, 1990) a given set O of entities is associated with the vertex set V of G, edges of G are weighted by *dissimilarities*, which express the degree in which pairs of entities differ one from the other, and one seeks homogeneous and/or well separated subsets of O, called *clusters*. The sum-of-diameters is then an homogeneity criterion, studied by Hansen and Jaumard, 1987. Avis, 1986 considers communication posts located in a large section of the plane and seeks how to bipartition them in order to minimize the transmitting power necessary for each pair of posts in the same group to communicate. Points of the plane are associated to vertices of G and weights of edges to Euclidean distances between them. Avis, 1986 provides an $O(n^2)$ algorithm for given-diameters bipartitionning. This clustering problem in the plane, pertains, as many others, to computational geometry (e.g. Preparata and Shamos, 1985). Hershberger and Suri, 1991 present an $O(n \log n)$ algorithm for Avis' problem. Monna and Suri, 1991 propose a $O(n^2)$ algorithm for minimum sum-of-diameters graph bipartitioning in the plane and Hershberger, 1992 reduces this complexity to $O(n \log^2 n / \log\ \log n)$. Capoyleas, Rote and Woeginger, 1991 show that minimizing any monotone function of the diameters of M clusters forming a partition of a set of n points in the plane can be done in $O(n^{6M})$ time. We consider here the general case of bipartitioning, where the $d_{k\ell}$ are dissimilarities and hence do not necessarily correspond to distances between points in the plane, and may even not satisfy the triangle inequality.

Brucker, 1978 shows that the minimum sum-of-diameters partitionning problem is NP-hard for $M \geq 3$ and mentions finding the complexity for $M = 2$ as an open problem. Hansen and Jaumard, 1987 provide an $O(n^3 \log n)$ algorithm for this last case, when, as usual in cluster analysis, G is a complete graph. This algorithm relies upon a property of the largest diameter of the two clusters (or diameter of the bipartition), recalled below, and on the solution of a series of given-diameters problems by the algorithm of Aspvall, Plass and Tarjan, 1979 for *2-satisfiability*, i.e., solution of a quadratic logical equation. Monma and Suri, 1991 propose to use within the same framework a labeling algorithm instead of the method of Aspvall *et al.* 1979. They claim to obtain an $O(mn \log n)$ algorithm which improves upon the previously best algorithm. The purpose of this paper is to examine this claim. A first remark is that the algorithm of Hansen and Jaumard, 1987 applied to a (possibly sparse) graph takes $O(mn \log n)$ time. Indeed, the algorithm of Aspvall, Plass and Tarjan, 1979 which is used $O(n \log n)$ times requires time in the order of the number of occurrences of literals in the 2-satisfiability problem. This is $O(m)$ for the given diameters problem on graph, which is equal to $O(n^2)$ for complete graphs (hence the reported complexity of $O(n^3 \log n)$ in Hansen and Jaumard, 1987), but not for sparse ones. However, a labeling algorithm to solve the given diameters problem might still be preferable, in order to make the general algorithm self-contained. It could also be more efficient, by some constant factor. It then turns out, as explained in the next section, that Monma and Suri's, 1991 labeling algorithm does not have a worst case complexity in $O(m)$, but in $O(m^2)$, leading to a complexity for their general algorithm in $O(m^2 n \log n)$. In the last section, an $O(m)$ labeling algorithm based on depth-first search of the graph G, is proposed. It immediately leads to an $O(mn \log n)$ algorithm for minimum sum-of-diameters bipartitioning and an $O(mn^2 \log n)$ algorithm for hierarchical divisive clustering with the sum-of-diameters criterion.

2. Monma and Suri's algorithm

A first result shows that there are at most n possible values for the largest diameter $d(V_1)$ of a minimum sum-of-diameters bipartition V_1, V_2 of V.

Theorem 1 *(Hansen and Jaumard, 1987) Let $G = (V, E)$ be an undirected weighted graph and $d_{k\ell}$ the weight associated with edge $\{v_k, v_\ell\} \in E$. Let V_1, V_2 be a bipartition of V such that $d(V_1) \geq d(V_2)$. Let T be a maximum weight spanning tree of G and $\{v_p, v_q\}$ a maximum weight edge which forms an odd cycle with T. Then*

(i) $d(V_1) \geq d_{pq}$;

(ii) if $d(V_1) > d_{pq}$ *there exists an edge* $\{v_k, v_\ell\} \in T$ *such that* $d(V_1) = d_{k\ell}$.

A corollary of Theorem 1, found independently by Guénoche, 1989, 1993 and by Monma and Suri, 1991 is that the bipartition induced by a bicoloring of T has minimum diameter. Guénoche, Hansen and Jaumard, 1991 use this result to obtain an $O(n^2 \log n)$ algorithm for hierarchical divisive clustering with the diameter criterion.

Theorem 1 suggests the following scheme for minimum sum-of-diameters bipartitioning:

(a) Determine T and $\{v_p, v_q\}$;

(b) Rank the weights $d_{k\ell}$ of the edges $\{v_k, v_\ell\}$ of T such that $d_{k\ell} > d_{pq}$ in order of decreasing values in a list L_1 and add d_{pq} to L_1;

(c) Rank the weights $d_{k\ell}$ of all edges $\{v_k, v_\ell\}$ of G by order of decreasing values in a list L_2;

(d) For each value of $d_{k\ell}$ in L_1, find by dichotomous search in L_2 the smallest $d_{rs} \leq d_{k\ell}$ corresponding to an edge such that V has a bipartition V_1, V_2 with $d(V_1) \leq d_{k\ell}$ and $d(V_2) \leq d_{rs}$. Keep the best bipartition found.

Both the algorithms of Hansen and Jaumard, 1987 and of Monna and Suri, 1991 follow this scheme. They differ in the way the given-diameters problem which occurs $O(n) \times O(\log n) = O(n \log n)$ times, is solved. Moreover, a few acceleration rules, which do not modify the complexity, are given in Hansen and Jaumard, 1987.

Monma and Suri, 1991 solve the subproblem of step *(d)*, for given values of $d_{k\ell}$ and d_{rs}, by labeling vertices 1 or 2, i.e., assigning vertices to V_1 or V_2, according to the two following propagation rules:

(R1) [applies if $d_{ij} > d_{k\ell}$] If v_i is labeled 1 (resp. 2) and v_j is unlabeled, then label v_j with 2 (resp. 1).

(R2) [applies if $d_{ij} > d_{rs}$] If v_i is labeled 2 and v_j is unlabeled then label v_j with 1. Otherwise v_j remains unlabeled.

A *label-violation* occurs at edge $\{v_i, v_j\}$ if either $d_{ij} > d_{k\ell}$ and both v_i and v_j are labeled 1 or if $d_{ij} > d_{rs}$ and both v_i and v_j are labeled 2. An arbitrary vertex v^* is first labeled 1 and rules **(R1)** and **(R2)** are used to label as many vertices as possible, until a label-violation occurs or a connected subgraph of G gets labeled without label-violation and no more vertices can be labeled. In the former case, labels are erased, v^* is labeled 2 and the propagation rules applied again. Then, if label-violation happens once more no bipartition of V has diameters $d(V_1) \leq d_{k\ell}$ and $d(V_2) \leq d_{rs}$, and the algorithm for the subproblem stops. In the latter case, if all vertices are labeled, the algorithm for the subproblem stops, the desired bipartition having been found. If unlabeled vertices remain, labeled vertices are deleted and the algorithm is applied to the subgraph so obtained.

The following example (illustrated in Figures 1 and 2) shows Monma and Suri's, 1991 labeling algorithm may require $O(m^2)$ operations in the worst case. Consider a path with $n-2$ vertices and n even, followed by a clique on 3 vertices, including the last one of the path. Assume the dissimilarities are greater than $d_{k\ell}$ for edges $\{v_{1+2k}, v_{2+2k}\}$ with $k = 0, 1, 2, \ldots, \dfrac{n-2}{2}$ and greater than d_{rs} but smaller than or equal to $d_{k\ell}$ for edges $\{v_{2+2k}, v_{3+2k}\}$ with $k = 0, 1, 2, \ldots, \dfrac{n-4}{2}$ and for $\{v_{n-2}, v_n\}$. As choice rule for v^* consider always the vertex with smallest index. Then v_1 is labeled 1 and **(R1)** and **(R2)** assign alternately labels of 2 and 1 to vertices v_2, v_3, etc. . The edge $\{v_{n-2}, v_n\}$ gets both end-vertices labeled 2, a label-violation. So all labelings of vertices are erased and v_1 is labeled 2. Then v_2 is labeled 1 and no more propagation occurs. Vertices v_1 and v_2 are erased, v_3 chosen as v^* and labeled 1. Again all vertices up to v_n are labeled and a label-violation occurs at edge $\{v_{n-2}, v_n\}$. Iterating shows that the total number of vertex labelings is $n + (n-2) + (n-4) + \ldots$, i.e., in $O(n^2) = O(m^2)$.

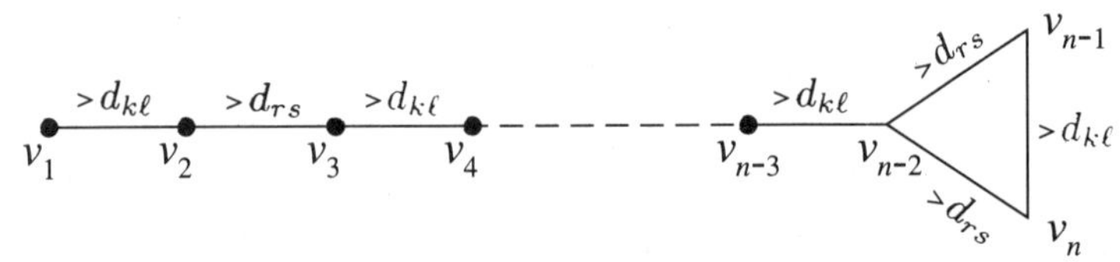

Figure 1: Graph requiring $O(m^2)$ operations

3. A linear labeling algorithm

The quadratic complexity of the labeling algorithm of Monma and Suri, 1991 is due, as illustrated by the example of the previous section, to multiple erasures of labels assigned to the same vertices. Such multiple erasures may be avoided

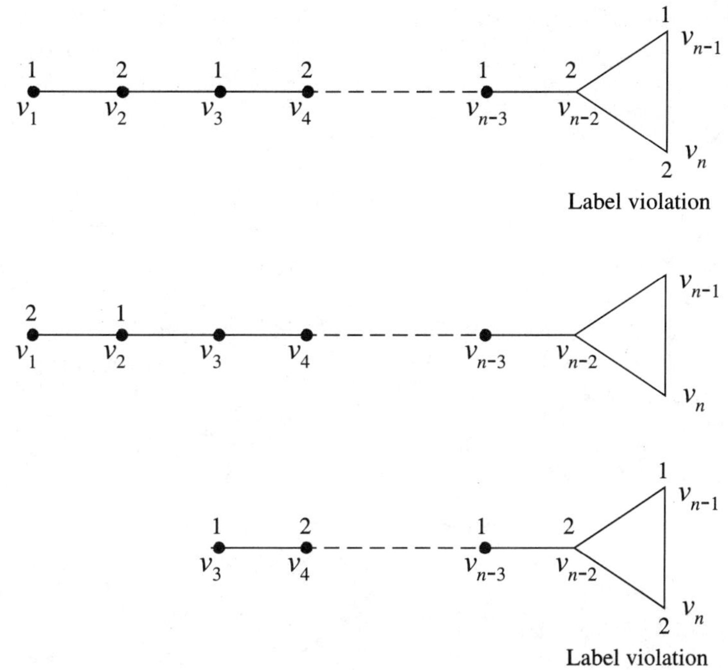

Figure 2: First labels for the vertices of graph G

by using temporary and permanent labels and specifying that a temporary label can only be replaced by a permanent one. Moreover, it is necessary to specify more precisely how to do the propagation of labels, i.e., by depth-first search of the graph G (possibly after removal of edges of weight $\leq d_{rs}$ which play no role in the subproblem). Let 1 and 2 denote permanent labels and 3 and 4 temporary ones. If a vertex v has a label 1 or 3 (resp. 2 or 4) when the algorithm ends without *permanent label-violation*, i.e., label violation for labels 1 or 2, it is assigned to V_1 (resp. V_2). A *temporary label-violation* occurs at edge $\{v_i, v_j\}$ if either $d_{ij} > d_{k\ell}$ and both v_i and v_j are labeled 3 or if $d_{ij} > d_{rs}$ and both v_i and v_j are labeled 4. Propagation rules for permament labels are **(R1)** and **(R2)**, stated above where, however, vertices with temporary labels are viewed as unlabeled (i.e., temporary labeled vertices may be assigned a permanent label which replaces the temporary one). For temporary labels, they are similar (and denoted by **(R3)** and **(R4)**) except that labels 3 and 4 are used instead of 1 and 2:

(R3) [applies if $d_{ij} > d_{k\ell}$] If v_i is labeled 3 (resp. 4) and v_j is unlabeled then label v_j with 4 (resp. 3).

(R4) [applies if $d_{ij} > d_{rs}$] If v_i is labeled 4 and v_j is unlabeled, then label v_j with 3. Otherwise v_j remains unlabeled.

Then the rules of the labeling algorithm are as follows:
Algorithm Bipartition

1. **Choice of an unlabeled vertex and test for ending:** Choose an unlabeled vertex v^* and label it 3. If there are none, stop: vertices labeled 1 or 3 (resp. 2 and 4), define V_1 (resp. V_2), $d(V_1) \leq d_{k\ell}$ and $d(V_2) \leq d_{rs}$.

2. **Temporary labels:** Set $V_3 = \emptyset$. Apply rules **(R3)** and **(R4)** in a depth-first search of G, beginning at v^*. If a temporary label-violation occurs at edge $\{v_i, v_j\}$ where v_j has been labelled before v_i, add v_j to V_3. When the search is completed proceed to Step 3 if $V_3 \neq \emptyset$ and return to Step 1 otherwise.

3. **Permanent labels:** Label each vertex v_j of V_3 with 1 if its temporary label was 4 and with 2 if its temporary label was 3. Then beginning with each vertex v_j of V_3 in turn, apply rules **(R1)** and **(R2)** in a depth-first search of G. If a permanent label-violation occurs, stop: there is no bipartition V_1, V_2 of V with $d(V_1) \leq d_{k\ell}$ and $d(V_2) \leq d_{rs}$. Otherwise, when all searches are completed return to Step 1.

We next prove the correctness and linear complexity of algorithm Bipartition.

Theorem 2 *Algorithm Bipartition provides in $O(m)$ time a bipartition V_1, V_2 of V such that $d(V_1) \leq d_{k\ell}$ and $d(V_2) \leq d_{rs}$ or the proof that no such bipartition exists.*

Proof.

1. Let G' be graph G after orientation from v_i to v_j of all edges $\{v_i, v_j\}$ for which a temporary label has been given to v_j propagating the label of v_i, during the first application of Step 2. Then if a temporary label-violation occurs at edge $\{v_i, v_j\}$ where v_j was labelled before v_i, there is a directed path from v_j to v_i in G'. Indeed suppose not. Then there is an edge $\{v_i, v_j\}$ in G with $d_{ij} > d_{rs}$ and v_i was not labelled from v_j due to this edge. If $d_{ij} > d_{k\ell}$, v_i should have been labeled from v_j following rule **(R3)** and the depth first search strategy; if $d_{ij} < d_{k\ell}$ and v_j is labelled 4, v_i should have been labeled from v_j according to rule **(R4)**; if $d_{ij} < d_{k\ell}$ and v_j is labelled 3 no label-violation may occur at edge $\{v_i, v_j\}$, a contradiction. Therefore the label of v_j implies that of v_i, and a label-violation. So v_j must receive a permanent label of different parity than that of its temporary label in any labeling without permanent label-violation. This justifies the labels given to the vertices of V_3 in Step 3, which remove all temporary label-violations (but may induce some permanent ones).

2. After Step 2 has been completed for the first time, any unlabeled vertex v_j may be adjacent to a temporarily labeled vertex v_i with $d_{ij} > d_{rs}$ only if the label of v_i is 3 and $d_{ij} \leq d_{k\ell}$. But then no temporary label-violation may occur at edge $\{v_i, v_j\}$. Similarly, after Step 3 has been completed for the first time any unlabeled vertex v_j may be adjacent to a permanently labeled vertex v_i with $d_{ij} > d_{rs}$ only if the label of v_i is 1 and $d_{ij} \leq d_{k\ell}$. Then no permanent label-violation may occur at edge $\{v_i, v_j\}$. Therefore labeling of vertices during the first application of Steps 1 and 2 imposes no constraints on the labelings of the remaining unlabeled vertices. So by induction the conclusions of point 1 and of this point apply for all successive applications of Steps 2 and 3.

3. As temporary label-violations are removed at the beginning of Step 3, permanent label-violations are checked for in that step, and Step 2 and 3 are iterated as long as unlabeled vertices remain, the algorithm finds a label-violation free labeling or proves there are none.

4. Regarding complexity it suffices to note that no edge can be checked more than 4 times (in both directions, for temporary and permanent labels). Hence the algorithm is in $O(m)$. ∎

References

1. Aspvall, B., M.F. Plass, and R.E. Tarjan, A Linear-time Algorithm for Testing the Truth of Certain Quantified Boolean Formulas, *Information Processing Letters* 8 (1979) 121–123.

2. Avis, D., Diameter Partitioning, *Discrete Computational Geometry* 1 (1986) 265–276.

3. Brucker, P., On the Complexity of Clustering Problems, in *Optimization and Operations Research, Lecture Notes in Economics and Mathematical Systems*, 157, eds. M. Beckmann and H.P. Kunzi, Heidelberg: Springer-Verlag, 1978, 45–54.

4. Capoyleas, V., G. Rote, and G. Woeginger, Geometric Clusterings, *Journal of Algorithms* 12 (1991) 341–356.

5. Gordon, A.D., *Classification: Methods for the Exploratory Analysis of Multivariate Data*, New York: Chapman and Hall, 1981.

6. Guénoche, A., Partitions with minimum diameter, paper presented at the International Federation of Classification Societies Conference, Charlottesville, USA, 1989.

7. Guénoche, A., Enumération des partitions de diamètre minimum, *Discrete Mathematics*, 111 (1993) 277–287.

8. Guénoche, A., P. Hansen, and B. Jaumard, Efficient Algorithms for Divisive Hierarchical Clustering with the Diameter Criterion, *Journal of Classification*, 8 (1991) 5–30.

9. Hansen, P., and B. Jaumard, Minimum Sum of Diameters Clustering, *Journal of Classification* 4 (1987) 215–226.

10. Hartigan, J.A., *Clustering Algorithms*, New York: Wiley, 1975.

11. Hershberger, J., Minimizing the Sum of Diameters Efficiently, *Computational Geometry: Theory and Applications* 2 (1992) 111–118.

12. Hershberger, J. and S. Suri, Finding Tailored Partitions, *Journal of Algorithms* 12 (1991) 431–463.

13. Kaufman, L., and P.J. Rousseeuw, *Finding Groups in Data: An Introduction to Cluster Analysis*, New York: Wiley, 1990.

14. Monma, C., and S. Suri, Partitioning Points and Graphs to Minimize the Maximum or the Sum of Diameters, in *Graph Theory, Combinatorics, and Applications, Proceedings of the Sixth Quadrennial International Conference on the Theory and Applications of Graphs*, eds., Y. Alavi, G. Chartrand, O.R. Oellerman, A.J. Schwenk, New York: Wiley, 1991, 899–912.

15. Preparata, F.P., and M.I. Shamos, *Computational Geometry, An Introduction*, New York: Springer, 1985.

Sylvie Gélinas
École Polytechnique de Montréal
Email address: sylvieg@crt.umontreal.ca
Pierre Hansen
GERAD and École des Hautes Études Commerciales, Montréal
Email address: pierreh@crt.umontreal.ca
Brigitte Jaumard
GERAD and École Polytechnique de Montréal
Email address: brigitt@crt.umontreal.ca

DIMACS Series in Discrete Mathematics
and Theoretical Computer Science
Volume 19, 1995

Agreement Subtrees, Metric and Consensus for Labeled Binary Trees

WAYNE GODDARD, EWA KUBICKA, GRZEGORZ KUBICKI,
and F. R. MCMORRIS

ABSTRACT. Analyzing methods for comparing trees is important when dealing with data that have been processed into trees in various ways by applying clustering algorithms, for instance. An interesting approach for doing this is based on the operation of *pruning* a (binary leaf labeled) tree by removing some leaves and suppressing all inner vertices of degree two which are formed as a result of this deletion. Given two trees T_1 and T_2, an agreement subtree is a tree that can be obtained from T_1, as well as from T_2, by pruning the fewest number of leaves from the two trees. We discuss the problem of finding agreement subtrees, recall two recently proposed metrics based on agreement trees, and present a method of consensus based on regrafting deleted leaves onto an optimal agreement subtree.

1. Introduction and Definitions

The development and analysis of different ways to compare various types of trees has been a concern of much research in the biological sciences (cf. [2], [3], [7], [11], [13], [15]). For example, trees representing hierarchical classifications of a collection of objects are usually produced as output of clustering algorithms, and when several algorithms are utilized it is desirable to determine the similarities among the outputs. Another situation is when multiple runs are made of one algorithm on the same data set. With the popular PAUP package [16] it is not uncommon to have thousands of output trees to analyze.

For simplicity, we focus our attention on ways to assess the similarity of two trees. In general, there are (at least) three approaches for comparing two discrete structures. The most obvious is to utilize a metric function. Structures a small distance apart are "more similar" than those a large distance apart. Of course it is important that the particular metric chosen for a study reflect some intuitive notion of dissimilarity appropriate for the application at hand. Other ways to measure the similarity of two

1991 *Mathematics Subject Classification.* Primary 92B10; Secondary 68Q25, 92-08.
The first author was supported by ONR grant N00014-91-J-1022; the last author was supported by ONR grant N00014-89-J-1643.
This paper is in final form and no version of it will be submitted for publication elsewhere.

structures involve some sort of maximum common substructure, or the production of a consensus structure.

The goal of this note is to briefly review recent developments on a particular type of maximum common substructure for trees, called the agreement subtree, and discuss associated metrics. We also propose a way of using the agreement subtree as a basis to form a consensus tree.

2. Agreement Subtrees

Suppose S is a set of n objects for which a hierarchical classification has been constructed. A common way to represent such a classification is by means of a tree T whose leaves (vertices of degree one) are labeled with the elements of S and the internal vertices are unlabeled. We will consider the case where T is a *labeled binary tree*: the internal vertices have degree three.

The operation of *pruning* on a tree T is the removal of a subset of the leaves of T and the suppression of all inner vertices of degree 2 that are formed by this removal. For example, a tree T_1, and the tree T_2 resulting after pruning the leaf 2 from T_1, are depicted in Figure 1.

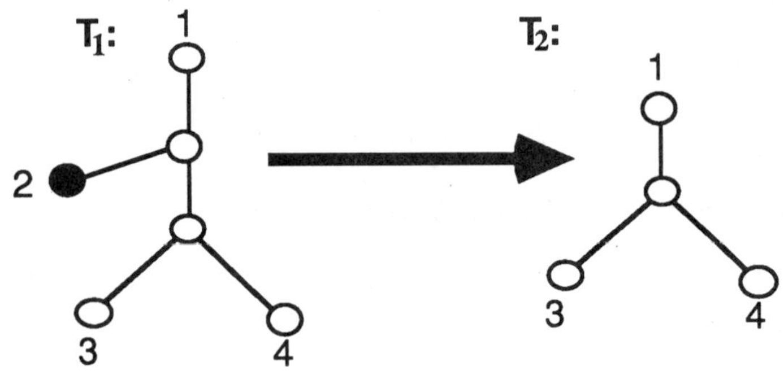

Figure 1.

Suppose T and U are two competing hierarchical classifications. A *common pruned subtree* of T and U is a tree that can be obtained from T as well as from U by pruning. An *agreement subtree* is a common pruned subtree with the most leaves. The idea of common pruned subtrees was introduced independently by Gordon [7] and Rosen [12] in order to ascertain the similarity of the original trees. This was further explored in Finden and Gordon [4] and mentioned in the review article of Swofford [15].

An important problem addressed in [4] and in [7] was the construction of an algorithm to find an agreement substructure. Unfortunately, this algorithm did not guarantee a common pruned subtree with the largest number of leaves. In [9] an exact $O(n^{\log n})$ algorithm was given and has since been implemented in COMPONENT [10]. Subsequently, $O(n^2)$ algorithms were produced in [5] and [14]. Additional computational considerations are found in [1] and [14].

Other aspects of the agreement subtree are explored in [6] and [8]. For example, the following bound on the size of an agreement subtree was obtained.

PROPOSITION [6]. *For two rooted complete binary trees with n leaves, an agreement tree has at least* $n^{0.17}$ *leaves.*

3. Metrics and Optimal Agreement Subtrees.

We let $\mathcal{A}(T,U)$ denote the set of all agreement subtrees for two trees T and U, and let $\#(T,U)$ denote the number of leaves in an agreement subtree. Let $\mathcal{T}_n$ denote the set of leaf-labeled binary trees with n leaves. The function $d_1: \mathcal{T}_n \times \mathcal{T}_n \to \mathbf{R}$ defined by $d_1(T,U) = n - \#(T,U)$ for all $T, U \in \mathcal{T}_n$ is easily seen to be a metric on $\mathcal{T}_n$. This metric is straightforward in the sense that it counts the number of leaves which have to be pruned from both trees to obtain a common substructure. Because of the polynomial algorithms in [5] and [14], d_1 is also easy to compute. However, this metric does not take into account the location of pruned leaves and therefore can possibly obscure some vital information about the intuitive "closeness" of two trees. As an illustration of this, let T_1, T_2, and T_3 be the binary labeled trees of size 10 with leaves labeled as in Figure 2.

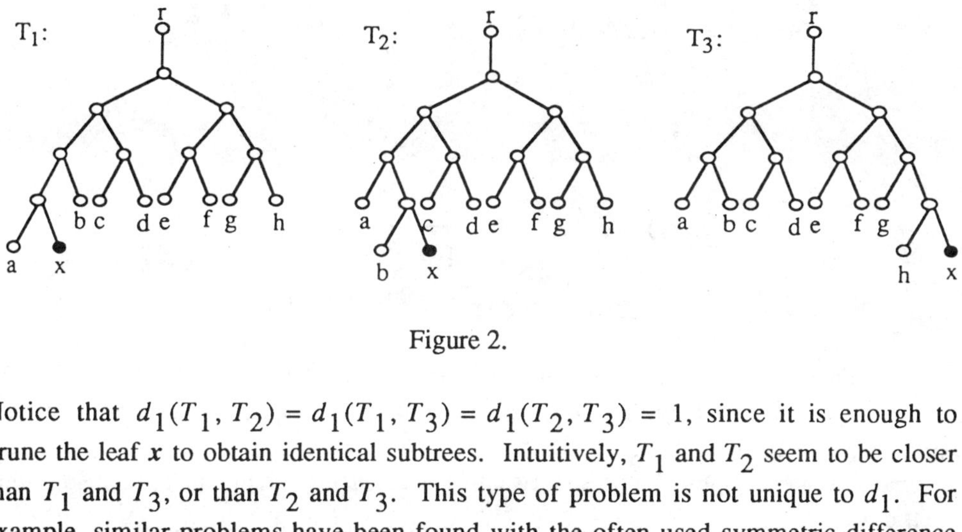

Figure 2.

Notice that $d_1(T_1, T_2) = d_1(T_1, T_3) = d_1(T_2, T_3) = 1$, since it is enough to prune the leaf x to obtain identical subtrees. Intuitively, T_1 and T_2 seem to be closer than T_1 and T_3, or than T_2 and T_3. This type of problem is not unique to d_1. For example, similar problems have been found with the often used symmetric difference metric ([3], [11]).

In order to address the situation above, the metric d_1 was modified in [5] by adding a fractional part, which reflects how far apart, with respect to an "optimal" agreement subtree, the pruned leaves are.

Suppose T_1 and T_2 belong to $\mathcal{T}_n$ and A is an agreement subtree for T_1 and T_2. Let P_A denote the set of all leaves pruned from T_1 and T_2 to obtain A. For each $x \in P_A$, let $A_i(x)$ denote the tree obtained from T_i by pruning all leaves in P_A except $x (i = 1, 2)$. Now let $A(x)$ be the tree formed by adding leaves x_1 and x_2 to A such that $A(x)$ is isomorphic to $A_1(x)$ after pruning x_2 and is isomorphic to $A_2(x)$ after pruning x_1. See Figure 3 for an illustration.

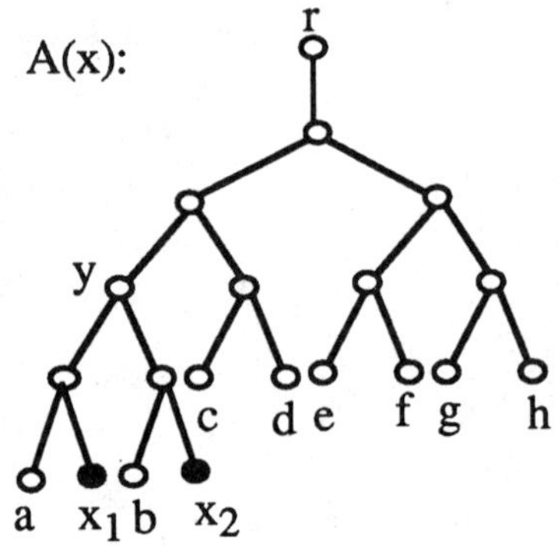

Figure 3.

Let $S(x,A)$ denote the length of the path between the leaves x_1 and x_2 in $A(x)$. In the example, $s(x,A) = 4$. Define

$$\lambda(A) = \sum_{x \in P_A} s(x,A),$$

and

$$L(T_1,T_2) = \min_{A \in \mathscr{A}(T_1,T_2)} \lambda(A)$$

Finally we set

$$d_2(T_1,T_2) = d_1(T_1,T_2) + \frac{1}{n\, d_1(T_1,T_2)} L(T_1,T_2) \text{ if } T_1 \neq T_2, \text{ and}$$
$$d_2(T_1,T_2) = 0 \text{ if } T_1 = T_2$$

We show in [5] that the function $d_2\colon \mathscr{T}_n \times \mathscr{T}_n \to \mathbf{R}$ is indeed a metric and also use the dynamic programming approach to produce an agreement subtree A which minimizes $\lambda(A)$ in polynomial time. Such an agreement subtree A is said to be *optimal*. Since an optimal agreement subtree is efficiently computable, then so is the metric d_2.

4. Regrafting for Consensus

Finden and Gordon [4] proposed a method for regrafting the pruned leaves from two rooted trees to form a "consensus" rooted tree on the original full set of leaves.

We now show how to use an optimal agreement subtree as a basis to regraft the pruned leaves so as to form a "consensus" labeled binary tree. With easy modification, our method could also be used for rooted trees.

Let $m(x,A)$ be the "midpoint" of the path between leaves x_1 and x_2 in $A(x)$. If the path has even length $m(x, A)$ is the middle vertex of the path. For example in Figure 3, $m(x, A)$ is vertex y. If the path between x_1 and x_2 has odd length then the midpoint $m(x,A)$ is the vertex formed by subdividing the middle edge of the path.

We define the *regrafted tree* R for the trees T_1 and T_2 from $\mathcal{T}_n$ with agreement subtree A as follows: start with A and attach each $x \in P_A$ as a leaf joined to $m(x,A)$. At first glance, R might appear to be a good candidate for a consensus tree. However, the example in Figure 4 shows that important substructure can be lost by doing regrafting in this fashion. We will now use a recursive procedure to eliminate this problem.

Define an *agreement consensus tree* $AC(T_1, T_2)$ of T_1 and T_2 as follows:

First find an optimal agreement tree A for T_1 and T_2.

For $x, y \in P_A$ define $x \sim y$ if x and y are joined by a path in T_1 that is vertex disjoint from A and are joined by a path in T_2 disjoint from A. Clearly, $\sim$ is an equivalence relation on P_A. Note also that if $x \sim y$ then $m(x,A) = m(y,A)$ since x_1 and y_1 are in the same place in $A(x)$ and $A(y)$, and similarly with x_2 and y_2.

Now let Q be an equivalence class of $\sim$, and let a_Q denote the vertex of A which is $m(x, A)$ for all $x \in Q$. Define $T_i(Q,a_Q)$ as the pruned subtree of T_i with leaves $Q \cup \{a_Q\}$. (Note that a_Q is not actually a leaf, but is temporarily considered as one.)

Finally, construct Z as the consensus tree for $T_1(Q,a_Q)$ and $T_2(Q,a_Q)$ recursively. Then attach Z to A by identifying the two copies of the vertex a_Q. Repeat this procedure for all equivalence classes Q. This procedure is depicted in Figure 4.

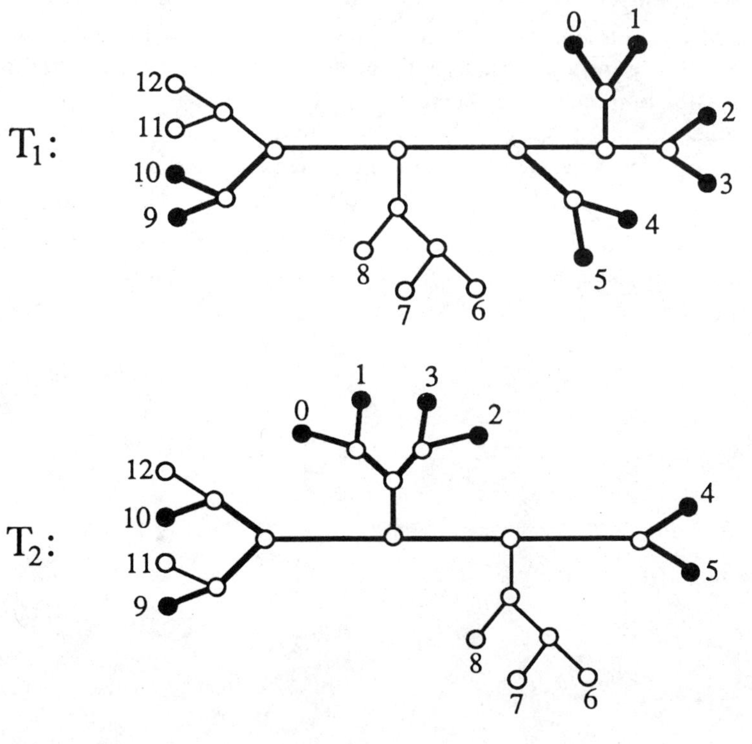

R:

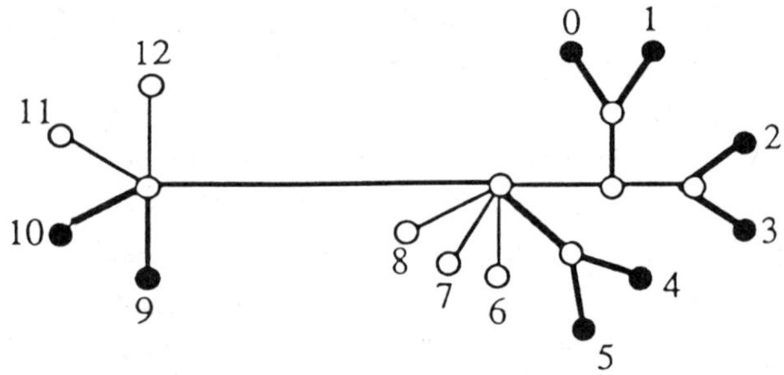

in $AC(T_1, T_2)$

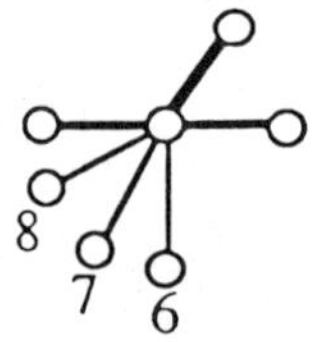

is replaced by

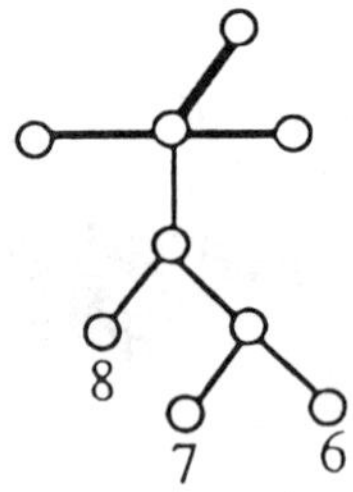

Figure 4 continued.

Clearly, the regrafted tree R can be constructed easily. However, the time complexity of constructing $AC(T_1, T_2)$ is unknown at this time. Also, this procedure awaits implementation.

The consensus tree satisfies the Pareto property of *preserving unanimity*. This means that if T_1 and T_2 both have an edge e whose removal gives the same bipartition of the vertices, then $AC(T_1, T_2)$ has this edge also. To see this, note that if the optimal agreement subtree contains leaves from both sets of the bipartition, then the result is true since regrafted vertices will go to the original sets of the bipartition. Otherwise, this situation must occur at some stage of the recursion.

In the same axiomatic spirit, we show in Figure 5 that AC is not *stable* in the following sense: It is possible that there exist trees T_1 and T_2 and a vertex w for which $AC(T_1, T_2) - w \neq AC(T_1 - w, T_2 - w)$.. This means that pruning vertex w before computing the consensus is not necessarily the same as computing the consensus and then pruning.

A formal axiomatic study of this consensus method remains an open problem.

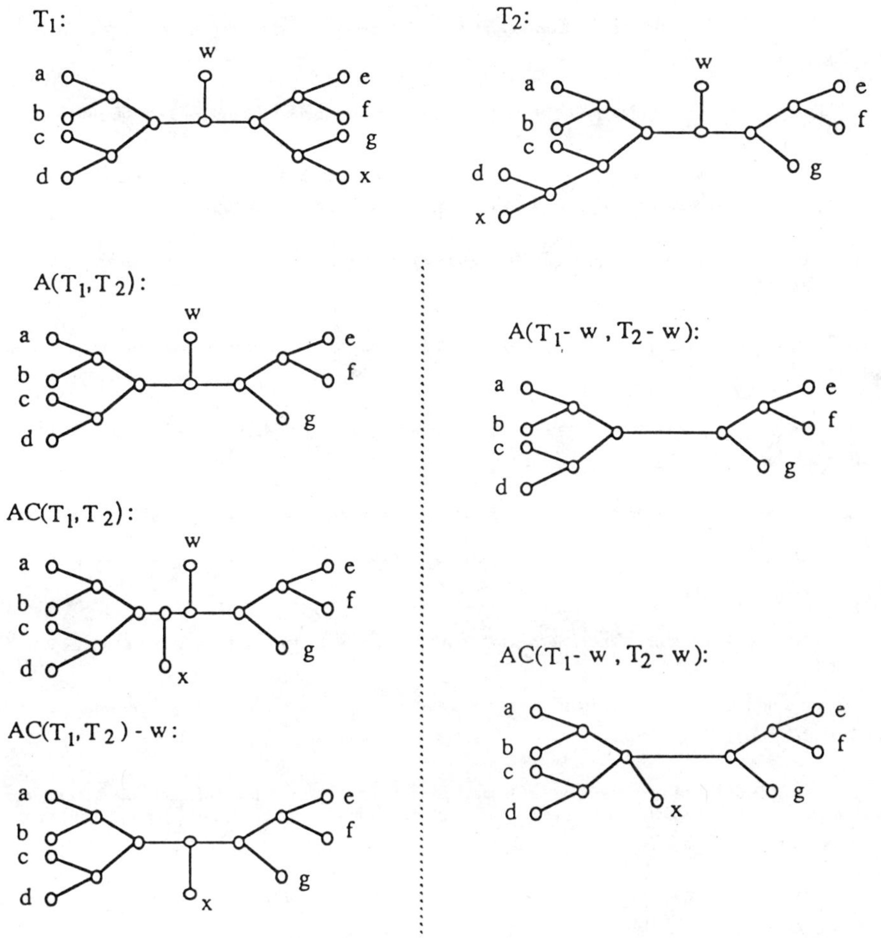

Figure 5.

References

1. A. Amir and D. Keselman, *Maximum Agreement Subtree in a Set of Evolutionary Trees*, preprint, College of Computing, Georgia Institute of Technology, 1994.

2. W.H.E. Day, *Optimal algorithms for comparing trees with labeled leaves*, J. Classif. **2** (1988), 7-28.

3. G.F. Estabrook, F.R. McMorris, and C.A. Meacham, *Comparison of undirected phylogenetic trees based on subtrees of four evolutionary units*, Syst. Zool. **34** (1985), 193-200.

4. C.R. Finden and A.D. Gordon, *Obtaining common pruned trees*, J. Classif. **2** (1985), 255-276.

5. W.D. Goddard, E. Kubicka, G. Kubicki, and F.R. McMorris, *The agreement metric for labeled binary trees*, to appear in Math. Biosci.

6. W.D. Goddard, G. Kubicki, *The minimum size of agreement subtrees of two binary trees*, Congr. Numer. **97** (1993), 131-136.

7. A.D. Gordon, *On the assessment and comparison of classifications*, in Analyse de Données et Informatique, R. Tomassone, ed., INRIA, Le Chesnay, France, (1980), 149-160.

8. E. Kubicka, G. Kubicki, and F.R. McMorris, *On agreement subtrees of two binary trees*, Congr. Numer. **88** (1992), 217-224.

9. E. Kubicka, G. Kubicki, and F.R. McMorris, *An algorithm to find agreement subtrees*, to appear in J. Classif.

10. R.D. Page, *COMPONENT 2.0 Tree Comparison Software for Microsoft Windows*, Natural History Museum, London, 1993.

11. D. Penny and M.D. Hendy, *The use of tree comparison metrics*, Syst. Zool. **34** (1985), 75-82.

12. D.E. Rosen., *Vicariant patterns and historical explanation in biogeography*, Syst. Zool. **27** (1978), 159-188.

13. M.A. Steel and D. Penny, *Distributions of tree comparison metrics – some new results*, Syst. Biol. **42** (1993), 126-141.

14. M. Steel and T. Warnow, *Kaikoura tree theorems: computing the maximum agreement subtree*, Inf. Proc. Lett. **48** (1993), 77-82.

15. D.L. Swofford, *When are phylogeny estimates from molecular and morphological data incongruent?*, in Phylogenetic Analysis of DNA Sequences, M.M. Miyamoto and J. Cracraft, eds., Oxford University Press, New York , 1991, pp. 295-333.

16. D.L Swofford, *PAUP: Phylogenetic Analysis Using Parsimony*, version 3.0, Illinois Natural History Survey, Champaign, Illinois, 1991.

DEPT. OF MATHEMATICS, UNIVERSITY OF PENNSYLVANIA, PHILADELPHIA, PA 19104

DEPT. OF MATHEMATICS, UNIVERSITY OF LOUISVILLE, LOUISVILLE, KY 40292

E-MAIL ADDRESS: FRMCMO∅1@HOMER.LOUISVILLE.EDU

DIMACS Series in Discrete Mathematics
and Theoretical Computer Science
Volume 19, 1995

How to Choose K Entities Among N

P. Hansen, B. Jaumard and N. Mladenovic

ABSTRACT. A new paradigm for cluster analysis is outlined. It is called *Sequential Clustering*. Given a set O of N entities a best cluster of K entities (where K is a parameter) is found and its entities removed from O. The procedure is iterated until all entities have been removed or the set of remaining ones presents no more structure. Various dissimilarity-based criteria for finding a best cluster are considered. Complexity of the resulting problems is examined and algorithms recalled, when some have already been proposed, or outlined when none appear to be yet available. Experiments with threshold type criteria and Ruspini's data are reported.

1. Introduction

Cluster Analysis aims mainly at solving the following general problem: given a set O of N entities, find subsets of O, called *clusters*, which are *homogeneous* and/or *well separated*. Homogeneity means that entities in the same cluster must resemble themselves. Separation means that entities in different clusters must differ one from the other. This general problem appears in many disciplines: biology, botany, medicine, psychology, geography, applied economics, marketing, computer science, image processing, flexible manufacturing, etc.

The multiple applications of cluster analysis may be grouped under the following headings:

(i) *classification* of large date sets, to facilitate their treatment;

1991 Mathematics Subject Classification. Primary 62H30; Secondary 90C29.
The first and second authors have been supported by the Office of Naval Research grant #N00014-92-J-1194 and FCAR (Fonds pour la Formation de Chercheurs et l'Aide à la Recherche) grant 92EQ1048. The first author has also been supported by NSERC (Natural Sciences and Engineering Council of Canada) grant to HEC and NSERC grant GP0105574. The second author has also been supported by NSERC grant GP0036426 and a FRSQ (Fonds pour la Recherche en Santé du Québec) fellowship.
An expanded version of this paper will be submitted for publication elsewhere.

(ii) *exploration* of data sets to find if they have some structure, which may suggest hypotheses and eventually lead to a theory;

(iii) *prediction* of properties of new entities, close to those of some cluster, from those of entities in that cluster;

(iv) *organization* in an efficient way of various economic or managerial activities, by applying the same measures to similar entities.

The concepts of homogeneity and separation may be made precise in many ways. Increasingly, this is done by specifying explicit criteria, in mathematical form. Such criteria are numerous and diverse: therefore, many *cluster analysis problems* have been studied. Moreover, these problems also differ in the way a predefined structure (such as a partition, or a hierarchy of partitions) is imposed *a priori* on the clusters to be found. In this way, various *paradigms* of cluster analysis, are defined. Each of them regroups many problems, and many more algorithms to solve them. The two best known and most studied paradigms are *hierarchical clustering* and *partitioning.* The purpose of the present paper is to outline a new paradigm for cluster analysis, called *sequential clustering.* It consists in finding a best cluster (in a well defined sense), removing its entities from O and iterating until either all entities have been removed or it can be shown that the remaining set of entities has no more structure (or, in other words, no remaining cluster with a good value as evaluated by the chosen criterion).

Contrary to the usual methods of cluster analysis which assign all entities to clusters (and usually provide some clustering even if the data has no structure), sequential clustering aims at finding within the data whatever structure there is (beginning with what is most apparent) and nothing more. Note that a similar procedure is often used in image processing where objects are recognized one after the other.

The paper is organized as follows. Steps of a cluster analysis study following the sequential clustering paradigm are formulated in Section 2, and the main questions to be solved are listed and briefly discussed. The remainder of the paper addresses the problem of finding a best cluster with K entities. Threshold type criteria in which the value depends on a single dissimilarity between a pair of entities are discussed in Section 3. Sum-type and sum-sum-type criteria in which such values depend on several dissimilarities are considered in Section 4. Illustration for a few criteria are provided in Section 5 and conclusions drawn in Section 6.

2. General Scheme for Sequential Clustering

Most algorithms of cluster analysis are based on dissimilarities between pairs of entities, obtained from the data before clustering. A general scheme for sequential clustering using such dissimilarities is the following:

Step 1. *Sample.* Select a sample $O = \{O_1, O_2, \ldots, O_N\}$ of N entities among which clusters are to be found.

Step 2. *Data.* Proceed to observe or measure p characteristics of the entities of O. This gives rise to an $N \times p$ data matrix X.

Step 3. *Dissimilarities.* Compute from the matrix X an $N \times N$ matrix $D = (d_{k\ell})$ of dissimilarities between pairs of entities. Such dissimilarities satisfy the properties $d_{k\ell} \geq 0$, $d_{kk} = 0$, $d_{k\ell} = d_{k\ell}$ for all $k, \ell = 1, 2, \ldots, N$; they do not necessarily satisfy the triangle inequality, i.e., may be or not distances.

Step 4. *Criterion.* Choose a criterion to evaluate the homogeneity and/or separation of the clusters to be obtained.

Step 5. *Choosing K entities among N.* Determine with a specialized algorithm an optimal cluster C^*, i.e., a best set of K entities of O according to the criterion of **Step 4**. (Usually K will be considered as a parameter, clusters will be determined for all values of $K = 2, 3, \ldots, N$ and the best value of K found together with the best cluster.)

Step 6. *Significance test.* Apply formal or informal tests to evaluate if the cluster C^* found in **Step 5** corresponds to some part of the structure inherent to O, or only to noise. In the former case, record the list of entities of C^*, remove them form O and return to **Step 5**. In the latter case proceed to **Step 7**.

Step 7. *Interpretation of results.* Describe the clusters found sequentially in **Step 5** by the lists of their entities and various techniques of descriptive statistics. Proceed to a substantive interpretation of these results.

To specify completely an algorithm of sequential clustering one must make precise to which problem it applies, how it determines candidate clusters among which to choose the best one (e.g. in function of the size parameter K), how this choice is made and which significance test applies.

Algorithms will vary greatly in complexity according to the type of criterion used. A few typical cases are discussed in the next two sections, focussing on how to choose K entities. Choice of the best clusters and significance tests will be briefly discussed in Section 5.

Dissimilarity based criteria for homogeneity of the entities of a cluster $C \subset O$ and/or separation of these entities from the remaining ones, in $O \setminus C$, may be classified into three groups according to the number of dissimilarities used in determining the value of the criterion under study:

(i) *threshold-type criteria,* in which a single dissimilarity determines this value. Examples of such criteria are the *diameter, radius* and *split,* defined in Section 3,

(ii) *sum-type criteria,* in which a sum of dissimilarities involving one entity determines this value. An example of such a criterion is the *star,* defined in Section 4,

(iii) *sum-sum-type criteria,* in which all dissimilarities between pairs of entities of the cluster are used to determine this value. Examples of such criteria are the *clique* and the *variance,* also defined Section 4.

3. Threshold-type criteria

We next discuss how to choose K entities among N for some threshold-type criteria, beginning with two criteria for homogeneity.

Problem 1. Find a set $C \subset O$ of K entities with minimum *diameter* $d(C)$, where the diameter of C is the largest dissimilarity between a pair of entities of C, i.e.,

$$d(C) = \max_{k|O_k \in C} \max_{\ell|O_\ell \in C} d_{k\ell}$$

for $K = 2, 3, \ldots, N$.

Problem 1 is NP-hard. Indeed consider the CLIQUE problem in graphs which is NP-hard (Garey and Johnson 1979): given a graph $G = (V, E)$ and an integer K does there exist a set of K pairwise adjacent vertices in G? Taking $d_{k\ell} = 0$ if $\{v_k, v_\ell\} \in E$ and $d_{k\ell} = 1$ otherwise, the answer to CLIQUE is yes if and only if Problem 1 has an optimal solution of value 0.

It is known however that branch-and-bound algorithms for cliques are quite efficient in practice, at least when $|V| \leq 100$ (e.g. Balas and Yu, 1987, Carraghan and Pardalos, 1991, Balas and Xue, 1993, Hansen and Mladenovic, 1994). Such algorithms are easily modified to solve Problem 1 for $K = 2, 3, \ldots, N$.

Step 1. *Initialization.* Rank the dissimilarities by order of non-increasing values. Set $C = O$ and $K = N$.

Step 2. *New clique.* Let $d_{k\ell}$ be the next dissimilarity. If $O_k \notin C$ or $O_\ell \notin C$ iterate this step. If $O_k \in C$ and $O_\ell \in C$, apply a maximum clique algorithm to the graph $G = (V, E)$ where the vertex set V is associated with O (i.e., v_j corresponds to O_j for $j = 1, 2, \ldots, N$) and $\{v_i, v_j\} \in E$ if and only if $d_{ij} \leq d_{k\ell}$. If a clique C' with $|C'| = |C|$ is found set $C \leftarrow C'$ and iterate this step (as soon as C' is found, i.e., interrupt the branch-and-bound algorithm used). Otherwise go to Step 3.

Step 3. *Output.* Print K, the cluster C and its diameter $d_{k\ell}$. Then set $C \leftarrow C \backslash O_k$, $K \leftarrow K - 1$ and return to Step 2 if $K > 1$. Otherwise stop.

Note that Problem 1 can be solved in low-order polynomial time in K and N in the particular case where the entities are points in the Euclidean plane: Aggarwal, Imai, Katoh and Suri, 1991 propose an $O(K^{2.5} N \log K + N \log N)$ algorithm.

Problem 2. Find a set $C \subset O$ with K entities and minimum *radius* $r(C)$, where the radius of C is the minimum over all entities of C of the maximum dissimilarity between that entity and another one of C, i.e.,

$$r(C) = \min_{k|O_k \in C} \max_{\ell|O_\ell \in C} d_{k\ell},$$

for $K = 2, 3, \ldots, N$.

This problem is polynomial and very easy to solve.

Step 1. *Ranking dissimilarities.* Construct two $N \times N$ matrices, the first containing in the jth column, for $j = 1, 2, \ldots, N$, the dissimilarities between O_j and all entities O_k of O ranked in order of non-decreasing values, and the second containing in the jth column the indices of the corresponding entities.

Step 2. *Output.* Find the minimum value in the second, third, $\ldots$, N^{th} line of the first matrix; output these values, which are the minimum radii of the best clusters for $K = 2, 3, \ldots, N$; using the second matrix, output also the list of corresponding entities.

The complexity of the above algorithm is $O(N^2 \log N)$.

The next problem has a criterion for separation.

Problem 3. Find a set $C \subset O$ of K entities with maximum *split* $s(C)$, where the split of C is the smallest dissimilarity between an entity of C and one outside of C, i.e.,

$$s(C) = \min_{k|O_k \in C} \min_{\ell|O_k \notin C} d_{k\ell},$$

for $K = 2, 3, \ldots, N$.

This problem is polynomial. It appears to have been first posed by Hsu and Nemhauser, 1979 in the context of location of obnoxious facilities. They propose an $O(N^2 \log N)$ algorithm for a given value of K, which is easily adapted into an $O(N^3)$ algorithm for $K = 2, 3, \ldots, N$. Let $G = (V, E)$ denote a complete graph with vertices $v_j \in V$ associated with the entities $O_j \in O$ for $j = 1, 2, \ldots, N$ and edges $(v_k, v_\ell) \in E$ weighted by the dissimilarities $d_{k\ell}$. Then a result of Rosenstielh, 1967 on trees and cocycles of graphs may be reformulated as follows: for any cluster $C \subset O$, the split $s(C)$ is equal to the weight associated with an edge of the minimum spanning tree T of G (or any such tree in case of non-uniqueness). Chang, Tang and Lee 1991 rediscover this result and use it to obtain a $\theta(N^2)$ algorithm for Problem 3. It is based on finding T, ranking its edges and considering forests induced by those edges of T with a weight less than a threshold value. For each forest a dynamic programming algorithm is used to find for which values of K clusters exist (and correspond to unions of trees in the current forest). Updating yields the desired complexity.

4. Sum and Sum-Sum Type criteria

The first criterion considered is for homogeneity, and of sum type.

Problem 4. Find a set $C \subset O$ of K entities with minimum *star* $st(C)$, where the star of C is the minimum over all entities of C of the sum of dissimilarities between that entity and all other ones of C, i.e.,

$$st(C) = \min_{k|O_k \in C} \sum_{\ell|O_\ell \in C} d_{k\ell},$$

for $K = 2, 3, \ldots, N$.

This problem is polynomial and very close to minimum radius sequential clustering discussed above.

Step 1. *Ranking and summing dissimilarities.* Construct two $N \times N$ matrices, the first containing in the jth column for $j = 1, 2, \ldots, N$, the sums of the smallest, two smallest, ..., N smallest dissimilarities between O_j and the entities of O, and the second containing in the jth column the indices of the entities corresponding to the dissimilarities last added to the sums.

Step 2. *Output.* Find the minimum value in the second, third, ..., N^{th} line of the first matrix; output these values which are the minimum stars of the best clusters for $K = 2, 3, \ldots, N$; using the second matrix output also the lists of corresponding entities.

The complexity of this algorithm is $O(N^2 \log N)$.

The next two criteria are for both homogeneity and separation, and of sum-sum type.

Problem 5. Find a set $C \subset O$ of K entities with minimum *clique* $cl(C)$, where the clique of C is the sum of dissimilarities between pairs of entities of C, i.e.,

$$cl(C) = \sum_{k|O_k \in C} \sum_{\ell|O_\ell \in C} d_{k\ell}$$

for $K = 2, 3, \ldots, N$.

This problem is NP-hard. The proof, by reduction to CLIQUE, is similar to that for minimum diameter sequential clustering given above.

Introducing 0–1 variables x_k, equal to 1 if $O_k \in C$ and to 0 otherwise, leads to reformulate Problem 5 (for a given value of K) as:

$$\begin{aligned} &\text{Min} \sum_{k=1}^{N-1} \sum_{\ell=k+1}^{N} d_{k\ell} x_k x_\ell \\ \text{s.t.} \quad &\sum_{k=1}^{N} x_k = K \\ \text{and} \quad &x_k \in \{0, 1\} \quad k = 1, 2, \ldots, N, \end{aligned}$$

for $K = 2, 3, \ldots, N$.

This problem can easily be reduced to a *quadratic knapsack problem* (Gallo, Hammer and Simeone, 1980). Indeed, let $\bar{d}$ denote the largest dissimilarity and let $d'_{k\ell} = \bar{d} - d_{k\ell}$ for $k, \ell = 1, 2, \ldots, N$, $k \neq \ell$. Then

$$\begin{aligned} \min \quad &\sum_{k=1}^{N-1} \sum_{\ell=k+1}^{N} d_{k\ell} x_k x_\ell = \min \sum_{k=1}^{N-1} \sum_{\ell=k+1}^{N} \left(\bar{d} - d'_{k\ell}\right) x_k x_\ell \\ = \quad &\frac{K(K-1)}{2}\bar{d} - \max \sum_{k=1}^{N-1} \sum_{\ell=k+1}^{N} d'_{k\ell} x_k x_\ell \end{aligned}$$

in view of the cardinality constraint and as the $d'_{k\ell} \geq 0$, the optimization problem becomes

$$\begin{aligned} \max \quad & \sum_{k=1}^{N-1} \sum_{\ell=k+1}^{N} d'_{k\ell} x_k x_\ell \\ \text{s.t.} \quad & \sum_{k=1}^{N} x_k \leq K \end{aligned}$$

in 0–1 variables.

Gallo, Hammer and Simeone, 1980 propose a branch-and-bound algorithm for the quadratic knapsack problem, which uses linearization of the objective function, and thus a usual knapsack problem to compute bounds. Chaillou, Hansen and Mahieu, 1989 present an algorithm using bounds based on network flows. Gallo, Grigoriades and Tarjan, 1989 provide a quicker method for computing such bounds.

The last problem we consider applies to sets of entities in Euclidean n-space.

Problem 6. Find a set $C \subset O$ of K entities with minimum sum of squared distances to the centroïd (or minimum *variance*).

Edwards and Cavalli-Sforza, 1965 observe that the variance of a cluster is equal to the sum of squared distances between all pairs of entities of the cluster divided by this cluster's cardinality. Problem 6 can therefore be written:

$$\begin{aligned} \min \quad & \frac{\displaystyle\sum_{k=1}^{N-1} \sum_{\ell=k+1}^{N} d_{k\ell} x_k x_\ell}{\displaystyle\sum_{k=1}^{N} x_k} \\ \text{s.t.} \quad & \sum_{k=1}^{N} x_k = K \end{aligned}$$

in 0–1 variables, for $K = 2, 3, \ldots, N$, where $d_{k\ell}$ is equal to the squared distance between points O_k and O_ℓ. Again this problem is NP-hard. Using the cardinality constraint to replace the denominator of the objective function by a constant reduces Problem 6 to Problem 5. Hence, it can also be reduced in turn to a quadratic knapsack problem.

5. Experiments

A few experiments have been made, in a preliminary attempt at validation of the proposed sequential clustering approach. We use the well-known data of Ruspini, 1970, i.e., 75 points in $\mathbb{R}^2$ divided into four clusters.

The three threshold-type criteria discussed above are adopted in turn. Each time, clusters are computed for all values of K between 2 and N. The cluster C^* chosen is that one for which decreasing K by one unit has involved the largest decrease in diameter or radius or increase in split. Following the general scheme, entities of C^* are removed and the procedure iterated. Change in the decrease (or increase) in objective function value is used to evaluate when to stop. If it drops sharply we consider no more clusters can be clearly distinguished.

Results for the diameter criterion are given on Figure 1. The first three iterations each isolate one of the clusters. Decreases Δ in diameter associated with these clusters are equal to 23.29., 28.26 and 45.58 respectively. A fourth iteration selects 13 out of the 15 entities of the fourth cluster. However, a value of Δ equal to 3.45 only shows such further clustering not to be warranted. The structure of Ruspini's data is clearly found. Total CPU time is 3.07 seconds on a SUN SPARC computer.

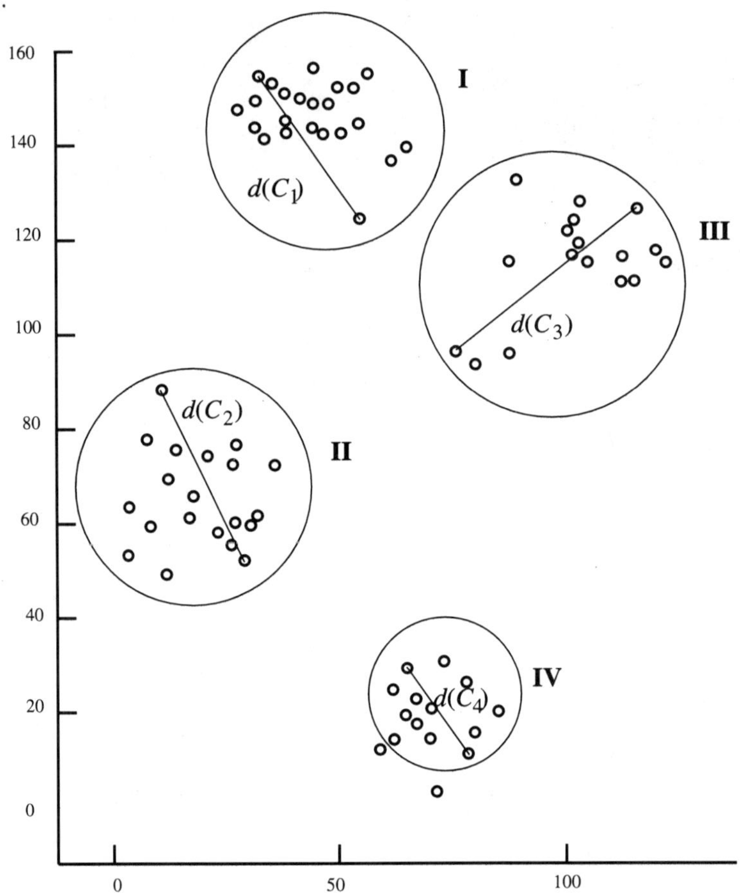

Figure 1. Sequential clustering with minimum diameter criterion: I — $|C_1| = 23$, $\Delta = 23.29$; II — $|C_2| = 20$, $\Delta = 28.26$; III — $|C_3| = 17$, $\Delta = 45.58$; IV — $|C_4| = 13, \Delta = 3.45$; CPU time = 3.07 sec.

Results for the radius criterion are given on Figure 2. Again the first three iteration recover three clusters, with values of the decrease Δ in radius associated with these clusters equal to 12.95, 16.46 and 32.04. The fourth iteration selects 14 out of 15 entities of the fourth cluster, but again a value of Δ equal to 2.83 only shows such further clustering to be inadequate. Again, the structure of the data is recovered. Total CPU time is only 0.07 seconds.

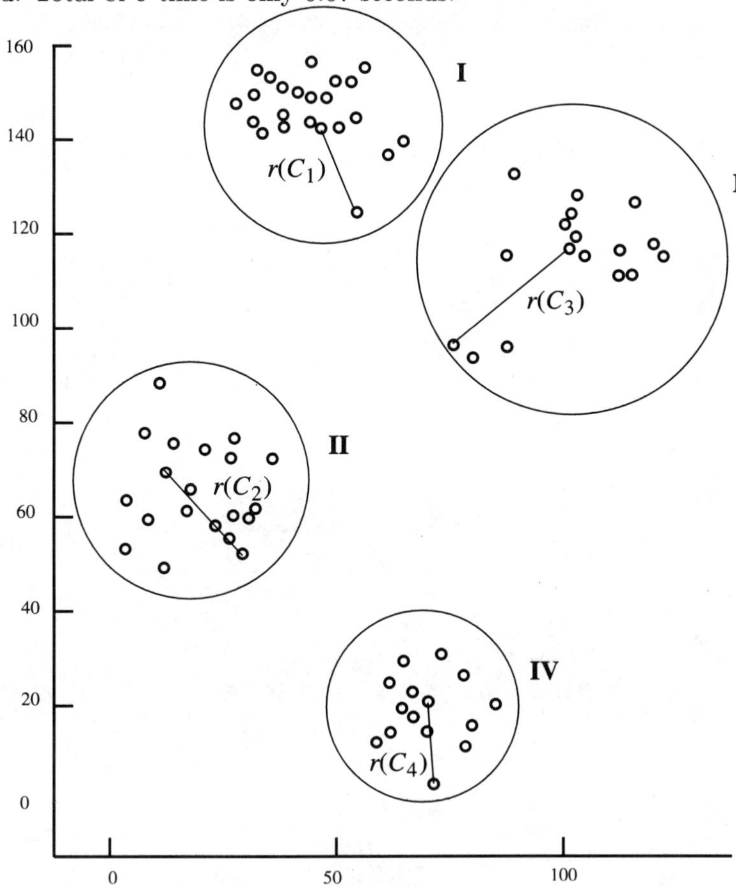

Figure 2. Sequential clustering with minimum radius criterion: I — $|C_1| = 23$, $\Delta = 12.95$; II — $|C_2| = 20$, $\Delta = 16.46$; III — $|C_3| = 17$, $\Delta = 32.04$; IV — $|C_4| = 15, \Delta = 2.83$; CPU time = 0.07 sec.

Results for the split criterion are given on Figure 3. The first iteration isolates the union of two clusters of the data with an increase Δ in split equal to 31.34. The second iteration isolates one further cluster, with a value of Δ equal to 29.87. The next three iterations isolate one entity each with Δ values of 11.40, 9.85 and 8.94. These values can be viewed as too low, in which case one should stop after the second iteration, or as indicative of smaller clusters within the last one in the data. So this time only part of the structure of the data is recovered. Total CPU time is 0.61 seconds.

Conclusions

The most used paradigms of cluster analysis, i.e., hierarchical clustering and partitioning imply all entities should be assigned to clusters. However, it is often the case that this is unnatural, some entities fitting badly into clusters, or being just noise. The proposed sequential clustering paradigm aims at finding within the data whatever structure there is and nothing more, by isolating clearly apparent clusters one at a time and stopping when this cannot be done anymore.

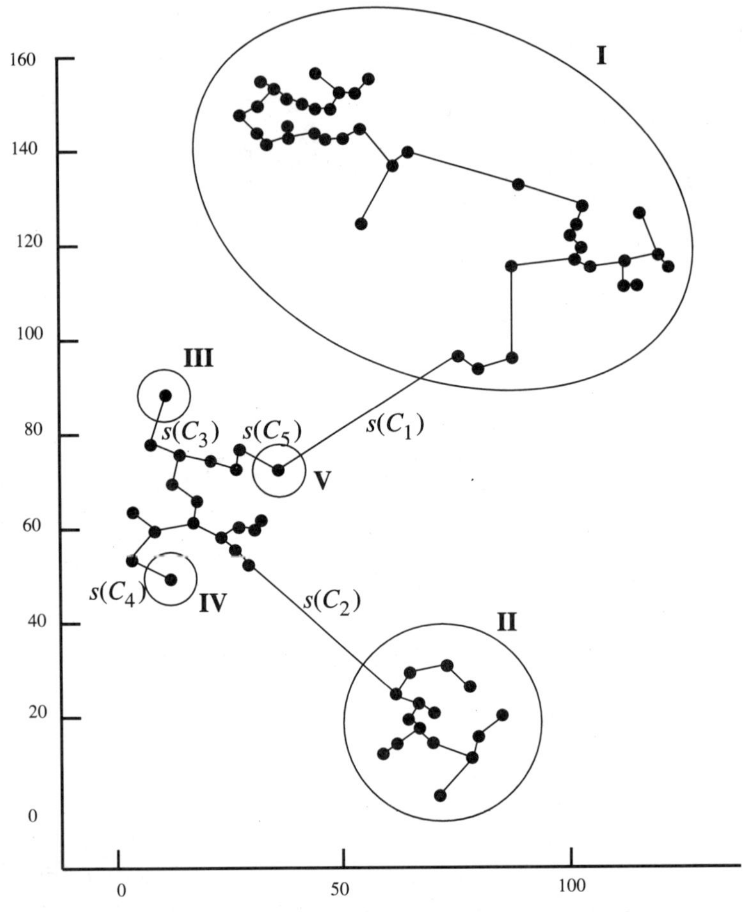

Figure 3. Sequential clustering with maximum split:
I — $|C_1| = 40$, $\Delta = 31.3$; II — $|C_2| = 15$, $\Delta = 29.87$;
III — $|C_3| = 1$, $\Delta = 11.40$; IV — $|C_4| = 1, \Delta = 9.85$;
V — $|C_5| = 1$, $\Delta = 8.94$; CPU time = 0.61 sec.

For various criteria the problem of choosing K entities among N, which is central to each iteration, can be solved either by a low-order polynomial algorithm or fairly well by a non-polynomial algorithm, for small or medium size data sets. Experiments with a standard data set (and a few others not reported on here) tend to show their structure can be clearly recovered. Further work to validate the sequential clustering approach will involve exploring more thoroughly which are the possible and relevant criteria (or combinations of criteria), developping algorithms for the most appealing ones and conducting empirical studies with a series of standard data sets.

References

1. Aggarwal, A., H. Imai, N. Katoh, and S. Suri, *Finding k Points with Minimum Diameter and Related Problems*, Journal of Algorithms **12**(1991), 38–56.

2. Balas, E., and Xue, J., *Weighted and Unweighted Maximum Clique Algorithms with Upper Bounds from Fractional Coloring*, Research Report #MSRR–590, Carnegie Mellon University (1993).

3. Balas, E., and Yu, C.S., *Finding a Maximum Clique in an Arbitrary graph*, SIAM J Computing **15**(1986), 1054–1068.

4. Carraghan R., and Pardalos, P.M., *An Exact Algorithm for the Maximum Clique Problem*, Operations Research Letters **9**(1990), 375–382.

5. Chaillou, P., Hansen, P., and Mahieu, Y., *Best Network Flow Bounds for the Quadratic Knapsack Problem*, in B. Simeone (ed.) Combinatorial Optimization, Lecture Notes in Mathematics **1403**(1989), 225–235.

6. Chang, M.S., Tang, C.Y., and Lee, R.C.T., *A Unified Approach for Solving Bottleneck k-Bipartition Problems*, Proceedings of the 19th Annual Computer Science Conference, San Antonio, Texas, March 5–7, 1991, ACM (1991), 39–47.

7. Gallo, G., Hammer, P.L., and Simeone, B., *Quadratic Knapsack Problem*, Math. Prog. Study **12**(1980), 132–149.

8. Gallo, G., Grigoriadis, M.D., and Trajan, R.E., *A Fast Parametric Maximum Flow Algorithm and Applications*, SIAM Journal on Computing **18**(1989), 30–55.

9. Hansen, P., and Mladenovic, N., *Two Algorithms for Maximum Cliques in Dense Graphs*, Les Cahiers du GERAD **G-92-18**(1992). To appear in European Journal of Operational Research.

10. Hsu, W.-L., and D.L. Nemhauser, *Easy and Hard Bottleneck Location Problems*, Discrete Applied Mathmatics **1**(1979), 209–215.

11. Preparata, F.P., and M.I. Shamos, *Computational Geometry, An Introduction*, New York: Springer, 1985.

12. Rosenstiehl, P., *L'arbre minimum d'un graphe*, in Théorie des Graphes, Rome, I.C.C. P. Rosenstiehl (ed.), Paris, Dunod, 1967, 357–368.

13. Ruspini, E.H., *Numerical Methods for Fuzzy Clustering*, Information Sciences **2**(1970), 319–350.

Pierre Hansen
GERAD and École des Hautes Études Commerciales, Montréal, Canada
Email address: pierreh@crt.umontreal.ca

Brigitte Jaumard
GERAD and École Polytechnique de Montréal, Canada
Email address: brigitt@crt.umontreal.ca

Nenad Mladenovic
GERAD and École des Hautes Études Commerciales, Montréal, Canada
Email address: nenad@crt.umontreal.ca

DIMACS Series in Discrete Mathematics
and Theoretical Computer Science
Volume 19, 1995

On the Classification of Monotone-Equivariant Cluster Methods

Melvin F. Janowitz and Rudolf Wille

ABSTRACT. Ordinal cluster methods are classified according to the significance they assign to the input data. This is determined by their interactions with isotone mappings on the nonnegative reals.

1. Background.

Cluster analysis is a class of exploratory data analysis techniques whose goal is to suggest possible internal structure for a given data set. Different techniques can of course give widely different internal structures. It is important to classify cluster techniques according to the assumptions that are being made about the input data. Typical assumptions might be that the data has ordinal significance, ratio significance, interval significance, etc. In the present work we shall be dealing with data that has at most ordinal significance in the sense that if one is given two numbers s and t such that $s \le t$, the actual values of the numbers are not important; indeed the only assertion we

1991 Mathematics Subject Classification. Primary 62H30; Secondary 06A15.
The first author was supported by ONR Grant N-00014-90-J-1008.

are making is that either $s < t$, or we may not be sure of the relative ordering between s and t. This can be expressed quantitatively by saying that an assertion of the form $s \leq t$ is valid only up to the action of a class of "admissible" isotone mappings on the nonnegative reals. So if $s \leq t$, but $\theta(s) = \theta(t)$ for some admissible mapping θ, then we are expressing doubt as to the relative ordering of s and t. This is very much in the spirit of modern measurement theory (See, for example, [1], pp. 64-67). We mention that there was earlier work along these lines using a more specialized kind of mapping, the *residuated* mappings of $[0,\infty)$. The primary references for this work are [3], [4] and [5]. Though a basic familiarity with cluster analysis is assumed, we shall introduce any needed terminology and notation in this opening section.

We shall be working with a finite nonempty set **P**. To avoid trivial cases, it will be assumed that **P** has at least 4 members. A *dissimilarity coefficient* (DC) on **P** is a mapping $d:\mathbf{P}\times\mathbf{P} \longmapsto [0,\infty)$ such that:

(DC1) $d(x,y) = d(y,x)$,

(DC2) $d(x,x) = 0$,

for all $x,y \in \mathbf{P}$. The DC d is called *definite* if it also satisfies

(DC0) $d(x,y) = 0$ implies $x = y$.

It is important to note that we are not assuming that any given input DC is definite. A dissimilarity coefficient is called an *ultrametric* if it also satisfies

(DC3) $d(x,y) \leq \max\{d(x,z),d(y,z)\}$ for all $x,y,z \in \mathbf{P}$.

This is clearly equivalent to the assertion that for any triple $\{x,y,z\} \subseteq \mathbf{P}$, $\max\{d(x,y),\ d(y,z),\ d(x,z)\}$ is taken on at least twice.

Notation: $\mathbf{D} = \mathbf{D}(\mathbf{P})$ = the set of DCs on $\mathbf{P}$, $\mathbf{U} = \mathbf{U}(\mathbf{P})$ = the set of ultrametrics on $\mathbf{P}$, $\mathbf{E} = \mathbf{E}(\mathbf{P})$ = the set of equivalence relations on $\mathbf{P}$, $\mathbf{T} = \mathbf{Tol}(\mathbf{P})$ = the set of reflexive symmetric relations on $\mathbf{P}$. These are sometimes called *tolerance relations* on $\mathbf{P}$.

Following notation originally introduced by Jardine and Sibson [2], we define a *numerically stratified clustering* to be a mapping $C:[0,\infty) \longmapsto \mathbf{T}$ such that:

(NSC1) $h \leq k$ implies $C(h) \subseteq C(k)$.

(NSC2) $C(h) = \mathbf{P}\times\mathbf{P}$ for some $h \in [0,\infty)$.

(NSC3) For each $h \in [0,\infty)$, there exists $\delta = \delta(h) > 0$ such that $C(h) = C(h+\delta)$.

A *dendrogram* is a numerically stratified clustering C such that $C(h) \in \mathbf{E}(\mathbf{P})$ for all h. The following facts were established in [2], and will be presented here without proof. When convenient, we shall follow the notation introduced in [2].

Fact 1.1. *There is a bijection between* DCs *and numerically stratified clusterings. It is given by*

$d \longmapsto Td$, *where* $Td(h) = \{(x,y): d(x,y) \leq h\}$,

$C \longmapsto d_C$, *where* $d_C(x,y) = \min\{h: (x,y) \in C(h)\}$.

Fact 1.2. *The* DC d *is an ultrametric iff* Td *is a dendrogram.*

Definition 1.3. A *cluster method* is a mapping $F:\mathbf{D} \longmapsto \mathbf{U}$; alternately, it may be viewed as a mapping $\underline{F}:\mathbf{D} \longmapsto$ dendrograms. The two notions are connected by the fact that $\underline{F}(d) = T[F(d)]$ for all $d \in \mathbf{D}$.

Definition 1.4. Let $d \in \mathbf{D}$. The numbers

$$0 = h_0 < h_1 < \ldots < h_n < h_{n+1}$$

in the image of d are called its *splitting levels*; corresponding to each h_i, the relation

$$R_i = Td(h_i) = \{(x,y): d(x,y) \leq h_i\}$$

is called a *splitting relation*. The DC d is clearly determined by its splitting levels and splitting relations. Since $\mathbf{P}\times\mathbf{P}$ is a splitting relation for every DC d, we shall normally not specifically mention either R_{n+1} or h_{n+1}; thus (unless otherwise specified) when we call R a splitting relation of d it will be understood that $R \subset \mathbf{P}\times\mathbf{P}$, and when we call h a splitting level of d, it will be understood that $Td(h) \subset \mathbf{P}\times\mathbf{P}$. This will be denoted by writing

$$d \leftrightarrow R_0 \subset R_1 \subset \ldots \subset R_n; h_0 < h_1 < \ldots h_n,$$

with the understanding that h_0 is necessarily 0. We agree to call this the *standard representation* of d. It is important to observe that d is completely specified by its standard representation together with the value of h_{n+1}.

Definition 1.5. (i) Let θ be a mapping on $[0,\infty)$, F a cluster method, and $d \in \mathbf{D}$. We let θ act on d by defining $\theta d \in \mathbf{D}$ by the rule

$$[\theta d](x,x) = 0, \text{ and}$$

$$[\theta d](x,y) = \theta(d(x,y)) \text{ for } x \neq y;$$

we say that F is *θ-compatible* in case $F(\theta d) = \theta F(d)$ for all $d \in \mathbf{D}$.

(ii) The cluster method F is called *monotone-equivariant* (ME) in case it is *θ-compatible* for every $\theta \in$ Aut, where Aut denotes the set of order automorphisms of $[0,\infty)$.

The standard well-known characterization of ME cluster methods is herewith presented without proof.

Theorem 1.6. *A cluster method* F *is monotone-equivariant if and only if*

(i) $\mathrm{im}(F(d)) \subseteq \mathrm{im}(d)$ *for all* $d \in \mathbf{D}$, *and*

(ii) *The splitting relations of* F(d) *depend only on the splitting relations of* d *and are independent of the splitting levels of* d.

Let's rephrase the result in a more specific manner. Consider the standard representations for d and F(d):

(1) $$d \leftrightarrow R_0 \subset R_1 \subset \dots \subset R_n; h_0 < h_1 < \dots < h_n,$$

(2) $$F(d) \leftrightarrow E_0 \subset E_1 \subset \dots \subset E_t; k_0 < k_1 < \dots < k_t.$$

To say that F is monotone-equivariant is equivalent to saying that by allowing repeats in the output splitting relations, we have

(3) $$F(d) \leftrightarrow S_0 \subseteq S_1 \subseteq \dots \subseteq S_n; h_0 < h_1 < \dots < h_n.$$

Thus F transforms a strictly increasing sequence of tolerance relations $R_0 \subset R_1 \subset \dots \subset R_n \subset \mathbf{P}\times\mathbf{P}$ into an increasing sequence $S_0 \subseteq S_1 \subseteq \dots \subseteq S_n \subseteq \mathbf{P}\times\mathbf{P}$ of equivalence relations. This shows that monotone-equivariant cluster methods are essentially graph-theoretic in nature.

2. An introductory Special Case.

In this section we restrict our attention to definite DCs, and monotone-equivariant cluster methods F whose domain consists of the definite DCs, and whose image is contained in the definite ultrametrics. We shall be working with the set M of isotone mappings on $[0,\infty)$ having the property that $\theta(h) > 0$ whenever $h > 0$. In view of this, we may ignore 0, and view d and F(d) as taking values in $(0,\infty)$. The idea is that we need not specify the value of any DC on a pair (x,x), so we shall just concentrate on the pairs (x,y) for which $x \neq y$. In view of this, we may adopt the temporary notation that $d \in \mathbf{D}$ may be specified by:

(4) $$d \leftrightarrow R_1 \subset \ldots \subset R_n; h_1 < \ldots < h_n \text{ with } h_1 > 0,$$

and F(d) by

(5) $$F(d) \leftrightarrow S_1 \subseteq \ldots \subseteq S_n; h_1 < \ldots < h_n.$$

For each $\theta \in M$, we seek to describe the set $\alpha(\theta)$ of all cluster methods F such that F is θ-compatible. We shall establish that there are only four possible values for α.

Suppose first that θ is injective. Then

$$\theta d \leftrightarrow R_1 \subset \ldots \subset R_n; \theta h_1 < \ldots < \theta h_n,$$

and by monotone equivariance,

$$F(\theta d) \leftrightarrow S_1 \subseteq \ldots \subseteq S_n; \theta h_1 < \ldots < \theta h_n.$$

Since $\theta F(d) \leftrightarrow S_1 \subseteq \ldots \subseteq S_n; \theta h_1 < \ldots < \theta h_n$, we see that F is θ-compatible. Similarly, if θ is a constant mapping, it is easy to see that F is θ-compatible. If θ is neither injective nor constant, then it is one of the following types:

<u>Type</u> A. Constant on an interval (0,h) and injective on (h,∞).

<u>Type</u> B. Constant on (h,∞) and injective on (0,h).

<u>Type</u> C. Constant on a pair of intervals (h,k) and (m,n) where $0 \leq h < k \leq m < n \leq \infty$ and $\theta(s) < \theta(t)$ for $s \in (h,k)$, $t \in (m,n)$.

Let d and F(d) be represented as in (4) and (5). For any $i \in \{1,2,\ldots,n\}$, we may choose θ so that:

For type A, $\theta d \leftrightarrow R_i \subset \ldots \subset R_n; \theta h_i < \ldots < \theta h_n$.

For type B, $\theta d \leftrightarrow R_1 \subset \ldots \subset R_i; \theta h_1 < \ldots < \theta h_i$.

For type C, $\theta d \leftrightarrow R_i;\theta h_i$.

Let F be θ-compatible. Using (5), we note that:

for type A, S_i depends at most on $\{R_i,\ldots,R_n\}$;

for type B, S_i depends at most on $\{R_1,\ldots,R_i\}$;

for type C, S_i depends at most on R_i.

The resulting cluster methods are called *divisive, agglomerative* and *flat,* respectively. It seems worth mentioning that this is one of the most common ways that practitioners of cluster analysis might group clustering algorithms.

3. The Formal Context Associated With Compatibility.

Recall [6] that a *formal context* is a triple $(\mathcal{G},\mathcal{M},\mathcal{I})$ where $\mathcal{G}$, $\mathcal{M}$ are sets and $\mathcal{I}$ is a binary relation from $\mathcal{G}$ to $\mathcal{M}$. The elements of $\mathcal{G}$ and $\mathcal{M}$ are respectively called *objects* and *attributes*. For $A \subseteq \mathcal{G}$, $B \subseteq \mathcal{M}$, one defines

$$A^{\mathcal{I}} = \{m \in \mathcal{M}\colon g\mathcal{I}m \text{ for all } g \in A\},$$

$$B^{\mathcal{I}} = \{g \in \mathcal{G}\colon g\mathcal{I}m \text{ for all } m \in B\}.$$

The mappings $A \longmapsto A^{\mathcal{I}}$, $B \longmapsto B^{\mathcal{I}}$ then form a Galois connection between the power sets of $\mathcal{G}$ and $\mathcal{M}$. A pair (A,B) is called a (*formal*) *concept* of the context $(\mathcal{G},\mathcal{M},\mathcal{I})$ if $A \subseteq \mathcal{G}$, $B \subseteq \mathcal{M}$, $A^{\mathcal{I}} = B$ and $B^{\mathcal{I}} = A$. A is the *extent* and B is the *intent* of the concept (A,B). The concepts of $(\mathcal{G},\mathcal{M},\mathcal{I})$ are customarily ordered by the rule:

$$(6) \qquad (A_1,B_1) \le (A_2,B_2) \Leftrightarrow A_1 \subseteq A_2.$$

It follows immediately from the theory of galois connections that $A_1 \subseteq A_2$ is equivalent to $B_2 \subseteq B_1$.

The notion of *compatibility* leads naturally to a context that will prove useful in classifying cluster methods. Unless otherwise specified, we shall continue with the notation we introduced earlier in the paper. Thus **P** is a finite set of objects that are to be clustered. We may let $\mathcal{G}$ denote a set of cluster methods on $\mathbf{P}$, and $\mathcal{M}$ a semigroup of mappings on $[0,\infty)$. The binary relation $\mathcal{I}$ is defined by the rule $(F,\theta) \in \mathcal{I}$ if and only if F is θ-compatible; i.e., if and only if $F(\theta d) = \theta F(d)$ for every DC d. The *Compatibility Problem* may then be taken to be the problem of determining the concepts of the context $(\mathcal{G},\mathcal{M},\mathcal{I})$. The *ME-Compatibility Problem* is defined as follows: Take $\mathcal{G}$ = ME = the set of monotone-equivariant cluster methods, and $\mathcal{M} = \mathrm{Map}_0$ = the set of 0-preserving isotone mappings on $[0,\infty)$. It is easy to show that the largest concept in this context is given by the pair $(\mathrm{ME},\mathrm{Inj}\cup\{\underline{0}\})$, where Inj is the semigroup of injective maps in Map_0, and $\underline{0}(h) = 0$ for all h. The problem now is to determine the concepts of the context $(\mathrm{ME},\mathrm{Map}_0,\mathcal{I})$. For convenience we agree to call $(\mathrm{ME},\mathrm{Map}_0,\mathcal{I})$ the *ME-compatibility context.*

We begin by introducing some added technical machinery. If the DC d has splitting levels $h_0,\ldots,h_n,h_{n+1}$ with $Td(h_{n+1}) = \mathbf{P}\times\mathbf{P}$, we say that d has *rank* n and write $\rho(d) = n$, while agreeing to assign d_0 rank -1, where d_0 is defined by $d_0(x,y) = 0$ for all $x,y \in \mathbf{P}$. We also let $I_n = \{0,1,\ldots,n\}$, and $\mathbf{D}_n = \{d \in \mathbf{D}\colon \rho(d) = n\}$. Assume that $d \in \mathbf{D}_n$, and $\theta \in \mathrm{Map}_0$. Associated with the pair (θ,d) there is a closure operator $\hat{\theta}_d\colon I_{n+1} \longmapsto I_{n+1}$ given by

$$\hat{\theta}_d(i) = \vee\{j \in I_{n+1}\colon \theta(h_i) = \theta(h_j)\}$$

We illustrate this closure operator with a concrete example. Suppose that when listed in ascending order, $\mathrm{im}(d) = \{h_i\colon 0 \le i \le 7\}$ where $Td(h_7) = \mathbf{P}\times\mathbf{P}$, and that $\theta \in \mathrm{Map}_0$ has the property that

$$\theta(h_1) = 0,\ \theta(h_2) = \theta(h_3) = \theta(h_4) = k_1,$$
$$\theta(h_5) = k_2,\ \theta(h_6) = \theta(h_7) = k_3,$$

where $0 < k_1 < k_2 < k_3$. Then $\hat{\theta}_d\colon I_7 \longmapsto I_7$ is given by

$$\hat{\theta}_d(0) = \hat{\theta}_d(1) = 1,\ \hat{\theta}_d(2) = \hat{\theta}_d(3) = \hat{\theta}_d(4) = 4,$$

$\hat{\theta}_d(5) = 5$, and $\hat{\theta}_d(6) = \hat{\theta}_d(7) = 7$.

Note that $\mathrm{im}(\hat{\theta}_d) = \{1,4,5,7\}$, and

$$\theta d \leftrightarrow R'_0 \subset R'_1 \subset R'_2; k_0 < k_1 < k_2,$$

where $R'_0 = R_1$, $R'_1 = R_4$, and $R'_2 = R_5$. It will prove useful to carefully consider what it means for $F(\theta d) = \theta F(d)$ for the specific pair (θ, d). The key observation to be made is that for $i \in I_n$, R_i is not a splitting relation for θd if and only if $i \neq \hat{\theta}_d(i)$. Alternately, one could say that R_i is necessarily a splitting relation for θd iff $i = \hat{\theta}_d(i)$. Suppose, for example, that

(7) $$d \leftrightarrow R_0 \subset R_1 \subset \ldots \subset R_n; h_0 < h_1 < \ldots < h_n,$$

(8) $$F(d) \leftrightarrow S_0 \subseteq S_1 \subseteq \ldots \subseteq S_n; h_0 < h_1 < \ldots < h_n.$$

If $\mathrm{im}(\hat{\theta}_d) = \{j_0 < j_1 < \ldots < j_t < n+1\}$, it is evident that

(9) $$\theta d \leftrightarrow R'_0 \subset R'_1 \subset \ldots \subset R'_t; \theta(h_{j_0}) < \theta(h_{j_1}) < \ldots < \theta(h_{j_t}),$$

where $R'_i = R_{j_i}$. Hence

(10) $$F(\theta d) \leftrightarrow S'_0 \subseteq S'_1 \subseteq \ldots \subseteq S'_t; \theta(h_{j_0}) < \theta(h_{j_1}) < \ldots < \theta(h_{j_t}).$$

To say that $F(\theta d) = \theta F(d)$ is equivalent to the assertion that for each index i in (10), $S'_i = S_{j_i}$. Thus $i \neq \hat{\theta}_d(i)$ implies that for $j = \hat{\theta}_d(j)$, the construction of S_j ignores the value or even the presence of R_i. For F to be compatible with θ requires of course that this assertion be valid for all $d \in \mathbf{D}$. In view of this it will be useful to define a *dependence function* $\mathbf{c}_\theta$ for θ by requiring that for $n \in \mathbb{W}$, and $j \in I_n$,

(11) $$\mathbf{c}_\theta(j,n) = \bigcap\{I_n \cap \mathrm{im}(\hat{\theta}_d) : j = \hat{\theta}_d(j) \in I_n,\ d \in \mathbf{D}_n\}.$$

Here $\mathbb{W}$ denotes the set of non-negative integers. This can be extended to a set A of mappings by requiring that

(12) $$\mathbf{c}_A(j,n) = \bigcap\{I_n \cap \mathrm{im}(\hat{\theta}_d) : j = \hat{\theta}_d(j) \in I_n,\ d \in \mathbf{D}_n,\ \theta \in A\}.$$

To obtain some intuition for what (12) is saying, assume $i \in \mathbf{c}_A(j,n)$, and that $\rho(d) = n$. If R_j is a splitting relation for d, then no matter what mapping $\theta \in A$ is chosen, if

R_j is a splitting relation for θd, then so necessarily is R_i. Recalling that for the consensus function F,

$$F^{\mathcal{J}} = \{\theta \in \mathrm{Map}_0 : F(\theta d) = \theta F(d) \text{ for all } d \in \mathbf{D}\},$$

we may define a *dependence function* $\mathbf{c}_F$ for F by the rule that $\mathbf{c}_F = \mathbf{c}_A$, where $A = F^{\mathcal{J}}$. Thus

(13) $\mathbf{c}_F(j,n) = \bigcap\{I_n \cap \mathrm{im}(\hat{\theta}_d) : j = \hat{\theta}_d(j) \in I_n,\ d \in \mathbf{D}_n,\ \theta \in F^{\mathcal{J}}\}.$

The dependence function defined in (13) is closely connected to the manner in which F treats input data. To see this, let $d \in \mathbf{D}_n$, and represent d and F(d) as in (7) and (8). Then when forming a given output relation S_j, one must consider the input relation R_i if and only if $i \in \mathbf{c}_F(j,n)$. Finally, we observe that

$$\mathbf{c}_F : \bigcup_{n \in \mathbb{W}} (I_n \times \{n\}) \longmapsto \mathcal{P}(\mathbb{W})$$

has the property that for each $n \in \mathbb{W}$, $\{j\} \subseteq \mathbf{c}_F(j,n) \subseteq I_n$. To summarize the above discussion, the dependence function $\mathbf{c}_F$ describes the operation of F by specifying that if

$$d \longleftrightarrow R_0 \subset R_1 \subset \ldots \subset R_n, \text{ and } F(d) \longleftrightarrow S_0 \subseteq S_1 \subseteq \ldots \subseteq S_n,$$

then for each $j \in I_n$, S_j depends *at most* on $\{R_i : i \in \mathbf{c}_F(j,n)\}$. In other words, when constructing S_j from $\{R_0, R_1, \ldots, R_n\}$, one simply ignores any R_k with $k \notin \mathbf{c}_F(j,n)$. It may prove illuminating here to consider an example. Following this, we shall present a more formal discussion of the connection between dependence functions and compatibility.

Example 3.1 If $\rho(d) \leq 2$, we define F(d) by taking $S_i = \sigma(R_i)$, where $\sigma(R)$ denotes the equivalence relation generated by the tolerance relation R; otherwise, we take $F(d) = d_0$. This clearly defines a monotone-equivariant cluster method F, so F is θ-compatible when θ is injective or $\theta = \underline{0}$. For any other $\theta \in \mathrm{Map}_0$, one can clearly construct $d \in \mathbf{D}_3$ such that $\theta d \in \mathbf{D}_2$. Then representing d, F(d), and θd

as in (7), (8) and (9), we have that $\theta F(d) = d_0$, while $F(\theta d) \leftrightarrow \sigma(R'_{j_0}) \subseteq \sigma(R'_{j_1}) \subseteq \sigma(R'_{j_2})$. Thus $F^{\mathcal{J}} = \mathrm{Inj} \cup \{\underline{0}\}$. It is evident that for θ injective, $\hat{\theta}_d(j) = j$ for all $j \in I_n$; for $\theta = \underline{0}$, $\hat{\theta}_d(j) = n+1$ for all $j \in I_n$, so $j \neq \hat{\theta}_d(j)$ for any $j \in I_n$. Hence $\mathbf{c}_F(j,n) = I_n$ for all $j \in I_n$. This does not say that S_j depends on each and every one of the splitting relations R_i for a given d; it merely says that the mappings in $F^{\mathcal{J}}$ do not induce any restriction on the construction of S_j. Indeed, when $\rho(d) > 2$, $F(d) = \underline{0}$, so F ignores the input splitting relations, though it must of course count them to determine the rank of d.

Remark 3.2 For a monotone equivariant cluster method F, $F^{\mathcal{J}}$ is a semigroup containing $\mathrm{Inj} \cup \{\underline{0}\}$. Thus in the ME-compatibility context, the definition of dependence functions may be viewed somewhat differently. For $A \subseteq \mathrm{Map}_0$, let $A^\wedge$ denote the semigroup generated by $A \cup \mathrm{Aut} \cup \{\underline{0}\}$, and note that $A^{\mathcal{J}} = (A^\wedge)^{\mathcal{J}}$. Observing that for any fixed $d' \in \mathbf{D}_n$

$$\{\mathrm{im}(\hat{\theta}_d) : j = \hat{\theta}_d(j) \in I_n,\ d \in \mathbf{D}_n,\ \theta \in A\}$$

coincides with

$$\{\mathrm{im}(\hat{\theta}_{d'}) : j = \hat{\theta}_{d'}(j) \in I_n,\ \theta \in A^\wedge\},$$

we may define a dependence function by considering a fixed $d \in \mathbf{D}_n$ and taking:

$$(14) \qquad \mathbf{c}_A(j,n) = \bigcap\{I_n \cap \mathrm{im}(\hat{\theta}_d) : j = \hat{\theta}_d(j) \in I_n,\ \theta \in A^\wedge\},$$

$$(15) \qquad \mathbf{c}_F(j,n) = \bigcap\{I_n \cap \mathrm{im}(\hat{\theta}_d) : j = \hat{\theta}_d(j) \in I_n,\ \theta \in F^{\mathcal{J}}\}.$$

The idea is to work with a single $d \in \mathbf{D}_n$, and for each splitting relation R_j, determine which splitting relations R_i can be removed by the action of some $\theta \in A^\wedge$, while simultaneously retaining R_j. The results that follow have obvious modifications for the ME-compatibility context, but will be stated in the more general setting so that they may be applied in situations that do not necessarily involve monotone equivariance.

We may order the set of dependence functions by the rule $\mathbf{c}_A \leq \mathbf{c}_B$ iff $\mathbf{c}_A(j,n) \subseteq \mathbf{c}_B(j,n)$ for all $j \in I_n$ and all $n \in \mathbf{W}$. We need to relate this ordering of dependence functions with the ordering of concepts that was given in (6). We shall emphasize the natural connection between compatibility and dependence functions. Let F be ME, and suppose that $F(\theta d) = \theta F(d)$. A glance at (10) shows that each S'_j depends at most on $\{R'_0, R'_1, \ldots, R'_t\}$. It follows that if $i \notin \mathrm{im}(\hat{\theta}_d)$, then $i \notin \mathbf{c}_F(j,n)$ for any $j \in [\mathrm{im}(\hat{\theta}_d)]\backslash\{n+1\}$. Suppose conversely that $i \notin \mathrm{im}(\hat{\theta}_d)$ implies that $i \notin \mathbf{c}_F(j,n)$ for any $j \in [\mathrm{im}(\hat{\theta}_d)]\backslash\{n+1\}$. If $j \in \mathrm{im}(\hat{\theta}_d)$, it is immediate that S_j can only depend on $\{R_{j_0}, R_{j_1}, \ldots, R_{j_t}\}$. If θd and $F(\theta d)$ are given as in (9) and (10), we consider $\theta F(d)$. Examining (8), it is clear that

(16) $$\theta F(d) \longleftrightarrow S_{j_0} \subseteq \ldots \subseteq S_{j_t}; \theta(j_0) < \ldots < \theta(j_t),$$

where $\theta(j_i)$ denotes $\theta(h_{j_i})$ for all i. Since S_{j_i} depends at most on $\{R_{j_0}, R_{j_1}, \ldots, R_{j_t}\}$, it must be true that $S_{j_i} = S'_i$ for $i = 0,1,\ldots,t$. But this says that $\theta F(d) = F(\theta d)$. In other words, we have established the following result.

Lemma 3.2 *The cluster method* F *is* θ*-compatible if and only if for each* DC d *of arbitrary rank* n, $i \neq \hat{\theta}_d(i)$ *implies that* $i \notin \mathbf{c}_F(j,n)$ *for any* $j \in \mathrm{im}(\hat{\theta}_d)\backslash\{n+1\}$.

Remark 3.3 Perhaps a more illuminating way of phrasing this is to say that F is θ-compatible if and only if for each DC d of rank n, $j = \hat{\theta}_d(j) < n+1$ implies that $\mathbf{c}_F(j,n) \subseteq \mathrm{im}(\hat{\theta}_d)$. In other words, what we are saying is that F is θ-compatible if and only if $\mathbf{c}_F \leq \mathbf{c}_\theta$.

Lemma 3.4 *Let* F,G *be* ME *cluster methods. Then* $\mathbf{c}_F \le \mathbf{c}_G$ *if and only if* $G^{\mathcal{J}} \subseteq F^{\mathcal{J}}$.

Proof: If $\mathbf{c}_F \le \mathbf{c}_G$, then by Lemma 3.2, G θ-compatible implies that F is θ-compatible. Suppose conversely that $G^{\mathcal{J}} \subseteq F^{\mathcal{J}}$. The fact that $\mathbf{c}_F \le \mathbf{c}_G$ is an immediate consequence of (13). ■

Theorem 3.5 *Let* A *be a nonempty subset of* Map_0, *and* G *a cluster method. Then* $\mathbf{c}_G \le \mathbf{c}_A$ *if and only if* $G \in A^{\mathcal{J}}$.

Proof: If $G \in A^{\mathcal{J}}$, then $A \subseteq (A^{\mathcal{J}})^{\mathcal{J}} \subseteq G^{\mathcal{J}}$, so by (13), $\mathbf{c}_G \le \mathbf{c}_A$. If on the other hand $\mathbf{c}_G \le \mathbf{c}_A$, then by Remark 3.3, G is θ-compatible for all $\theta \in A$. But this puts $G \in A^{\mathcal{J}}$. ■

Corollary 3.6 *For each nonempty subset* A *of* Map_0, *if* $A^* = (A^{\mathcal{J}})^{\mathcal{J}}$, *then*

$$\mathbf{c}_{A^*} = \vee\{\mathbf{c}_G\colon G \in A^{\mathcal{J}}\}.$$

Proof: By Theorem 3.5, $\mathbf{c}_{A^*}$ is an upper bound for $\{\mathbf{c}_G\colon G \in A^{\mathcal{J}}\}$. If $\mathbf{c}_B$ is any other upper bound for this set, then $G \in A^{\mathcal{J}}$ implies that $\mathbf{c}_G \le \mathbf{c}_B$, so a second application of the Theorem produces the fact that $G \in B^{\mathcal{J}}$. But this says that $A^{\mathcal{J}} \subseteq B^{\mathcal{J}}$, so $B \subseteq A^*$. It is immediate that $\mathbf{c}_{A^*} \subseteq \mathbf{c}_B$, thus establishing the result. ■

Corollary 3.7 *If* A,B *are nonempty subsets of* Map_0, *then*:

(i) $\mathbf{c}_A \le \mathbf{c}_B$ *implies* $A^{\mathcal{J}} \subseteq B^{\mathcal{J}}$.

(ii) $A^{\mathcal{J}} \subseteq B^{\mathcal{J}}$ *implies that* $\mathbf{c}_{A^*} \le \mathbf{c}_B$.

(iii) *If* $A = A^*$, *then* $\mathbf{c}_A \le \mathbf{c}_B$ *iff* $A^{\mathcal{J}} \subseteq B^{\mathcal{J}}$.

Proof: (i) is an immediate consequence of Theorem 3.5. To establish (ii), note that by Theorem 3.5, $G \in A^{\mathcal{J}}$ implies

$\mathbf{c}_G \le \mathbf{c}_B$. By Corollary 3.6, $\mathbf{c}_{A*} \le \mathbf{c}_B$. (iii) is now immediate. ■

Working in the *ME-compatibility context,* the compatibility question may now be rephrased. For each nonempty subset A of Map_0, let $\alpha(A) = \mathbf{c}_A$. For singletons, we agree to write $\alpha(\theta)$ in place of the more correct $\alpha(\{\theta\})$. The problem now is to describe the image of α.

4. Kernels of isotone mappings.

We begin with some needed terminology. For $\theta \in \mathrm{Map}_0$, the *kernel* of θ is the equivalence relation defined on $[0,\infty)$ by the rule that $x \equiv y$ iff $\theta(x) = \theta(y)$. The equivalence classes of the kernel of θ are either singletons or proper intervals on which θ is constant. There is a second partition of $[0,\infty)$ naturally associated with this equivalence relation. It is formed by taking the set $\{J_\alpha\}_{\alpha\in I}$ of proper interval classes of the kernel together with the maximal intervals of $[0,\infty)\setminus\bigcup_{\alpha\in A}J_\alpha$ This second partition will be called the *interval kernel* of θ, and will be denoted by the symbol $ik(\theta)$.

This idea is relevant because if $ik(\theta) = ik(\psi)$, then $\alpha(\theta) = \alpha(\psi)$. It is important to note that if ψ is an order automorphism of $[0,\infty)$, then applying ψ to the right of θ shifts the intervals in $ik(\theta)$, but does not change their nature. On the other hand, applying ψ to the left of θ shifts the levels of θ, but leaves the intervals in its interval kernel intact. Since θ and $\psi\theta$ have the same points of discontinuity, it should be noted that it is possible to have $ik(\theta) = ik(\theta')$ without the existence of order automorphisms ψ_1, ψ_2 such that $\theta = \psi_1 \circ \theta' \circ \psi_2$.

Remark 4.1. Observations about interval kernels.

(i) The classes of the interval kernel of θ consist of at most a countable number of intervals. Each of these intervals consists either of a single point, a proper interval on which θ is constant, or a proper interval on which θ is one-one. Denote these three types of intervals, respectively, by p, q, and r. Thus the interval kernel of θ can be represented by a finite or countably infinite word consisting of p's, q's and r's. There will be no confusion if we simply refer to these as *k-words*.

(ii) For a given k-word W, we shall say that the cluster method F is W-*compatible* in case F is compatible with some mapping (hence all mappings) $\theta \in \mathrm{Map}_0$ whose interval kernel is W. This follows from the fact that compatibility with respect to θ is determined entirely by which relations R_i of a DC d that can be collapsed by θ when forming θd. It will be convenient to denote the corresponding dependence function by $\mathbf{c}_W$ in place of $\mathbf{c}_F$.

(iii) By the construction of the interval kernel, the following sub-words are illegal: pp, pr, rp, rr

(iv) A k-word may begin with any of the symbols p,q,r but cannot end with p. This follows from the fact that $[0,\infty)$ has no largest element.

(v) A k-word consists of at most a countable number of letters, but it need not have a second letter or a last letter. If W does not have a second letter, then it clearly must have infinitely many q-intervals. Hence it may be regarded as being of type qW_1qW_2, $pW_1qW_2qW_3$, or $rW_1qW_2qW_3$ where each W_i is an arbitrary (possibly empty) word.

(vi) The possible 1 letter k-words are q (constant map) and r (injective map).

(vii) The possible 2 letter k-words are pq, qq, qr, rq.

(vii) The possible 3 letter k-words are pqq, pqr, qpq, qqq, qqr, qrq, rqq, rqr.

(ix) The mapping α described at the end of Section 3 naturally induces a mapping β from k-words into dependence functions. If W is a k-word that describes the interval kernel of θ, one just defines $\beta(W) = \alpha(\theta) = \alpha([\theta])$, where

$$[\theta] = \{\psi \in \mathrm{Map}_0 : \mathrm{ik}(\psi) = \mathrm{ik}(\theta)\}.$$

We define a quasiorder on the set of k-words by the rule $W_1 \leq W_2$ iff $\beta(W_1) \leq \beta(W_2)$. Thus $W_1 \equiv W_2$ iff $\beta(W_1) = \beta(W_2)$. Evidently, determining the image of β is equivalent to determining the image of α, so this leads to an alternate view of the solution of the ME-compatibility problem. Note that when we say that $W_1 \equiv W_2$, we are saying that if $\mathrm{ik}(\theta_1) = W_1$ and $\mathrm{ik}(\theta_2) = W_2$, then $[\theta_1]^{\mathcal{G}} = [\theta_2]^{\mathcal{G}}$. Before proceeding, we present a technical lemma that will be useful in the sequel.

Lemma 4.2 *Let* F *be compatible with respect to all* $\theta \in \mathrm{Map}_0$ *whose interval kernel is* W. *Suppose that for each* DC d, *and every mapping* ψ *whose interval kernel is* V, *it is true that there is a mapping* θ *with interval kernel* W *such that the restriction of* θ *to* im(d) *coincides with the restriction of* ψ *to* im(d). *Then* F *is* ψ*-compatible with all* $\psi \in \mathrm{Map}_0$ *whose interval kernel is* V. *It follows that* $\mathbf{c}_W \leq \mathbf{c}_V$.

Proof: Let $\mathrm{ik}(\psi) = V$, and let d be a fixed DC. By hypothesis, we can find $\theta \in \mathrm{Map}_0$ with interval kernel W such that $\psi d = \theta d$. Since F is ME, it follows that $\psi F(d) = \theta F(d)$. Hence

$$F(\psi d) = F(\theta d) = \theta F(d) = \psi F(d).$$

Since this works for any fixed DC, it follows that F is ψ-compatible. ■

Our next goal is to describe the equivalence classes of k-words. This will be done in a number of steps. We shall use the standard representations given in (1) and (3) for d and F(d), sometimes suppressing the actual splitting levels, so that for $\rho(d) = n$, we have

(17) $$d \leftrightarrow R_0 \subset R_1 \subset \ldots \subset R_n,$$

(18) $$F(d) \leftrightarrow S_0 \subseteq S_1 \subseteq \ldots \subseteq S_n.$$

<u>Step</u> 1. It is easy to show that every monotone-equivariant cluster method is compatible with all injective members of Map_0, so $\beta(r)$ is the compatibility function defined by $\mathbf{c}_r(i,n) = I_n$ for all $i \in I_n$. The only way that the interval kernel of θ could be q is for $\theta = \underline{0}$, the zero mapping. F is compatible with $\underline{0}$ iff $F(d_0) = d_0$. Since this is true for all monotone-equivariant cluster functions, we have that $\beta(q) = \beta(r)$.

<u>Step</u> 2. We consider $\beta(qq)$, so let $ik(\theta) = qq$. Then θ can be chosen so that $\theta d \leftrightarrow R_i$ for any choice of $i \in I_n$. It follows that the corresponding dependence function is $\mathbf{c}_{qq}(i,n) = \{i\}$ for all $i \in I_n$. Hence $\mathbf{c}_{qq} \leq \mathbf{c}_W$ for any k-word W. Such cluster methods are commonly called *flat*, and are defined by a mapping $\sigma:\mathbf{Tol}(\mathbf{P}) \longmapsto \mathbf{E}$, in the sense that in (18), $S_i = \sigma(R_i)$ for all i. If $ik(\theta) = qW_1qW_2$, where W_1, W_2 are arbitrary (possibly empty) words, then for any DC d, the W_1 and W_2 portions of θ can be shifted by an order automorphism so that they miss the image of d. By Lemma 4.2, $\mathbf{c}_{qW_1qW_2} \leq \mathbf{c}_{qq}$. Thus $\mathbf{c}_{qW_1qW_2} = \mathbf{c}_{qq}$, and we have $qq \equiv qW_1qW_2$.

<u>Step</u> 3. Let $ik(\theta) = pW_1qW_2qW_3$ where again W_1, W_2, W_3 are arbitrary (possibly empty) words. As in Step 2, for any input DC d, we can apply an order automorphism so that we may assume that W_1, W_2 and W_3 miss all values in the image

of d. Thus we may choose θ so that $\theta d \leftrightarrow R_0 \subset R_i$ or so that $\theta d \leftrightarrow R_0$. Since $\theta(h) > 0$ whenever $h > 0$, there is no way to choose θ so as to eliminate R_0. It follows that $\beta(pW_1qW_2qW_3) = \mathbf{c}_{pqq}$, where $\mathbf{c}_{pqq}(i,n) = \{0,i\}$. Cluster methods having this dependence function are often called *semi-flat*. Note that $pqq \equiv pW_1qW_2qW_3$.

Step 4. If $ik(\theta) = rW_1qW_2qW_3$, then by an argument similar to that given in Step 3, we can show that $\beta(rW_1qW_2qW_3) = \mathbf{c}_{pqq}$, and we also have $pqq \equiv rW_1qW_2qW_3$.

This then disposes of all words that have more than one "q" interval, all words not having a second letter, as well as the words only having a single letter, so we temporarily drop them from our considersations. Suppose then that the interval kernel of θ has in it at most one "q" interval. The possibilities are: pq, pqr, qr, rq, and rqr. We continue our step by step discussion.

Step 5. If $ik(\theta) = pq$, then $\theta(0) = 0$ and there exists a nonzero constant k such that $\theta(h) = k$ for all $h > 0$. It follows that $\beta(pq) = \mathbf{c}_{pq}$, where $\mathbf{c}_{pq}(0,n) = \{0\}$, and $\mathbf{c}_{pq}(i,n) = I_n$ for $i > 0$. We agree to call such cluster methods 0-*stable*.

Step 6. Let $ik(\theta) = qr$. For $i \in I_n$, θ can be chosen so that $\theta d \leftrightarrow R_i \subset R_{i+1} \subset \ldots \subset R_n$. Clearly no R_j $(j > i)$ can be removed by θ without removing R_i. Thus $\beta(qr) = \mathbf{c}_{qr}$, where

$$\mathbf{c}_{qr}(i,n) = \{i, i+1, \ldots, n\}.$$

This produces the class of *divisive* cluster methods.

Step 7. Let $ik(\theta) = rq$. This is dual to Step 6, and θ can be chosen so that for $i \in I_n$, $\theta d \leftrightarrow R_0 \subset R_1 \subset \ldots \subset R_i$.

Hence $\beta(rq) = \mathbf{c}_{rq}$, where $\mathbf{c}_{rq}(i,n) = \{0,1,...,i\}$, and we have the class of *agglomerative* cluster methods.

Step 8. We now examine $ik(\theta) = pqr$. Here θ may be chosen so that $\theta d \longleftrightarrow R_0$ or $\theta d \longleftrightarrow R_0, R_i \subset R_{i+1} \subset ... \subset R_n$. Thus $\beta(pqr) = \mathbf{c}_{pqr}$, where

$$\mathbf{c}_{pqr}(0,n) = \{0\},\ \mathbf{c}_{pqr}(i,n) = \{0,i,i+1,...,n\} \text{ for } i > 0.$$

Step 9. We are left with the case where θ has interval kernel rqr. Here we must consider the fact that we may compose a pair of rqr mappings. The easiest way to see what is happening is to view this schematically as follows:

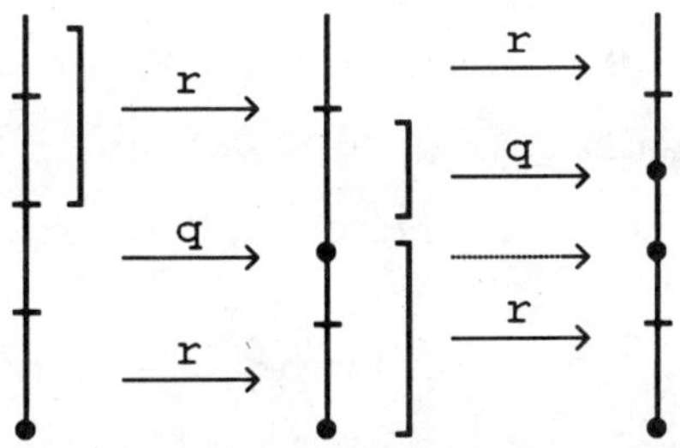

This shows that the composition of a pair of rqr-mappings can yield a mapping whose interval kernel is rqqr. By Step 4, $\beta(rqr) = \beta(pqq)$.

We summarize the situation in the following diagram:

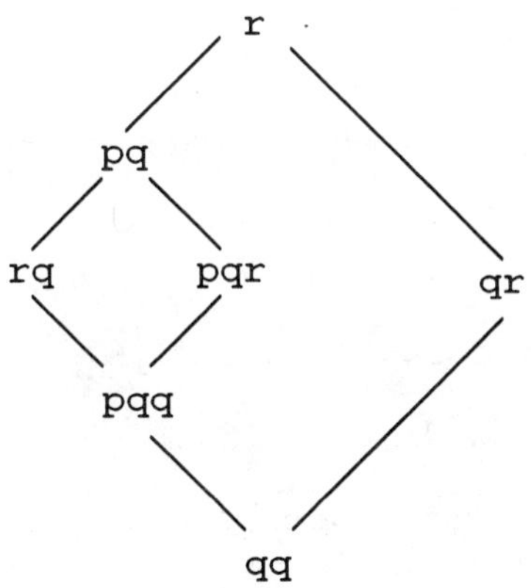

r $\longrightarrow$ monotone-equivari

pq $\longrightarrow$ 0-stable

rq $\longrightarrow$ agglomerative

pqr $\longrightarrow$ 0-divisive

pqq $\longrightarrow$ semiflat

qr $\longrightarrow$ divisive

qq $\longrightarrow$ flat

Figure 1.

Figure 1 displays the image of α on singletons. The ordering is that of compatibility implication. In order to establish that this is also the image of α on arbitrary subsets of Map_0, we need to argue that the meet operation in the above diagram is intersection. To do this, we need only establish two assertions:

Item 1. If F is both pq and qr compatible, then F is qq compatible.

Proof: Let d and F(d) be represented as in (17) and (18). Given $i \in I_n$, choose $\theta \in Map_0$ so that $ik(\theta) = qr$, and $\theta d \leftrightarrow R_i \subset \ldots \subset R_n$. By θ-compatibility, $F(\theta d) \leftrightarrow S_i \subsetneq \ldots \subsetneq S_n$. Now choose $\theta_1 \in Map_0$ with $ik(\theta_1) = pq$. Then $\theta_1\theta d \leftrightarrow R_i$, and by θ_1-compatibility, $F(\theta_1\theta d) \leftrightarrow S_i$. But this implies that F is flat, thus showing that $\beta(pq \wedge rq) = \beta(qq)$.

Item 2. If F is both rq and pqr compatible, then F is pqq compatible.

Proof: Using the same notation as in Item 1, for $i \in I_n$, we may choose $\theta \in Map_0$ so that $ik(\theta) = pqr$, and $i \in I_n$. Choose $\theta \in Map_0$ so that $ik(\theta) = qr$, and

$$\theta d \leftrightarrow R_0 \text{ or } \theta d \leftrightarrow R_0 \subset R_i \subset \ldots \subset R_n.$$

If $\theta d \leftrightarrow R_0$, this shows that S_0 depends only on R_0. If $\theta d \leftrightarrow R_0 \subset R_i \subset \ldots \subset R_n$, then $\theta_1 \in Map_0$ can be chosen so that $ik(\theta_1) = rq$ and $\theta_1 d \leftrightarrow R_0 \subset R_1 \subset \ldots \subset R_i$. Then using θ-compatibility, S_i depends at most on $\{R_0, R_i, \ldots, R_n\}$; by θ_1-compatibility, S_i depends at most on $\{R_0, R_1, \ldots, R_i\}$. It is immediate that S_i can only depend on

$$\{R_0, R_i, \ldots, R_n\} \cap \{R_0, R_1, \ldots, R_i\} = \{R_0, R_i\},$$

thus showing that F is semiflat. This then completes the proof. ∎

The actual formal concepts of the ME-compatibility context can easily be read off from Figure 1. The intents are the principal filters of the diagram that appears to the left of the figure, and the extents are the principal ideals of the ordered set formed by the classes of cluster methods that appear to the right of Figure 1. For example, the smallest concept is the pair (flat,Map_0). If

$$M_1 = \{\theta \in Map_0 : h > 0 \text{ implies } \theta(h) > 0\},$$

then (flat $\cup$ semiflat,M_1) is a concept, etc. Note that M_1 consists of the class pqq and all classes that lie above it in Fig. 1.

5. Some examples.

We present some examples here to indicate that the seven classes of cluster methods displayed in Figure 1 are both non-trivial and distinct.

Example 5.1 Let F be defined as in Example 3.1. We have already shown there that $F^{\mathcal{J}} = \{r,q\}$.

Example 5.2 The canonical example of a flat cluster method is provided by single-linkage clustering. This operates by transforming the input sequence

$$d \leftrightarrow R_0 \subset R_1 \subset \ldots \subset R_n$$

into the sequence

$$F_S(d) \leftrightarrow S_0 \subseteq S_1 \subseteq \ldots \subseteq S_n,$$

where for each index i, $S_i = \sigma(R_i)$ is the transitive closure of R_i.

Example 5.3 Here is an example of a semiflat cluster method that is not flat. For d definite, take $F(d) = d_0$; otherwise, $F(d) = F_S(d)$.

Example 5.4 For an example of an agglomerative method that is not semiflat, we consider a version of complete-linkage clustering. We shall denote this by F_C. Briefly here is how it operates. If $d \leftrightarrow R_0 \subset R_1 \subset \dots \subset R_n$, we let $S_0 = \sigma(R_0)$. To form S_1, we merge any classes A, B of S_0 having the property that all possible links have been made between them. If necessary, the transitive closure of the resulting relation is formed to obtain an equivalence relation. This is clearly an agglomerative cluster method since S_i depends at most on $\{R_0, R_1, \dots, R_i\}$. This also shows that F_C is neither divisive nor is it 0-divisive. We have yet to verify that it is not semiflat. To see this, we let $\mathbf{P} = \{a,b,c,e\}$ and form DCs d and d′ as follows:

$$d \leftrightarrow R_0 \subset R_1 \subset R_2 \subset R_3,\quad d' \leftrightarrow R_0' \subset R_1' \subset R_2' \subset R_3',$$

where

$R_0 = R_0' = \Delta$, $R_1 = \Delta \cup \{(a,b),(b,a)\}$, $R_1' = R_1 \cup \{(c,e),(e,c)\}$,

$R_2 = R_2' = R_1 \cup \{(a,c),(c,a)\}$, $R_3 = R_3' = \mathbf{P} \times \mathbf{P}$. Letting

$$F_C(d) \leftrightarrow S_0 \subseteq S_1 \subseteq S_2 \subseteq S_3,\quad F_C(d') \leftrightarrow S_0' \subseteq S_1' \subseteq S_2' \subseteq S_3',$$

we have $S_0 = S_0' = \Delta$, $S_1 = R_1$, $S_1' = R_1'$. Since all links between {a,b} and {c} are made by R_2 the partition associated with S_2 is {{a,b,c},{e}}. On the other hand, not all links are made by R_2' between {a,b} and {c,e}, so $S_2' = S_1' \neq S_2$. Thus even though $R_i = R_i'$ for $i \in \{0,2\}$, we have $S_2 \neq S_2'$, thus showing that F_C is not semiflat. Obvious modifications of this construction must be made when $\mathbf{P}$ has more than 4 members.

Example 5.5 We need a cluster method F_D that is divisive but not flat. For a given $d \in \mathbf{D}$, choose a, b so that d(a,b) is maximal. Let E be the equivalence relation whose associated partition consists of the three classes {a}, {b}, and **P**\{a,b}. Then with d given its standard representation, let

$$F_D(d) \leftrightarrow S_0 \subseteq S_1 \subseteq \ldots \subseteq S_n,$$

where $S_i = E \cap \sigma(R_i)$. Since S_i depends on R_i and R_n, F_D is divisive, but is not flat. The motivation behind this cluster method is that it is unlikely that a,b should be in the same cluster for any S_i $(i \leq n)$ if d(a,b) is maximal.

Example 5.6 A similar technique may be used to construct a 0-divisive cluster method that is not semiflat. Let $d \leftrightarrow R_0 \subset R_1 \subset \ldots \subset R_n$, and $F_{D_0}(d) \leftrightarrow S_0 \subseteq S_1 \subseteq \ldots \subseteq S_n$. If $R_0 = \Delta$, take $S_0 = R_0$; otherwise, choose a pair (x,y) such that xR_0y, and let S_0 be the equivalence relation formed by merging the pair {x,y}. For i > 0, let S_i be the output from $F_D(d)$. Thus S_0 depends only on R_0, and for i > 0, S_i depends on R_i and R_n. This cluster method is clearly not semiflat. On forming the dependence function associated with F_{D_0}, one sees, however, that it is 0-divisive. Another way or arguing this would be to observe that S_0 depends only on R_0, and for i > 0, S_i depends *at most* on $\{R_0, R_i, \ldots, R_n\}$.

Example 5.7 Let $d \leftrightarrow R_0 \subset R_1 \subset \ldots \subset R_n$. Take $S_0 = \sigma(R_0)$. If $\rho(d) = 2$, let $S_i = \sigma(R_1)$ for i > 0; otherwise, take $S_i = \sigma(R_i)$ for all i. Since this cluster method is θ-compatible only for θ with interval kernel of type pq, q or r, it is 0-stable but neither 0-divisive nor agglomerative.

6. Conclusion

We have shown how monotone-equivariant cluster methods may be classified by means of the data types upon which they are designed to act. Here a *data type* is a class of dissimilarity coefficients closed under the action of some fixed semigroup of isotone mappings of $[0,\infty)$. Under the assumption that all of the mappings fix 0, a formal concept was defined that allowed us to produce 7 classes of monotone equivariant cluster methods. Though we restricted our attention to the situation where no overlap was allowed between clusters, the same 7 classes appear in the more general situation where some mild form of overlap between clusters at the same level is allowed.

If mappings are allowed to move 0, or if dissimilarity coefficients take values in the extended nonnegative reals, or if we work with cluster methods that are not necessarily monotone-equivariant, the situation becomes much more complex, and many more classes of cluster methods arise. The results will be presented in a paper to appear elsewhere.

References

1. F. S. Roberts, *Measurement Theory with applications to Decision making, Utility, and the Social Sciences,* Encyclopedia of Mathematics and its Applications, Vol. 7, Addison-Wesley, Reading, 1979.

2. N. Jardine and R. Sibson, *Mathematical Taxonomy*, Wiley, New York, 1971.

3. F. Baulieu, *Order theoretic classification of monotone equivariant cluster methods*, Algebra Universalis 20 (1985), 351-267.

4. F. Baulieu, *Classification of normalized cluster methods in an order theoretic model*, Discrete Applied Math. 32 (1991), 1-29.

5. R. C. Powers, *Order theoretic classification of percentile cluster methods*, Ph.D. dissertation, University of Massachusetts, Amherst, MA (1988).

6. R. Wille, *Restructuring lattice theory: an approach based on hierarchies of concepts.* In: I. Rival (ed.), Ordered Sets, NATO ASI Series 83, Reidel, Dordrecht, 1982.

Department of Mathematics and Statistics, University of Massachusetts, Amherst, MA 01003
E-mail address: melj@math.umass.edu

Fachbereich Mathematik, Technische Hochschule Darmstadt, 75670 Darmstadt, Federal Republic of Germany
E-mail address: wille@mathematik.th-darmstadt.de

DIMACS Series in Discrete Mathematics
and Theoretical Computer Science
Volume 19, 1995

CONTIGUITY-CONSTRAINED HIERARCHICAL CLUSTERING

FIONN D. MURTAGH

ABSTRACT. Approaches for unsupervised regionalization and contiguity-constrained clustering are discussed in the context of 2-dimensional image data and planar-referenced multidimensional data. In particular we consider the property of reversals or inversions in hierarchies. In regard to the latter, we discuss fast nearest neighbor based approaches, which facilitate parallel implementations.

1. INTRODUCTION: THE DATA

It is often necessary to segment 2-dimensional image data, or to regionalize multivariate data which can be represented naturally in the plane. Examples of the latter include geographic and astronomical data. Generalizion to data which is more than 2-dimensional is possible. So also is generalization to 1-dimensional data, where the problem is then one of segmentation of a time series or of a spectrum.

We can understand limitations or peculiarities of segmentation or regionalization algorithms if we study them in some common framework. In this paper, a geometric framework will be used to discuss a family of stepwise agglomerative methods.

Image data can be handled with such methods (but note that faster, more heuristic methods may well score over what are often $O(n^2)$, or computationally even more demanding, methods discussed in this paper). Point pattern data can also be treated by such methods, if "cartographic representability" is relevant.

A 2-dimensional image is a mapping

$$f : X \times Y \longrightarrow I\!R^+$$

where $(x, y), x \in X, y \in Y$ is a pixel; where $f(x, y)$ is the pixel intensity value, or vector; and where $X, Y \subset Z^+$. A simple way to incorporate cartographically referenced point pattern data into this framework is to consider such data to be overlaid with a regular pixel grid which is sufficiently fine that each object or case is located in an individual pixel. A binary image is arrived at: a pixel has a 1 value if it contains a point, and otherwise a 0 value. Alternatively, for object sets of sufficient density, the number of points falling within the pixel boundaries yields a frequency which becomes the pixel value. Imposing a pixelated grid on a planar point set is commonly done for large numbers of points: see, e.g., Upton and Fingleton (1988)

1991 *Mathematics Subject Classification.* 62-07, 62H30, 68Q25, 68U10, 85-04.

Key words and phrases. multiresolution analysis, spatially-constrained clustering, hierarchical segmentation, hierarchical quantization.

This paper is in final form and no version of it will be submitted for publication elsewhere.

or Slezak et al. (1992). A reasonable definition of (binary) contiguity when using 2-dimensional image data is the set of four nearest neighbor pixels (north, south, east, west) of a given pixel; or the eight nearest neighbor pixels.

Cartographically-referenced data may adopt a different perspective, replacing the regular pixel increments in x and y with continuous values. Thus a vector f is located at position (x, y) in the real plane. Contiguity may be explicitly defined by the user: location (x, y) is externally defined by the user to be contiguous to location (x', y'). Other stratagems may be specified, perhaps based on a distance in the representational plane between (x, y) and (x', y').

As an alternative to binary contiguity, continuous contiguity constraints have been studied (e.g. Perruchet, 1983). In this case, we consider a compound dissimilarity formed from the product of the continuous contiguity, δ, and the dissimilarity defined in intensity space, d: $D(i, j) = \delta(i, j)d(i, j)$. However discrete ("either-or") contiguity is more mathematically tractable: in particular, it does not raise the problem of the appropriate relative weighting (or normalization) of d relative to δ.

Various generalizations of the frameworks sketched out so far are feasible, and in fact may be immediate. These include data associated with locations on the sphere rather than the Euclidean plane; 3-dimensional, or 1-dimensional (i.e. spectral), image data; etc.

2. Hierarchical Regionalization

Hierarchical clustering algorithms are widely-used for the following reasons: (i) they allow for a multiresolution or multiscale analysis of the data; (ii) they order clusters not just by set inclusion but also by agglomeration criterion value, and therefore impose a particular model on the data; (iii) each partition may be regarded as an approximation to the original data if data values are replaced by the cluster mean; and (iv) even if a partition of the data is desired, an agglomerative algorithm presents us with one, stepwise optimal or greedy, way to do this. Unconstrained hierarchical clustering methods have been "amended" to cater for constraints of various types (Murtagh, 1985). Although contiguity-constrained clustering is a leading problem, constraints other than contiguity ones have been considered by some authors (Ferligoj, 1993).

A problemsome characteristic of stepwise agglomerative procedures, whether constrained or not, is that they may give rise to inversions or reversals. This is when there is non-monotonic variation in the cluster (region) criterion value. In Figure 1, q is merged with r, and later s joins this cluster. But if the dissimilarity between s and the cluster is less than the dissimilarity between q and r then we have a representational dilemma, as shown in Figure 1. This has particular relevance if we cut the hierarchy in order to derive a partition: there is no longer an unequivocal way to do this.

Among hierarchical algorithms which are motivated by widely- and long-used unconstrained hierarchical algorithms there are only two quite distinct contiguity-constrained algorithms which are guaranteed to avoid inversions. These are a contiguity-constrained complete link method, and a differently-premised contiguity-constrained single link method. Batagelj and Ferligoj (1982) adopt a powerful perspective on this theme. They take the Lance-Williams dissimilarity update formula (which embraces many well-known agglomerative criteria) and show that the

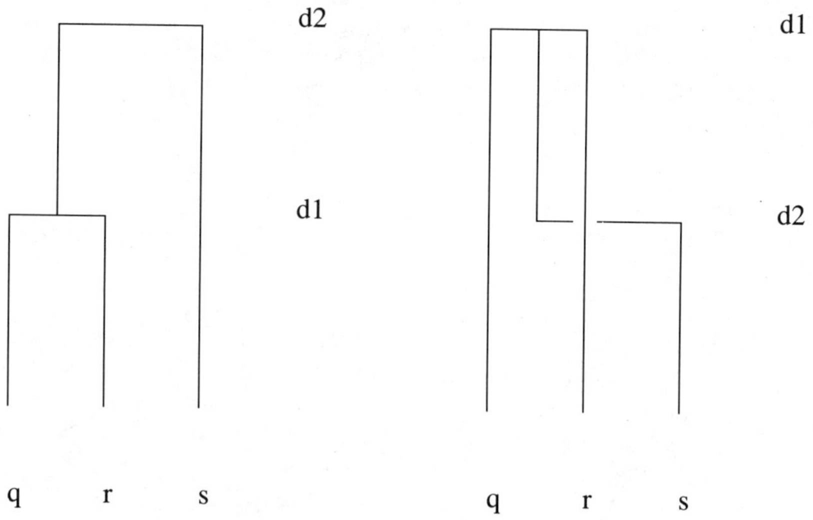

FIGURE 1. An example of an inversion or reversal in a sequence of agglomerations. We do not have: $d1 < d2$.

contiguity-constrained complete link method is the only widely-used criterion not to give rise to inversions when weakly-constrained in the manner defined here. They furthermore give conditions for a variant of the Ward minimum variance method to be similarly inversion-free. It has not been shown, however, if the latter method is meaningful in practice.

Consider an object set, I. (Thus the indices of I, $i \in I$, range over all positional couples (x, y) in the 2-dimensional case considered above.) Define G as a contiguity graph: i.e. $G : I \times I \longrightarrow \{0, 1\}$ with $G(i, i') = 1$ if object i is contiguous to i', and 0 otherwise. Now, initially the members of I are singleton clusters, but in the course of agglomerative clustering they become regions. How is contiguity propagated? We consider two cases, leading to two broad clustering strategies.

Objects i or i' may be non-singleton clusters or regions. Let a member of region i be denoted by p, and a member of region i' be denoted by q. Then an agglomerative criterion will be based on $d(p, q)$ where d is some distance or even dissimilarity. (A *dissimilarity* is a positive semi-definite mapping of the arguments onto the non-negative reals; a *distance* in addition satisfies the triangular inequality.)

Case 1: $\min d(i, i') = d(p, q)$ for some $p \in i$ and for some $q \in i'$, such that $(p, q) \in G$. In this case we require that the link between clusters which define their closest dissimilarity be a contiguous link. Thus the definition and redefinition of clusters or regions is based on a privileged edge in the contiguity graph, G, at each agglomeration; this is an additional contiguity-related constraint which comes into play when dealing with regions.

Case 2: $\min d(i, i')$ implies that

$$\exists p \in i, \exists q \in i' \quad s.t. \quad (p, q) \in G$$

In this case we do not require that the link between clusters which define their closest dissimilarity be in the contiguity graph, but we do require that some other link be in this graph.

The first of these is a "strong" contiguity constraint, and the latter is a "weak" contiguity constraint. In the next section we will associate these, respectively, with the single link agglomerative criterion, and with a complete link criterion. To avoid misunderstanding, note that "strong" contiguity will be associated with single link clustering, and "weak" contiguity with complete link and other methods.

We assume that any reference to "traditional" clustering algorithms will be sufficiently clear: background reading may be found in Murtagh (1985b).

3. Inversions or Reversals

Whenever stepwise agglomerations in hierarchical clustering are constrained to be between clusters which are contiguous, the problem of inversions is likely. This problem is also referred to as reversals, or non-monotonic increase in cluster criterion value. Such a situation is when $d(q \cup r, s) \not\geq d(q, r)$ for three clusters q, r, and s; where q and r agglomerate to form $q \cup r$; and where the cluster criterion value (e.g. minimal compactness) is related to the dissimilarity d between clusters. In using the common agglomerative clustering criteria, with the restriction that only contiguous clusters can merge, inversions tend to arise since a previously forbidden merger between two similar clusters may be permitted by changes in the contiguity relation. The presence of inversions in a hierarchy is disadvantageous: it makes difficult the interpretation of partitions and the definition of dissimilarity between clusters. It also may prevent unique solutions when differing algorithmic implementations are used.

The strong contiguity-constrained clustering method, or the contiguity-constrained single link method, is as follows. At each agglomeration, fuse together the two regions (or clusters) of least interconnecting dissimilarity, such that this dissimilarity is between a pair of contiguous *objects* (Case 1 above). Initially all regions are singletons. Each agglomeration in this method is necessarily between a pair of contiguous objects. Hence, given the contiguity graph G where each edge connecting a pair of objects is weighted by the dissimilarity between the objects, we see that this method may be specified as follows: obtain the minimal spanning tree (MST) of the weighted graph, G; and subsequently transform this into the single link hierarchy. A simple proof that this method cannot present inversions is to replace the dissimilarities between all pairs of non-contiguous objects by some arbitrarily large ("infinite") value. The construction of the single link hierarchy on this amended set of dissimilarities is well-defined, in the sense that, at all stages, the traditional algorithm can be employed and, assuming the contiguity graph is connected, infinite dissimilarities will never be used as cluster criterion (i.e. connectivity) values.

References to the use of this contiguity-constrained single link and minimal spanning tree method include: Monestiez (1977) and Fischer (1980). Closely related methods under the headings of "mode estimation", "mode-seeking clustering" or "density-contour tree" include: Hartigan (1975, pp. 205–212), Kittler (1976), Johnston et al. (1979), and Shaffer et al. (1979).

An alternative to propagating the contiguity constraint permits agglomerations between a pair of clusters such that there is a contiguity link between at least one member of each of the clusters. Of the major hierarchical methods, only the complete link method excludes the possibility of inversions when constrained in this manner. We have:

Proposition: *The complete link method, with the constraint that at least one member of each of the two clusters to be agglomerated be contiguous, is guaranteed not to give rise to inversions.*

For proof, see Ferligoj and Batagelj (1982). For another proof, see Murtagh (1985b).

Many traditional agglomerative methods use the Lance-Williams dissimilarity update formula to propagate dissimilarities defined for objects onto clusters; i.e. they define $d(q \cup r, s)$ as a function of $d(q, r)$, $d(q, s)$ and $d(r, s)$. Thus one implementation strategy is to keep all relevant dissimilarities constantly updated as the agglomerations proceed. An alternative implementation strategy, valid for certain methods, is to replace clustered objects with some average cluster representative. For image data, this strategy is thus to define an average pixel value whenever a cluster is formed. These strategies have been referred to as "stored dissimilarity" versus "stored data" approaches. Further details, including which methods equally well support either strategy, may be found in Murtagh (1985b).

The centroid and minimum variance (Ward) methods, for instance, specify the new cluster representative value as $(n_1 g_1 + n_2 g_2)/(n_1 + n_2)$ where g_1 and g_2 are the vector values associated the agglomerands; and where n_1 and n_2 are the cardinalities associated with the agglomerands. The dissimilarities defining the agglomeration criterion value are respectively: $\| g_1 - g_2 \|^2$, and the latter multiplied by $n_1 n_2/(n_1+n_2)$. These results hold for the Euclidean metric. Contiguity-constrained agglomerative regionalization, based on the minimum variance or Ward method, is used by Beaulieu and Goldberg (1989) and by Tilton (1990).

4. Reciprocal Nearest Neighbor Strategies

The use of "nearest neighbor chains" and "reciprocal nearest neighbors" are implementation strategies which have proven useful for a range of commonly-used hierarchical clustering methods. For certain criteria, they provide alternative algorithms which yield *identical* results (Murtagh, 1992). They often give rise to $O(n^2)$ time complexity, in some cases coupled with $O(n)$ storage complexity, and the potentially biggest gain of all is that they may facilitate parallel implementation.

The nearest neighbor chain, and reciprocal nearest neighbors, are defined as follows. For any set of points, x_1, x_2, ..., x_k, consider $x_2 = NN(x_1)$, $x_3 = NN(x_2)$, ... , $x_k = NN(x_{k-1})$, where NN is the nearest neighbor (for computational convenience, required to be unique). Note that no condition whatsoever is set on the starting point. Clearly we will always have at least two members in such a nearest neighbor chain; and such a chain will always end with a reciprocal nearest neighbor pair (RNN), i.e. $x_k = x_{k-2}$. If an agglomerative clustering criterion respects the *reducibility property*, i.e.

$$d(p,q) < d(p,r) \text{ and } d(p,q) < d(q,r) \Rightarrow d(p,q) < d(p \cup q, r)$$

then *inversions cannot arise.* That is to say, agglomerations may be carried out *locally*, i.e. whenever and wherever a RNN is found. For unconstrained hierarchical clustering, the single link method, the complete link method, and Ward's minimum variance method, all adhere to this reducibility property. When we consider such criteria subject to the weak contiguity constraint, then we believe that the complete link hierarchy is the only one which, if constructed using RNNs, is guaranteed to be inversion-free. Thus, an identical hierarchy is obtained compared to the use of any alternative implementation strategy; and we are free to use the RNN approach if we have good computational reasons to do so.

Parenthetically, we remark that the NN-chain and RNNs open up two broad lines of attack on the problem of constructing a hierarchy:

(1) Build a NN-chain; carry out an agglomeration whenever a RNN pair is found; continue the NN-chain from the unagglomerated member just before the RNN pair.

(2) Determine the NNs of all points; thereby determine all RNNs; agglomerate these (using the "stored data" or "stored dissimilarity" approaches discussed by Anderberg, 1973); continue.

We now return to the issue of the contiguity-constrained complete link method offering the possibility of implementation using RNNs.

Proposition: *The complete link method, with the constraint that at least one member of each of the two clusters to be agglomerated be contiguous, and the clusters of which are agglomerated RNNs, is guaranteed not to give rise to inversions.*

Proof: Consider three clusters, possibly singletons, in a NN-chain as follows: $i \longrightarrow j \longrightarrow k \longrightarrow j$, where this notation implies that $j = NN(i)$ etc. We see that k and j form a RNN pair.

To have obtained this set of relationships, subject to the contiguity constraint, we must have: (i,j) is a contiguous couple, as is (j,k); and $d(i,j) > d(j,k)$.

But is (i,k) contiguous? Maybe not. What if this couple were not contiguous, and furthermore what if the contrary (i.e. (i,k) being contiguous) could have differently affected the hierarchy? Such an effect on the hierarchy would be the case if $d(i,k) < d(j,k)$, implying that i and k should have been agglomerated before j and k. In the presence of the contiguity constraint, can this scenario come back to haunt us?

The one potentially difficult situation which we are examining has the following dissimilarity tie-up:

$$d(i,k) < d(j,k) < d(i,j)$$

An inversion is possible if: $d(j,k) \not\leq d(j \cup k, i)$. By choosing $d(j \cup k, i) = \max\{d(i,j), d(i,k)\}$, i.e. the complete link criterion, we bypass the difficulty. Under our various assumptions, we have here that $d(j \cup k, i) = d(i,j)$.

In carrying out an agglomeration using the complete link criterion, we "cover over the tracks" sufficiently well to rule out any infelicities unexpectedly arising through the enforcement of the contiguity relation.

End of proof

The NN-chain approach ensures $O(n^2)$ time performance in the non-constrained case (Murtagh, 1985a). For 2-dimensional image data, the basic algorithm is as follows. A label array, of the same dimensions as the input image array, is used, which is initialized to unique integer sequence numbers.

(1) Build a NN-chain from an arbitrary pixel, allowing a NN to be found among the 4 or 8 NNs of a given pixel. This involves: (i) selecting a pixel; (ii) determining all neighbors; (iii) excluding neighbors which are similarly labeled to the initial pixel; (iv) for all unique neighbor labels, determine dissimilarities between the corresponding regions; (v) find the minimum such dissimilarity.

(2) When a RNN pair is found, agglomerate them: i.e. remove them from the NN-chain; put the information relative to the agglomeration into the hierarchy data structure; and relabel appropriately the label array.

The time complexity of the above algorithm is studied as follows. The number of growths and contractions of the NN-chain, combined, is $O(n)$ (Murtagh, 1985b, p. 85). For each growth, a NN must be found. How many neighbors can a region have, at any stage of the processing? Even for planar graphs, this is $O(n)$. Thus a worst-case $O(n^2)$ complexity ensues for the overall algorithm, even in the contiguity-constrained case.

However we have a further problem when treating large datasets. The efficient implementations of unconstrained hierarchical methods which are based on the above NN-chain algorithm (and which, moreover, constitute the most efficient, general-purpose applications of these algorithms to date) additionally make use of storage of dissimilarities. This is so, since otherwise there is much redundancy in repeated dissimilarity calculation at various stages of the hierarchy-building process. Image data, or large point pattern sets, do not afford such a luxury. A small image of, say, 100×100 gives 10000 pixels, and dissimilarities among these would necessitate specific memory management procedures.

5. Application

The Kohonen self-organizing feature map (SOFM) can be regarded as a generalization of k-means partitioning. A set of cluster centers are iteratively determined, such that they best represent a set of input vectors. The homogeneity criterion used is a minimal distance one. In addition, the cluster centers are arranged by proximity. One way to do this is to have them associated with the intersections of a regular grid in the plane. The grid dimensions, e.g. 50×50, are fixed by the user. A grid of such dimensions implies a partitioning into 2500 clusters. Various links between the SOFM method, on the one hand, and partitioning methods and dimensionality-reduction methods, on the other hand, are explored in Murtagh and Hernández-Pajares (1994).

With such a large number of clusters, it may be desirable to use some further, complementary analysis method on the SOFM output. Ultsch (1993a, 1993b) uses

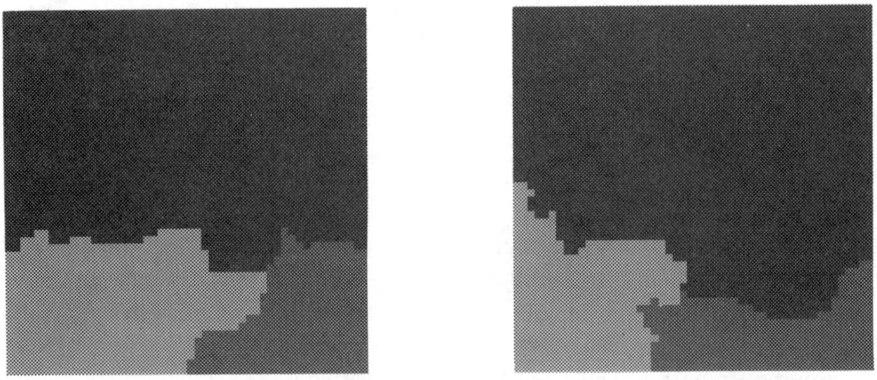

FIGURE 2. Results of two contiguity-constrained clusterings of a 50×50 image. This image represents the output of a Kohonen SOFM, applied to a large astronomical data array. Shown in both cases is a segmentation into 3 clusters.

a graphical method to do this. Given that the locations of the clusters are already related to their proximities, we find that contiguity-constrained clustering is an appropriate way to automate this phase of the analysis.

An application is discussed in Murtagh (1994) of the analysis of about a quarter million 4-dimensional object vectors. A 50×50 SOFM output regular grid is used. The 2500 4-dimensional cluster centers associated with this are then analyzed using a contiguity-constrained clustering method.

For consistency with the SOFM method, the agglomerative criterion used is the centroid one. Other possibilities are also investigated, such as a minimum variance homogeneity criterion. Figure 2 shows two segmentations of the 50×50 cluster centers, into three final clusters. One of the results in Figure 2 is based on the contiguity-constrained clustering of 2500 equi-weighted cluster centers. The other result in Figure 2 is based on a weighted version of this method, which additionally takes into account the numbers of input objects associated with each of the 2500 cluster centers. Doing this is made possible by use of the formulas discussed at the end of section 3, for cluster representative value, and for dissimilarity.

The centroid and minimum variance methods employed for the agglomerative criteria are both subject to inversions (cf. section 3). To preempt problems arising with RNN or NN-chain strategies, we instead used the more traditional approach.

This involved a global search for the minimum dissimilarity at each agglomeration. Various other operations were, however, based on array operations. In particular the calculation of all dissimilarities (from which a new agglomerable pair would be chosen) was carried out in an efficient way. Murtagh (1994) gives more details.

6. Conclusions

We have reviewed some major directions in the literature on stepwise hierarchical regionalization. We investigated whether or not reversals occurred in unconstrained hierarchical clustering, and in the case of their contiguity-constrained cousins. We sought implementations based on notions of reciprocal nearest neighbors and nearest neighbor chains, which have proven very effective in the unconstrained case.

We have shown how a complete link criterion is apparently the only one to guarantee no inversions when agglomerative clustering is subject to a weak contiguity constraint. However this method does not allow for a natural region representative. The centroid and minimum variance criteria do allow a natural definition of cluster and region representative, and this facilitates efficient implementation based on use of array operations.

References

1. M.R. Anderberg, *Cluster Analysis for Applications*, Academic Press, New York (1973).
2. J.-M. Beaulieu and M. Goldberg, Hierarchy in picture segmentation: a stepwise optimization approach, *IEEE Transactions on Pattern Analysis and Machine Intelligence*, **11** (1989) 150–163.
3. A. Ferligoj and V. Batagelj, Clustering with relational constraint, *Psychometrika*, **47** (1982) 413–426.
4. A. Ferligoj, Constrained clustering: theory and application, unpublished ms., 20 pp. (1993).
5. M.M. Fischer, Regional taxonomy, *Regional Science and Urban Economics*, **10** (1980) 503–537.
6. J. Hartigan, *Clustering Algorithms*, Wiley, New York (1975).
7. B. Johnston, T. Bailey and R. Dubes, A variation on a nonparametric clustering method, *IEEE Transactions on Pattern Analysis and Machine Intelligence*, **1** (1979) 400–408.
8. J. Kittler, A locally sensitive method for cluster analysis, *Pattern Recognition*, **8** (1976) 23–33.
9. P. Monestiez, Méthode de classification automatique sous contraintes spatiales, *Statistique et Analyse des Données*, **2** (1977) 75–84.
10. F. Murtagh, A survey of algorithms for contiguity-constrained clustering and related problems, *The Computer Journal*, **28** (1985a) 82–88.
11. F. Murtagh, *Multidimensional Clustering Algorithms*, Physica-Verlag, Würzburg (1985b).
12. F. Murtagh, Comments on 'Parallel algorithms for hierarchical clustering and cluster validity', *IEEE Transactions on Pattern Analysis and Machine Intelligence*, **14** (1992) 1056–1057.
13. F. Murtagh, Interpreting the Kohonen self-organizing feature map using contiguity-constrained clustering, *Pattern Recognition Letters* (1994), submitted.
14. F. Murtagh and M. Hernández-Pajares, The Kohonen self-organizing feature map method: an assessment, *Journal of Classification* (1994), forthcoming.
15. P.M. Narendra and M. Goldberg, Image segmentation with directed trees, *IEEE Transactions on Pattern Analysis and Machine Intelligence*, **2** (1980) 185–191.
16. Ch. Perruchet, Constrained agglomerative hierarchical classification, *Pattern Recognition*, **16** (1983) 213–217.
17. E. Shaffer, R. Dubes and A.K. Jain, Single-link characteristics of a mode-seeking clustering algorithm, *Pattern Recognition*, **11** (1979) 65–70.
18. E. Slezak, A. Bijaoui and G. Mars, Identification of structures from galaxy counts – use of the wavelet transform, *Astronomy and Astrophysics*, **227** (1990), 301–316.

19. J.C. Tilton, Image segmentation by iterative parallel region with application to data compression and image analysis, in *Proc. Symposium on the Frontiers of Massively Parallel Computation*, NASA, Greenbelt, MD (1988) 357–360.
20. J.C. Tilton, Image segmentation by iterative parallel region growing, *Information Systems Newsletter* (Feb. 1990) 50–52.
21. A. Ultsch, Knowledge extraction from self-organizing neural networks, in O. Opitz, B. Lausen and R. Klar, Eds., *Information and Classification*, Springer-Verlag, Berlin (1993a), 301–306.
22. A. Ultsch, Self-organizing neural networks for visualization and classification, in O. Opitz, B. Lausen and R. Klar, Eds., *Information and Classification*, Springer-Verlag, Berlin (1993b), 307–313.
23. G.J.G. Upton and B. Fingleton, *Spatial Data Analysis by Example. Volume 1. Point Pattern and Quantitative Data*, Wiley, New York (1988).
24. S.W. Wharton, A generalized histogram clustering scheme for multidimensional data, *Pattern Recognition*, **16** (1983) 193–199.

Astrophysics Division, Space Science Department, European Space Agency

Current address: Space Telescope – European Coordinating Facility, European Southern Observatory, Karl-Schwarzschild-Str. 2, D-85748 Garching, Germany

E-mail address: fmurtagh@eso.org

Part II

Target Tracking

DIMACS Series in Discrete Mathematics
and Theoretical Computer Science
Volume 19, 1995

IMAGE SEGMENTATION BASED ON OPTIMAL LAYERING FOR PRECISION TRACKING

ANIL KUMAR, YAAKOV BAR-SHALOM AND ELIEZER ORON

ABSTRACT. We present a method for precision tracking of a low observable target based on data obtained from imaging sensors. The image is assumed to consist of gray level intensities in each pixel. The intensity range is divided into a target layer and background layers. A binary image is obtained and grouped into clusters using image segmentation techniques. Using the centroid measurements of the clusters, the Probabilistic Data Association Filter (PDAF) is employed for tracking the centroid of the target.

The boundaries of the target layer are optimized by minimizing the Bayes risk. A closed-form analytical expression is obtained for the single frame based centroid measurement noise variance.

The simulation results presented validate both the expression for the measurement noise variance as well as the performance predictions of the proposed tracking method.

The method is first illustrated on a dim synthetic target occupying about 80 pixels within a 64×64 frame in the presence of noise background which can be stronger than the target. The usefulness of the method for practical applications is demonstrated for a highway traffic surveilance problem by considering a sequence of real target images (a moving car) of about 20 pixels in size, in a noisy urban environment.

1. INTRODUCTION

This work considers the problem of tracking a small target (of the order of tens of pixels), where the variables of interest are the location and velocity of its centroid.

The tracking of an extended hot spot target using imaging sensors was reported in [5,9,10,13,14]. These methods assume that the target covers about 10 pixels. In [3], the utility of nonlinear filtering approach for extracting motion (rotational and translational) parameters of a known rigid body, based on a sequence of images was demonstrated. In our previous work [11] we proposed a technique for precision centroid tracking for small targets (up to about 100 pixels) which is less sensitive to target intensity. This technique is robust in the sense that the noise background can be stronger than the target.

In [11], the image was divided into predetermined layers of gray level intensities and thresholded. Typically the target is predominantly in one layer and the back-

---

1991 *Mathematics Subject Classification.* Primary 68U10.

Research sponsored by ONR/BMDO-IST Contract N00014-91-J-1950 and by AFOSR Grant F049620-92-J-0150

ground has a wider spread. A binary image was obtained and grouped into clusters using image segmentation techniques. The association of the various clusters to the track to be estimated relies on both the motion and pattern (object) recognition characteristics of the target. Using the centroid measurements of the clusters, the Probabilistic Data Association Filter (PDAF) [2,4] is employed for state estimation yielding improved centroid position estimates as well as velocity estimates.

In the above method the critical step is the division of gray level intensity range into several layers because the binary image and the clusters obtained from it depend on the layer boundaries. In this paper, we address the problem of layering the image optimally by minimizing the misclassification probability. As shown later, this will reduce the single frame based centroid measurement noise variance at least by a factor of 2 compared to [11]. The variance of the single frame based centroid position is given by a closed-form analytical expression. This is an important result because the measurement centroid variance is a vital tracking filter design parameter. Even though statistical models for computation of the target centroid of an image were presented in [15], they were based on the assumption that there is a predetermined target region within which the centroid is computed from binarized pixel intensities. In our case there is no well defined target region — the segmentation (which consists of binarization and clustering) yields the target region. In other words, the method presented here is substantially more general.

In Section 2, the general expression for the single frame based centroid measurement noise variance, based on the target and background intensity statistical characterizations is presented. The optimal layering algorithm used for image segmentation and the clustering algorithm used to reduce the binary image to clusters are the topic of Section 3. In Section 4, the closed form expression for the single frame based centroid measurement noise of variance of a random target cluster is presented. Also, for different target and noise background statistics, the single frame based centroid measurement noise variance obtained by analytical computations is compared to the values obtained through simulations. The state space model used for tracking and the simulation results validating the performance of the tracking filter are presented in Section 5 for both a synthetic as well as a real target. Finally, conclusions are given in Section 6.

2. Centroid of a Random Cluster

The General Case. Consider a cluster of N points (pixels) in a Cartesian coordinate system. Each point is denoted by a single index $i, i = 1, ..., N$. If each point has an intensity I_i, then the centroid of cluster is defined as

$$x_{n_c} = \frac{\sum_{i=1}^{N} x_{n_i} I_i}{\sum_{i=1}^{N} I_i} \tag{2.1}$$

where x_{n_i} is the nth coordinate of pixel i. When I_i are independent random variables (RVs) then x_{n_c} given by (2.1) being a ratio of two RVs (n^{th} coordinate first moment of the intensity and the total intensity) , is also a RV. It is convenient to use the true centroid as the coordinate system origin [5]. That is,

$$\sum_{i=1}^{N} x_{n_i} \bar{I}_i = 0 \tag{2.2}$$

where $\bar{I}_i$ denotes the mean of I_i. Then the variance of the centroid reduces to [11]

$$(2.3) \qquad var(x_{n_c}) = \frac{\sum_{i=1}^{N} x_{n_i}^2 var(I_i)}{\left(\sum_{i=1}^{N} \bar{I}_i\right)^2}$$

and x_{n_i} is the nth coordinate of pixel i w.r.t. the center of mass. In spite of above assumption (2.2), expression (2.3) turns out to be very useful as shown later.

Centroid of a Binary Image. In our model, the clustering points are the image pixels. The gray level image is transformed to a binary image, with new binary value β_i, in pixel i, by a hard limiter according to:

$$\beta_i = 1 \ \{I_L' \leq I_i \leq I_H'\}$$

$$(2.4) \qquad \beta_i = 0 \ \{\text{otherwise}\}$$

where I_L' and I_H' are the lower and upper threshold limit for the intensity for the initial choice of the "target layer". The probability p_i of a unit value in pixel i, called detection probability of this pixel by the threshold rule, is

$$(2.5) \qquad P\{\beta_i = 1\} = p_i$$

$$(2.6) \qquad P\{\beta_i = 0\} = 1 - p_i$$

For example, if I_i is $\mathcal{N}(\mu, \sigma^2)$, i.e., Gaussian with mean μ and variance σ^2, then

$$(2.7) \qquad p_i = \frac{1}{\sqrt{2\pi}\sigma} \int_{I_L'}^{I_H'} e^{-\frac{(x-\mu)^2}{2\sigma^2}} dx = \Phi\left(\frac{I_H' - \mu}{\sigma}\right) - \Phi\left(\frac{I_L' - \mu}{\sigma}\right)$$

where Φ is the standard error function.

The mean and the variance of the binary value of a single pixel is

$$(2.8) \qquad \mu(\beta_i) = p_i$$

$$(2.9) \qquad var(\beta_i) = p_i(1 - p_i)$$

If the binary image so obtained is grouped into clusters using an appropriate clustering algorithm, then the analytical expression for the variance of the centroid of a cluster of N pixels, using (2.3) with I_i replaced by β_i is, for the nth coordinate

$$(2.10) \qquad var(x_{n_c}) = \frac{\sum_{i=1}^{N} x_{n_i}^2 p_i(1 - p_i)}{\left(\sum_{i=1}^{N} p_i\right)^2}$$

3. Image Segmentation and Clustering

Pixel Classification. In the method presented, it is assumed that the pixel intensity is discretized into 256 gray levels. The image is divided into several layers of gray level intensities. Each layer is characterized by upper and lower gray level limits. The threshold values I_L' and I_H' needed for implementing (2.4) are obtained based on the prior information about the target and noise pixels gray level intensities. Using the upper and lower limits of the "target layer" as threshold limits, target and noise pixel detection probabilities (p_t', p_ν') are obtained. If the target size (N_t) is known, then the *a priori* probabilities of the two classes of pixels (target and noise) are given by

$$\pi_1 = \frac{N_t p_t'}{N_t p_t' + (S - N_t)p_\nu'} \tag{3.1}$$

$$\pi_2 = 1 - \pi_1 \tag{3.2}$$

where S is the size of the image in pixels.

If the probability density of the gray level intensity value for each class, and the *a priori* probabilities of the two classes are known, then new threshold levels I_L and I_H based on a criterion that minimizes the misclassification probability can be derived [12] as follows.

In the case of a single target, the pixels can be classified into two classes (class 1 is target and class 2 is noise). The probability densities for the two classes of pixels, $p(I|1)$ and $p(I|2)$ and the *a priori* probabilities of the classes π_1 and π_2, are assumed available (this is discussed further in the next section).

If the pixels are classified according to the hard limiter rule (2.4), then the probability of misclassifying a class 2 (background) pixel as class 1 (target), i.e., the type I error, is

$$P_{e_I} = P(\mathcal{I}_t|2) \triangleq \int_{I_L}^{I_H} p(I|2)dI \tag{3.3}$$

where $\mathcal{I}_t$ is the event $I \in [I_L, I_H]$. Similarly, the probability of misclassifying a class 1 pixel as class 2 (type II error) is

$$P_{e_{II}} = 1 - P(\mathcal{I}_t|1) \triangleq 1 - \int_{I_L}^{I_H} p(I|1)dI \tag{3.4}$$

Since it is desirable to have $P_{e_I} < P_{e_{II}}$ we assign the following Bayesian costs.

$$C_{miss} = C_{21} < C_{12} = C_{fa} \tag{3.5}$$

The Bayesian risk is the weighted sum of P_{e_I} and $P_{e_{II}}$. That is,

$$\bar{C} = C_{12}\pi_2 P_{e_I} + C_{21}\pi_1 P_{e_{II}} \tag{3.6}$$

The Bayesian risk $\bar{C}$ will be minimized by differentiating (3.6) with respect to I_L and I_H and setting the result to zero. This yields the same equation for I_L and I_H, namely,

$$C_{12}\pi_2 p(I|2) = C_{21}\pi_1 p(I|1) \tag{3.7}$$

The two solutions of the above are the optimal layer boundaries. If the mathematical forms of the probability densitities of the classes are known, Eq. (3.7) can be solved analytically, else numerical techniques have to be employed.

Assume the probability densities $p(I|1)$ and $p(I|2)$ are Gaussian, namely, $p(I|1) = N(\mu, \sigma^2)$ and $p(I|2) = N(\eta, \kappa^2)$. It is convenient now to define the target layer thresholds as

$$I_L = \mu - \delta_1 \tag{3.8}$$

$$I_H = \mu + \delta_2 \tag{3.9}$$

Then the optimal δ_1 and δ_2 can be obtained by solving the set of nonlinear equations

$$\frac{C_{12}\pi_2}{\kappa} e^{-\frac{(\mu-\delta_1-\eta)^2}{2\kappa^2}} - \frac{C_{21}\pi_1}{\sigma} e^{-\frac{-\delta_1^2}{2\sigma^2}} = 0 \tag{3.10}$$

$$\frac{C_{12}\pi_2}{\kappa} e^{-\frac{(\mu+\delta_2-\eta)^2}{2\kappa^2}} - \frac{C_{21}\pi_1}{\sigma} e^{-\frac{-\delta_2^2}{2\sigma^2}} = 0 \tag{3.11}$$

Using I_L and I_H given by (3.8,3.9), with δ_1 and δ_2 from (3.10,3.11) optimal target and noise pixel detection probabilities p_t and p_ν are obtained.

For multiple targets having different probability densities the pixels can be classified into more classes.

Clustering and Centroid Estimation. Most of the standard clustering methods are difficult to implement in real time. Because of this, a clustering algorithm, which can be implemented in real time, was chosen. The clustering method that is used in the implementation is the single linkage or nearest neighbor technique [1,6,8,16]. In this technique, a pixel belongs to the cluster if it is linked to at least one other pixel of the cluster by a distance which is less or equal to a certain proximity distance. By choosing an optimum proximity distance as shown below, fewer noise clusters are obtained.

The proximity distance, denoted by d_p, affects the size, shape and number of the clusters. As shown in [11], the average distance between neighboring pixels in the target and the noise is $d_t \approx \sqrt{\frac{1}{p_t}}$ and $d_\nu \approx \sqrt{\frac{1}{p_\nu}}$ and the optimal d_p must be between these two values. That is,

$$d_t < d_p < d_\nu \tag{3.12}$$

This guarantees that most of the target pixels will be in the same cluster while most of the noise pixels stay isolated from each other. While most of the target pixels will be in the same cluster there will be a small number of noise pixels attached around its edge. Such a noisy edge increases the centroid measurement variance. As shown in [11], in practice one can use

$$d_p^* = \frac{d_t + d_\nu}{2} \tag{3.13}$$

as the approximate optimal proximity distance which gives minimum centroid measurement noise variance. This takes care of covering the gaps in the target image while minimizing the noise pixels' interference.

4. Centroid of a Random Target Cluster in Noise Background

Consider an image of size S, consisting of a target of size (area) N_t pixels within a noise background of $S - N_t$ pixels, with pixel detection probabilities p_t, p_ν for the target and the noise respectively. The intensity of the pixels is given by

$$\begin{aligned} I_i &= s_i \quad 1 \leq i \leq N_t \\ I_i &= \nu_i \quad \text{otherwise} \end{aligned} \tag{4.1}$$

In this case the image is a combination of N_t pixels of the target and $S - N_t$ pixels of the background which might have noise detections. Then the variance of the centroid estimate is, using (2.10)

$$var(x_{n_c}) = \frac{p_t(1-p_t)\sum_{i\in \mathbf{T}} x_{n_i}^2 + p_\nu(1-p_\nu)\sum_{i\in \mathbf{N}} x_{n_i}^2}{(p_t N_t + p_\nu N_\nu)^2} \tag{4.2}$$

where $\mathbf{T}$ is the set of target pixels and $\mathbf{N}$ is the set of noise pixels attached to the target following the segmentation. The key to have (4.2) useful is to use in it the set $\mathbf{N}$ - the "noise ring" around the target. This is obtained next.

Consider a circular target of radius r_t. Since the clustering algorithm is based on the proximity distance d_p, the target region can be defined as a circular region of radius $r_t + d_r$ where d_r is the expected width of the noise ring. This width, as explained in the next subsection, is calculated by considering the noise pixels as random Poisson points in two dimensions [7]. In this case, the noise background (ring of width d_r around the target) contains N_ν pixels, where

$$N_\nu = \pi\left((r_t + d_r)^2 - r_t^2\right) \tag{4.3}$$

For a circular target we have

$$\sum_{i\in \mathbf{T}} x_{n_i}^2 = 2\int_0^{r_t}\int_0^{\pi} r_t^3 cos^2(\theta) d\theta dr_t = \frac{r_t^4 \pi}{4} \tag{4.4}$$

$$N_t = \pi r_t^2 \tag{4.5}$$

Using (4.4), (4.5) and (4.2) we get the centroid variance as

$$var(x_{n_c}) = \frac{(1-p_t)p_t r_t^4 + (1-p_\nu)p_\nu\left((r_t + d_r)^4 - r_t^4\right)}{4\pi\left(p_t r_t^2 + p_\nu(r_t + d_r)^2 - p_\nu r_t^2\right)^2} \tag{4.6}$$

Expression (4.6) was used in evaluating the measurement noise variance of the target cluster. This differs from (2.40) in [11] by having the realistic (Poisson process based) d_r instead of a heuristic value denoted as r_ϵ.

Expected Width of the Noise Ring. The expected width of the noise ring d_r can be calculated as follows. The nearest neighbor clustering technique links all the pixels within the proximity distance d_p to the target cluster. The number of noise pixels within the distance d_p from the target edge is given by

$$N_r = \pi\left((r_t + d_p)^2 - r_t^2\right) \tag{4.7}$$

All the pixels in this "first ring layer" of area N_r are attached to the target cluster and constitute the first link in the chain of attachments. Any pixel outside this first layer can be linked to the target cluster if it is within d_p distance from any of the pixels in this first ring layer. This amounts to considering all the pixels in a semicircle (facing outward) of radius d_p centered at each pixel in the ring N_r with a detection. The number of pixels in the semicircle is given by its area

$$N_s = \frac{1}{2}\pi d_p^2 \tag{4.8}$$

The average number of noise pixels with detections in the first ring layer of area N_r is (with p_ν being the noise pixel detection probability)

$$\rho = N_r\, p_\nu \tag{4.9}$$

and in a semicircle of area N_s is

$$s = N_s\, p_\nu \tag{4.10}$$

If the noise pixels around the the target are considered as random Poisson points in two dimensions, then the probability of the number of links n_l being zero is given by the probability that the number of detections in the first ring layer is zero. That is,

$$P\{n_l = 0\} = P\{\text{zero detections in the first ring layer}\} = P\{n_\nu(N_r) = 0\} \tag{4.11}$$

where $n_\nu(N)$ denotes the number of noise-due detections in area N.

The probability of the number of links being 1 is the probability that the number of pixels in the first ring layer is not equal to zero times the probability that the number of pixels in a semicircle at each detection in the first layer is zero (i.e., the chain stopped after the first link). That is

$$P\{n_l = 1\} = [1 - P\{n_\nu(N_r) = 0\}]\,[P\{n_\nu(N_s) = 0\}]^\rho \tag{4.12}$$

In general, the probability of the number of links being k is

$$\begin{aligned} P\{n_l = k\} &= [1 - P\{n_\nu(N_r) = 0\}]\,[1 - P\{n_\nu(N_s) = 0\}]^\rho \\ &\quad [1 - P\{n_\nu(N_s) = 0\}]^{\rho s}\,[1 - P\{n_\nu(N_s) = 0\}]^{\rho s^2} \ldots \\ &\quad [1 - P\{n_\nu(N_s) = 0\}]^{\rho s^{k-2}}\,[P\{n_\nu(N_s) = 0\}]^{\rho s^{k-1}} \end{aligned} \tag{4.13}$$

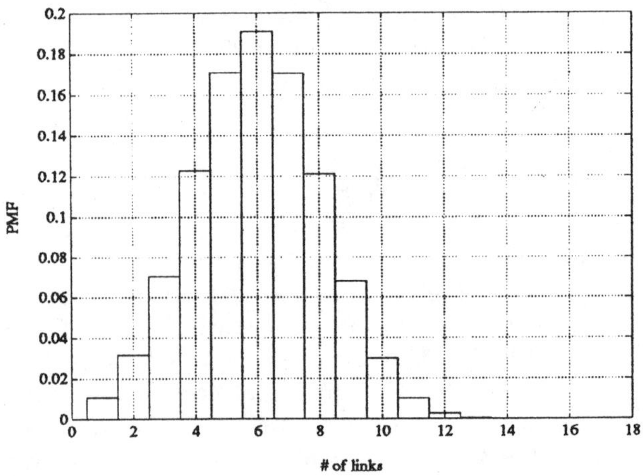

FIG. 1. Probability mass function of the length of a link around the target cluster

The above expression is meaningful for $s < 1$, i.e., when the chain tapers off.

The probability mass function (PMF) of the length of a link is shown in Fig. 1 for case A of Table 1 below. It can be observed that this PMF peaks for a chain length of 6 and the sum adds up to 1 satisfying the normalization condition for a PMF. Also the series defined by (4.13) is absolutely convergent.

The expected "outward" distance of a uniformly distributed point in a semicirlce (the expected value of the y coordinate of the points in a semicircle at the origin and pointing upward) is

$$\bar{d} = \frac{1}{N_s} \int_{x=-d_p}^{d_p} \int_{y=0}^{\sqrt{d_p^2 - x^2}} y \, dy \, dx = \frac{4 \, d_p}{3 \, \pi} \tag{4.14}$$

Once the PMF (4.13) and the expected distance (4.14) of a point in semicircle are calculated from the given parameters, the expected width of the (total) noise ring around the target can be taken as

$$d_r = \sum_{k=0}^{\infty} k \, \bar{d} \, P\{n_l = k\} \tag{4.15}$$

Examples. To demonstrate the practical utility of the analytical expression for the centroid variance (4.6) and the optimal thresholding (3.7), four different cases are considered. In Table 1, the initial threshold levels (I_L' and I_H') and optimal threshold levels (I_L and I_H) obtained using (3.7) are shown along with the target and noise pixel detection probabilities. Cases A through C correspond to a synthetic image with the following target and noise statistics:

Target pixel intensity : $\mathcal{N}(100, 10^2)$

TABLE 1. Threshold levels, pixel detection probabilities and proximity distances.

Case	I_L'	I_H'	I_L	I_H	p_t'	p_ν'	p_t	p_ν	$d_p^{*'}$	d_p^*
A	90.0	110.0	88.94	115.22	0.68	0.097	0.80	0.121	2	2
B	95.0	105.0	88.02	116.15	0.38	0.048	0.83	0.131	3	2
C	97.5	102.5	88.01	116.15	0.19	0.024	0.83	0.131	4	2
D	140.5	159.5	139.03	190.05	0.71	0.023	0.87	0.028	2	2

$$\text{Noise pixel intensity} : \mathcal{N}(50, 50^2) \tag{4.16}$$

and they differ only in the initial threshold levels prior to the optimization. The gray scale synthetic image with noise for a single scan is shown in Fig. 2(a). Case D in Table 1 corresponds to a real target image (a moving car). The real gray image is shown in Fig. 3(a). This image shows two cars moving in opposite directions crossing each other during daytime. The car moving from left to right (marked in the Fig. 3(a)) with an unknown velocity was chosen as the target. The segmented synthetic and real images are shown in Fig. 2(b) and 3(b) respectively. The statistics calculated from the first frame of the real image are as follows:

$$\text{Target pixel intensity} : \mathcal{N}(150, 9.5^2)$$

$$\text{Noise pixel intensity} : \mathcal{N}(120, 10^2) \tag{4.17}$$

For a given circular target, the measurement noise variance and the best proximity distances $d_p^{*'}$ and d_p^* obtained from initial and optimal thresholds (I_L',I_H') and (I_L, I_H), from 300 Monte-Carlo runs for different target and noise pixel detection probabilities and *a priori* probabilities π_1 and π_2 are shown in Table 1. The proximity distances shown are rounded to nearest integer. The *a priori* probabilities are obtained using (3.1), (3.2). Also, when the noise pixels are distributed around the target pixel forming a ring like structure, the analytically calculated measurement noise variances using (4.6) are shown in Table 2. From this table, it can be observed that the theoretical and experimental variances are close for all the cases. Compared with [11], the noise variances are smaller at least by a factor of 2. The variance depends on the size of the target N_t, statistical characteristics of the target and noise pixel intensities and the target pixel detection probability p_t. It also depends on the type of the clustering algorithm used in the implementation.

In [11], the results presented for the theoretical and experimental variance depend heavily on the pixel detection probabilities p_t', p_ν' which are fixed once the threshold levels I_L' and I_H' are chosen. Whereas in our case, the (image based) *a priori* information employed and the optimal thresholding give rise to almost the same pixel detection probabilities p_t, p_ν irrespective of the initial threshold levels. Since, in general, $p_t > p_t'$, the centroid variance is lower than presented in [11]. Also, note that the optimal target layer is not symmetric about the target mean intensity.

TABLE 2. Computed value and sample average of the measurement noise variance.

Case	Size	*Variance* (Theory)	(Estimated)
A	78	0.151	0.142
B	78	0.143	0.129
C	78	0.143	0.131
D	20	0.106	N/A

5. PRECISION IMAGE TRACKING

State Space Model. A nearly constant velocity state model [4] was employed to track a target moving with constant velocity. In the constant velocity model the acceleration is taken as a white process noise with variance q_x and q_y in the two cartesian coordinates. The change in velocity over a sampling period T is of the order of $\sqrt{q}T$. In the state space model only the centroid position measurements in two dimensional Cartesian coordinates were considered.

The measurement noise variance (same for each coordinate) is given by (4.6). The measurement and process noise sequences are uncorrelated. The Probabilistic Data Association Filter (PDAF) [4], was used to obtain the state estimates.

Numerical Results. The results are presented for two cases of target pixel detection probabilities and the state estimation was done for a period of 20 scans. The initial state vector of the target in the image (pixel) frame was

$$x(0) = [11.0 \quad 1.5 \quad 15.0 \quad 1.0]' \tag{5.1}$$

The initial target position was near the bottom of the image and moderate constant velocities were considered (1.5 and 1.0 pixel/scan in x and y coordinates, respectively) so that the target stays in the same frame for 20 scans. The target and noise parameters for the synthetic and real images, used in the simulations are given in (4.16) and (4.17) and the threshold levels and detection probabilities were as in Table 1.

Two cases (A & D of Table 2) with different target and noise background statistics were considered for tracking. In an image of size 64×64, Case A corresponds to a (synthetic) circular target of 78 pixels (radius of 5 pixels) and case D corresponds to a real target of 20 pixels (radius of ≈ 3 pixels). The minimum cluster size I_{min}, for cases A and D were chosen as 20 and 3, respectively. Also, as given in Table 2, the optimal proximity distances for these cases are 2 for both cases.

The PDAF was used with the following parameters. The measurement noise had a variance of 0.15 (cf. Table 2) for case A and 0.071 for case D. The target detection probability was taken as 0.97. A low level process noise $q_x = q_y = 0.025$ was assumed by the tracking filter.

Table 3 summarizes the results of 100 Monte Carlo runs in terms of achieved accuracies for case A. The results are given only for the x coordinate but the normalized state estimation error squared (NEES) pertains to the entire four dimensional state. From the last columns of Table 3, it is clear that the NEES is within the 95% probability region (based on the chi-square distribution [4]) indicating that the filter is consistent, i.e., the calculated variances match the actual errors. The results show that it is possible to achieve subpixel accuracy in the range of 0.2 pixel RMS error. Compared to 0.3 to 0.4 pixel RMS error achieved in [11] this is a significant improvement.

Note that same results are valid for cases B and C also because the optimal target and noise pixel detection probabilities are same for the first three cases. In [11] the pixel RMS error depends on detection threshold levels and hence target and noise pixel detection probabilities. Whereas in our case, optimal thresholds obtained using *a priori* information give rise to almost same pixel detection probabilities irrespective of initial threshold levels.

TABLE 3. Position and velocity variances (in steady state) for each coordinate and filter consistency verification from 100 runs.

Case	Position Variance	Velocity Variance	Average NEES	95% Region for NEES
A	0.034	0.00086	4.11	3.5-4.5

For case D, the real images were obtained by a FPA based platinum-silicide camera (resolution 256 × 256) in infrared light. The 256 × 256 images were reduced to 64 × 64. The images show two cars moving in opposite directions crossing each other during daytime. The car moving from left to right (with an unknown velocity) was chosen for tracking. The target parameters were shown in last rows of Tables 1 and 2.

Using the PDAF, the estimated values of the centroid position and velocity of the target along with the estimated accuracies are given in Figs. 4 and 5. The filter-calculated RMS estimation errors were 0.2 for each coordinate in position and 0.031 pixel/frame in velocity. Whereas in [11] the position and velocity RMS errors were 0.4 and 0.09 pixel/frame respectively. Since the ground truth is not known, in order to check the filter's consistency (validity of the filter-calculated estimation accuracy), the 10-sample time-averaged Normalized Innovation Squared (NIS) [4] was calculated and was found to be 2.24, which is within the chi-square limits [0.96, 3.14] based on $\frac{\chi^2_{20}}{10}$. Thus the filter-calculated accuracies are remarkably reliable.

These results show that in spite of the non-white and non-Gaussian noise in the image, the method was successful for this real image. The above results were obtained with the interactive software IMDAT [2].

6. CONCLUSIONS

A precise method for tracking low observable targets obtained from imaging

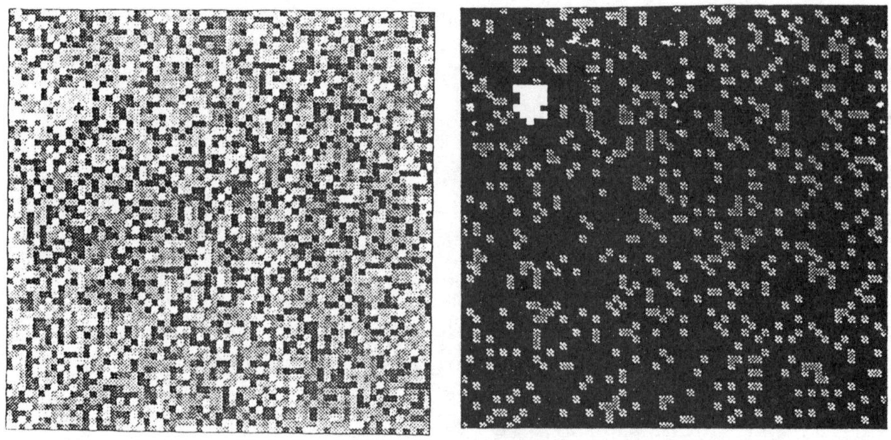

FIG. 2. (a) Synthetic gray scale image with target marked "+" (b) The segmented areas with the marked target in white and others in gray.

sensors was presented. Using segmentation and clustering techniques the image is binarized and grouped into small number of clusters. Using the optimal thresholds obtained by minimizing the Bayes risk, the image is segmented into several layers of gray level intensities.This optimal layering of the image has the following properties: (a) following the segmentation, a closed-form analytical expression is obtained for the single frame based centroid measurement noise variance; (b) in comparison to [11], the measurement noise variance is smaller by at least a factor of 2, thus improving the performance of the tracker.

The simulation results presented validate both the expression for the measurement noise variance as well as the performance predictions of the proposed tracking method. The method was first illustrated on a dim synthetic target occupying about 80 pixels within a 64 × 64 frame in the presence of noise background which can be stronger than the target. The usefulness of the method for practical applications is demonstrated by considering a sequence of real target images (a moving car) of about 20 pixels in size, in a noisy urban environment where the measurement noise variance was calculated as having 0.32 pixel RMS value.

REFERENCES

1. M. R. Anderberg (eds.), *Cluster analysis and applications*, Academic Press, 1973.
2. Yaakov Bar-Shalom, *IMDAT – Image Data Association Tracker 2.0*, Interactive software package, 1993.
3. T. J. Broida and R. Chellappa, *Estimation of object motion parameters from noisy images*, IEEE Trans. Pattern Analysis and Machine Intelligence **PAMI-8** (1986), 90–99.
4. Yaakov Bar-Shalom and T. E. Fortmann, *Tracking and data association*, Academic Press, 1988.

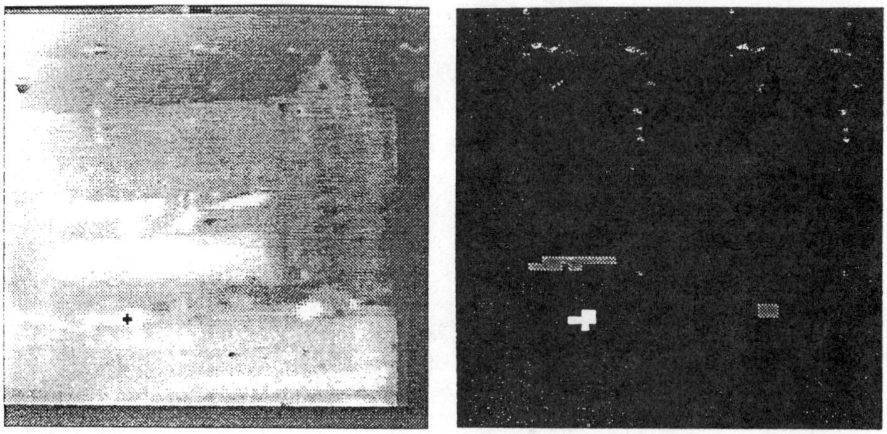

FIG. 3. (a) Real gray scale image with target marked "+" (b) The segmented areas with the marked target in white and others in gray.

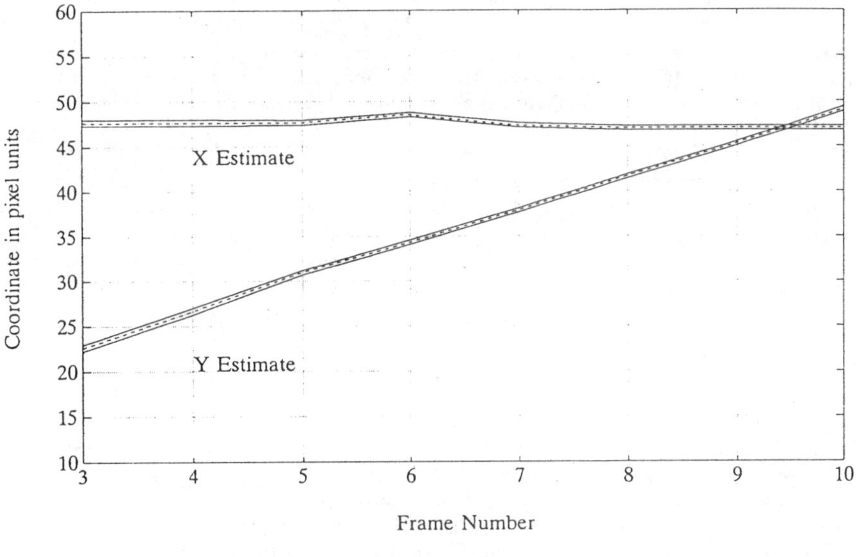

FIG. 4. Estimate with $\pm 1\sigma$ error for position.

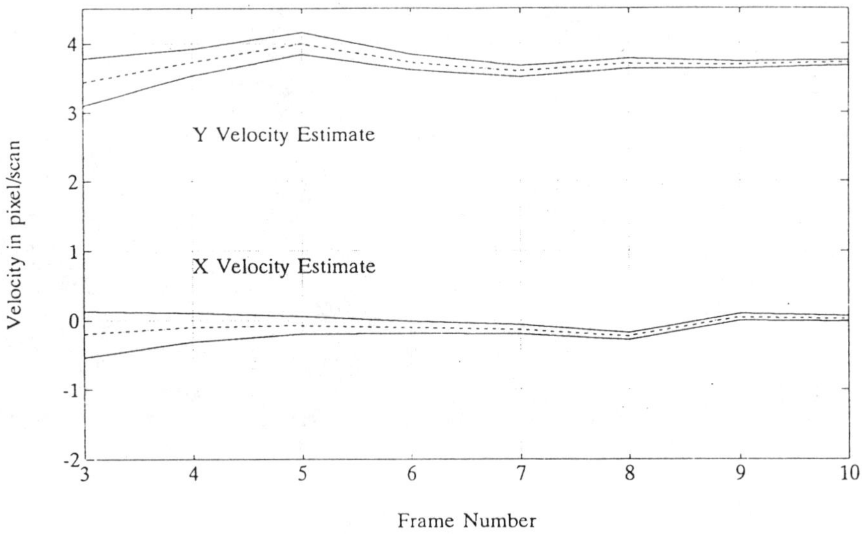

FIG. 5. Estimate with $\pm 1\sigma$ error for velocity.

5. Yaakov Bar-Shalom, H. M. Shertukde and K. R. Pattipati, *Use of measurements from an imaging sensor for precision target tracking*, IEEE Trans. Aerospace and Electronic Systems **AES-25** (1989), 863-871.
6. Anil K Jain and Richard C. Dubes, *Algorithms for clustering data*, Prentice Hall, 1988.
7. Anil Kumar, Yaakov Bar-Shalom and Eliezer Oron, *Precision tracking based on segmentation with optimal layering for imaging sensors*, submitted to IEEE Trans. Pattern Analysis and Machine Intelligence (1993).
8. P. R. Krishnaiah and L. N. Kanal, *Handbook of statistics* II, North-Holland, 1982.
9. P. S. Maybeck, R. L. Jensen and D. A. Harnly, *An adaptive extended kalman filter for target image tracking*, IEEE Trans. Aerospace and Electronic Systems **AES-17** (1981), 173-180.
10. P. S. Maybeck and R. I. Suizu, *Adaptive tracker field-of-view variation via multiple model filtering*, IEEE Trans. Aerospace and Electronic Systems **AES-21** (1985), 529-539.
11. Eliezer Oron, Anil Kumar and Yaakov Bar-Shalom, *Precision tracking with segmentation for imaging sensors*, IEEE Trans. Aerospace and Electronic Systems **AES-29** (1993).
12. A. Rosenfeld and A. C. Kak, *Digital picture processing* I *& (II)*, Academic Press, 1982.
13. H. M. Shertukde and Yaakov Bar-Shalom, *Tracking of crossing targets with imaginig sensors*, IEEE Trans. Aerospace and Electronic Systems **AES-27** (1991), 582-592.
14. D. M. Tobin and P. S. Maybeck, *Enhancements to a multiple model adaptive estimator/image tracker*, IEEE Trans. Aerospace and Electronic Systems **AES-24** (1988), 417-426.
15. Don R. VanRheeden and Richard A. Jones, *Noise effects on centroid tracker aim point estimation*, IEEE Trans. Aerospace and Electronic Systems **AES-24** (1988).
16. J. Zupan, *Clustering of large data sets*, Research Studies Press, 1982.

DEPARTMENT OF ELECTRICAL AND SYSTEMS ENGINEERING, UNIVERSITY OF CONNECTICUT, STORRS, CT 06269-3157

E-mail address: ybs@ee.uconn.edu

MAYFLOWER COMMUNICATIONS COMPANY, INC., 80 MAIN STREET, READING, MA 01867

IAI, ISRAEL

DIMACS Series in Discrete Mathematics
and Theoretical Computer Science
Volume 19, 1995

Multidimensional Assignments and Multitarget Tracking

AUBREY B. POORE

ABSTRACT. The ever-increasing demand in surveillance is to produce highly accurate target and track identification and estimation in real-time, even for dense target scenarios and in regions of high track contention. The use of multiple sensors, through more varied information, has the potential to greatly enhance target identification and state estimation. For multi-target tracking, the processing of multiple scans all at once yields high track identification. However, to achieve this accurate state estimation and track identification, one must solve an NP-hard data association problem of partitioning observations into tracks and false alarms in real-time. An objective in this work is to formulate a general class of these data association problems as multidimensional assignment problems to which new, fast, near optimal, Lagrangian relaxation based algorithms are applicable. The dimension of the formulated assignment problem corresponds to the number of data sets being partitioned with the constraints defining such a partition. The linear objective function is developed from Bayesian estimation and is the negative log posterior or likelihood function, so that the optimal solution yields the maximum a posteriori estimate. After formulating this general class of problems, the equivalence between solving data association problems by these multidimensional assignment problems and by the currently most popular method of multiple hypothesis tracking is established. Track initiation and track maintenance using an N-scan sliding window are then used as illustrations. After a brief discussion of the Lagrangian relaxation based algorithms, numerical simulations for a model problem are presented to demonstrate the potential effectiveness of the solution procedures.

1991 *Mathematics Subject Classification.* 90C08, 90C27.

This work was partially supported by the Federal Systems Company of the IBM Corporation in Boulder, CO and Owego, NY and by the Air Force Office of Scientific Research through AFOSR Grant Numbers AFOSR-91-0138 and AFOSR-93-0133.

1. Introduction

The central problem in any multitarget/multisensor surveillance system is the data association problem of partitioning the observations into tracks and false alarms [**4,7,34,38**]. Current methods for multitarget tracking generally fall into two categories: sequential and deferred logic. Methods for the former include nearest neighbor, one-to-one or few-to-one assignments, and all-to-one assignments as in the joint probabilistic data association (JPDA) [**4**]. For track maintenance, the nearest neighbor method is valid in the absence of clutter when there is little or no track contention, i.e., when there is little chance of misassociation. Problems involving one-to-one or few-to-one assignments are generally formulated as (two dimensional) assignment or multi-assignment problems for which there are some excellent optimal algorithms [**5,6,18**]. This methodology is real-time but can result in a large number of partial and incorrect assignments in dense or high contention scenarios, and thus incorrect track identification. The difficulty is that decisions, once made, are irrevocable, so that there is no mechanism to correct misassociations. The use of all observations in a scan or within a neighborhood of a predicted track position (e.g., JPDA) to update a track moderates the misassociation problem and has been highly successful for tracking a few to a moderate number of targets in dense clutter [**4**].

Deferred logic techniques consider several data sets or scans of data all at once in making data association decisions. At one extreme is batch processing in which all observations (from all time) are processed together, but this is computationally too intensive for real-time applications. The other extreme is sequential processing. Deferred logic methods between these two extremes are of primary interest in this work. The principal deferred logic method used to track large numbers of targets in low to moderate clutter is called multiple hypothesis tracking (MHT) in which one builds a tree of possibilities, assigns a likelihood score to each track, develops an intricate pruning logic, and then solves the data association problem by an explicit enumeration scheme. The use of these enumeration schemes to solve this NP-hard combinatorial optimization problem in real-time is inevitably faulty in dense scenarios since the time required to solve the problem optimally can grow exponentially in the size of the problem.

Another important aspect in surveillance systems is the growing use of multisenor data fusion [**2,3,9,38**] in which one associates reports from multiple sensors together. Once matched, this more varied information has the potential to greatly enhance target identification and state estimation. The central problem is again that of data association and the principal method employed is multiple hypothesis tracking. For these problems I highly recommend the work of Deb, Pattipati, Bar-Shalom, Washburn, and Tsaknakis [**9,10,11,23**]. Although the current setting is that of multitarget tracking, these data association problems occur in other areas such as the recent work of Cox, Rehg, and Hingorani on edge grouping and contour segmentation [**8**].

These data association problems of partitioning multiple data sets are fundamentally important NP-hard combinatorial optimization problems. Thus an objective in this work is to formulate a large class of these problems as multidimensional assignment problems and to give an example from multitarget tracking. The development in Section 2 extracts the salient features and assumptions that occur in a large class of these problems; a concrete example is developed in Sections 3 and 4. Our reason for this specific formulation is the recent development of a class of real-time and near-optimal algorithms [**9-11,23-33**] for solving these multidimensional assignment problems to the noise level in the problem. We do not discuss algorithms for solving the multidimensional assignment problems such as multiple hypothesis tracking, greedy, branch and bound techniques, and the Lagrangian relaxation methods currently under development [**25,29**]; however, we do present an overview of the salient features and discuss some of the difficulties that one must overcome to obtain fast near-optimal solutions via Lagrangian relaxation in Section 5. Section 6 presents a numerical study of a model problem to demonstrate the effectiveness of the problem formulation and solution techniques. The current work differs from our earlier work [**24-27**] in the generality of the formulation, derivation of the cost coefficients, and the allowable dynamics, and represents to some extent an overview of problem formulation, algorithms, and numerical results.

The use of combinatorial optimization in multitarget tracking is not new and dates back to the mid-sixties and the pioneering work of Sittler [**37**], who used maximum likelihood estimation to evaluate all possible track updates and employed track splitting (several hypotheses were maintained for each track) and pruning (when their probabilities fell below certain threshold). Maximum likelihood estimation was further investigated by Stein and Blackman [**7**], who developed a comprehensive probability for track initiation, track length expectancy, missed detections and false alarms. Morefield [**21**] pioneered the use of the integer programming to solve a set packing and covering problems arising from a data association problem. Multiple hypothesis tracking has been popularized by the fundamental work of Reid [**34**]. These works are further discussed in the books of Blackman [**7**] and Bar-Shalom and Fortmann [**4**], which also serve as excellent introductions to the field of multitarget tracking.

There are a host of topics not considered in this work. We do not discuss the design of a tracking system and the host of associated issues [**2,3,7**], the use of parallel computation, the various nonlinear estimation techniques for nonlinear dynamics, the integration of multisensor data into a multitarget tracking system. Many of these topics are discussed in the aforementioned references and the many references therein.

The outline of the paper is as follows. General conditions under which linear multidimensional assignment problems are applicable to data association are established in Section 2. Utilizing the results of this section, data association problems in multitarget tracking formulated for multiple hypothesis tracking methods are posed as multidimensional assignment problems in Section 3. This development includes track initiation and track maintenance using an N-scan sliding window as developed by the authors in earlier work [**27**]. After presenting a model problem in Section 4, we briefly present the salient features of a class of Lagrangian relaxation algorithms in Section 5, and the results of a numerical study in Section 6 to demonstrate the potential effectiveness of the algorithms.

2. The Linear Multidimensional Assignment Problem

The goal of this section is to formulate the data association problem for a large class of multiple target tracking and sensor fusion problems as a multidimensional assignment problem. This development extracts the salient features and assumptions that occur in a large class of these problems; a concrete example will be given in a subsequent section for multitarget tracking. The plan for this section is to begin with a simple description of a model multitarget tracking. The path to these assignment problems begins with a brief review of the work of Morefield [**21**], who formulated data association problems as set covering and packing problems. After this review in Subsection 2.1 and a refinement of the definition of a partition of the data, the assignment problems is derived in Subsection 2.2. This development generalizes the various special and example cases in our earlier work on track initiation and maintenance [**24-27,30-32**].

In tracking, a common surveillance problem is to estimate the past, current, or future state of a collection of objects (e.g., airplanes) moving in three dimensional space from a sequence of measurements made of the surveillance region by one or more sensors. The objects will be called targets. The dynamics of these targets are generally modeled from physical laws of motion, but there may noise in the dynamics and certain parameters of the motion may be unknown. (The dynamics are often modeled as stochastic differential equations.) At time $t = 0$ a sensor (or sensors) is turned on to observe the region. In an ideal situation measurements are taken at a finite sequence of times $\{t_k\}_{k=0}^{n}$ where $0 = t_0 < t_1 < \cdots < t_n$. (Due to the finite amount of time required for a sensor to sweep the surveillance region, measurements are generally made asynchronously, i.e., not at the same time, so that a time tag is associated with each measurement.) At each time t_k the sensor produces a sequence of measurements $Z(k) = \{z_{i_k}^k\}_{i_k=1}^{M_k}$ where each $z_{i_k}^k$ is a vector of noise contaminated measurements. The actual type of measurement varies with the sensor. Some of the measurements may be false, and the number of targets and which measurement emanates from which target are not known a priori.

The problems then are to determine the number of targets, which measurements go with which targets and which are false (i.e., the data association problem), and to estimate the state of each target given a sequence of measurements that emanate from that target. As part of posing the data association problem, one must estimate the state of a potential target given a particular sequence of measurements from the data sets $\{Z(k)\}_{k=1}^{N}$. Thus, the two problems of data association and state estimation are inseparable parts of the same problem. Before proceeding to the data association problem, we briefly discuss the form of the governing equations of motion and the relation of the measurements to the state of the target.

Given a sequence of measurements $\{z(k)\}_{k=0}^{n}$ postulated to belong to a particular target, the discrete form of the dynamics of the target and measurement equation are generally assumed to be governed by a state-space system of the form

$$\begin{aligned} x(k+1) &= f_k(x(k)) + g_k(x(k))w(k) \\ z(k) &= h_k(x(k)) + v(k) \end{aligned} \tag{2.1}$$

where $x(k)$ is a vector of n state variables at time t_k, $\{w(k)\}$ and $\{v(k)\}$ are independent white noise sequences of normal random variables with zero mean and covariance matrices $Q(k)$ and $R(k)$, respectively, $z(k)$ represents the measurement at time k associated with this particular target and $h_k(x(k))$ maps the state $x(k)$ to the measurement space $z(k)$. Although we shall not make explicit use of this discrete system, one generally uses Kalman filtering for linear problems and a nonlinear filtering technique such as extended Kalman filtering for nonlinear problems [**1**].

2.1 Data Association Problem. The combinatorial optimization problem that governs a large number of data association problems in multitarget tracking and multisensor data fusion [**2-4,7-11,19,23-32**] is generally posed as

$$\text{Maximize}\left\{\frac{P(\Gamma=\gamma \mid Z^N)}{P(\Gamma=\gamma^0 \mid Z^N)} \mid \gamma\in\Gamma^*\right\} \tag{2.2}$$

where Z^N represents N data sets (2.3), γ is a partition of indices of the data ((2.4) and (2.5)), Γ^* is the finite collection of all such partitions (2.5), Γ is a discrete random element defined on Γ^*, γ^0 is a reference partition, and $P(\Gamma = \gamma \mid Z^N)$ is the posterior probability of a partition γ being true given the data Z^N. Each of these terms must be defined. Having done so, the objective will be to formulate a reasonably general class of these data association problems (2.2) as multidimensional assignment problems (2.16).

In the above surveillance example, the data sets were measurements made of the surveillance region. To allow for more general types of data such as tracks and measurements, we shall call the elements in the data sets *reports* [**34**]. Thus, let $Z(k)$ denote a data set of M_k reports $\{z_{i_k}^k\}_{i_k=1}^{M_k}$ and let Z^N denote the cumulative data set of N such sets defined by

$$Z(k) = \{z_{i_k}^k\}_{i_k=1}^{M_k} \quad \text{and} \quad Z^N = \{Z(1),\dots,Z(N)\}, \tag{2.3}$$

respectively. In multisensor data fusion and multitarget tracking the data sets $Z(k)$ may represent different classes of objects. For track initiation in multitarget tracking the objects are measurements that must be partitioned into tracks and false alarms. In our formulation of track maintenance (Section 3.3) one data set will be tracks and remaining data sets will be measurements which are assigned to existing tracks, as false measurements, or to initiating tracks. In sensor level tracking, the objects to be fused are tracks [4]. In centralized fusion, the objects may all be measurements that represent targets or false reports, and the problem is to determine which measurements emanate from a common source.

The next task is to define what is meant by a partition of the cumulative data set Z^N in (2.3). Since this definition is to be independent of the actual data in the cumulative data set Z^N, we first define a partition of the indices in Z^N. Let

$$I^N = \{I(1), I(2), \ldots, I(N)\} \text{ where } I(k) = \{i_k\}_{i_k=1}^{M_k} \tag{2.4}$$

denote the indices in the data sets (2.3). A partition γ of I^N and the collection of all such partitions Γ^* is defined

$$\begin{aligned}
\gamma &= \{\gamma_1, \ldots, \gamma_{n(\gamma)} \mid \gamma_i \neq \emptyset \text{ for each } i\}, && (2.5a)\\
\gamma_i \cap \gamma_j &= \emptyset \text{ for } i \neq j && (2.5b)\\
I^N &= \cup_{j=1}^{n(\gamma)} \gamma_j, && (2.5c)\\
\Gamma^* &= \{\gamma \mid \gamma \text{ satisfies } (2.5a) - (2.5c)\}. && (2.5d)
\end{aligned}$$

Here, γ_i in (2.5a) will be called a track, so that $n(\gamma)$ denotes the number of tracks (or elements) in the partition γ. A $\gamma \in \Gamma^*$ is called a set partitioning of the indices I^N if properties (2.5a) - (2.5c) are valid, a set covering of I^N if property (2.5b) is omitted but the other two properties (2.5a) and (2.5c) are retained, and a set packing if (2.5c) is omitted but (2.5a) and (2.5b) are retained [22]. A partition $\gamma \in \Gamma^*$ of the index set I^N induces a partition of the data Z^N via

$$Z_\gamma = \{Z_{\gamma_1}, \ldots, Z_{\gamma_{n(\gamma)}}\} \text{ where } Z_{\gamma_i} = \{\{z_{i_k}^k\}_{i_k \in \gamma_i}\}_{k=1}^N. \tag{2.6}$$

Clearly, $Z_{\gamma_i} \cap Z_{\gamma_j} = \emptyset$ for $i \neq j$ and $Z^N = \cup_{j=1}^{n(\gamma)} Z_{\gamma_j}$. Each Z_{γ_i} will be called a track of data.

We now come to the key independence assumptions that allow one to convert the objective function in the optimization problem (2.2) to an equivalent linear one. Assume $(\hat{Z}^N, \Gamma)$ is a random element defined on the space $(S \times \Gamma^*)$ with probability "density" $p(Z^N, \Gamma = \gamma)$ where Γ^* is defined above, S is the range space of the data Z^N, $\hat{Z}^N$ is a continuous random variable and Γ is discrete. The assumptions are made on the form of the density function $p(Z^N \mid \Gamma = \gamma)$ and the probability $P_\Gamma(\Gamma = \gamma)$ in Bayes' formula $P(\Gamma = \gamma \mid Z^N) = \frac{1}{p(Z^N)} p(Z^N \mid \Gamma = \gamma) P_\Gamma(\Gamma = \gamma)$ and can be summarized as

$$p(Z^N \mid \Gamma = \gamma) = \prod_{\gamma_i \in \gamma} p(Z_{\gamma_i} \mid \Gamma = \gamma) \tag{2.7a}$$

$$p(Z_{\gamma_i} \mid \Gamma = \gamma) = p(Z_{\gamma_i} \mid \Gamma = \omega) \quad \text{for all } \gamma \text{ and } \omega \in \Gamma^* \tag{2.7b}$$

$$P_\Gamma(\Gamma = \gamma) = C \prod_{i=1}^{n(\gamma)} G(\gamma_i) \tag{2.7c}$$

where C is a constant independent of the partition $\gamma \in \Gamma^*$ and G is a probability distribution on the set of tracks γ_i in (2.5). (If $T = \{\lambda_i \mid \lambda_i$ is a track occurring in some partition $\lambda \in \Gamma^*\}$ and $|T|$ denotes the cardinality of T, then $G(\lambda_i) \geq 0$ and $\sum_{i=1}^{|T|} G(\lambda_i) = 1$.) We also note that since for each $\gamma \in \Gamma^*$, Z_γ corresponds to a partition of the data into $n(\gamma)$ feasible tracks of data, assumption (2.7a) says that the $n(\gamma)$ tracks of data, $\{Z_{\gamma_1}, \ldots, Z_{\gamma_{n(\gamma)}}\}$, are independent if γ is the true track. Of course, there may be dependence between the reports within a single track. Equation (2.7b) further states that these tracks are independent across all partitions of the data.

By equation (2.7a), the irrelevant conditioning on the partition γ in the expression for $p(Z_{\gamma_i} \mid \Gamma = \gamma)$ can be dropped and replaced by $p(Z_{\gamma_i})$. This, along with Bayes' formula, and (2.7) yield the following sequence of decompositions

$$\begin{aligned} P(\Gamma = \gamma \mid Z^N) =& \frac{1}{p(Z^N)} p(Z^N \mid \Gamma = \gamma) P_\Gamma(\Gamma = \gamma) && (2.8a) \\ =& \frac{1}{p(Z^N)} \left[\prod_{i=1}^{n(\gamma)} p(Z_{\gamma_i}) \right] P_\Gamma(\Gamma = \gamma) && (2.8b) \\ =& \frac{C}{p(Z^N)} \prod_{\gamma_i \in \gamma} p(Z_{\gamma_i}) G(\gamma_i) && (2.8c) \end{aligned}$$

Equation (2.8a) is Bayes' formula. If $P_\Gamma(\Gamma = \gamma) = C$, a constant, over all partitions $\gamma \in \Gamma^*$, then (2.8b) is proportional to the likelihood function. Otherwise, the expansion (2.8c) follows from (2.7c). Thus (2.8c) includes both the likelihood function with the identification $C = 1$ and $G(\gamma_i) = 1$ and the posterior function for a more general probability distribution G on the tracks. This completes the review of the required parts Morefield's formulation [**21**]. The next objective is to refine this formulation in a way that is amenable to the assignment problem.

2.2 Assignment Formulation. For notational convenience in representing tracks, we add a *zero index* to each of the index sets $I(k)$ $(k = 1, \ldots, N)$ in (2.4), a *dummy report* z_0^k to each of the data sets $Z(k)$ in (2.3), and require that each

$$\begin{aligned} \gamma_i =& (i_1, \ldots, i_N) \\ Z_{\gamma_i} =& Z_{i_1 \cdots i_n} \equiv \left(z_{i_1}^1, \ldots, z_{i_N}^N\right) \end{aligned} \tag{2.9}$$

where i_k and $z_{i_k}^k$ can now assume the values of 0 and z_0^k, respectively. The dummy report z_0^k serves several purposes in the representation of missing data, false reports, initiating tracks, and terminating tracks. If Z_{γ_i} is missing an actual report from the data set $Z(k)$, then $\gamma_i = \left(i_1, \ldots, i_{k-1}, 0, i_{k+1}, \ldots, i_N\right)$ and $Z_{\gamma_i} = \{z_{i_1}^1, \ldots, z_{i_{k-1}}^{k-1}, z_0^k, z_{i_{k+1}}^{k+1}, \ldots, z_{i_N}^N\}$. A false report $z_{i_k}^k$ $(i_k > 0)$ is represented by $\gamma_i = (0, \ldots, 0, i_k, 0, \ldots, 0)$ and $Z_{\gamma_i} = \{z_0^1, \ldots, z_0^{k-1}, z_{i_k}^k, z_0^{k+1}, \ldots, z_0^N\}$ in which there is but one actual report. The partition γ^0 of the data in which all reports are declared to be false reports is defined by

$$\begin{aligned} Z_{\gamma^0} =& \big\{ Z_{0 \cdots 0 i_k 0 \cdots 0} \equiv \left(z_0^1, \ldots, z_0^{k-1}, z_{i_k}^k, z_0^{k+1}, \ldots, z_0^N\right) \mid \\ & \qquad i_k = 1, \ldots, M_k; k = 1, \ldots, N \big\}. \end{aligned} \tag{2.10}$$

If each data set $Z(k)$ represents a scan of measurements, a track that initiates on scan $m > 1$ will contain only the dummy report z_0^k from each of the data sets $Z(k)$ for each $k = 1, \ldots, m-1$. Likewise, a track that terminates on scan m would have only the dummy report from each of the data sets for $k > m$. These representations are discussed further in Section 3 for both track initiation and track maintenance.

The use of the 0-1 variable

$$z_{i_1 \cdots i_N} = \begin{cases} 1 & \text{if } (i_1, \ldots, i_N) \in \gamma \\ 0 & \text{otherwise} \end{cases} \tag{2.11}$$

yields an equivalent characterization of a partition (2.5) and (2.9) as a solution of the equations

$$\sum_{(i_1,\ldots,i_{k-1},i_{k+1},\ldots,i_N)=(0,\ldots,0)}^{(M_1,\ldots,M_{k-1},M_{k+1},\ldots,M_N)} z_{i_1 \cdots i_N} = 1 \text{ for } i_k = 1, \cdots, M_k \text{ and } k = 1, \cdots, N. \tag{2.12}$$

With this characterization of a partition of the cumulative data set Z^N as a set of equality constraints (2.12), we now proceed to the formulation of the multidimensional assignment problem.

For the derivation of the assignment problem, observe that for $\gamma_i = (i_1 \cdots i_n)$ as in (2.9) and the reference partition (2.10), the expansion (2.8c) can be written as

$$\frac{P(\Gamma = \gamma \mid Z^N)}{P(\Gamma = \gamma^0 \mid Z^N)} \equiv L_\gamma \equiv \prod_{(i_1 \cdots i_N) \in \gamma} L_{i_1 \cdots i_N} \tag{2.13a}$$

where

$$L_{i_1 \cdots i_N} = \frac{p(Z_{i_1 \cdots i_N}) G(Z_{i_1 \cdots i_N})}{\prod_{k=1, i_k \neq 0}^{N} p(Z_{0 \cdots 0 i_k 0 \cdots 0}) G(Z_{0 \cdots 0 i_k 0 \cdots 0})}. \tag{2.13b}$$

Here the index i_k in the denominator corresponds to the kth index of $Z_{i_1 \cdots i_N}$ in the numerator. Next define

$$c_{i_1 \cdots i_N} = -\ln L_{i_1 \cdots i_N}, \tag{2.14}$$

so that

$$-\ln\left[\frac{P(\gamma | Z^N)}{P(\gamma^0 | Z^N)}\right] = \sum_{(i_1, \ldots, i_N) \in \gamma} c_{i_1 \cdots i_N}. \tag{2.15}$$

Thus in view of the of the characterization of a partition (2.5) and (2.6) specialized by (2.9) as a solution of the equation (2.12), the independence assumptions (2.7) and the expansion (2.8), problem (2.2) is equivalently characterized as the following N-dimensional assignment problem:

$$
\begin{aligned}
\text{Minimize} \quad & \sum_{i_1=0}^{M_1} \cdots \sum_{i_N=0}^{M_N} c_{i_1\cdots i_N} z_{i_1\cdots i_N} \\
\text{Subject To:} \quad & \sum_{i_2=0}^{M_2} \cdots \sum_{i_N=0}^{M_N} z_{i_1\cdots i_N} = 1, \quad i_1 = 1,\ldots, M_1, \\
& \sum_{i_1=0}^{M_1} \cdots \sum_{i_{k-1}=0}^{M_{k-1}} \sum_{i_{k+1}=0}^{M_{k+1}} \cdots \sum_{i_N=0}^{M_N} z_{i_1\cdots i_N} = 1, \\
& \qquad \text{for} \quad i_k = 1,\ldots, M_k \text{ and } k = 2,\ldots, N-1, \\
& \sum_{i_1=0}^{M_1} \cdots \sum_{i_{N-1}=0}^{M_{N-1}} z_{i_1\cdots i_N} = 1, \quad i_N = 1,\ldots, M_N \\
& z_{i_1\cdots i_N} \in \{0,1\} \text{ for all } i_1,\ldots, i_N,
\end{aligned}
\tag{2.16}
$$

where $c_{0\cdots 0}$ is arbitrarily defined to be zero. Note that the definition of a partition (2.5) and (2.9) and the 0-1 variable $z_{i_1\cdots i_N}$ in (2.11) imply $z_{0\cdots 0} = 0$. (If $z_{0\cdots 0}$ is not preassigned to zero and $c_{0\cdots 0}$ is defined arbitrarily, then $z_{0\cdots 0}$ is determined directly from the value of $c_{0\cdots 0}$, since it does not enter the constraints other than being a zero-one variable.) Also, each cost coefficient with exactly one nonzero index (i.e., $c_{0\cdots 0 i_k 0\cdots 0} = 0$ for all $i_k = 1,\ldots, M_k$ and $k = 1,\ldots, N$) is zero due to use of the normalizing partition γ^0 in the likelihood ratio in (2.2) and (2.13b).

Having completed the derivation of the assignment problem (2.16), several remarks are in order. The definition of a partition in (2.5) and (2.6) implies that each actual report belongs to at most one track of reports Z_{γ_i} in a partition Z_γ of the cumulative data set. One can modify this to allow multi-assignments of one, some, or all the actual reports. The assignment problem (2.16) is changed accordingly. For example, if $z^k_{i_k}$ is to be assigned no more than, exactly, or no less than $n^k_{i_k}$ times, then the " $= 1$" in the constraint (2.16) is changed to " $\leq, =, \geq n^k_{i_k}$," respectively. In making these changes, one must pay careful attention to the independence assumptions (2.7). (These assumptions need not be valid in many potential applications such as sensor level tracking [**3**], and this may lead to the possibility of nonlinearity in the objective function in (2.16).)

The assignment problem (2.16) can be replaced by an equivalent one with all zero indices removed and the equality replaced by a "less than or equal" sign. To see this, first observe that any assignment problem of the form (2.16) can be converted to one in which all cost coefficients with at exactly one nonzero index have value of zero. (The case of $c_{0\cdots 0}$ and $z_{0\cdots 0}$ are of course excepted from this conclusion; however, this one variable is determined independent of the constraints in (2.16) and was included in the problem for notational convenience.) For the latter problem, all zero-one variables with exactly one nonzero index can be treated as slack variables, which when removed yields the desired form. The opposite conversion is even easier.

Next, the normalization of $P(\Gamma = \gamma \mid Z^N)$ by dividing by $P(\Gamma = \gamma^0 \mid Z^N)$ where the partition γ_0 of all false reports is useful for at least two reasons. First it allows the cancellation of the term $\frac{C}{p(Z^N)}$ in (2.8c). Since $p(Z^N)$ can have dimensions, the result is that the ratio $L_{i_1 \cdots i_N}$ in (2.13) is dimensionless. This normalization also yields a convenient preprocessing technique called fine gating as discussed in the next paragraph.

The complexity of this problem (2.16) makes its formulation and solution formidable. We have already mentioned that it is NP-hard [**14**]. Note also that the computation of all the cost coefficients in (2.16) can be a time consuming task. For example, six scans of measurements with one hundred measurements per scan requires the computation of one trillion cost coefficients for track initiation. Thus preprocessing is essential for reducing the complexity. One class of methods for tracking and sensor fusion problems is called gating [**4,7**], which is generally composed two parts - coarse and fine gating. The philosophy of the former is to compute only those cost coefficients that are feasible for the underlying physical problem, thereby removing unlikely pairings of measurements. Techniques vary widely and are discussed in the book by Blackman [**7**]. If a sequence of points $\{z^1_{i_1}, \ldots, z^N_{i_k}\}$ passes this coarse gating procedure, then the corresponding coefficients $c_{i_1 \cdots i_N}$ is computed. If $c_{i_1 \cdots i_N}$ is positive, then one preassigns (before solving (2.16)) the corresponding 0-1 variable $z_{i_1 \cdots i_N} = 0$, since the assignment of the reports in the corresponding $\{z^1_{i_1}, \ldots, z^N_{i_k}\}$ as false reports leads to a lower cost. (The normalization in (2.2) where γ^0 is the partition of all false reports is fundamental to this conclusion.) We emphasize that this does not mean that each of the actual reports in $\{z^1_{i_1}, \ldots, z^N_{i_k}\}$ is declared to be a false report in advance of solving the problem. Another commonly used complexity reducing technique is that of clustering [**2,7**] which is used to decompose the problem into a number of smaller independent problems. Thus all preprocessing techniques used to reduce the computational costs associated with formulating and solving data association problems in tracking and sensor fusion can also be used to reduce the complexity of the assignment problem (2.16).

3. Assignment Formulation of Multiple Hypothesis Tracking for Multitarget Tracking

The principal method for multiscan processing in multitarget tracking and for sensor fusion is an explicit enumeration scheme called multiple hypothesis tracking (MHT) [**7,34**]. The goal in this section is to demonstrate that a data association problem formulated for multitarget tracking in terms of multiple hypothesis tracking can be formulated as a multidimensional assignment problem as developed in the previous section. In particular, we convert the general data association problem as posed by Reid [**34**] and extended by Kurien [**19**] to include maneuvering targets to a multidimensional assignment problem in Subsection 3.1. The underlying assumptions in the work of Reid [**34**], Kurien [**19**], and Blackman [**7**] are the same as those independence assumptions expressed in (2.7).

Track initiation and a sliding window formulation for track maintenance, as developed in our previous work [**24-27,30-32**] for a special case, are presented in the second and third subsections, respectively, for this more general formulation.

Before beginning this conversion, we briefly discuss some similarities and differences in the multiple hypothesis tracking and assignment formulations. First, one can view multiple hypothesis tracking enumeration algorithm as a method of solving the multidimensional assignment problem (2.16). As discussed at the end of the previous section, all the preprocessing that is generally used for multiple hypothesis tracking can be used for other methods for solving the assignment problem (2.16). The particular algorithms that we have developed for obtaining near-optimal solutions of the assignment problem (2.16) [**24,26,29**] make use of all of these preprocessing techniques, except the explicit pruning of noncompetitive partitions. We do prune tracks of reports from all partitions via the gating techniques discussed at the end of the previous section. Since the Lagrangian relaxation based algorithms [**10,11,23-33**] are not enumerative in nature, pruning of whole partitions is not done explicitly. The algorithms instead compute both the correct partition and score during the course of solving the problem. Thus the pruning of the unlikely partitions is done implicitly in the course of solving the problem.

3.1 Likelihood Calculations. The development of an expression for the posterior probability $P(\Gamma = \gamma \mid Z^N)$ in (2.2) for the problem described by Reid [**34**] and Kurien [**19**] is quite tedious and thus we shall not go through the details of the derivation. The purpose here is to develop the framework of the derivation in sufficient detail to explain the meaning of the various terms in the posterior or likelihood function. The posterior probability, which is developed sequentially (scan by scan) [**34**], will be stated in equation (3.4) and is then converted to the required form (2.13).

Recall from (2.3) that the kth data set of reports is $Z(k) = \{z_i^k\}_{i_k=1}^{M_k}$ and the cumulative set of k such data sets is $Z^k = \{Z(1), \ldots, Z(k)\}$ so that $Z^k = Z(k) \cup Z^{k-1}$. The customary notation here [**34**] is to define $\Omega^k = \{\Omega_j^k \mid j = 0, 1, \ldots, I_k\}$ to be the set of hypotheses about the feasible partitions of the cumulative set of measurements Z^k into tracks and false alarms. In the notation of the previous section, each Ω_j^k represents the event that a partition $\gamma^k \in \Gamma^{k*}$ is true. (Here, the dependence of the γ on the number of scans k has been explicitly included as a superscript.) We shall follow this customary notation. The hypothesis Ω_0^k is that all reports are false. Let $\Omega_{\psi_l}^{k-1}$ denote that specific hypothesis in Ω^{k-1} that produces Ω_l^k, and let $\psi_l(k)$ denote the hypothesis that indicates the specific status of all targets postulated by $\Omega_{\psi_l}^{k-1}$ at the scan time t_k and the specific origin of all reports received at scan time t_k. Thus

$$\Omega_l^k = \psi_l(k) \cup \Omega_{\psi_l}^{k-1} \text{ and } Z^k = Z(k) \cup Z^{k-1} \tag{3.1}$$

Using this and Bayes' rule, one can write

$$P(\Omega_l^k|Z^k) = p(Z(k)|\Omega_l^k, Z^{k-1})P(\psi_l(k)|\Omega_{\psi_l}^{k-1}, Z^{k-1}) \times P(\Omega_{\psi_l}^{k-1}|Z^{k-1})\left[\frac{p(Z^{k-1})}{p(Z^k)}\right] \tag{3.2}$$

The term $\left[\frac{p(Z^{k-1})}{p(Z^k)}\right]$ is a normalizing factor, which will be removed when $P(\Omega_l^k|Z^k)$ is normalized by dividing this expression by the probability of all false reports, i.e., $P(\Omega_0^k|Z^k)$.

We partition the reports just as in the work of Reid **[34]** and Kurien **[19]** in the following way. Assume the parent hypothesis $\Omega_{\psi_l}^{k-1}$ postulates that τ^k targets that were extended from scan $k-1$ to scan k. Of the total number of measurements M_k, ν^k reports originate from new targets with an associated probability mass function $\mu_\nu^k(\nu^k)$, f_k reports are false reports with probability mass function $\mu_f^k(f^k)$, and the remaining δ_k reports are associated with existing targets. Of the τ^k targets that exist after scan $k-1$, χ^k of these τ^k targets are terminated and are not observed (on scan k); δ^k of the $\tau^k - \chi^k$ nonterminated targets are detected; and, $\tau^k - \chi^k - \delta^k$ nonterminated targets are not detected. Of the δ^k detected targets, m^k executed a maneuver and the remaining $\delta^k - m^k$ detected targets continued with the existing model (2.1). (Thus, the total number of targets that exist after scan k is $\tau_{k+1} = \tau^k - \chi^k + \nu^k$.)

For ease of reference, here are these numbers, probabilities, and the definition of some required indicator functions:

(3.3)

P_χ^k	probability of termination on scan k
P_d^k	probability of detection on scan k
P_m^k	probability of a maneuver on scan k
z_i^k	measurement i from scan k
δ^k	number of measurements on scan k originating from previously established tracks
ν^k	number of new targets detected on scan k
f^k	number of false alarms on scan k
M_k	total number of measurements on scan k, i.e., M_k
τ^k	number of targets that were extended from scan $k-1$ to scan k
χ^k	number of terminated (and nondetected) targets on scan k
m^k	number of maneuvering (and detected) targets on scan k

$$f_i^k = \begin{cases} 1, & \text{if } z_i^k \text{ is a false alarm;} \\ 0, & \text{otherwise;} \end{cases}$$
$$\nu_i^k = \begin{cases} 1, & \text{if } z_i^k \text{ is a new target;} \\ 0, & \text{otherwise;} \end{cases}$$
$$m_i^k = \begin{cases} 1, & \text{if } z_i^k \text{ originated from a maneuvering target;} \\ 0, & \text{otherwise;} \end{cases}$$
$$\delta_i^k = \begin{cases} 1, & \text{if } z_i^k \text{ belongs to an existing track;} \\ 0, & \text{otherwise;} \end{cases}$$
$$\Delta_{ij} = \begin{cases} 1, & \text{if } i = j; \\ 0, & \text{otherwise.} \end{cases}$$

In addition, let $p_t^k(z_i^k|\Omega_l^k, Z^{k-1})$ represent the likelihood that the report originated from a target maintaining the state space model (2.1), $p_m^k(z_i^k|\Omega_l^k, Z^{k-1})$ represent the likelihood that the report originated from a maneuvering target, $p_\nu^k(z_i^k|\Omega_l^k, Z^{k-1})$, the likelihood that the report originated from a new source, and $p_f^k(z_i^k|\Omega_l^k, Z^{k-1})$, the likelihood the report represents a false alarm.

The development of the formula for (3.2) can be found in the work of Kurien **[19]** and is given by

$$\begin{aligned} P(\Omega_l^k|Z^k) = & P(\Omega_{\psi_l}^{k-1}|Z^{k-1})\left\{\frac{p(Z^{k-1})}{p(Z^k)}\right\}\left\{\frac{\nu^k!f^k!}{M_k!}\mu_f^k(f^k)\mu_\nu^k(\nu^k)\right\} \\ & \left\{(P_\chi^k)^{\chi^k}\left[(1-P_\chi^k)(1-P_d^k)(1-P_m^k)\right]^{\tau^k-\delta^k-\chi^k}\right\} \\ & \prod_{i=1}^{M_k}\left\{\left[(1-P_\chi^k)P_d^k(1-P_m^k)p_t^k(z_i^k|\Omega_l^k, Z^{k-1})\right]^{\delta_i^k(1-m_i^k)}\right. \\ & \left[(1-P_\chi^k)P_m^k P_d^k p_m^k(z_i^k|\Omega_l^k, Z^{k-1})\right]^{\delta_i^k m_i^k} \\ & \left.\left[p_\nu^k(z_i^k|\Omega_l^k, Z^{k-1})\right]^{\nu_i^k}\left[p_f^k(z_i^k|\Omega_l^k, Z^{k-1})\right]^{f_i^k}\right\} \end{aligned} \tag{3.4}$$

We normalize this expression by dividing by $P(\Omega_l^0|Z^k)$ where Ω_0^k is the event that all reports are false and assume $\mu_f^k(f^k) = \exp(-\lambda_f^k)\frac{(\lambda_f^k)^{f^k}}{f^k!}$ and $\mu_\nu^k(\nu^k) =$

$\exp(-\lambda_\nu^k)\frac{(\lambda_\nu^k)^{\nu^k}}{\nu^k!}$ are Poisson probability mass functions, so that

$$\frac{P(\Omega_l^k|Z^k)}{P(\Omega_0^k|Z^k)}=\frac{P(\Omega_{\psi_l}^{k-1}|Z^{k-1})}{P(\Omega_0^{k-1}|Z^{k-1})}\left\{(P_\chi^k)^{\chi^k}\left[(1-P_\chi^k)(1-P_d^k)(1-P_m^k)\right]^{\tau^k-\delta^k-\chi^k}\right\}$$
$$\prod_{i=1}^{M_k}\left\{\left[\frac{(1-P_\chi^k)P_d^k(1-P_m^k)p_t^k(z_i^k|\Omega_l^k,Z^{k-1})}{\lambda_f^k p_f^k(z_i^k|\Omega_l^k,Z^{k-1})}\right]^{\delta_i^k(1-m_i^k)}\left[\frac{(1-P_\chi^k)P_m^kP_d^kp_m^k(z_i^k|\Omega_l^k,Z^{k-1})}{\lambda_f^kp_f^k(z_i^k|\Omega_l^k,Z^{k-1})}\right]^{\delta_i^km_i^k}\left[\frac{\lambda_\nu^kp_\nu^k(z_i^k|\Omega_l^k,Z^{k-1})}{\lambda_f^kp_f^k(z_i^k|\Omega_l^k,Z^{k-1})}\right]^{\nu_i^k}\right\} \tag{3.5}$$

The formula for $\frac{P(\Omega_l^k|Z^k)}{P(\Omega_0^k|Z^k)}$ has been developed recursively for $k \geq 1$; however, we now consider N scans of measurements as a single entity. For a track $Z_{i_1\cdots i_N}$ define

$$P_\phi^k=\begin{cases}P_\chi^k, & \text{if track } Z_{i_1\cdots i_N} \text{ terminates on scan } k;\\ (1-P_\chi^k)(1-P_d^k)(1-P_m^k), & \text{if track } Z_{i_1\cdots i_N} \text{ has a missed detection on scan } k;\\ 1, & \text{otherwise.}\end{cases} \tag{3.6}$$

Each track of reports $Z_{i_1\cdots i_N}$ in a feasible partition $\gamma \in \Gamma$ is scored as

$$\begin{aligned}L_{i_1i_2\cdots i_N} \equiv& L(Z_{i_1i_2\cdots i_N}) \equiv L(z_{i_1}^1,\dots,z_{i_N}^N)\\ =&\prod_{k=1}^N\left\{\left[\frac{(1-P_\chi^k)P_d^k(1-P_m^k)p_t^k(z_{i_k}^k|Z_{i_1\cdots i_N})}{\lambda_f^kp_f^k(z_{i_k}^k|Z_{0\cdots 0i_k0\cdots 0})}\right]^{\delta_{i_k}^k(1-m_{i_k}^k)}\right.\\ &\left[\frac{(1-P_\chi^k)P_m^kP_d^kp_m^k(z_{i_k}^k|Z_{i_1\cdots i_N})}{\lambda_f^kp_f^k(z_{i_k}^k|Z_{0\cdots 0i_k0\cdots 0})}\right]^{\delta_{i_k}^km_{i_k}^k}\\ &\left.\left[\frac{\lambda_\nu^kp_\nu^k(z_{i_k}^k|Z_{i_1\cdots i_N})}{\lambda_f^kp_f^k(z_{i_k}^k|Z_{0\cdots 0i_k0\cdots 0})}\right]^{\nu_{i_k}^k}\right\}^{(1-\Delta_{0i_k})}\left\{P_\phi^k\right\}^{\Delta_{0i_k}}\end{aligned} \tag{3.7a}$$

provided at least two of the indices in $\{i_1, i_2, \dots, i_N\}$ are nonzero and

$$L_{0\cdots 0i_k0\cdots 0} \equiv 1 \text{ provided } i_k \neq 0 \tag{3.7b}$$

where the conditioning now is only on the assumption that the collection of reports in $Z_{i_1\cdots i_N} \equiv \{z_{i_1}^1,\dots,z_{i_N}^N\}$ represent a true event. The conditioning in the probability density for a false report $p_f^k(z_{i_k}^k|Z_{0\cdots 0i_k0\cdots 0})$ in the denominator is on the assumption that $z_{i_k}^k$ is a false report which is represented by $Z_{0\cdots 0 i_k 0\cdots 0}$. Using (3.7), (3.5) can be expressed as

$$\frac{P(\Omega_l^k|Z^k)}{P(\Omega_0^k|Z^k)} = \prod_{\{i_1 i_2 \cdots i_N\}\in\gamma} L_{i_1 i_2 \cdots i_N}, \tag{3.8}$$

which is needed in the assignment formulation (2.13).

Several remarks about this expression are in order. Each track of reports $Z_{i_1\cdots i_N} = \{z_{i_1}, \ldots, z_{i_N}\}$ carries with it such information as the time at which the track initiates, maneuvers or terminates. For example, if this particular target initiates on scan $r > 1$ and terminates on scan $s < N$, then $i_k = 0$ and $\Delta_{0i_k} = 1$ for $k = 1, \ldots, r-1$ and for $k = s, \ldots, N$, so that

$$\{P_\phi^k\}^{\Delta_{0i_k}} = \begin{cases} 1 & \text{if } k = 1, \ldots, r-1 \\ & \text{and } k = s+1, \ldots, N; \\ (1-P_\chi^k)(1-P_d^k)(1-P_m^k) & \text{if } r < k < s \text{ and there} \\ & \text{is a missed detection} \\ & \text{on scan } k \\ P_\chi^k & \text{if } k = s \\ 1 & \text{otherwise.} \end{cases} \tag{3.9}$$

Next, consider a particular sequence of reports $\{z_{i_1}, \ldots, z_{i_N}\}$. If there is only one nonzero index, then this point must be a false report by definition. If there are two or more points assigned together as a track, then the underlying assumption is that there are no false reports in the list. Thus each point z_{i_k} is either a dummy, associated with a the track defined by the collection, or may correspond to a new target. This last situation may cause some ambiguity in the definition of $L_{i_1 i_2 \cdots i_N}$. This is generally decided from auxiliary information. Here is one example of how such auxiliary reasoning is frequently used. If z_{i_k} is the first non-dummy report and the subsequent reports in the list $\{z_{i_1}, \ldots, z_{i_N}\}$ are dynamically feasible with it, then $\nu_{i_k}^k = 1$ and $\delta_{i_k}^k = 0$. For this same list of reports, one would not consider any subsequent points in the list as new sources. Another ambiguous situation is that of a maneuvering target. Without additional information, one cannot decide whether a report arising from an existing track corresponds to a maneuvering target or non-maneuvering target. *These issues will be discussed in future work.* Finally, track terminations can also be based on a track length model as in the work of Stein and Blackman [7].

3.2 Track initiation. If one specializes the development in subsection 3.1 to the situation in which each object in the data set $Z(k)$ is a measurement or observation for $k = 1, \ldots, N$, then the problem of partitioning the observations into tracks and false alarms can be posed as the multidimensional assignment problem (2.16) where

$$c_{i_1\cdots i_N} = -\ln L_{i_1\cdots i_N} \tag{3.10}$$

and $L_{i_1 \cdots i_N}$ is defined by equation (3.7).

3.3. Track maintenance using a sliding window. Suppose now that the observations on P *previous* scans (of observations) have been partitioned into tracks and false alarms and that K *new* scans of observations are to be added. One approach to solving the resulting data association problem is formulate the problem as a track initiation problem with $P+K$ scans. This is the previously mentioned *batch* approach. The *deferred logic approach* adopted here is to treat the track extension problem within the framework of a window sliding over the observation sets. First assume that the scans of observations are partitioned into three components: D *discarded* scans of observations, R *retained* scans of observations from the P *previously* processed scans, and K *new* scans of observations. Thus the number of scans in the sliding window is $N = R + K$ while the number of *discarded* scans is $D = P - R$.

Let M_0 denote the number of confirmed tracks previously constructed from the discarded and retained regions that are present at the beginning of the window of N scans. Tracks that initiate within the window are not included in this list. Tracks that have been declared as terminated at some scan within the present window are still included in this number M_0. Thus a tentative decision about a terminated track in the window may be changed given new information. A track is actually terminated only if the termination point occurs prior to the current window, i.e., the termination occurs in the discarded region. Other than these restrictions, the M_0 tracks are obtained as the solution of a previous problem assignment problem. Another possibility is to include all tracks in the best K such solutions or even all feasible tracks. For $i_0 = 1, \ldots, M_0$ the i_0^{th} such track is denoted by T_{i_0} and the $(N+1)$-tuple $\{T_{i_0}, z_{i_1}^1, \ldots, z_{i_N}^N\}$ will denote a track T_{i_0} plus a set of observations or measurements $\{z_{i_1}^1, \ldots, z_{i_N}^N\}$, actual or dummy, that are feasible with the track T_{i_0}. The $(N+1)$-tuple $\{T_0, z_{i_1}^1, \ldots, z_{i_N}^N\}$ will denote a track that initiates in the sliding window. Note the use once again of a dummy variable T_0. A false report in the sliding window is one with all but one non-zero index in the $(N+1)$-tuple $\{T_0, z_{i_1}^1, \ldots, z_{i_N}^N\}$

The hypothesis about a partition $\gamma \in \Gamma^*$ being true is now conditioned on the truth of the M_0 tracks entering the N-scan window. (Thus the assignments prior to this sliding window are fixed.) The likelihood function is given by $L_\gamma = \prod_{\{T_{i_0}, z_{i_1}, \ldots, z_{i_N}\} \in \gamma} L_{i_0 i_1 \cdots i_N}$, where $L_{i_0 i_1 \cdots i_N} = L_{T_{i_0}} L_{i_1 \cdots i_N}$, $L_{T_{i_0}}$ is the composite likelihood from the discarded scans just prior to the first scan in the window for $i_0 > 0$, $L_{T_0} = 1$, and $L_{i_1 \cdots i_N}$ is defined as in (3.8) for the N-scan window. ($L_{T_0} = 1$ is used for any tracks that initiate in the sliding window.) Thus the track extension problem can be formulated as Maximize $\{L_\gamma \mid \gamma \in \Gamma^*\}$. With the same convention as in Section 2, a feasible partition is one which is defined by the properties in (2.4) and (2.6). Analogously, the definition of the zero-one variable

$$z_{i_0 i_1 \cdots i_N} = \begin{cases} 1 & \text{if } \{T_{i_0}, z_{i_1}^1, \ldots, z_{i_N}^N\} \text{ is assigned as a unit,} \\ 0 & \text{otherwise,} \end{cases}$$

and the corresponding cost for the assignment of the sequence $\{T_{i_0}, z_{i_1}, \ldots, z_{i_N}\}$ to a track by $c_{i_0 i_1 \cdots i_N} = -\ln L_{i_0 i_1 \cdots i_N}$ yield the following multidimensional assignment formulation of the data association problem for track maintenance:

$$
\begin{aligned}
\text{Minimize} \quad & \sum_{i_0=0}^{M_0} \cdots \sum_{i_N=0}^{M_N} c_{i_0 \cdots i_N} z_{i_0 \cdots i_N} \\
\text{Subject To} \quad & \sum_{i_1=0}^{M_1} \cdots \sum_{i_N=0}^{M_N} z_{i_0 \cdots i_N} = 1, \quad i_0 = 1, \ldots, M_0, \\
& \sum_{i_0=0}^{M_0} \sum_{i_2=0}^{M_2} \cdots \sum_{i_N=0}^{M_N} z_{i_0 i_1 \cdots i_N} = 1, \quad i_1 = 1, \ldots, M_1, \\
& \sum_{i_0=0}^{M_0} \cdots \sum_{i_{k-1}=0}^{M_{k-1}} \sum_{i_{k+1}=0}^{M_{k+1}} \cdots \sum_{i_N=0}^{M_N} z_{i_0 \cdots i_N} = 1, \\
& \qquad \text{for} \quad i_k = 1, \ldots, M_N \ \text{ and } k = 2, \ldots, N-1, \\
& \sum_{i_0=0}^{M_0} \cdots \sum_{i_{N-1}=0}^{M_{N-1}} z_{i_0 \cdots i_N} = 1, \quad i_N = 1, \ldots, M_N, \\
& z_{i_0 \cdots i_N} \in \{0, 1\} \ \text{ for all } i_0, \ldots, i_N.
\end{aligned}
\tag{3.11}
$$

Note that the association problem involving N scans of observations is an N-dimensional assignment problem for track initiation and an $(N+1)$-dimensional one for track maintenance.

4. An example problem

Consider the problem of many targets traveling in two dimensional space according to the constant acceleration model $x(t, \alpha) = (x_1(t, \alpha), x_2(t, \alpha))$ where

$$
\begin{aligned}
x_1(t, \alpha) &= x_{01} + t v_1 + \frac{t^2}{2} a_1 \\
x_2(t, \alpha) &= x_{02} + t v_2 + \frac{t^2}{2} a_2
\end{aligned}
\tag{4.1}
$$

where the parameters $\alpha = (x_{01}, v_1, a_1, x_{02}, v_2, a_2)$ identify a particular target. At a discrete set of *scan times* $\{t_k\}_{k=1}^N$ ($t_1 \leq t_2 \leq \ldots \leq t_N$) a radar located at the origin in this Cartesian space observes error contaminated ranges and angles of the targets in the measurement space which is a circle with radius R. Some measurements are spurious and some measurements of true targets are missed. At time t_k, the radar is assumed to return the set of noise contaminated measurements $\{z_{i_k}^k\}_{k=1}^{M_k}$, where M_k is the number of measurements and $z_{i_k}^k = (r_{i_k}^k, \theta_{i_k}^k)$. To every scan, a dummy measurement z_0^k is added to represent missed detections. Each measurement $(r_{i_k}^k, \theta_{i_k}^k)$ is related to the true observable

$H(x(t_k,\alpha))$ by

$$\begin{pmatrix} r^k_{i_k} \\ \theta^k_{i_k} \end{pmatrix} = H(x(t_k,\alpha)) + \begin{pmatrix} e^k_r \\ e^k_\theta \end{pmatrix} \tag{4.2}$$

where e^k_r and e^k_θ are independent zero-mean Gaussian random variables with standard deviations σ^k_r and σ^k_θ, respectively. The true observable $H(x(t,\alpha))$ is related to the track $x(t,\alpha) \equiv (x_1(t,\alpha), x_2(t,\alpha))$ by

$$H(x(t,\alpha)) = \begin{pmatrix} \sqrt{x_1(t,\alpha)^2 + x_2(t,\alpha)^2} \\ \arctan\left[\frac{x_2(t,\alpha)}{x_1(t,\alpha)}\right] \end{pmatrix}. \tag{4.3}$$

(If the definition of arctan is based on the principal angle, then the appropriate shifts in $\theta^k_{i_k}$ must be made.) If the measurement corresponds to a false alarm or new target, then

$$\begin{pmatrix} r^k_{i_k} \\ \theta^k_{i_k} \end{pmatrix} = \begin{cases} \begin{pmatrix} w^k_r \\ w^k_\theta \end{pmatrix} & \text{if the measurement is spurious,} \\ \begin{pmatrix} u^k_r \\ u^k_\theta \end{pmatrix} & \text{if the measurement arises from a new source,} \end{cases} \tag{4.4}$$

where the random sequences w^k and u^k have some assumed densities p^k_f and p^k_ν, respectively. A common assumption is that each of w^k_r, w^k_θ, u^k_r, u^k_θ is uniformly distributed so that

$$p^k_f(w^k) = \begin{cases} \frac{1}{2\pi R} & \text{if } 0 \le w^k_r \le R \text{ and } -\pi < w^k_\theta \le \pi \\ 0 & \text{otherwise} \end{cases} \tag{4.5}$$

A similar expression would hold for p^k_ν. The number of false alarms and new targets are assumed to be generated at each time interval $[t_{k-1}, t_k]$ according to a Poisson distributions with expected numbers λ^k_f and λ^k_ν, respectively. For this example, we also assume the targets do not maneuver so that $m^k = 0$ and $P^k_m = 0$ for all k in (3.7).

Under these assumptions the likelihood expression in (3.8) simplifies to

$$L_{i_1 i_2 \cdots i_N} = \prod_{k=1}^{N} \left\{P^k_\phi\right\}^{\Delta_{0 i_k}} \left\{ \left[\frac{(1-P^k_\chi)P^k_d p^k_t(z^k_{i_k}|Z_{i_1\cdots i_n})}{\lambda^k_f p^k_f(z^k_{i_k}|Z_{i_1\cdots i_n})}\right]^{\delta^k_{i_k}} \left[\frac{\lambda^k_\nu p^k_\nu(z^k_{i_k}|Z_{i_1\cdots i_n})}{\lambda^k_f p^k_f(z^k_{i_k}|Z_{i_1\cdots i_n})}\right]^{\nu^k_{i_k}} \right\}^{(1-\Delta_{0 i_k})} \tag{4.6a}$$

provided at least two of the indices in $\{i_1, i_2, \ldots, i_N\}$ are nonzero and

$$L_{0\cdots 0 i_k 0 \cdots 0} \equiv 1 \text{ provided } i_k \neq 0. \tag{4.6b}$$

where

$$P^k_\phi = \begin{cases} P^k_\chi, & \text{if track } Z_{i_1\cdots i_N} \text{ terminates at scan } k; \\ (1-P^k_\chi)(1-P^k_d), & \text{if track } Z_{i_1\cdots i_N} \text{ has a missed detection on scan } k; \\ 1, & \text{otherwise.} \end{cases} \tag{4.7}$$

In writing down this score, one assumes that the targets dynamics (4.1) are known, so that the measurement errors can be computed using the densities

$$p_t^k(z_{i_k}^k | Z_{i_1 \cdots i_n}) = \frac{1}{2\pi\sigma_r(t_k)\sigma_\theta(t_k)} \exp\Big\{ -\frac{1}{2} \left[z_{i_k}^k - H_k(x(t_k,\alpha))\right]^T \Sigma_{r\theta}^{-1}(t_k) \left[z_{i_k}^k - H_k(x(t_k,\alpha))\right] \Big\} \tag{4.8}$$

However, it is precisely these tracks that are to be identified, so that they must be estimated from the given the measurements $\{z_{i_1}^1, \ldots, z_{i_k}^N\}$. We do this by replacing the parameters α by the maximum likelihood estimate

$$\hat{\alpha} = \text{Arg Max } L_{i_1 \cdots i_n}(x(\cdot,\alpha)) = \text{Arg Min } c_{i_1 \cdots i_n}(x(\cdot,\alpha)) \tag{4.9}$$

which can be equivalently characterized as the solution to the nonlinear least squares problem

$$\begin{aligned} \hat{\alpha} &= Arg\ Min\ c_{i_1 \cdots i_N}(x(\cdot,\alpha)) \\ &= Arg\ Min \sum_{k=1}^{N} (1 - \Delta_{0 i_k}) \left[z_{i_k}^k - H_k(x(t_k,\alpha))\right]^T \\ &\qquad \Sigma_{r\theta}^{-1}(t_k) \left[z_{i_k}^k - H_k(x(t_k,\alpha))\right], \end{aligned} \tag{4.10}$$

where $\Sigma_{r\theta}(t) = diag(\sigma_r^2(t), \sigma_\theta^2(t))$. Although the estimation problem (4.9) or (4.10) is expressed as a batch nonlinear least squares problem, the parameters could be estimated recursively (i.e., sequentially) as the scans of measurements become available. Also, the model (4.1) model fits the state space representation (2.1) with no noise in the state equation. If one includes noise as in (2.1), then an extended Kalman filter, iterated extended Kalman filter, or some other nonlinear estimation technique [1] must be used to estimate the state of the system.

5. A Prototype Algorithm

The algorithm used to construct near-optimal solutions of the multidimensional assignment problem in real-time is based on Lagrangian relaxation, which originally gained prominence for producing efficient bounds in the traveling salesman problem [**16,17**]. Reviews of this methodology can be found in [**12,15,22**]. The algorithm that motivates those for the multidimensional assignment problem is that of Frieze and Yadegar [**13**] who developed a Lagrangian relaxation based algorithm for constructing good solutions for the three dimensional assignment problem. The current algorithm is recursive in that a N-dimensional assignment problem is relaxed to a $(N-1)$-dimensional one by incorporating one set of constraints into the objective function using a Lagrangian relaxation of this set. This relaxed problem is then maximized with respect to the Lagrange multipliers and the corresponding solution is used to recover a feasible solution of the N-dimensional problem. Each $(N-1)$-dimensional problem is solved in a similar manner, and the process is repeated until it reaches the two-dimensional problem, which is solved by the forward/reverse auction algorithm [**5,6**]. Explicit descriptions of this algorithm can be found in the work of Poore and Rijavec [**24-27,29,32**], particularly [**29**], and that of Deb, Pattipati, Bar-Shalom, Washburn, and Tsaknakis [**9-11,23**]. We shall not describe these algorithms here. Instead, we content ourselves with an overview of the algorithms from the viewpoint of a more general integer programming problem, and then discuss some of the discuss some of the difficulties associated with the multidimensional assignment problem (2.16).

Consider the integer programming problem

$$\begin{aligned} \text{Minimize} \quad & c^T z = V(z) \\ \text{Subject To} \quad & Az \leq b \\ & Bz \leq d \\ & z_i \text{ is an integer for } i \in I \end{aligned} \tag{5.1}$$

where the partitioning of the constraints is natural in some sense and the inequalities are to be interpreted componentwise. The Lagrangian relaxation of (5.1) relative to the constraints $Bz \leq d$ is defined to be

$$\begin{aligned} \Phi(u) = \text{Minimize} \quad & \phi(z,u) \equiv \{c^T z + u^T(Bz - d)\} \\ \text{Subject To} \quad & Az \leq b \\ & z_i \text{ is an integer for } i \in I \\ & u \geq 0. \end{aligned} \tag{5.2}$$

If the constraint set $Bz \leq d$ is replaced by $Bz = d$, the nonnegativity constraint on u is removed. $L = c^T z + u^T(Bz - d)$ is the Lagrangian relative to the constraints $Bz \leq d$, and hence the name *Lagrangian relaxation.* Next, if $\bar{z}$ is an optimal solution to (5.1), problems (5.1) and (5.2) imply

$$\Phi(u) \leq V(\bar{z}) \quad \text{for all } u \geq 0. \tag{5.3}$$

Given a specific multiplier u, let $z_r(u)$ and $\bar{z}_r(u)$ denote suboptimal and optimal solutions of the relaxed problem (5.2), respectively. Generally, $\bar{z}_r(u)$ is not feasible for the relaxed constraint $Bz \leq d$; however, if $\bar{z}_r(u)$ *is* feasible, then it is then also optimal for (5.1). Thus one must develop a recovery procedure for constructing a feasible solution of (5.1) from either of $z_r(u)$ or $\bar{z}_r(u)$. There are several reasons why the resulting feasible solution z_f might be a good solution of (5.1). First, if the multiplier $u = \bar{u} \geq 0$ is chosen as the maximizer of the problem

$$\text{Maximize } \{\Phi(u) \ : \ u \geq 0\} \tag{5.4}$$

and the duality gap $[\Phi(\bar{u}), V(\bar{z})]$ is small, then the recovered feasible solution z_f of (5.1) from the solution $z_r(\bar{u})$ of (5.2) may be close to $\bar{z}$. (The experience of many researchers is that this duality gap tends to be much smaller for equality constrained problems than for corresponding inequality constrained problems [**12**].) Secondly, the term $u^T(Bz - d)$ in (5.2) acts like a penalty for violating the constraint $Bz - d \leq 0$, thereby forcing $z_r(\bar{u})$ closer to the optimal solution $\bar{z}$ of

(5.1). Furthermore, the recovery procedure for the multidimensional assignment problem is designed to utilize any remaining flexibility in optimizing the objective function [**24,25,29**]. Given this rationale, the following prototype algorithm abstracts the salient features of our Lagrangian relaxation based algorithm for the multidimensional assignment problems (2.16) and incorporates the salient features of the Frieze-Yadegar algorithm for the three dimensional algorithm [**13**]:

ALGORITHM OUTLINE. *Construct a sequence of multipliers* $\{u_k\}_{k=0}^{\infty}$ *in the course of maximizing* $\Phi(u)$ *defined in (5.2) and a corresponding sequence of feasible solutions* $\{z_k\}_{k=0}^{\infty}$ *of (5.1) as follows:*

A. Choose an initial approximation u_0.

B. Given u_k*, determine a new multiplier* u_{k+1} *from a step in the maximization problem (5.4), so that* $\Phi(u_k) \le \Phi(u_{k+1})$.

C. Given u_{k+1} *and a solution* $z_r(u_{k+1})$ *of (5.2), recover a feasible solution* z_{k+1} *of the integer programming problem (5.1).*

In the absence of any prior knowledge of the initial multiplier u_0, a good neutral choice in Part A is $u_0 = 0$. Part B of this algorithm is the nonsmooth optimization phase and one of the most widely used methods for non-smooth optimization is the subgradient algorithm [**12,15**], which is the nonsmooth analog to the steepest ascent method. Analogous to conjugate gradient methods for smooth optimization is the class called "bundle methods" [**20**]. This includes the space dilation method of Shor [**36**], the "bundle-trust region" method due to Schramm and Zowe [**35**], and the conjugate subgradient method of Wolfe [**39**]. Part C of this algorithm is the aforementioned recovery procedure.

Note that $\Phi(u_k) \le \Phi(\bar{u}) \le V(\bar{z}) \le V(z_k)$, so that we have a bound on the optimal solution with this algorithm. It is this bound that guides several of the relaxation algorithms. One of the major difficulties with the multidimensional assignment problem (2.16) occurs with this inequality in that the relaxed problem must be solved optimally for it to be valid. This causes no problem when a three dimensional problem is relaxed to a two dimensional one, since there are many excellent (polynomial time) algorithms for solving this problem optimally. For the four and higher dimensional assignment problems, the only known method for ensuring an optimal solution is branch and bound, but this is insufficient for real-time needs. To resolve this difficulty, we use a merit function [**29**] for the algorithm described at the beginning of this section. Other methods for dealing with this difficulty are the subject of a forthcoming paper [**33**].

6. Numerical Simulations

This section contains parametric studies of the tracking problem described in Section 4 to demonstrate algorithm behavior and potential effectiveness. There are several points to be made with these numerical simulations. Tables 1 - 6 demonstrate the computation to or below the noise level in the problems and the corresponding solution quality. These tables also compare the Lagrangian relaxation algorithms with a greedy algorithm to demonstrate the superior track identification within the same order of magnitude of time of this greedy algorithm. Tables 7, 8, and 9 demonstrate the effect of changing target dynamics on the solution times and quality. A more complete numerical study can be found in the work of Poore and Rijavec [**28**].

In each study, 20 randomly generated problems were solved and the results were averaged. The simulations were conducted on an IBM RS/6000-550 workstation. The tracking problems were precisely those presented in Section 4 with the following parameters: 100 targets were initiated on the first scan and never terminated, not even by leaving the space. New targets were distributed uniformly in space, with the initial velocities and accelerations being uniformly distributed in direction and magnitude. The sensor returned instantaneous scans of the target space every 10 seconds. The targets were observed in polar coordinates, with both the range and angle being subject to zero-mean Gaussian white noise with the standard deviations of σ_r and σ_θ, respectively. The sensor also allowed for the missed detection and clutter (false reports) that were uniformly distributed in space. Length of each simulation was 15 scans or 150 seconds.

All tables are based on a *fast scenario* in which the targets were generated with the initial velocities between 200 and 900 mph. The acceleration is less than 44064 miles per hour squared, for a total of no more than 1800 mph velocity gain during the simulation. Tables 7, 8, and 9 compare some problem characteristics between this fast scenario and a *slow scenario*, in which the targets were generated with the initial velocities between 200 and 500 mph. The acceleration is less than 4795 miles per hour squared, for a total of no more than 200 mph velocity gain during the simulation.

The solution quality is computed in two stages. The first compares each true track with all the solution tracks. Using the knowledge about the measurement errors in the tracking problem, we then determine which pairs of true and solution tracks are feasible with each other, i.e., which solution tracks are good models of each true track. The list of feasible pairings is then passed to the second stage, a two dimensional assignment problem solver which selects a maximum number of the pairings from the list, bearing in mind that each solution track can model only one true track and that each true track can be modeled by at most one solution track. All the true tracks that are represented in the final list of pairings are considered correctly identified. The true tracks that are *not* represented in the final list are considered missed.

Tables 1 and 2 compare the assignment problem solutions obtained by the Lagrangian relaxation and greedy algorithms on track initiation problems. For each combination of problem parameters, lower bound on the optimal solution, true solution value, relaxation solution value and the greedy solution value are shown. The true solution value was obtained by adding the costs of the variables that were known to represent the true tracks.

σ_r	σ_θ	lower bound	true value	relaxation value	greedy value
0.33	0.5	1137.61	1168.32	1160.96	1612.56
0.33	1.0	1105.34	1173.35	1143.52	1584.22
0.66	0.5	1102.55	1145.27	1136.40	1598.78
0.66	1.0	1072.71	1145.20	1118.57	1564.89

Table 1: Values for the Measurement Error Study

PD	PFA	Window	lower bound	true value	relaxation value	greedy value
0.70	0.05	5	1434.61	1553.89	1451.05	1575.71
0.70	0.05	7	1139.46	1250.50	1194.32	1797.96
0.70	30.00	5	2116.52	2253.20	2145.59	2242.52
0.70	30.00	7	2209.77	2449.43	2336.52	2770.06
0.95	0.05	5	918.48	937.08	927.40	1108.30
0.95	0.05	7	1398.32	1441.10	1429.46	1995.20
0.95	30.00	5	2184.92	2239.24	2221.50	2354.16
0.95	30.00	7	3118.69	3241.66	3226.00	3626.97

Table 2: Values for the PD/PFA/Window Study

Tables 1 and 2 show that the value of the solution computed by the Lagrangian relaxation is invariably below the true solution value. This implies that the relaxation algorithm computes to below the noise level in the problem and that further computations would not yield better solutions of the tracking problems, even though the assignment problem solution might improve. In contrast, the solution value computed by the greedy algorithm is almost always *above* the true solution value. Tables 3 and 4 give the tracking problem solution quality (expressed as a percentage of the correctly identified tracks), as computed by the greedy and relaxation algorithms.

σ_r	σ_θ	6 sc.	7 sc.	8 sc.	9 sc.	10 sc.	11 sc.	12 sc.	13 sc.	14 sc.	15 sc.
0.33	0.50	96	88	92	94	95	96	97	97	97	97
0.33	1.00	95	90	92	94	95	96	97	98	97	98
0.66	0.50	94	87	91	94	95	96	97	97	97	98
0.66	1.00	93	89	91	93	94	95	96	96	96	96

Table 3: Percentage of Correctly Identified Tracks (greedy)

σ_r	σ_θ	6 sc.	7 sc.	8 sc.	9 sc.	10 sc.	11 sc.	12 sc.	13 sc.	14 sc.	15 sc.
0.33	0.50	100	100	100	100	100	100	100	100	100	100
0.33	1.00	100	100	100	100	100	100	100	100	100	100
0.66	0.50	100	100	100	100	100	100	100	100	100	100
0.66	1.00	99	99	99	99	99	99	100	100	100	100

Table 4: Percentage of Correctly Identified Tracks (relaxation)

Clearly, the solutions computed by the greedy algorithm are not below the noise level. As a result, the quality of the corresponding tracking solution suffers. The superior performance of the relaxation algorithm does carry a time penalty, as shown by Tables 5 and 6.

σ_r	σ_θ	6 sc.	7 sc.	8 sc.	9 sc.	10 sc.	11 sc.	12 sc.	13 sc.	14 sc.	15 sc.
0.33	0.50	0.14	0.70	0.71	0.62	0.55	0.51	0.47	0.44	0.42	0.39
0.33	1.00	0.20	1.66	1.77	1.50	1.29	1.19	1.07	0.99	0.92	0.88
0.66	0.50	0.20	1.25	1.33	1.14	1.04	0.97	0.89	0.86	0.84	0.80
0.66	1.00	0.35	3.86	4.35	3.88	3.24	2.93	2.81	2.62	2.48	2.37

Table 5: Greedy Algorithm Solution Times

σ_r	σ_θ	6 sc.	7 sc.	8 sc.	9 sc.	10 sc.	11 sc.	12 sc.	13 sc.	14 sc.	15 sc.
0.33	0.50	0.16	1.12	1.37	0.89	0.77	0.60	0.60	0.48	0.49	0.42
0.33	1.00	0.24	4.47	7.10	5.14	4.41	3.30	2.49	2.43	2.06	1.76
0.66	0.50	0.24	1.92	2.57	1.88	1.49	1.35	1.15	1.00	1.00	0.99
0.66	1.00	0.37	8.54	13.11	12.41	9.69	7.85	7.98	6.78	6.10	5.35

Table 6: Lagrangian Relaxation Solution Times

It should be noted, however, that the relaxation solution times are still compatible with real time requirements. Moreover, the implementation of the Lagrangian relaxation was not designed with speed as a primary requirement. More efficient coding would certainly yield significant speedups. In contrast, the greedy algorithm is very simple and thus offers only a limited scope for improvement.

Tables 7, 8, and 9 compare the impact of changing the bounds on target dynamics in the tracking problems for a problem with the probability of detection 0.95, probability of false alarms 0.05, σ_r of 0.66 miles and σ_θ of $1°$. The tracking window had six scans. If a faster target movement is allowed, observations that are further apart start to interact. In Table 7, this can be seen in the first few scans where the data association problems for the fast scenario have 2-3 times more variables than the problems for the slow scenario. As the simulation progresses and the information is accumulated, tighter gates can be constructed and the difference lessens.

scenario	6 sc.	7 sc.	8 sc.	9 sc.	10 sc.	11 sc.	12 sc.	13 sc.	14 sc.	15 sc.
slow	5088	39586	47267	52074	54389	53868	55367	42526	42524	41036
fast	11896	108936	121229	104064	88532	76053	69403	64349	61412	58729

Table 7: Assignment Problem Sizes

scenario	6 sc.	7 sc.	8 sc.	9 sc.	10 sc.	11 sc.	12 sc.	13 sc.	14 sc.	15 sc.
slow	0.18	1.57	1.71	1.98	1.90	1.80	1.72	1.44	1.47	1.43
fast	0.37	8.54	13.11	12.41	9.69	7.85	7.98	6.78	6.10	5.35

Table 8: Assignment Problem Solution Times

scenario	6 sc.	7 sc.	8 sc.	9 sc.	10 sc.	11 sc.	12 sc.	13 sc.	14 sc.	15 sc.
slow	98	100	100	100	100	100	100	100	100	100
fast	99	99	99	99	99	99	100	100	100	100

Table 9: Percentage of Correctly Identified Tracks

Table 8 shows that the problems for the fast scenario also take significantly more time to solve. However, the difference in the solution quality is minor, since the six scan tracking window provides enough information in both cases to obtain excellent solutions.

7. Concluding Remarks

In this work we have formulated some general classes data association problems in multitarget tracking and multisensor data fusion as multidimensional assignment problems. The objective function was derived from a composite negative log posterior or likelihood function for each of the tracks of reports from the different the data sets. This formulation is of sufficient generality to cover our earlier work [**24-27,31,32**], the popular multiple hypothesis tracking method introduced by Reid [**34**] and modified by Kurien [**19**] to include maneuvering targets and terminations. Track initiation and track maintenance are both posed as multidimensional assignment problems. In the area of centralized multisensor fusion and track initiation, Deb, Pattipati, Bar-Shalom [**9-11,23**] have developed similar multidimensional assignment problems and algorithms for their solution.

Although these multidimensional assignment problems are NP-hard [**14**], they are fundamentally important and central to multitarget and multisensor data fusion tracking systems. The only known method for solving these problems optimally is branch and bound; however, such a method is too computationally intensive to produce real-time solutions, especially in dense scenarios. Since the objective function is generally noisy due to various sources of errors (e.g., plant noise, observation errors, and uncertainty about exact values of various probability parameters), one only needs to solve these problems to or just below the noise level in the problem. Thus high quality and near optimal solutions computed to the noise level in the problem is the goal of algorithm development. Motivated by the work of Frieze and Yadegar [**13**], several algorithms based on Lagrangian relaxation have been and continue to be developed [**9-11, 23-33**]. These algorithms are near-optimal, provide a lower and upper bound on the optimal value, and are real-time for many scenarios. This work has presented a brief discussion of some of the salient features of these algorithms and a numerical study to demonstrate their potential effectiveness. A more complete development of our algorithms can be found in [**24-33**] and those of Deb, Pattipati, Bar-Shalom, Washburn, and Tsaknakis in references [**9-11,23**]. The quality of these algorithms will, no doubt, continue to improve for some time. Future work [**33**] will address the different relaxation techniques and their effectiveness.

References

1. B. D. O. Anderson and J. B. Moore, *Optimal Filtering*, Prentice-Hall, Englewood Cliffs, New Jersey, 1979.
2. Y. Bar-Shalom, *Multitarget-Multisensor Tracking: Advanced Applications*, Artech House, Dedham, MA., 1990.
3. ______, *Multitarget-Multisensor Tracking: Applications and Advances*, Artech House, Dedham, MA., 1992.
4. Y. Bar-Shalom and T. E. Fortmann, *Tracking and Data Association*, Academic Press, Boston, 1988.
5. D. P. Bertsekas, *Linear Network Optimization: Algorithms and Codes*, MIT Press, Cambridge, MA, 1991.
6. D. P. Bertsekas and D. A. Castañon, *A Forward/Reverse Auction Algorithm for Asymmetric Assignment Problems*, Computational Optimization and Applications **1** (1992), no. 3, 277 - 298.
7. S. S. Blackman, *Association and fusion of multiple sensor data*, Multitarget-Multisensor Tracking: Applications and Advances (Y. Bar-Shalom, ed.), Artech House, Dedham, MA, 1992.
8. I. J. Cox, J. M. Rehg, and S. Hingorani, *A Bayesian Multiple Hypothesis Approach to Contour Grouping and Segmentation*, Int. J. of Computer Vision **11** (1993), no. 1, 24 - 55.
9. S. Deb, K. R. Pattipati, and Y. Bar-Shalom, *A multisensor-multitarget data association algorithm for heterogeneous systems*, IEEE Transactions on Aerospace and Electronic Systems **29** (1993), no. 2, 560 - 568.
10. S. Deb, K. R. Pattipati, and Y. Bar-Shalom, *A S-dimensional assignment algorithm for track initiation*, Proc. IEEE Systems Conf., Kobe, Japan, 1992 (to appear).
11. S. Deb, K. R. Pattipati, Y. Bar-Shalom, and H. Tsaknakis, *A new algorithm for the generalized multidimensional assignment problem*, Proc. IEEE International Conference on Systems, Man, and Cybernetics, 1992 (to appear).
12. M.L. Fisher, *The Lagrangian Relaxation Method for Solving Integer Programming Problems*, Management Science **27** (1981), no. 1, 1 - 18.
13. A. M. Frieze and J. Yadegar, *An Algorithm for Solving 3-Dimensional Assignment Problems with Application to Scheduling a Teaching Practice*, Journal of the Operational Research Society **32** (1981), 989-995.

14. M. R. Garvey and D. S. Johnson, *Computers and Intractability: A Guide to the Theory of NP-Completeness*, W. H. Freeman & Co, CA, 1979.

15. A. M. Geoffrion, *Lagrangean relaxation for integer programming*, Mathematical Programming Study 2: Approaches to Integer Programming (M. L. Balinski, eds.), North Holland Publishing Company, Amsterdam, 1974.

16. M. Held and R. M. Karp, *The traveling salesman problem and minimal spanning trees*, Operations Research **18** (1970), 1138 - 1162.

17. M. Held and R. M. Karp, *The Traveling-Salesman Problem and Minimum Spanning Trees: Part II*, Mathematical Programming **1** (1971), 6-25.

18. R. Jonker and T. Volgenant, *A Shortest Augmenting Path Algorithm for Dense and Sparse Linear Assignment Problems*, Computing **38** (1987), 325 - 430.

19. T. Kurien, *Issues in the designing of practical multitarget tracking algorithms*, Multitarget-Multisensor Tracking: Advanced Applications (Y. Bar-Shalom, eds.), Artech House, Dedham, MA, 1990.

20. C. Lemarechal, *Bundle Methods in Nonsmooth Optimization*, Nonsmooth Optimization, IIASA Proceedings 3 (C. Lemarechal and R. Mifflin, eds.), Pergamon, Oxford, 1978, pp. 79 - 102.

21. C. L. Morefield, *Application of 0-1 Integer Programming to Multitarget Tracking Problems*, IEEE Transactions on Automatic Control **AC-22** (1977), no. 3, 302-312.

22. G. L. Nemhauser and L.A. Wolsey, *Integer and Combinatorial Optimization*, Wiley-Interscience, New York, 1988.

23. K. R. Pattipati, S. Deb, Y. Bar-Shalom, and R. B. Washburn, Jr., *A new relaxation algorithm and passive sensor data association*, IEEE Transactions on Automatic Control **37** (1992), no. 2, 198-213.

24. A. B. Poore and N. Rijavec, *A New Class of Methods for Solving Data Association Problems Arising from Multi-target Tracking*, Proceedings of the 1991 American Automatic Control Conference, vol. 3, IEEE, 1991, pp. 2303-2304.

25. A. B. Poore and N. Rijavec, *A Lagrangian Relaxation Algorithm for Multi-dimensional Assignment Problems Arising from Multi-target Tracking*, SIAM Journal on Optimization **3** (1993), no. 3, 645-663.

26. A. B. Poore and N. Rijavec, *Multitarget Tracking and Multidimensional Assignment Problems*, Signal and Data Processing of Small Targets 1991 (Oliver E. Drummond, ed.), SPIE, Bellingham, WA, 1991, pp. 345 - 356.

27. A. B. Poore, N. Rijavec, and T. Barker, *Data association for track initiation and extension using multiscan windows*, Signal and Data Processing of Small Targets (O. Drummond, ed.), SPIE, Bellingham, WA, 1992, pp. 432-441.

28. A. B. Poore and N. Rijavec, *A Numerical Study of Some Data Association Problems Arising in Multitarget Tracking*, Large Scale Optimization: State of the Art (W. W. Hager, D. W. Hearn and P. M Pardalos, eds.), Kluwer Academic Publishers, Boston, 1994, pp. 347 - 370.

29. A. B. Poore and N. Rijavec, *Partitioning multiple data sets: multidimensional assignments and Lagrangian relaxation*, DIMACS Series in Discrete Mathematics and Theoretical Computer Scienc (to appear).

30. A. B. Poore, N. Rijavec, M. Liggins, and V. Vannicola, *Data association problems posed as multidimensional assignment problems: problem formulation*, Signal and Data Processing of Small Targets 1954 (Oliver E. Drummond, ed.), vol. 1954, SPIE, Bellingham, WA, 1993, pp. 552-561.

31. A. B. Poore, N. Rijavec, T. Barker, and M. Munger, *Data association problems posed as multidimensional assignment problems: numerical simulations*, Signal and Data Processing of Small Targets 1954 (Oliver E. Drummond, ed.), vol. 1954, SPIE, Bellingham, WA, 1993, pp. 564-573.

32. A. B. Poore, N. Rijavec, T. Barker, and M. Munger, *Data association problems posed as multidimensional assignment problems: algorithm development*, Signal Processing, Sensor Fusion, and Target Recognition II 1955 (I. Kadar and V. Libby, ed.), vol. 1955, SPIE, Bellingham, WA, 1993, pp. 172-183.

33. A. B. Poore, *A comparison of several relaxation algorithms for multidimensional assignment problems*, in preparation.
34. D. B. Reid, *An Algorithm for Tracking Multiple Targets*, IEEE Transactions on Automatic Control **AC-24** (1979), no. 6, 843-854.
35. H. Schramm and J. Zowe, *A Version of the Bundle Idea for Minimizing a Nonsmooth Function: Conceptual Idea, Convergence Analysis, Numerical Results*, SIAM Journal on Optimization **2** (1992), 121-152.
36. N. Z. Shor, *Minimization Methods for Non-Differentiable Functions*, Springer-Verlag, New York, 1985.
37. R. W. Sittler, *An Optimal Data Association Problem in Surveillance Theory*, IEEE Transactions on Military Electronics **MIL-8** (1964), 125-139.
38. E. Waltz and J. Llinas, *Multisensor Data Fusion*, Artech House, Dedham, MA, 1990.
39. P. Wolfe, *A Method of Conjugate Subgradients for Minimizing Nondifferentiable Functions*, Mathematical Programming Study **3** (1975), 147-173.

DEPARTMENT OF MATHEMATICS, COLORADO STATE UNIVERSITY FORT COLLINS, COLORADO 80523

E-mail address: poore@math.colostate.edu

Part III

Computer Vision

DIMACS Series in Discrete Mathematics
and Theoretical Computer Science
Volume 19, 1995

GROUPING EDGES: AN EFFICIENT BAYESIAN MULTIPLE HYPOTHESIS APPROACH

INGEMAR J. COX, JAMES M. REHG, SUNITA L. HINGORANI,
AND MATT L. MILLER

ABSTRACT. A contour segmentation algorithm is presented that takes an edge map and extracts continuous curves of arbitrary smoothness, correctly handling curve intersections and capable of extrapolating over significant measurement gaps. The algorithm incorporates noise models of the edge detection process and limited scene statistics. It is based on an explicit contour model and employs a statistical distance measure to quantify the likelihood of each segmentation hypothesis. A Bayesian multiple hypothesis tree organizes possible segmentations, making it possible to postpone grouping decisions until a sufficient amount of information is available. We have demonstrated its performance on real and synthetic images.

The algorithm is based on the multiple hypothesis tracking (MHT) algorithm of Reid [32], developed in the context of surveillance and target tracking. In fact, it was this interdisciplinary research that partially motivated the organization of this DIMACS workshop. Our implementation of the MHT exploits results from combinatorial optimization that allow the k-best hypothesis to be determined in polynomial time without explicit enumeration of all legal hypotheses. A very efficient approximation to the MHT is thereby achieved which is likely to significantly extended the practical application of the algorithm.

1. INTRODUCTION

The set of image contours produced by objects in a scene encode important information about their shape, position, and orientation. Image contours arise from discontinuities in the underlying intensity pattern, due to the interaction of surface geometry and illumination [38]. A large body of work, from such areas as model-based object recognition [17, 20, 22, 25], contour motion flow [18, 39] and shape from contour [19], depend critically on the reliable extraction of image contours.

Contour extraction is a two step process consisting of edge detection followed by contour grouping. Edge detectors [7, 4] are sensitive to significant changes in

---

1991 *Mathematics Subject Classification.* Primary 68T10, 68U10.

An earlier version of this paper appeared in the Int. J. of Computer Vision, **11**, 1, 5-24, (1993). This paper includes new material describing an efficient implementation and additional experimental results. Thanks to Y. Bar-Shalom, H. Durrant-Whyte, T. Kurien and J. J. Leonard for valuable discussion on issues related to target tracking. Special thanks to David Lowe for providing the "Totem Pole" image and making his source code available for our comparison.

image intensity over a small neighborhood. They output the position (and possibly orientation) of detected edge points (edgels) in the form of an edge map. The grouping process, with which this paper is concerned, involves collecting individual edge points together to form continuous curves. Once the assignment of edge points to curves has been made, they can be processed further to obtain, for example, curvature statistics or model parameters.

Associating edge points with a contour is difficult because the input data (from edge detectors) is noisy; there is uncertainty in the position of the edge, there may be false and/or missing points, and contours may intersect and interfere with one another. These difficulties are illustrated in Figure (1). An algorithm that operates by walking along the contour from left to right, classifying edge pixels as it finds them, will be unable to differentiate between the three possible assignments for the black pixel, at the point when it is first encountered.

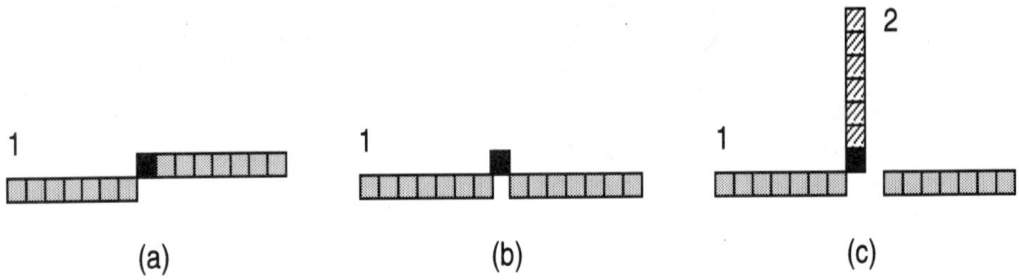

FIGURE 1. Three possible contour scenarios. In (a), the black pixel belongs to contour 1, in (b) it is a false edge due to noise, and in (c) it belongs to contour 2 (and contour 1 continues to the right across the gap).

The scenarios in Figure (1) illustrate four basic requirements for a successful contour segmentation algorithm. First, there must be a mechanism for integrating information in the neighborhood of an edgel to avoid making irrevocable grouping decisions based on insufficient data. Second, there must be a prior model for the smoothness of the curve to base grouping decisions on. This model must have an intuitive parameterization and sufficient generality to describe arbitrary curves of interest. Third, it must incorporate noise models for the edge detector, in order to optimally incorporate noisy measurements, and detect and remove spurious edges. And finally, since intersecting curves are common, the algorithm must be able to handle these as well. As we argue in Section (2), our algorithm is the first unified framework to incorporate these four requirements.

We formulate contour grouping as a Bayesian multiple hypothesis "tracking" problem. Our work is based on an algorithm originally developed by Reid[32] in the context of military and surveillance tracking of multiple targets (aircraft) in the presence of noise. The algorithm has three main components: A "dynamic" contour model that encodes the smoothness prior, a measurement model that incorporates edge detector noise characteristics, and a Bayesian hypothesis tree that encodes the likelihood of each possible edge assignment and permits multiple hypotheses to develop in parallel until sufficient information is available to make a decision.

Our edge grouping algorithm has been implemented in a system that takes an intensity image and segments it into smooth curves. The system has three parts: edge detection and thinning, edge grouping, and post-processing. The output of the first stage is an eight-connected edge image. The tracking algorithm outlined above is applied to the processed edge image in the second stage. Finally, tracks that correspond to the same underlying contour are merged in a post-processing step. These stages are explained in detail in the sections that follow.

Our algorithm is novel in that the edge grouping and segmentation processes are tightly coupled, rather than pipelined. This coupling is seen to significantly improve the final segmentation, providing longer contours than would otherwise be obtained. In addition, the algorithm is unique in explicitly distinguishing between the geometric uncertainty and data association uncertainty. The geometric uncertainty present in both the measurements and contours is represented by conventional covariance matrices and the grouping and segmentation uncertainty, i.e. the data association uncertainty, is represented by a hypothesis tree with Bayesian probabilities associated to each leaf. This decoupling of the data association and noise uncertainties is a distinguishing feature of this approach that should allow for more robust interpretation and integration by later processing modules.

Section (2) reviews previous work in contour grouping. Section (3) describes the Bayesian multiple hypothesis contour grouping algorithm. Section (4) describes important implementation techniques for constraining the memory and computational requirements of the algorithm. Section (5) describes experimental results on real and synthetic images, and is followed by a discussion of the novel properties of our algorithm and some directions for future work in Section (6).

2. Previous work

Contour segmentation is a classical problem with a long history. Many previous attempts share with our system the basic approach of edge labeling through a sequential search of the edge map. The Multiple Hypothesis Tracking (MHT) approach provides a sound theoretical framework for this grouping paradigm. The key concepts of track, measurement, and hypothesis nicely describe the main components of the segmentation task. Furthermore, algorithms based on alternative paradigms [24] still often require eight-connected "chains" of pixels as inputs. The performance of this initial, sequential segmentation can impact the quality of the final grouping, as we illustrate in Section (5). In the remainder of this section, we briefly review other segmentation approachs.

The early segmentation systems described in [28, 26] employ search trees analogous to our hypothesis trees. However, these approaches lack two other key components of our system: dynamic models for contours and track initiation from new measurements. Montanari [28] uses dynamic programming [3] to detect the path through a set of pixels that globally optimizes an objective function emphasizing high edge strength and low overall curvature. Martelli [26] uses the A^* algorithm to speed up the search through the application of heuristics. Like our method, these algorithms perform tree search and can follow arbitrary smooth curves. However, they generally lack the ability to handle intersections and exploit statistical models.

Lowe [24] describes an algorithm for contour grouping based on multiscale smoothing and local curvature analysis. His approach is similar to ours in its use of an implicit model of local curve smoothness, but it has several important differences. The most important distinction is the interval-based, multiresolution nature of the algorithm. However, this formulation dictates that the input to the system be a set of edge chains, requiring an additional algorithm to perform an initial segmentation. In some situations, errors in the initial segmentation will be propagated to the final output, as we illustrate in Section (5).

The Hough transform is probably the most popular algorithm for detecting *parameterized shapes* (typically line segments and arcs) within an image [14]. However this approach has several deficiencies, including an inability to detect arbitrary smooth curves and the difficulty of efficiently discretizing the parameter space to form the accumulator array. Hough transforms are typically used in domains where straight lines are desired and some prior knowledge is available. In fact, many of the algorithms described below perform fairly well for straight contours. Our interest, however, is in recovering smooth contours without explicit parameterization into, for example, circles and arcs.

David and Zucker [13] present an algorithm that covers salient curves with an overlay of short oriented edge elements. These elements are attracted to the minima of a potential field generated from an edge image, and perform segmentation by propagating labels to their neighbors. One basic problem with the algorithm is that it doesn't treat intersecting curves explicitly, and it's unlikely that their simple labeling process will segment them correctly (the paper focuses on the problem of generating the covering, and doesn't present segmentation results.) It also lacks a framework for incorporating noise models of the edge detection process.

Most recently, Shashua and Ullman [33] describe preliminary results in identifying disjoint groupings of edge elements by optimizing an overall saliency measure favouring long smooth curves. They present some interesting results, but is unclear whether their optimization criteria has the flexibility to produce a wide variety of segmentations. In particular, explicitly optimizing for contour length may not be appropriate since a short high contrast contour may be perceptually more significant than a longer but much noisier contour.

In an accompanying paper, Feldman [16] develops a Bayesian approach to characterizing whether 3 points are interpreted as colinear by human observers. This work is somewhat complimentary to our own. In particular, Feldman seeks to determine the conditions under which an *isolated* set of 3 dots is seen as either a curvilinear contour (chain) or an unordered random cluster. In order to do this, Feldman proposes a generative model for contour formation based on a linear translation process. Coincidentally, we assume a similar contour formation model, as described in Section (3.4). It would be interesting to include Feldman's perceptual model within our framework, perhaps using his model as a source of motivated estimates for some of the free parameters in our model. Feldman also includes a brief discussion of how these 3 point clusters could be grouped into larger clusters using a sliding window of triplets. It is unclear how sucesssive triplets ought to be combined into edges, since (as Feldman points out [16]) one would not expect successive triplets in an edge to be independent. Moreover, it is unclear how isolated spurious edgels could be handled if such points must be categorized as an unordered random cluster, with minimum size of 3 edgels (dots). Most significantly, Feldman's model ignores the issue of computational efficiency - all possible permutations of the edgels must be examined in order to determine the most likely configuration, an exponential process - since it is motivated as a competence theory for human perception rather than as a method for finding contours. While the MHT framework is also exponential in theory, a computationally efficient approximation exists in practice, see Section (4).

3. Multiple Hypothesis Framework for Contour Grouping

In this section we present a technique for the extraction of contours from an edge image, based heavily on the Multiple Hypothesis Tracking (MHT) algorithm originally developed by Reid[32]. The three main components of the algorithm are measurements, tracks, and hypotheses. In our context, measurements are edge elements, samples of an ideal image curve that are obtained from the input edge map. Tracks provide an evolutionary description of the image curves. Each track has a dynamic model that specifies curve smoothness, and an estimator that gives the contour shape as a function of the edge points assigned to it. In our case, the estimator is a Kalman Filter with edge points as its measurements. An association hypothesis is an assignment of an edge point to either a track or the set of noise edgels not belonging to any track (called false alarms in the tracking literature.) Possible assignments of measurements to tracks are organized into a hypothesis tree. Each node of the tree contains an event, an assignment of a group of measurements to a set of tracks.

The hypothesis tree represents all possible interpretations of the edge image as sequences of events. Each hypothesis node is assigned a probability, computed from track estimators and priors. If all measurements were processed and the tree grown to its full depth, each path through the tree from the root to a leaf would describe a unique labeling of all edgels in the image, visited in a certain order. The optimal maximum a posteriori (MAP) segmentation would then be the maximum probability path. In practice, the tree must be pruned at each iteration to control its growth.

Figure (2) outlines the basic operation of the contour grouping algorithm. An iteration begins with the set of current hypotheses (at iteration $k-1$). Each hypothesis (leaf) contains a set of active tracks, and becomes a parent hypothesis node in the current iteration. For each active contour in each parent node, a predictor is applied to compute its next location in the image (in our case this is the prediction step of a Kalman Filter). An adaptive search region is centered at each of these predicted locations as shown in Figure (3) and detailed in Section (4.2). Measurements are extracted from these surveillance regions and matched to contour predictions using the statistical Mahalanobis distance. A threshold on the distance measure (called a validation gate) produces a set of candidate assignments of measurements to contours. The resulting enumeration of associations produces a set of children (events) for each parent node, extending the depth of the tree by another level. Associated with each new leaf is a probability whose computation is described in Section (3.3). In the final step of the iteration, the tree is pruned to remove unlikely segmentations.

The three subsections that follow describe the algorithm in more detail, drawing from [2, 8, 23]. The next subsection defines the segmentation problem and introduces some useful terminology. It is followed by a description of the hypothesis generation process. The final subsection outlines the probability computations.

3.1. Problem Statement. The goal of the segmentation algorithm is to generate an assignment of edge points to estimated contours satisfying two properties: *disjointness*, each edgel belongs to either a unique contour or the set of false alarms, and *completeness*, there is an assignment for each edgel in the input image. To

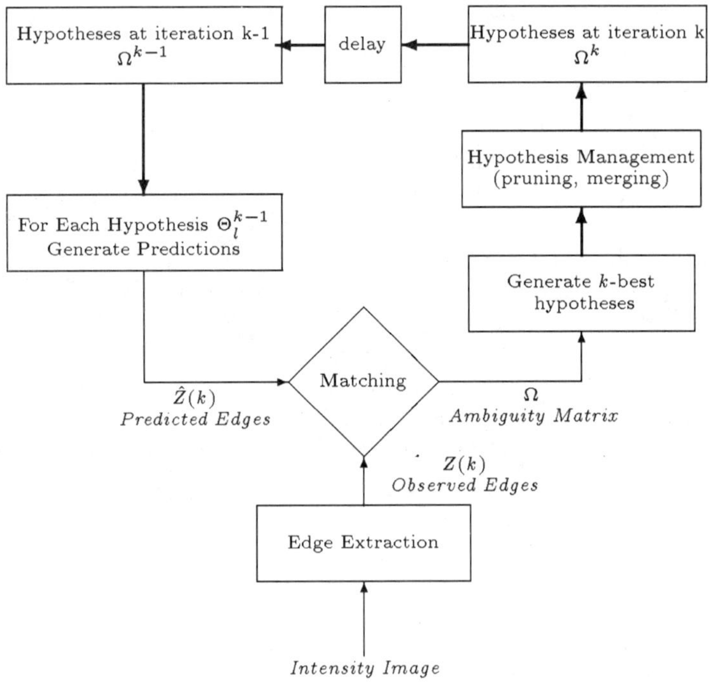

FIGURE 2. Outline of the contour grouping algorithm

accomplish this, the algorithm must perform three simultaneous tasks: the extraction of edge measurements from the image, the generation of contours from groups of measurements, and the enumeration of the likelihoods of all possible edge assignments, so that an optimal segmentation can be selected. Since there are any number of possible segmentations for a given set of edgels, the generation of likelihoods for assignments must encode the definition of a "correct" segmentation. To define the problem statement in detail, we must introduce some basic notation.

Suppose that at iteration $k-1$ the system contains a set of tracks (contours) denoted by $Q(k-1)$. Through a process described in Section (4.2), a set of edge measurements is obtained from their surveillance regions. Let $Z(k)$ denote the set of m_k measurements found at iteration k,

(1) $$Z(k) \triangleq \{z_i(k)\}_{i=1}^{m_k}$$

Each $z_i(k)$ is a vector $\{x_i, y_i, \dot{x}_i, \dot{y}_i\}$ obtained from the edge image, where (x_i, y_i) are the coordinates of an edge point and $(\dot{x}_i, \dot{y}_i)$ are obtained from its slope.

Once a measurement set is obtained, the hypothesis tree must be updated to include these measurements in the total segmentation. Each node in the tree contains a set of associations (an event) for all of the measurements found at that level (iteration). There are three possible hypotheses for each edge measurement:

- The measurement starts a new track.
- The measurement is a false alarm.
- The measurement belongs to an existing track.

In addition to these associations, each tree node contains the status of a set of tracks. The tracks present in a given node are a combination of tracks from the node's parent and any new tracks resulting from the set of measurement associations. For any track in existence at time $k-1$, there are three possible track hypotheses at time k:

- The track is assigned one of the new measurements and continues.
- The track does not receive a measurement, but still continues.
- The track does not receive a measurement, and terminates.

We denote by $\theta_m(k)$ a particular set of measurement and track assignments at the kth iteration, called an *event*. In our implementation, each $\theta_m(k)$ is represented by a matrix stored in the mth leaf of the kth level of the hypothesis tree.

Now let Z^k denote the total set of measurements up to and including iteration k,

$$Z^k \triangleq \{Z(i)\}_{i=1}^{k} \tag{2}$$

A particular hypothesis, Θ_m^k, denotes a particular disjoint partitioning of the measurements, Z^k, and has a likelihood $P\{\Theta_m^k|Z^k\}$, whose computation is described in Section (3.3). Let $\Theta_{l(m)}^{k-1}$ denote the parent hypothesis from which Θ_m^k is derived. Then $\theta_m(k)$ specifies the status of all contours postulated by $\Theta_{l(m)}^{k-1}$ at iteration k, as well as the origins of the measurement set $Z(k)$.

Finally, suppose that at some iteration k_f all of the edge measurements in the image have been processed, resulting in a set of M_f possible segmentations. The optimal contour segmentation is then that $\Theta_o^{k_f}$ which maximizes $P(\Theta_o^{k_f} \mid Z^{k_f})$. This conditional probability will be examined in detail in Section (3.3), where a recursive update formula for the association hypotheses is derived.

3.2. Hypothesis Generation. The goal of hypothesis generation is to enumerate all valid assignments of measurements to tracks and false alarms. This is an exponential process that can quickly overwhelm available computational resources. Therefore, in practice, we choose to enumerate only the k-best hypotheses. Section (4.3.3) describes how the k-best hypothesis can be efficiently generated in polynomial time. However, first, it is important to understand the properties and constraints associated with a valid hypotheses, since these characteristics are essential to the k-best algorithm.

At each iteration, hypotheses are generated for the current measurement set, given the previous set of tracks. This corresponds to growing the hypothesis tree another level. Each event from the previous time step generates children based on its ability to explain the current measurement set. In the terminology of the previous subsection, the hypothesis generation routine processes $Z(k)$ and $\{\Theta_m^{k-1}\}$ to obtain $\{\theta_m(k)\}$. In generating each event, it must satisfy two requirements:

(1) Within a single event, each measurement is assigned to only one contour. This guarantees that the disjointness requirement described at the beginning of Section (3.1) will be met.
(2) Within a single event, each track is assigned only one measurement. This requirement is imposed by the sequential nature of the tracking algorithm.

In more traditional target tracking applications, such as radar, the second condition is enforced physically by the sensor: a single plane cannot produce more than one radar return in a given time instant. For contour segmentation, this requirement must be enforced in the measurement acquisition stage to ensure the validity of the model of tracks evolving in steps of unit arc length along image contours[1]. Section (4.2) shows how the second requirement can be met through careful selection of search regions.

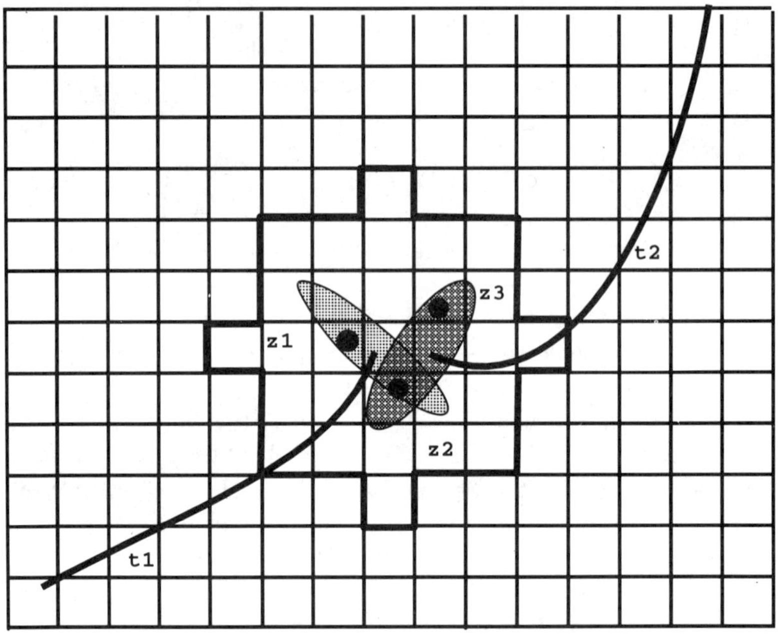

FIGURE 3. Predicted contour locations, a surveillance region and statistical Mahalanobis (elliptical) regions for a situation with two known contours (T_1 and T_2) and three new measurements ($z_1(k)$, $z_2(k)$ and $z_3(k)$).

The set of all possible assignments of measurements to tracks is represented by an array of ones and zeros called a *ambiguity matrix*. A particular assignment is represented by a related matrix called the *event* matrix. Thus hypothesis generation can be posed as the construction of a set of event matrices from a single ambiguity matrix. Given the hypothesis Θ_m^{k-1} and the measurement set $Z(k)$, the ambiguity matrix is generated as follows. First, the set $Z(k)$ is tested against the list of tracks

[1]The assignment of multiple measurements to a contour can be accommodated within Reid's theoretical framework. However, in practice, the number of possible assignments that must be considered for each contour at a given iteration would rise dramatically, making it impractical. Moreover, we would be unable to generate the k-best hypotheses, as described in Section (4.3.3), because essential properties of the event matrices would no longer be true.

in Θ_m^{k-1}. The subset of measurements that validate one or more tracks makes up the valid list for that event. Second, this list is used to construct a matrix that has the contours of Θ_m^{k-1} as its columns and the total measurements from $Z(k)$ as its rows. A non-zero element at matrix position $c_{i,j}$ indicates that measurement $z_i(k)$ validated contour t_j. In addition to columns from known contours, the ambiguity matrix has an initial column denoting false alarms and a final column denoting new contours. As an example, consider Figure (3), which depicts a situation in which there are two known contours (T_1 and T_2) and three new measurements, $\{z_1(k), z_2(k)$ and $z_3(k)\}$, at a given iteration. The corresponding ambiguity matrix is shown in Figure (4), where T_F denotes the false alarm column and T_N the new target column.

$$\begin{array}{c} \\ \Omega = \end{array} \begin{array}{c} \begin{array}{cccc} T_F & T_1 & T_2 & T_N \end{array} \\ \left(\begin{array}{cccc} 1 & 1 & 0 & 1 \\ 1 & 1 & 1 & 1 \\ 1 & 0 & 1 & 1 \end{array} \right) \end{array} \begin{array}{c} \\ z_1(k) \\ z_2(k) \\ z_3(k) \end{array}$$

FIGURE 4. The Corresponding Ambiguity Matrix for the configuration illustrated in Figure (3).

The ambiguity matrix provides a concise representation of the possible assignments between measurements and contours. For example, the four non-zero entries in the middle row reveal that measurement $z_2(k)$ could have originated from either contour T_1 or T_2 or be a false alarm or the start of a new contour. A set of event matrices is generated, constituting the children of the parent hypothesis. An event matrix is constructed by taking each row of the ambiguity matrix and selecting one nonzero element, subject to the constraint that there be no more than one nonzero element per existing track column. There may be any number of non-zero elements in the false alarm and new contour columns. Two possible event matrices are illustrated in Figure (5). Each event matrix is stored in a leaf node of the hypothesis tree.

A complication arises when validating measurements to contours: there is a π ambiguity in the gradient of the measurement. This is dealt with by the concept of a measurement pair; every measurement is treated as two measurements with opposing velocities. The measurement set $Z(k)$ then becomes

$$Z(k) = \{z_1^+(k), z_1^-(k), \ldots, z_{m_k}^+(k), z_{m_k}^-(k)\} \tag{3}$$

and there are now $2m_k$ measurements. A ambiguity matrix is constructed with pairs of rows representing the measurement pairs, as shown in Figure (6). Conventional ambiguity matrices are generated from this *compound* ambiguity matrix by selecting one row from each pair. In practice, it is often the case that one direction or the other will not validate any of the existing contours, thus ruling out a large number of the multiple hypotheses.

$$\Theta_1(k) = \begin{array}{c} \begin{array}{cccc} T_F & T_1 & T_2 & T_N \end{array} \\ \left(\begin{array}{cccc} 0 & 1 & 0 & 0 \\ 0 & 0 & 1 & 0 \\ 1 & 0 & 0 & 0 \end{array} \right) \end{array} \begin{array}{c} \\ z_1(k) \\ z_2(k) \\ z_3(k) \end{array}$$

$$\Theta_2(k) = \begin{array}{c} \begin{array}{cccc} T_F & T_1 & T_2 & T_N \end{array} \\ \left(\begin{array}{cccc} 1 & 0 & 0 & 0 \\ 0 & 1 & 0 & 0 \\ 0 & 0 & 0 & 1 \end{array} \right) \end{array} \begin{array}{c} \\ z_1(k) \\ z_2(k) \\ z_3(k) \end{array}$$

FIGURE 5. Two of many Event Matrices derived from the one Ambiguity Matrix of Figure (4).

3.3. Probability calculations. Once the candidate hypotheses have been enumerated, they can be assigned likelihoods through a recursive formula. The following derivation is based on [2, 23]. Consider a given global hypothesis Θ_m^k, made up of a current event and a previous hypothesis using all measurements up to $k-1$, i.e.

(4) $$\Theta_m^k = \left\{\Theta_{l(m)}^{k-1}, \theta_m(k)\right\}$$

Let T denote the total number of contours postulated by the parent hypothesis $\Theta_{l(m)}^{k-1}$. Next, we define the event $\theta_m(k)$, based on current measurements, to consist of τ measurements from known contours, ν measurements from new contours, ϕ false alarms and χ terminated (or ended) contours from the parent hypothesis.

For all the current measurements, $z_i(k)$, $i = 1, \ldots, m_k$ we define the indicator variables

(5) $$\tau_i \triangleq \begin{cases} 1 & z_i(k) \text{ came from a known contour} \\ 0 & \text{otherwise} \end{cases}$$

(6) $$\nu_i \triangleq \begin{cases} 1 & z_i(k) \text{ is a new contour} \\ 0 & \text{otherwise} \end{cases}$$

(7) $$\delta_t \triangleq \begin{cases} 1 & \text{if contour } t \text{ (in } \Theta_{l(m)}^{k-1}\text{) is detected at iteration } k \\ 0 & \text{otherwise} \end{cases}$$

(8) $$\chi_t \triangleq \begin{cases} 1 & \text{if contour } t \text{ (in } \Theta_{l(m)}^{k-1}\text{) is deleted at iteration } k \\ 0 & \text{otherwise} \end{cases}$$

$$\Omega = \begin{array}{c} \begin{array}{cccc} T_F & T_1 & T_2 & T_N \end{array} \\ \left(\begin{array}{cccc} 1 & 1 & 0 & 1 \\ 1 & 1 & 0 & 1 \\ 1 & 1 & 1 & 1 \\ 1 & 1 & 1 & 1 \\ 1 & 0 & 1 & 1 \\ 1 & 0 & 1 & 1 \end{array} \right) \end{array} \begin{array}{l} \\ z_1^+(k) \\ z_1^-(k) \\ z_2^+(k) \\ z_2^-(k) \\ z_3^+(k) \\ z_3^-(k) \end{array}$$

FIGURE 6. A ambiguity matrix in which pairs of rows represent measurement pairs.

Now consider the conditional probability of the global hypothesis Θ_m^k, written $P\{\Theta_m^k|Z^k\}$. Using Bayes' rule, we have

$$\begin{aligned} (9)\quad P\left\{\Theta_m^k|Z^k\right\} &= P\left\{\theta_m(k), \Theta_{l(m)}^{k-1}|Z(k), Z^{k-1}\right\} \\ &= \frac{1}{c} p[Z(k)|\theta_m(k), \Theta_{l(m)}^{k-1}, Z^{k-1}] P\left\{\theta_m(k)|\Theta_{l(m)}^{k-1}, Z^{k-1}\right\} \times \\ &\quad P\{\Theta_{l(m)}^{k-1}|Z^{k-1}\} \end{aligned}$$

where c is a normalization constant. The last term of this equation, $P\{\Theta_{l(m)}^{k-1}|Z^{k-1}\}$, represents the probability of the parent global hypothesis and is therefore available from the previous iteration. The remaining two terms may be evaluated as follows.

If the number of false alarms and new features are assumed to be Poisson distributed with densities λ_F and λ_N, respectively, then the second factor of Equation (9) is obtained by combining results from [2] and [23] to yield

$$\begin{aligned} (10)\quad P\{\theta_m(k)|\Theta_{l(m)}^{k-1}, Z^{k-1}\} &= \frac{\phi!\nu!}{m_k!}\mu_F(\phi)\mu_N(\nu)\prod_t \left(P_D^t\right)^{\delta_t}\left(1-P_D^t\right)^{1-\delta_t} \times \\ &\quad \left(P_\chi^t\right)^{\chi_t}\left(1-P_\chi^t\right)^{1-\chi_t} \end{aligned}$$

where P_D^t and P_χ^t are the probabilities of detection and termination of contour t and $\phi = m_k - \tau - \nu$ and $\nu = \sum_{i=1}^{m_k} \nu_i$ are the total number of false alarms and new contours respectively hypothesized by $\theta_m(k)$.

To determine the first term on the right hand side of Equation (9) we assume that a measurement $z_i(k)$ has a Gaussian probability density function (PDF)

$$\begin{aligned} N_{t_i} = N[z_i(k)] &\triangleq N[z_i(k); \hat{z}_i(k|k-1), S^{t_i}(k)] \\ (11)\qquad &= \left|2\pi S^{t_i}(k)\right|^{-\frac{1}{2}} e^{-\frac{1}{2}\{(z(k)-\hat{z}(k|k-1))^T\{S^{t_i}(k)\}^{-1}(z(k)-\hat{z}(k|k-1))\}} \end{aligned}$$

if it is associated with contour t_i, where $\hat{z}_i(k|k-1)$ denotes the predicted measurement for contour t_i and $S^{t_i}(k)$ is the associated innovation covariance. If the measurement is a false alarm, then its PDF is assumed uniform in the observation volume, i.e. V^{-1}. The value of V is discussed in more detail in the next section. Likewise, the probability of a new contour is also taken to be uniform with PDF V^{-1}. Under these assumptions, we have that

$$(12) \qquad p[Z(k)|\theta_m(k), \Theta^{k-1}_{l(m)}, Z^{k-1}] = \prod_{i=1}^{m_k} [N_{t_i}[z_i(k)]]^{\tau_i} V^{-(1-\tau_i)} = V^{-\phi-\nu} \prod_{i=1}^{m_k} [N_{t_i}[z_i(k)]]^{\tau_i}$$

Substituting Equations (12) and (10) into Equation (9) yields the final expression for the conditional probability of an association hypothesis

$$(13) \qquad P\{\Theta^k_m|Z^k\} = \frac{1}{c}\frac{\phi!\nu!}{m_k}\mu_F(\phi)\mu_N(\nu)V^{-\phi-\nu}\prod_{i=1}^{m_k}[N_{t_i}[z_i(k)]]^{\tau_i} \times \left\{\prod_t (P^t_D)^{\delta_t}(1-P^t_D)^{1-\delta_t}\left(P^t_\chi\right)^{\chi_t}\left(1-P^t_\chi\right)^{1-\chi_t}\right\} \times P\{\Theta^{k-1}_{l(m)}|Z^{k-1}\}$$

As assumed above, the probability mass functions $\mu_F(\phi)$ and $\mu_N(\nu)$ are Poisson distributed,

$$(14) \qquad \mu_F(\phi) = \exp\{-\lambda_F V\}\frac{(\lambda_F V)^\phi}{\phi!} \quad \textit{and} \quad \mu_N(\nu) = \exp\{-\lambda_N V\}\frac{(\lambda_N V)^\nu}{\nu!}$$

where λ_F and λ_N are the mean rates of false alarms and new contours. Inserting Equations (14) in (13), finally results in

$$(15) \qquad P\{\Theta^k_m|Z^k\} = \frac{1}{c'}\lambda_N^\nu\lambda_F^\phi\prod_{i=1}^{m_k}[N_{t_i}[z_i(k)]]^{\tau_i} \times \left\{\prod_t (P^t_D)^{\delta_t}(1-P^t_D)^{1-\delta_t}\left(P^t_\chi\right)^{\chi_t}\left(1-P^t_\chi\right)^{1-\chi_t}\right\} \times P\{\Theta^{k-1}_{l(m)}|Z^{k-1}\}$$

where c' is a normalization constant.

One of the parameters in the above expression is the probability of termination, P_χ. It controls the distance over which a contour will be extended if no measurements are present. In order to ensure that tracks which receive no measurements eventually terminate, we set

$$(16) \qquad P_\chi = 1 - \exp(-m/\lambda_\chi)$$

where λ_χ is a control parameter and m is specific to each contour and is the number of consecutive iterations in which no measurement was assigned to the contour. Thus, P_χ increases with each missed measurement, guaranteeing the eventual termination of a contour.

The previous derivation made two sets of assumptions about the statistical distribution of false alarms and new features: Poisson arrival rates for the detection of false alarms and new features as the contour is traversed, and uniform distributions for the location of false alarms and new features within the surveillance volume. These assumptions simplify the derivation of the PDF, and the parameters in the distributions provide a set of control values that can be adjusted to modify the outcome of the segmentation. The role of these parameters is analogous to the gains of a Kalman filter: they provide a convenient way to index a useful family of filters. In Section (5), we give an example of how model parameters can be varied to achieve different segmentations. Finally, we note that the Poisson distributions in Equation (14) can be replaced by uniform distributions, if desired.

3.4. Contour model. The contour model has two components: a dynamic model that describes the evolution of the curve, and an estimator that processes measurements to obtain curve shape. An appropriate dynamic model can be formed from a wide variety of parametric curve models including polynomials, B-splines and SNAKEs [21], as well as constant gradient (piecewise linear) and constant curvature models. The choice of contour model is not critical to the performance of the algorithm. However, a model should be efficient in both computation and memory since this will have a significant impact on the performance of the overall algorithm.

We employ a Kalman Filter in our estimation step because its recursive nature makes it computationally efficient. This is an important issue, since the algorithm can have hundreds of tracks in existence simultaneously. Moreover, the combination of a Kalman Filter and a noise-driven linear dynamic model has been used successfully in a wide variety of classical tracking applications to follow targets with smoothly varying trajectories at a reasonable computational cost. (In addition, see [15] for further discussion of alternative Kalman filter models of line segments and [34] for an interesting comparison between Kalman Filters and the SNAKE model.)

The Kalman Filter fits nicely into the framework described in the previous sections. In particular, the quantity in the exponential of Equation (11) is exactly the innovation term calculated by the Kalman filter. A first order (piecewise linear) Kalman filter is described in the next section. Although the contour is represented *locally* by a straight line, the process noise in the Kalman formulation permits curves that are *globally* nonlinear to be represented. This is illustrated in Section (5) by the shapes of the curves we were able to model and extract.

3.4.1. *The measurement vector and associated Kalman filter.* The edge image under analysis is assumed to consists of edge points with both position and gradient information, i.e. the measurement vector $z(k)$ is given by

$$z(k) = (x, \dot{x}, y, \dot{y})' \tag{17}$$

where (x, y) are the coordinates of the edge point, $\dot{x}$, $\dot{y}$ are the x and y gradients with respect to arc length and $'$ denotes transpose.

As the contour is traversed, we wish to estimate the position and gradient of the contour's endpoint, i.e. the Kalman filter state vector is

$$x(k) \stackrel{\Delta}{=} (x, \dot{x}, y, \dot{y})' \tag{18}$$

As a result of this choice, the observation function $H(k)$ is a 4x4 identity matrix.

Contour dynamics are described with a noise driven acceleration model common in the tracking literature ([2], Chapter 2). Each curve is driven by a continuous white Gaussian acceleration with autocorrelation q. Sampling in increments, Δs, of arc length produces a discrete state transition matrix $F(k)$, given by

$$F(k) = \begin{pmatrix} 1 & \Delta s & 0 & 0 \\ 0 & 1 & 0 & \\ 0 & 0 & 1 & \Delta s \\ 0 & 0 & 0 & 1 \end{pmatrix} \tag{19}$$

This model has the property that curves of arbitrary smoothness can be obtained by appropriate setting of the process noise parameter q. For example, a very small value forces contour segments to be straight lines. The increment of arc length, Δs, specifies the distance traveled along the curve in one iteration. This in turn is a function of the size of the surveillance region, as discussed in Section (4.2).

The sampling process that produces the discrete dynamic model also specifies the discrete process noise covariance used in the Kalman Filter. In our case, this covariance matrix is given by

$$Q(k) = \begin{pmatrix} \Delta s^3/3 & \Delta s^2/2 & 0 & 0 \\ \Delta s^2/2 & \Delta s & 0 & 0 \\ 0 & 0 & \Delta s^3/3 & \Delta s^2/2 \\ 0 & 0 & \Delta s^2/2 & \Delta s \end{pmatrix} q \tag{20}$$

where q is the variance of the continuous white noise process [2].

The final step is the specification of the measurement noise covariance. In principle, the measurement noise model can be obtained from an analysis of the edge detector. This has been done, for example, by Matthies *et al* [27] in the case of the Canny detector. In our case, we used a simple diagonal covariance matrix,

$$R(k) = \begin{pmatrix} \sigma_p & 0 & 0 & 0 \\ 0 & \sigma_g & 0 & 0 \\ 0 & 0 & \sigma_p & 0 \\ 0 & 0 & 0 & \sigma_g \end{pmatrix} \tag{21}$$

where σ_p and σ_g are the variance in the measurement of position and gradient, respectively. These parameters represent the extent of deviation of the edge detector output from the ideal, due to image noise and discretization effects.

4. Implementation

This section gives more detail on the implementation of the edge grouping algorithm, and its incorporation into a contour segmentation system. In the subsections that follow, we discuss the initialization process, the surveillance region, efficient hypothesis tree implementation, and the post-processing contour fusion step.

4.1. Initialization. After the edge detection and thinning stage, the input edge image is scanned from left to right, top to bottom, until an edge element is found. When the first edge measurement is located, the grouping algorithm is invoked. It immediately initiates two contours (a pair moving in opposite directions), since the edge measurement contains both location and gradient information[2]. A (trivial) hypothesis matrix is constructed which represents the possibilities that the edge is a false alarm or the start of a new contour (initially there are no existing contours). A hypotheses tree is generated with three branches, one representing the assignment of the edge as a false alarm and two hypotheses assigning the edge to be the start of a new contour (one for each direction). These new contours predict the location of new edgels, obtain new measurements and spawn new contours, thereby propagating the grouping process across the image. Eventually all of these contours will terminate and the hypothesis tree will be void of any active contours. At this point, the edge image is scanned for the next unlabeled edge point, and the initialization process is repeated. When the entire image has been scanned, all of the edge points will be labeled.

4.2. The surveillance region. The surveillance region is an adaptive variable-sized window that travels with the tip of the contour and is used to extract measurements from the edge map. Every iteration, each contour searchs for edge points in a series of circles of increasing diameter centered at the predicted contour endpoint. The search halts as soon as at least one measurement is found, or the maximum search radius is reached. The size of the surveillance region determines the distance the curve must travel in that time period, and is reflected in the step size, Δs, for the curve. The use of a set of windows of increasing size ensures that no more than one measurement from the given contour will be found in a single time period (see assumption (2) in Section (3)). We also assume that a thinning algorithm has been applied to reduce sequences of pixels to eight-connected chains.

The search for measurements takes place after the prediction phase of the state estimator generates an extrapolated endpoint location, (x, y), for the contour. This location determines the discrete image coordinates, (x_i, y_j), at which the surveillance region is centered. If there is no edge at the predicted location, circles of radius 1, $\sqrt{2}$, 2, $\sqrt{5}$, are searched. These surveillance regions are labeled 1 through 5 in Figure (7). It should be noted that the surveillance region of a contour is *not* equivalent to its validation region, which is defined by the Mahalanobis distance and depicted in Figure (7) as an ellipse.

4.3. Implementation of the Bayesian hypothesis tree. Successful practical implementation of the MHT algorithm requires careful control of memory usage and tree size. This section describes several techniques that have proven invaluable in practice: track trees, disjoint clusters, a novel hypothesis generation algorithm and pruning.

4.3.1. *Track Trees.* The same track may appear in more than one global hypothesis. Rather than duplicate a track for each hypothesis containing it, thereby incurring additional computational and storage overheads, track trees are formed[23]. Each branch of a track tree represents a different assignment of a measurement to the

[2]If only location information is available, contour initiation is somewhat more complex - a second neighboring measurement is needed to define the contour line.

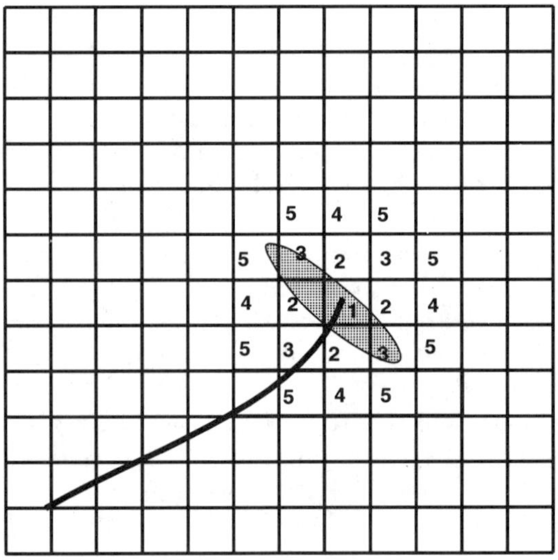

FIGURE 7. A contour, its surveillance region (labeled 1 through 5) and its validation region

track. Each global hypothesis then contains pointers to leaves of the track trees. An explicit hypothesis tree is not represented since it is unnecessary. Only the leaf nodes of the hypothesis tree need to be represented, parent hypotheses can be reconstructed by tracing the pointers up the corresponding track trees, though, in practice, parent hypotheses are not needed.

4.3.2. *Spatially disjoint hypothesis trees.* A considerable reduction in the combinatorics can be achieved by realizing that it is not necessary to form a single global hypothesis tree if there are tracks that do not compete for common measurements. Instead, tracks can be partitioned into separate *clusters* as proposed by Reid[32].

4.3.3. *Generating the k-best hypotheses for each disjoint hypothesis tree.* A brute force implementation of the MHT would, at each iteration, enumerate *all* possible global hypotheses, calculate the probability of each hypothesis and prune so as to keep only the k-best. This enumeration is impractical.

Recently, several papers have proposed approximate solutions to determining a ranked set of assignments. Nagarajan *et al* [30] present an algorithm in which the k-best hypotheses are generated by an "easy search process instead of going through an extensive enumeration". The authors do not provide a theoretical analysis of the computational complexity of their branch and bound scheme. However, while evidence is presented to demonstrate that in some cases a very significant reduction in computation is achieved, in the worst case the cost may still be exponential. Brogan [6] provides an algorithm for determining a ranked set of p assignments. However, this set is not guaranteed to be the p-best, i.e. it is possible to miss certain good combinations. A sufficient condition is provided to determine $q \leq p$ such that the first q assignments are optimal. Once again, no formal analysis of the computational complexity is provided.

Consider first the problem of finding the single most probable hypothesis. This can be cast as a weighted bipartite matching problem by constructing a bipartite graph in which each node on one side represents one of the measurements, each node on the other represents one of the targets, and each arc, $< z_i, y_j, l >$, gives the log likelihood, l, that measurement z_i should be assigned to target y_j. The log of the likelihood of a given assignment can be found by summing the log likelihoods of all the arcs that it specifies. Finding the best hypothesis, then, is a matter of finding the assignment that maximizes this sum. This is an instance of the classical assignment problem from combinatorial optimization, and can be solved very efficiently in polynomial time [31].

Danchick and Newnam [12] recognize that finding the best hypothesis can be formulated as a classical linear assignment problem and then show how modifications to the cost matrix followed by repeated solutions to these new assignment problems allow the k-best assignments to be computed. Danchick and Newnams' algorithm has two disadvantages. First, in the worst case, Danchick and Newnams' algorithm requires the solution of $k!$ linear assignment problems. Though the average case is expected to be considerably better, it is highly desirable to reduce the order of this dependency. Second, at the end of each iteration (or "sweep") Danchick and Newnams' algorithm must identify and eliminate duplicate assignments.

These problems can be avoided by using Murty's algorithm [29] for producing assignments in order of decreasing likelihood. This algorithm is guaranteed to find the k-best assignments in $O(N^4)$ polynomial time and linear in k. In fact, "the computations required at each stage are the solving of at most $(n-1)$ assignment problems, each of sizes $2, 3, \ldots, n$"

Given a solution, S, to an assignment problem, P, we can partition the assignment problem into a list of new problems with the following properties:

(1) The set of valid solutions for any one of the problems in the list doesn't intersect with the set of solutions for any other problem in the list. That is, there are no duplicate problems.
(2) The union of the sets of valid solutions for all the problems in the list is exactly the set of solutions for problem P, minus solution S.

Murty gives a method for computing this partitioning in $O(N^2)$ time.

For the k-best algorithm, a list of problem/solution pairs is kept. Each pair consists of an assignment problem and its best solution. The list is initialized with the initial problem to be solved. In each iteration, the best solution is found, then removed from the list, and replaced with its partitioning. So, in the first iteration, the single best solution, S_0, is found to the problem, and the list is altered so the set of possible solutions no-longer contains S_0. The next iteration gives the next-best solution, S_1, and changes the list so that possible subsequent solutions no-longer include S_1 or S_0; and so on. Figure (8) outlines the algorithm. The partitioning is performed by the loop in step 4.4. The reader is directed to [29, 10] for more detail.

We implemented Murty's algorithm using the Hungarian method [31] to solve each assignment problem. The Hungarian method has time complexity $O(N^3)$. There are faster weighted bipartite matching algorithms but the Hungarian method was most accessible. The complexity of Murty's algorithm using the Hungarian method is $O(kN^4)$. This is the worst case time, and the average time is considerably better.

(1) Find the best solution, S_0, to P_0 (this can be done using a standard algorithm like the Hungarian method)
(2) Initialize the list of problem/solution pairs with $< P_0, S_0 >$
(3) Clear the list of solutions to be returned
(4) For i = 1 to k, or until the list of problem/solution pairs is empty
    4.1 Search through the list of problem/solution pairs, and find the pair, $< P, S >$ that has the best solution value
    4.2 Remove $< P, S >$ from the list of problem/solution pairs
    4.3 Add S to the list of solutions to be returned
    4.4 For each triple, $< t, z, l >$, found in S
        4.4.1 Let $P' = P$
        4.4.2 Remove the triple $< t, z, l >$ from P'
        4.4.3 Look for the best solution, S', to P'
        4.4.4 If S' exists
            4.4.4.1 Add $< P', S' >$ to the set of problem/solution pairs
        4.4.5 From P, remove all triples that include t, and all triples that include z, except $< t, z, l >$ itself. (This reduces the dimension of the problem by one).

FIGURE 8. Murty's algorithm for finding the k- best solutions to an assignment problem, P_0.

In each iteration of the MHT algorithm, we want to find the k-best descendents of the k-best hypotheses from the previous iteration. This involves finding the k-best solutions to k different assignment problems. The k solutions could come from any subset of this set of problems. The obvious way of using Murty's algorithm for this task is to run it k times, once for each assignment problem, and then choose the k best solutions from the resulting set of k^2 solutions. However, if instead, we simply change steps 1 and 2 of Murty's algorithm to initialize the list of problem/solution pairs with all k problems at the beginning, and then run the rest of the algorithm only once, the correct result will be obtained for considerably less computation.

4.3.4. *Pruning.* Pruning is essential to any practical implementation of this algorithm. Pruning is based on a combination of an "N-scan-back" algorithm[23] and ratio pruning, i.e. a simple lower limit on the ratio of the probabilities of the current and best hypotheses.

The "N-scan-back" algorithm assumes that any ambiguity at time k is resolved by time $k+N$, i.e. it defines the number of frames to look ahead in order to resolve an ambiguity. Then, if hypothesis Θ_l^k at time k has q children, the sum of the probabilities of the the leaf nodes is calculated for each of the q branches. Whichever branch has the greatest probability is retained and all other branches are pruned. The result is an irrevocable decision regarding the assignment of measurements to tracks based on looking ahead N time steps.

Murty's algorithm can be viewed as a ranked queue of solutions, allowing for more sophisticated methods of deciding when to stop generating new hypotheses. We have had good results terminating hypothesis generation whenever the likelihood of the next hypothesis is below some fixed percentage of the likelihood of the first (most likely) hypothesis. Nagarajan et al use a similar termination condition [30].

4.4. Contour "fusion". The algorithm may encounter a new contour at any point along the contour, not just at end points. It may even encounter the contour simultaneously at different points along it depending on the topology of the curves. When this happens, a single contour is usually erroneously partitioned into two or more segments. Therefore, at the end of the Bayesian data association stage a contour "fusion" stage, analogous to multisensor track-to-track fusion[2], is applied. The fusion step joins tracks whose initial or final states are compatible into single curve.

Let i and j denote two contours with corresponding state vectors given by $\hat{x}^i(k)$ and $\hat{x}^j(k)$. Remember that each state vector provides an estimate of the current position and gradient of the end point of the contour. Following Bar-Shalom[2], let

$$\hat{\Delta}^{ij}(k) = \hat{x}^i(k) - \hat{x}^j(k) \tag{22}$$

denote the estimate of

$$\Delta^{ij}(k) = x^i(k) - x^j(k) \tag{23}$$

where $x^i(k)$ and $x^j(k)$ are the true states of the contours. We need to test whether $x^i(k) = x^j(k)$, or alternatively, whether

$$\Delta^{ij}(k) = \begin{cases} 0 & i.e.\ same\ contour \\ non-zero & i.e.\ different\ contour \end{cases} \tag{24}$$

Probability of detection	P_D	0.85
Contour termination parameter	λ_χ	20
Mean rate of false alarms	λ_F	0.05
Mean rate of new contours	λ_N	0.05
N-scan-back pruning depth	N_SCAN	3
Lower limit		0.01
Process noise variance	q	0.01
Measurement variances	σ_p, σ_g	0.3, 0.2
Validation size	γ	9.49

TABLE 1. Bayesian hypothesis tree and contour model (Kalman filter) parameters.

If the errors in the state estimates of the two contour are independent, then the covariance $T^{ij}(k)$ of Equation (22) is

$$T^{ij}(k) = P^i(k) + P^j(k) \tag{25}$$

If the estimation errors are assumed Gaussian, then the Mahalanobis distance can once again be used, this time to merge contours if

$$\hat{\Delta}^{ij}(k)'[T^{ij}(k)]^{-1}\hat{\Delta}^{ij}(k) \leq \delta \tag{26}$$

where δ is obtained from χ^2 tables.

A caveat is in order here. If i and j are two segments of the same contour then strictly the state estimates of each contour are not for the same point k along the contour, since the two segments are disjoint. It is therefore necessary to predict the end point of contour i to the end point of contour j.

5. Experimental Results

We present experimental results on real and synthetic images that demonstrate the performance of the system[3]. Since the definition of a correct segmentation varies with the application, it is difficult to quantify the general performance of a contour grouping algorithm. The results presented here serve two purposes: to demonstrate qualitatively that our algorithm produces meaningful segmentations of complicated real images, and to show that the set of algorithm parameters provide sufficient flexibility to adjust the performance to the requirements of a particular task domain. In addition, we illustrate the advantage our system provides over simple edge chaining approachs, and the implication for systems that use such a chain code as input. Table (1) gives the parameter settings that were used to obtain all of the experimental results that follow, except where explicitly noted otherwise.

We begin with a series of synthetic images that demonstrate the variety of different segmentations that are possible by varying the process noise variance, q, and the contour termination parameter, λ_χ. The input image, shown in Figure (9), consists of the edges extracted from a set of stripes. It was generated by the Boie-Cox edge detector[5, 4], which provides both position and coarse gradient information (horizontal, vertical and two diagonals).

[3] A subset of these experiments are described in [11]

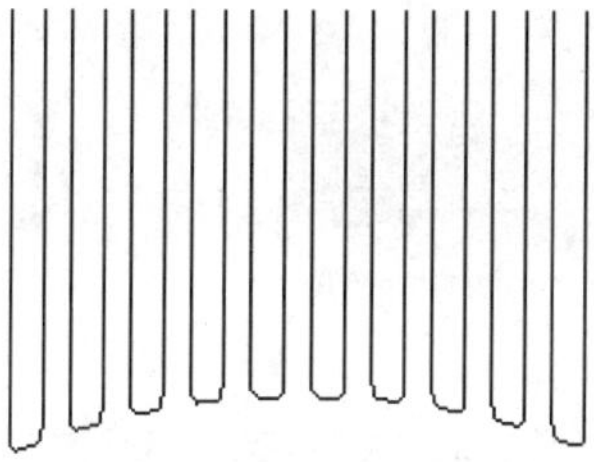

FIGURE 9. Edges of a set of stripes

Figure (24) shows the result of applying the contour grouping algorithm to this edge map using the parameter values of Table (1), but with the process noise variance q changed to 0.001.

Each stripe is segmented into three contours because the large changes in gradient at the corners do not allow points beyond the corner to validate to the previous contour segment, i.e. the points fail the Mahalanobis test. For the next segmentation, the process noise is increased to 0.055, decreasing the smoothness constraint. As a result, each "stripe" is grouped into a single closed contour as shown in Figure (25).

The next two images show the effect of varying the termination parameter, λ_χ. Contours can apply their dynamic models to extrapolate over gaps in the measurements. The termination parameter affects the probability of termination of a contour, as described in Section (3.3), and therefore determines the maximum gap size that a contour can link. Any gap that exceeds this maximum distance results in a segmentation into two distinct curves. By increasing λ_χ to a value of 250, and thereby decreasing the probability of termination, the algorithm can extend contours across the gaps between the central stripes, as shown in Figure (26). The stripes to the left and right of the pattern are not linked because their spacing is slightly larger than those at the center. Finally, increasing λ_χ to 500 allows the entire subjective contour to be extracted, as illustrated in Figure (27).

The next set of figures illustrate the differences between our algorithm and traditional edge linking approachs. Figure (11) shows the edges extracted from the synthetic image of Figure (10) consisting of a black and white circle against a white and black background. Figure (28) shows the result of applying a standard edge linking algorithm (see [1], Chapter 4) to the edge map. The algorithm uses the eight-connected neighborhood property of the edge map to link edges into chains through sequential search. Because of the failure of the edge detector at curve in-

FIGURE 10. Synthetic "Circle" image.

tersections, the simple chaining routine incorrectly[4] groups the semicircles with the lines. Figure (29) illustrates the implications of this failure for higher level curve analysis. It shows the result of applying David Lowe's contour grouping scheme [24] to the edge chains from Figure (28). In places where a single chain spanned one or more curves, Lowe's algorithm segmented the individual curves successfully. But when the initial linking process broke a single curve into one or more parts (as it did with both the circle and the line), the mistake is passed on to the final segmentation. In contrast, Figure (30) shows the result of applying our algorithm to the same edge map. While interval-based approachs like Lowe's are often very effective in practice, they are limited by the accuracy of the linking stage that proceeds them. Our framework provides a mechanism for bridging this gap. Note, however, that the segmentation of Figure (30) required three parameter changes. First, the probability of termination was decreased ($\lambda_\chi = 200$) in order to cross a 4 pixel gap in the vertical line. Second, the measurement noise had to be doubled to deal with the fact that the edge detector placed the central line segment one pixel to the right of the upper and lower segments. Lastly, the process noise was halved to prevent edge points on the circle validating to the line segment.

We now present results on real images. Figure (12) shows the edges extracted from the image of a fork. Figure (31) shows the corresponding contour groupings for the parameter settings of Table (1). Note that while the curvature at the end of the handle was smooth enough to allow the contour to continue around, the curvature at the ends of the prongs is too great and the contour is broken.

Figures (13), (14) and (32) show the results of applying the algorithm to the image of a face. Figures (15), (16) and (33) demonstrate the output of the algorithm

[4]We use the terms "correct" and "incorrect" loosely here since the correct grouping depends strongly on the desired purpose or use of such segments in any higher level task such as object recognition. Section (6) discusses this point further.

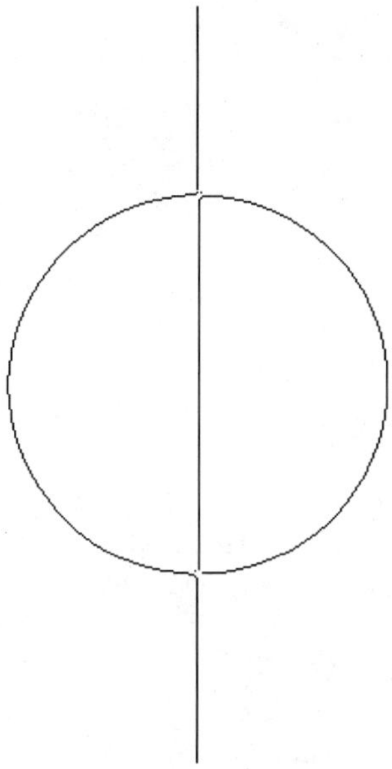

FIGURE 11. Edge map from test image of circle against background.

applied to an image of industrial tools. Figures (17), (18) and (34) demonstrate the output of the algorithm applied to a traffic scene. Figures (19), (20) and (35) demonstrate the output of the algorithm applied to an aerial scene. Figures (21) and (36) show the edges and associated contour groupings obtained from the image of a computer mouse.

Finally, Figure (22) is the image of a totem pole provided by courtesy of David Lowe. Figure (23) are the edges extracted and Figures (37) and (38) are the edge groupings obtained by applying the multiple hypothesis approach (with the same parameter settings as Figure (30)) and Lowe's [24] approach respectively.

Although it is very difficult to compare the two segmentations, careful inspection reveals that the MHT algorithm almost invariably produces longer contours, because of its ability to bridge gaps. For example, consider the yellow almost vertical contour in the bottom left of Figure (37) - this is broken into three segments by Lowe. Similarly, consider the red-coded contour beginning in the lower right of Figure (37) - this is broken into four segments in Figure (38). Similar examples are prevalent throughout the image, for example in the upper lip, the rim of the headdress above the eyes and where the headdress meets the shoulder. Lowe's algorithm is good but it's performance is fundamentally limited by the 8-connected analysis which precedes it. The MHT algorithm does not have this limitation.

6. DISCUSSION

We have presented a contour segmentation algorithm that takes an edge image and extracts continuous curves of arbitrary smoothness, correctly handling curve intersections and capable of extrapolating over gaps in the measurement sequence. The algorithm incorporates noise models of the edge detection process and limited scene statistics, and outputs a covariance for each segmented curve. We have demonstrated its performance on real and synthetic images.

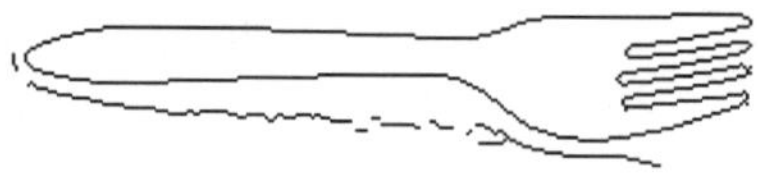

FIGURE 12. Edges of a fork

The algorithm is novel in that the edge grouping and segmentation processes are coupled, in contrast to the traditional approach of grouping followed by segmentation. The coupling is seen to improve the final segmentation, providing longer contours than would otherwise be obtained. This is a significant advantage since the fewer contours there are, the smaller the search space of any recognition stage that follows.

The uncertainty in both the origins *and* values of measurements is one reason why perception is different from traditional estimation and control problems. We modeled the uncertainty in the data association or grouping process by probability values for each association hypothesis, while the geometric uncertainty due to noise is represented by covariances. This decoupling of the data association and noise uncertainties is a distinguishing feature of this approach that should allow for more robust interpretation and integration by later processing modules.

There are at least five important aspects of the algorithm that contribute to its uniqueness and success:

(1) *Explicit dynamic contour models.* Providing a parametric model of the contour whose state estimate changes with each new measurement allows tracking of arbitrarily smooth contours in a manner that is independent of the underlying sampling pattern of the image. The Kalman filter based model permits an implicit specification of the desired smoothness of a curve, without requiring explicit parameterization into quadratics, cubics, etc. As was demonstrated in the previous section, the variation of a few parameters permits a variety of segmentation behaviors.

(2) *Measurement noise.* The usefulness of a measurement increases dramatically when an accurate estimate of its certainty is available. Edge certainties are available from edge detectors[4, 27] and can be incorporated into our framework through the Kalman filter measurement covariance. A

FIGURE 13. "Lena"

FIGURE 14. Edges of "Lena" Figure (13).

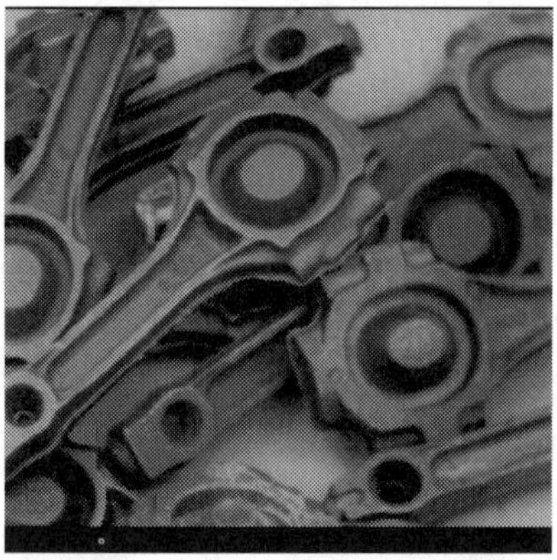

FIGURE 15. Machine tools.

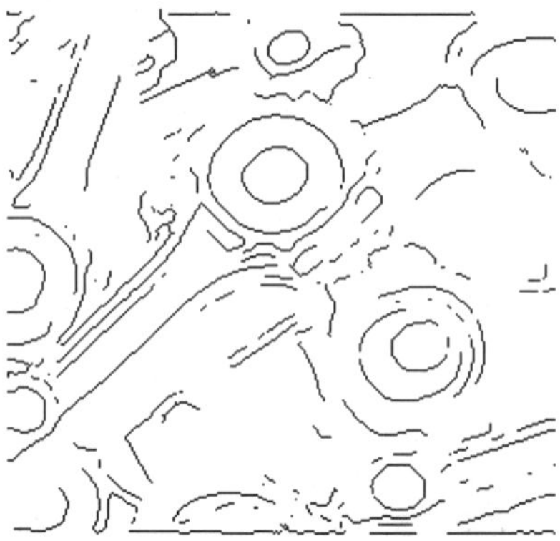

FIGURE 16. Edges of machine tools of Figure (15).

FIGURE 17. Traffic scene

FIGURE 18. Edges of traffic scene Figure (17).

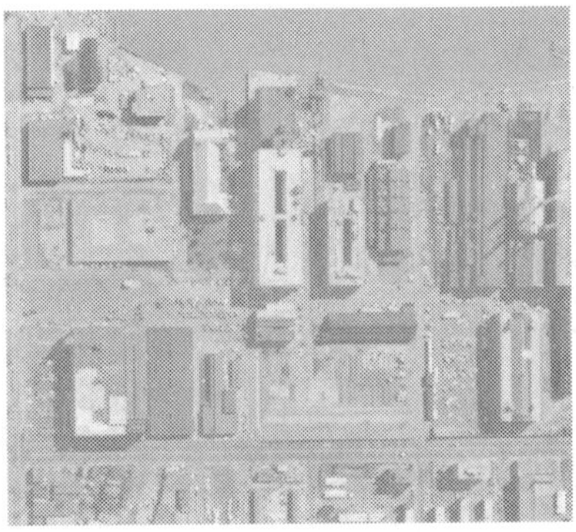

FIGURE 19. Aerial scene

FIGURE 20. Edges of aerial scene Figure (19).

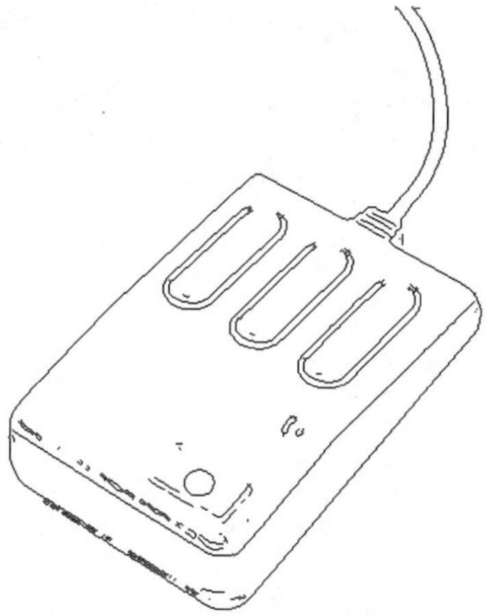

FIGURE 21. Edges of a computer mouse

FIGURE 22. Totem pole (Courtesy of David Lowe).

FIGURE 23. Edges of totem pole.

description of edge certainty as a function of image noise would permit the algorithm to optimally weight each measurement: low contrast, very noisy edges would have small weightings while high contrast edges would be given large weightings.

(3) *Contour initiation.* There are two key factors in the ability of the algorithm to correctly segment intersections and bifurcations. The first is the contour model, which permits extrapolation across curve intersections, where edge detector measurements are typically missing or incorrect. The second is the fact that each new measurement can initiate a new contour, which permits the simultaneous generation of all possible segmentations in the neighborhood of an intersection. Corners, bifurcations and intersections are all reasonably partitioned, even though these concepts are *not* explicitly represented in the algorithm. This is intuitively appealing since it is probably impractical to enumerate all intersection topologies.

(4) *Mahalanobis distance measure.* The Mahalanobis distance utilizes statistical information to *quantify* closeness and similarity properties of curves.

(5) *Hypothesis Tree.* The ability (through N-Scan Pruning) to grow the hypothesis tree to variable depths makes it possible to postpone the interpretation of edge measurements until sufficient evidence has been accumulated.

While other algorithms have contained subsets of these features, we believe that our framework is the first to incorporate all of these important concepts in a unified, synergistic manner.

There are three other observations that can be made about the application of the multiple hypothesis approach to the contour segmentation problem. First, it should be noted that *all* free parameters of the system are physical quantities of the sensor and/or scene that can be physically measured, i.e. it is physically grounded. Thus it is possible in principle to estimate the correct parameter settings from a representative sample of images.

Second, the multiple hypothesis filter can be considered to provide a unified top-down/bottom-up approach to unsupervised learning[9], the contour model (Kalman filter) providing a tool for *looking for the expected* and the hypothesis generation and associated Bayesian probabilities providing a mechanism for *explaining the unexpected.*

Third, the use of Murty's algorithm together with other implementation techniques [23, 35, 36, 37], allows a computationally efficient (low order polynomial) approximation to Reid's original MHT algorithm.[5] This implementation has low overhead when applied to images (or regions thereof) with little or no ambiguity - in these cases, the total number of hypotheses is very small - while in ambiguous regions, e.g. highly-textured images, the amount of computational significantly increases. This is a useful property of the implementation that should allow it to be applicable over a wide class of scenes.

Experimental results demonstrated good performance. Corners, bifurcations and intersections are all reasonably partitioned, even though these concepts are *not* explicitly represented in the algorithm. Instead, use of the Mahalanobis distance to quantify closeness and similarity properties suffices. Experiments on a simple circle and line figure demonstrated the weaknesses of using an 8 connected component analysis and showed how the coupling of the grouping and segmentation processes can lead to superior results. Experimental results were also provided that suggest that the final segmentations provided by the MHT algorithm are usually longer and less fragmented. A preliminary investigation of our segmentation results reveals that many errors are due to the post processing stage that merges some contours. These errors could be avoided if the merging was incorporated into the main algorithm.

There are five main avenues for future work. First, alternative contour models, such as piecewise constant curvature, should be explored. Second, it would be interesting to tightly couple the edge detection and contour grouping stages, with the latter providing the former with expectations of where to look for edges. Hopefully, a synergy between the two stages is possible. At the very least, this approach is expected to improve upon relaxation[40] and hysteresis[7] techniques for line and curve enhancement. Third, measurements should be performed to verify or refute the statistical assumptions made here and to determine better models of measurement and process noise that would improve performance. Forth, experiments should be performed with edge measurements that have multiple attributes other than x, y position and gradient. For example, edgels could be labeled with whether they are bright-to-dark or dark-to-bright transitions, contrast information could be included etc. The framework outlined here can easily accommodate these extra dimensions and it would be interesting to evaluate the usefulness of additional information for segmentation. Finally, to obtain a quantitative evaluation of segmentation performance, it is necessary to consider a higher level objective such as recognition. With the Bayesian formulation, it should be possible to incorporate higher level tasks into the segmentation process, by modifying the prior probabilities in the model. Further research is needed to support this conjecture.

---

[5]We also believe that our implementation significantly extends the practicability of Reid's MHT to motion correspondence.

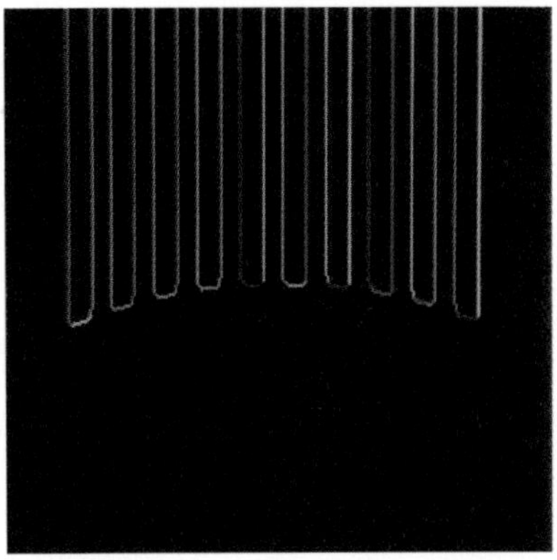

FIGURE 24. Color coded contour groupings for Figure (9)

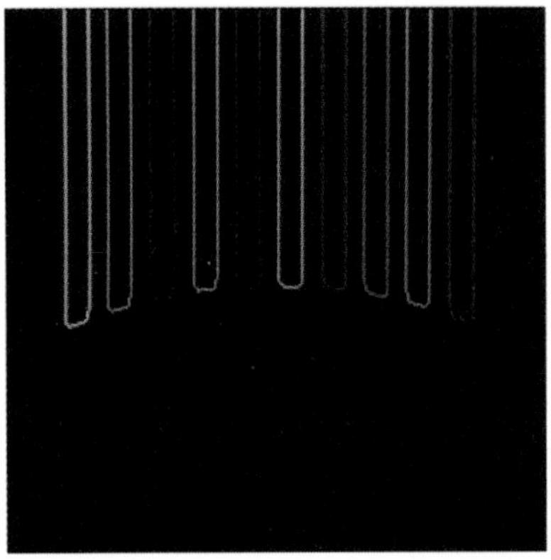

FIGURE 25. Contour grouping for Figure (9) with process noise set to 0.055

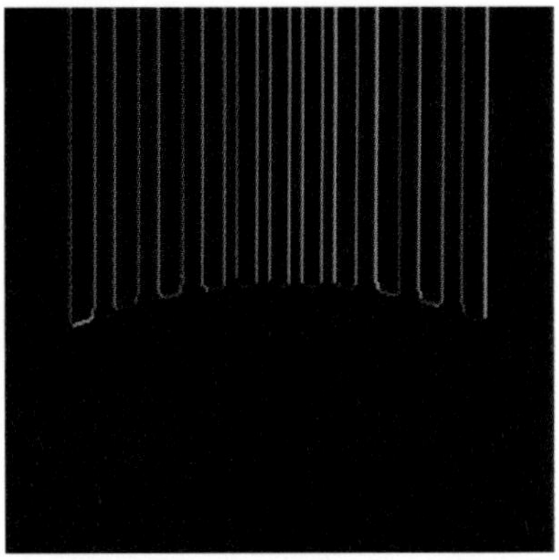

FIGURE 26. Contour grouping for Figure (9) with $\lambda_\chi = 250$.

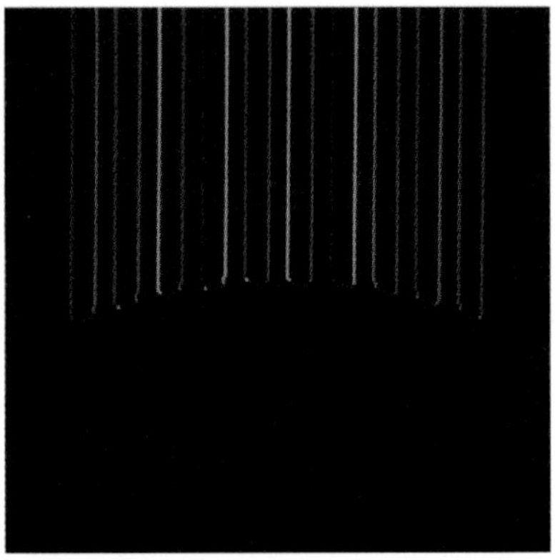

FIGURE 27. Contour grouping for Figure (9) with $\lambda_\chi = 500$.

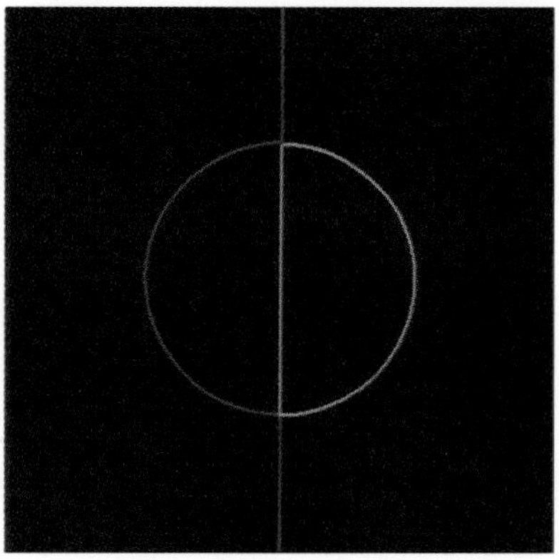

FIGURE 28. Result of linking previous edge map using simple neighborhood properties.

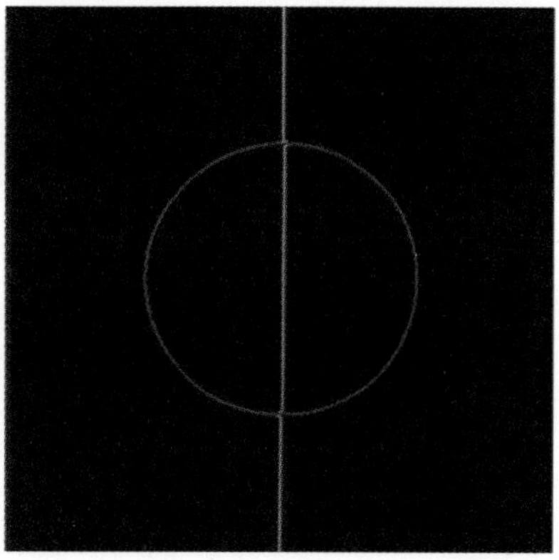

FIGURE 30. Result of applying MHT segmentation algorithm to the edge map of Figure (11).

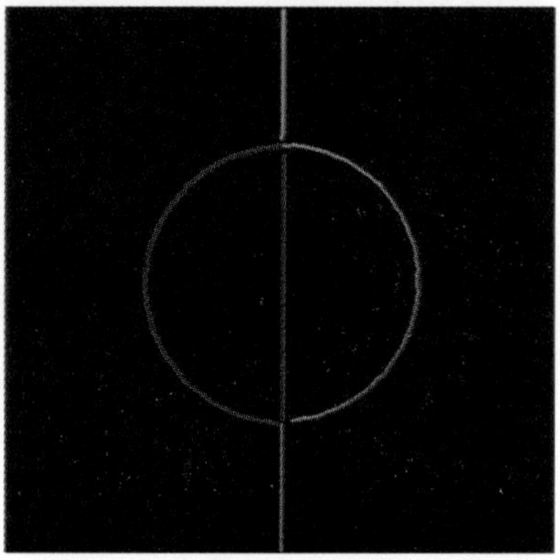

FIGURE 29. Result of applying Lowe's multiscale contour grouping algorithm to the edge chains of Figure (28).

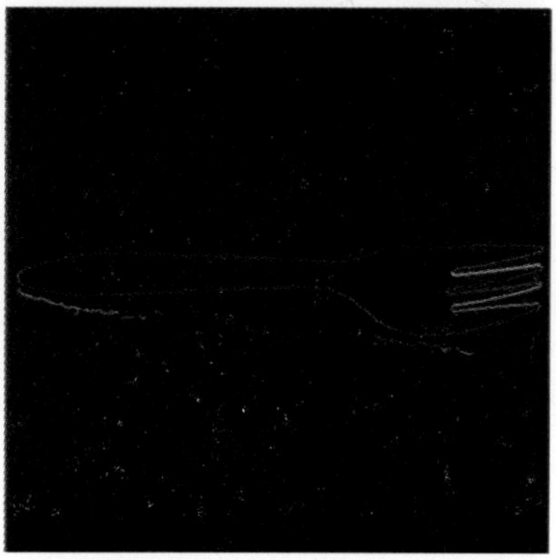

FIGURE 31. Contour grouping for "fork" of Figure (12)

FIGURE 32. Contour grouping for "Lena" image of Figure (14)

FIGURE 34. Contour groupings obtained for traffic scene of Figure (17).

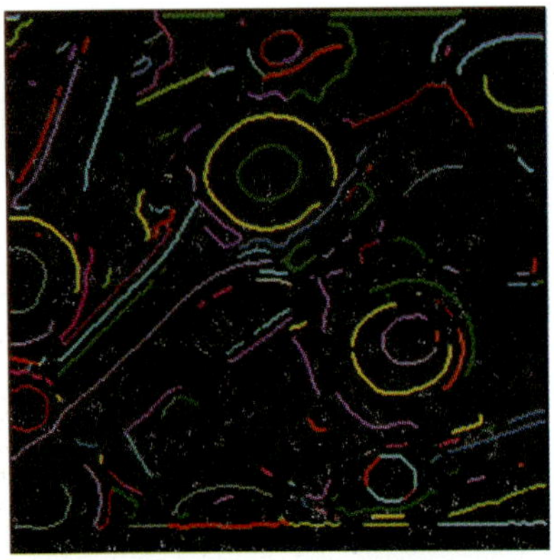

FIGURE 33. Contour grouping for the tools of Figure (16)

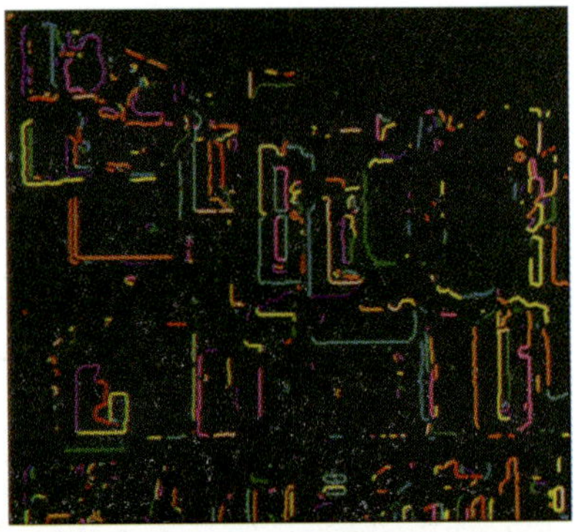

FIGURE 35. Contour groupings obtained for aerial scene of Figure (19).

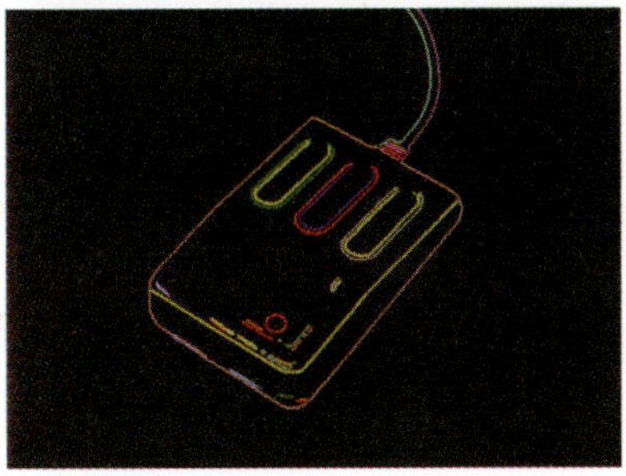

FIGURE 36. Contour grouping for Figure (21)

FIGURE 38. Edge groupings obtained by Lowe's algorithm on totem pole.

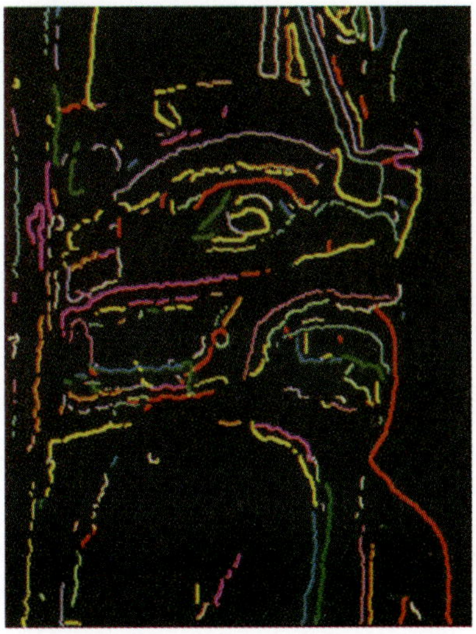

FIGURE 37. Edge groupings obtained by MHT on totem pole.

References

1. D. Ballard and C. Brown. *Computer Vision.* Prentice-Hall, 1982.
2. Y. Bar-Shalom and T. E. Fortmann. *Tracking and Data Association.* Academic Press, 1988.
3. R. Bellman and S. Dreyfus. *Applied Dynamic Programming.* Princeton University Press, Princeton, NJ, 1962.
4. R. A. Boie and I. J. Cox. Two dimensional optimum edge detection using matched and wiener filters for machine vision. In *IEEE First International Conference on Computer Vision*, pages 450–456. IEEE, June 1987.
5. R. A. Boie, I. J. Cox, and P. Rehak. On optimum edge recognition using matched filters. In *IEEE Computer Society Conference on Computer Vision and Pattern Recognition*, pages 100–108. IEEE, June 1986.
6. W. L. Brogan. Algorithm for ranked assignments with applications to multiobject tracking. *IEEE J. of Guidance*, 12(3):357–364, 1989.
7. J. Canny. A computational approach to edge detection. *IEEE Trans. Pattern Analysis and Machine Intelligence*, 8(6):34–43, 1986.
8. C. Chong, S. Mori, and K. Chang. Distributed multitarget multisensor tracking. In Y. Bar-Shalom, editor, *Multitarget-Multisensor Tracking: Advanced Applications*, pages 247–295. Artech House, 1990.
9. I. J. Cox and J. J. Leonard. Unsupervised learning for mobile robot navigation using probabilistic data association. In *Workshop on Computer Learning and Natural Learning*, Berkeley, California, USA, 1991.
10. I. J. Cox and M. L. Miller. On finding ranked assignments with application to multi-target tracking and motion correspondence. Technical report, NEC Research Institute, 1994.
11. I. J. Cox, J. M. Rehg, and S. Hingorani. A bayesian multiple hypothesis approach to contour segmentation. In *Proc. 2nd European Conf. on Computer Vision*, pages 72–77, Italy, 1992.
12. R. Danchick and G. E. Newnam. A fast method for finding the exact n-best hypotheses for multitarget tracking. *IEEE Trans. Aerospace and Electronic Systems*, 29(2):555–560, 1993.
13. C. David and S. Zucker. Potentials, valleys, and dynamic global coverings. *Int. J. Computer Vision*, 5(3):219–238, 1990.
14. E. Davies. *Machine Vision: Theory, Algorithms, Practicalities.* Academic Press, 1990.
15. R. Deriche and O. Faugeras. Tracking line segments. *Image and Vision Computing*, 8(4):251–270, 1990.
16. J. Feldman. Perceptual models of small dot clusters. In I. J. Cox, P. Hansen, and B. Julesz, editors, *Proc. DIMACS Workshop on Partitioning Data Sets: With applications to psychology, vision and target tracking*, 1994.
17. W. Grimson. On the recognition of curved objects. *IEEE Trans. Pattern Analysis and Machine Intelligence*, 11(6):632–643, June 1989.
18. E. Hildreth. Computation underlying the measurement of visual motion. *Artificial Intelligence*, 27(3):309–355, 1984.
19. R. Horaud and M. Brady. On the geometric interpretation of image contours. In *Proc. First Int. Conference on Computer Vision*, pages 374–382, London, England, 1987.
20. D. Huttenlocher and S. Ullman. Recognizing solid objects by alignment with an image. *Int. J. Computer Vision*, 5(2):195–212, November 1990.
21. M. Kass, A. Witkin, and D. Terzopoulos. Snakes: Active contour models. *Int. J. Computer Vision*, pages 321–331, 1988.
22. D. Kriegman and J. Ponce. On recognizing and positioning 3-d objects from image contours. *IEEE Trans. Pattern Analysis and Machine Intelligence*, 12(12):1127–1137, December 1990.
23. T. Kurien. Issues in the design of practical multitarget tracking algorithms. In Y. Bar-Shalom, editor, *Multitarget-Multisensor Tracking: Advanced Applications*, pages 43–83. Artech House, 1990.
24. D. Lowe. Organization of smooth image curves at multiple scales. *Int. J. Computer Vision*, 3:119–130, 1989.
25. D. G. Lowe. Three-dimensional object recognition from single two-dimensional images. *Artificial Intelligence*, 31:355–395, 1987.
26. A. Martelli. An application of heuristic search methods to edge and contour detection. *Communications of the ACM*, 19(2):73–83, 1976.

27. L. Matthies, T. Kanade, and S. Shafer. Kalman filter-based algorithms for estimating depth from image sequences. *Int. J. Computer Vision*, 3:209–236, 1989.
28. U. Montanari. On the optimal detection of curves in noisy pictures. *Communications of the ACM*, 15(5):335–345, May 1971.
29. K. G. Murty. An algorithm for ranking all the assignments in order of increasing cost. *Operations Research*, 16:682–687, 1968.
30. V. Nagarajan, M. R. Chidambara, and R. N. Sharma. Combinatorial problems in multitarget tracking - a comprehensive survey. *IEE Proceedings, Pt F*, 134(1):113–118, 1987.
31. C. H. Papadimitriou and K. Steiglitz. *Combinatorial Optimization*. Prentice Hall, 1982.
32. D. B. Reid. An algorithm for tracking multiple targets. *IEEE Transactions on Automatic Control*, AC-24(6):843–854, December 1979.
33. A. Shashua and S. Ullman. Grouping contours by iterated pairing network. In Richard P. Lippmann, John Moody, and David S. Touretzky, editors, *Advances in Neural Information Processing Systems 3*. Morgan Kaufmann, 1991. Proc. NIPS'90, Denver CO.
34. R. Szeliski and D. Terzopoulos. Physically-based and probabilistic modeling for computer vision. In *SPIE Vol. 1570 Geometric Methods in Computer Vision*, pages 140–152, San Diego, July 1991. Society of Photo-Optical Instrumentation Engineers.
35. J. K. Uhlmann. Adaptive partitioning strategies for ternary tree structures. *Pattern Recognition Letters*, 12:537–541, 1991.
36. J. K. Uhlmann. Satisfying general proximity/similarity queries with metric trees. *Information Processing Letters*, 40:175–179, 1991.
37. J. K. Uhlmann. Algorithms for multiple-target tracking. *American Scientist*, 80:128–141, 1992.
38. A. Witkin. Intensity-based edge classification. In *Proc. AAAI-82*, pages 36–41, Pittsburgh, PA, 1982.
39. K. Wohn, J. Wu, and R. Brockett. A contour-based recovery of image flow: Iterative transformation method. *IEEE Trans. Pattern Analysis and Machine Intelligence*, 13(8), August 1991.
40. S. W. Zucker, R. A. Hummel, and A. Rosenfeld. An application of relaxation labeling to line and curve enhancement. *IEEE Trans. on Computers*, C-26(4):394–403, 1977.

NEC Research Institute, 4 Independence Way, Princeton, NJ 08540
E-mail address: ingemar@research.nj.nec.com

Dept. of Electrical and Computer Eng., Carnegie Mellon University, Pittsburgh, PA
E-mail address: jimr@cs.cmu.edu

NEC Research Institute, 4 Independence Way, Princeton, NJ 08540
E-mail address: sunita@research.nj.nec.com

Pilies 32-10, Vilnius, LITHUANIA

DIMACS Series in Discrete Mathematics
and Theoretical Computer Science
Volume 19, 1995

FINDING SALIENT CONVEX GROUPS

DAVID W. JACOBS

ABSTRACT

This paper describes an algorithm that robustly locates convex collections of line segments in an image. The algorithm is guaranteed to find all convex sets of line segments in which the length of the line segments accounts for at least some fixed proportion of the length of their convex hull. This enables the algorithm to find convex groups whose contours are partially occluded or missing due to noise. We perform an expected case analysis of the algorithm's performance that shows that its run-time is $O(n^2 \log(n) + nm)$, when we wish to find the m most salient groups in an image with n line segments. We support this analysis with experiments on real data. Our analysis not only reveals the circumstances under which our algorithm is efficient, but also tells us when the groups found are unlikely to occur at random, and so are likely to capture the underlying structure of a scene. We also demonstrate the grouping system as a module in an efficient recognition system that combines grouping with indexing.

1. INTRODUCTION

In this paper, we consider *Grouping* (or *Perceptual Organization*) as a process that clusters image features into higher level organizations, each likely to come from a single object. As in any clustering problem, two important questions arise. First, which features should we cluster together? Second, how can we locate these clusters efficiently? We approach the first problem by looking for *non-accidental properties.* In this approach, we group together features that have some property that is frequently shared by features originating in a single object, but that is unlikely to arise at random. In particular, we consider convexity, which is a common characteristic of object parts. We show that convex sets of line segments with relatively small gaps rarely occur at random. We then show that this contributes to a solution of the second problem; the very fact that these groups are rare enables us to find them efficiently.

We approach grouping as a process that is done bottom-up, using general clues applicable to wide classes of objects and images, as opposed to a top-down process that organizes an image by recognizing the objects in it. Grouping can greatly reduce the combinatorics of visual tasks that match features across multiple images,

1991 *Mathematics Subject Classification.* Primary 68U10.

This paper describes research supported at the Artificial Intelligence Laboratory of the Massachusetts Institute of Technology, and further supported at the NEC Research Insititue. Research at MIT was funded in part by an Office of Naval Research University Research Initiative grant under contract N00014-86-K-0685, and in part by the Advanced Research Projects Agency of the Department of Defense under contract N00014-91-J-4038, Army contract number DACA76-85-C-0010, and under Office of Naval Research contract N00014-85-K-0124.

1052-1798/95 $1.00 + $.25 per page

or between an image and a model. Groups of features contain a good deal more information than individual features; we can use this information to determine correct matches efficiently.

While grouping has proven a difficult problem, it deserves attention because even partial progress can be quite valuable. For example, a grouping system can focus and improve a search for an object by a recognition system by collecting together features that are more likely to come from a single object than are a random collection of features. Many recognition systems now exploit simple grouping techniques. By extending our understanding of grouping, we extend the domain in which recognition is feasible.

One approach to perceptual organization is to identify common properties of image groups that originate in a single object or process, and that occur relatively infrequently in groups generated by a random process. We describe the past work that has developed this approach in Section 2. In this paper we combine proximity and convexity as grouping clues, forming what we call *salient, convex* groups. We present an algorithm that finds all collections of convex line segments where the length of the line segments accounts for at least some fixed proportion of the length of their convex hull. The *salience* of a group is therefore measured by the percentage of the convex boundary that is identified as a brightness discontinuity. By taking account of the gap between the end of the group and its beginning, this salience measures also has the effect of enforcing a *closure* constraint on the groups that it finds. We show that a random process is unlikely to produce such groups. At the same time, salient convex groups can be very useful in domains in which objects frequently have convex parts. And in such domains, salient convexity will be a strong clue that a group of lines originate in a convex part of a single object, as we will discuss.

We present an algorithm for finding salient convex groups that has several desirable qualities. First, the algorithm is simple and we demonstrate that it is efficient, on average. Second the algorithm is robust to noise and occlusion because it allows for gaps to appear in groups. Third the algorithm is robust to clutter, because it finds all image groups that meet a simple test of salience. Since salience is measured by the overall properties of a group, the presence of extraneous lines cannot divert the system into connecting lines that appear good locally, but do not lead to groups with strong overall salience. Therefore, for images of a wide variety of objects, the grouping system we describe will reliably find groups that come from a single object, while producing groups unlikely to be due to a random process.

Our analysis and experiments on real data also show that precisely because salient convex groups are unlikely to occur at random, they can be found efficiently using a simple, backtracking search. We show that the expected time required to find the m most salient convex groups in an image containing n line segments is $O(n^2\log(n)+mn)$. This leads to a grouping algorithm that is efficient and robust in practice. We demonstrate this by showing the actual performance of the algorithm on a number of real images. We also show sample applications of the algorithm to object recognition.

2. Past Work

We will emphasize three aspects of past work on perceptual organization. First, we describe past use of non-accidental properties as grouping clues. We describe how our current work derives from and differs from previous approaches. Second, we consider past efforts at making grouping systems efficient. Third, we describe the role that convexity has played in a variety of approaches to image understanding.

The use of non-accidental properties in grouping has been developed by Witkin and Tenenbaum[44], Binford[3], Kanade[27], Lowe[30] and Richards and Jepson[36]. These authors argue that the visual system is sensitive to properties commonly produced by a single object or process, and rarely occurring at random. They show how this approach can explain a number of well-known Gestalt grouping phenomena, including grouping due to symmetry, parallelism, smooth continuation, and proximity (see Kohler[28] for a discussion of the Gestalt work).

Lowe in particular has contributed to this approach by showing how to derive probability distributions that reflect the strength of grouping clues. Lowe, for example, shows that parallel image lines are relatively likely to come from a single object. This derivation assumes that objects tend to contain parallel lines, and that lines coming from different objects have random relative orientations. Because these types of clues are probabilistic, one expects to achieve better performance by combining many clues together, and this seems to be the experience of many researchers (Lowe[30], Jacobs[23], Shashua and Ullman[40], Denasi et al.[12], Sarkar and Boyer[37], Mohan and Nevatia[33], Williams[43], and Nitzburg and Mumford[34]).

Lowe[30] also stressed the importance of grouping to object recognition. He pointed out that by grouping together features that are likely to have been all produced by a single object, one can intelligently order one's search for that object. If one looks first at these groups, one can achieve significant speedups in a recognition system, compared to performing a random search. Others have also explicitly used grouping, or the formation of more complex features, to speed up recognition systems (Jacobs[22, 23], Califano and Mohan[7], Clemens[9], Syeda-Mahmood[41], Burns and Riseman[6], Huttenlocher and Wayner[21] and Wayner[42]). The interaction between grouping and the computational complexity of recognition is treated more theoretically by Grimson[16] and Clemens and Jacobs[10].

Most past work on contour grouping has focused on using a small set of properties identified by Gestalt psychologists: proximity, symmetry, parallelism, collinearity and smooth continuation (although Williams[43] has explored the role of topological constraints in grouping). Work in computer vision has stressed that there is zero probability that a perfect instance of any of these properties will occur as a result of some appropriate, uniformly distributed random process. For example, lines positioned by a uniform random distribution have zero probability of being parallel, or having a common end point.

Convexity differs from Gestalt properties in that lines positioned with a uniform random distribution have a small, but non-zero likelihood of being convex. We discuss this probability in greater detail in Section 4.2.2. We feel that it is very important to explore properties such as convexity, which are not among the classic Gestalt non-accidental properties. Images frequently do not contain sufficient Gestalt clues to produce correct edge groupings, and in such cases additional cues such as convexity may distinguish the groups of edges that do come from a single object. Also, even when Gestalt cues are available, additional probabilistic information about the origin of edges may help a grouping system. This is particularly true since even non-accidental properties with zero chance of occurring at random still do not provide certain grouping information. Parallelism, for example, does frequently occur between lines from different objects in many real situations. Moreover, due to noise in feature detection, we must always allow for some deviation from the pure form of a non-accidental property. For example, grouping systems based on parallelism must consider grouping lines that are a bit non-parallel. There is a non-zero probability that random orientations will lead to nearly parallel lines. Therefore, as Lowe points out, clues like parallelism in practice provide only probabilistic information. If we intend to perform grouping by combining probabilistic information, it makes sense to attempt to understand how to use as many different kinds of information as possible. This can provide coverage of a wider variety of images, and more accurate grouping information in general.

So far we have discussed non-accidental properties as a motivation for using particular grouping clues. A second important issue in perceptual grouping has been to understand how to efficiently form relevant groups. Approaches have spanned a wide spectrum from efficiently computing simple local relationships between image features to more complex algorithms that attempt to integrate more global image information.

Lowe's system, SCERPO, provides an example of simple and efficient computation of pairwise relationships. It groups together pairs of lines that are parallel, collinear, or nearby. Indexing is used so that pairs of lines with these properties can be found without having to explicitly examine all pairs of lines in the image. Scerpo can then form larger groups by finding chains of features connected by these pairwise relations. The system has low complexity, but the decision about what lines to group together rests essentially on the relationship between just two lines. As another example, Cox, Rehg, and Hingorani[11] describe a system that uses a Bayesian approach to finding smooth curves based on ideas from target tracking. This method makes smoothness decisions based on the recent history of a curve. Huttenlocher and Wayner[21] also use local relationships in forming convex groups. They begin with each side of each line segment as a convex group, and then extend a group by adding the nearest neighbor that will preserve its convexity. By only making the best local extension to each group, they guarantee that the output will be linear in the size of the input. And by using a Delauney triangulation of the line segments, they guarantee that nearby neighbors are found efficiently, and that total run time is $O(n \log(n))$. Their system is quite different from ours because they achieve efficiency by making purely local decisions about the best way to extend a group. This can make their algorithm sensitive to small local perturbations in the image, and can cause it to miss useful groups that are found by optimizing a measure of a group's value that is based on its overall, global structure.

Other approaches have attempted to efficiently integrate some more global information into grouping. For example, a hierarchical approach can be used that makes local decisions at a variety of scales, allowing for the integration of information from more spatially separated parts of the image. Boldt, Weiss and Riseman[4] take this approach in detecting collinear groups, while Dolan and Riseman[13] and Saund[38] apply hierarchical methods to cocircularity detection. Mahoney[31] uses parallel processing to gain efficiency in an algorithm that extracts smooth, connected curves. Other researchers have used methods such as convolution or relaxation to integrate smoothness information (Zucker[45], Finkel and Sajda[15], Heitger and Van der HeydtHV, Hérault and Horaud[18] and Guy and Medioni[17]). Other network approaches include Hopfield nets (Mohan and Nevatia[33]) and Bayes nets (Sarkar and Boyer[37]). Typically these systems are not able to formulate and then optimize a simple measure of the the value of a group. Rather, clever methods are used to find groups that bear a strong, though heuristic relationship to some clearly desirable grouping criteria.

In contrast, Shashua and Ullman[40] find the image curve that will explicitly optimize a cost function based on the total curvature and number and size of gaps in a curve. This is computed using dynamic programming, in a network. The system requires $O(Nkn^2)$ computation, where the image is of size n^2, the system represents k discrete orientations at each pixel ($k = 16$ in their experiments), and the system finds a curve of length N. At the end of the computation, the system computes a salience measure for each orientation at each pixel, which reflects the overall salience of the most salient curve of length N ending at that pixel. From this, they can extract the most salient curve in the image.

This system produces interesting results, and finds groups that optimize a more global criteria than some previously mentioned systems. It is similar to our current work in that it finds groups that optimize a global function of a curve that includes gap size. It differs from our work in several respects. First, of course, their system uses smoothness, rather than convexity as a grouping clue. Second, it can be implemented using dynamic programming because the cost of each addition to a curve is based on only a small nearby portion of the curve. By contrast, convexity cannot be checked by comparing a line only to the previous line in the group. In spite of this, as we will show, the systems require similar amounts of computation, on average. Third, Shashua and Ullman's system does not take account of the gap between the beginning and end of a curve, as our system does. There is some evidence that such gaps are of considerable psychological significance (Elder and Zucker[14], Kovacs and Julesz[29]. See also Julesz' article in this volume). Perhaps the most significant difference, however, is that Shashua and Ullman's system is designed to find the most perceptually salient contour, which initially attracts our attention. Our system is designed to find all groups that satisfy a salience criteria, which may be useful in systems that may search through a number of possible groupings, applying some higher level knowledge to identify the correct ones.

Overall, efficiency has clearly emerged as an important issue in many grouping systems. Most systems address this issue by efficiently making local grouping decisions, perhaps using methods such as hierarchical processing or relaxation to incorporate some non-local information into the process. Shashua and Ullman find the optimal solution to a grouping criterion by restricting themselves to one that can be solved using dynamic programming. In contrast, our method finds all globally acceptable groups according to a simple fitness criterion. Moreover, we examine clues such as convexity and closure that appear difficult to enforce with more local methods. We achieve this by showing that a brute force search in fact has a reasonable expected running time.

Convexity and closure are not classic Gestalt grouping clues, but there are two reasons for exploring their effectiveness: first, convexity combined with closure can be strong clues that lines share a common origin, and second many approaches to object recognition rely on interpreting objects in terms of their convex parts. The results here and in Jacobs[23, 22] show that convexity and closure can provide significant inferential leverage for grouping in domains that contain objects with convex parts. These are common in many natural and man-made environments. And while Gestalt grouping clues have been used as the basis of many grouping systems, it has not been demonstrated that they are sufficient to produce robust and effective grouping systems across a wide variety of domains. We expect that by identifying and showing how to make use of additional grouping clues, we can find ways to enhance the performance of existing systems. It has also not been demonstrated that Gestalt principles can explain all aspects of human perceptual organization. Jacobs[23] shows some examples of grouping performance that appear due to convexity, although this possibility has not been explored in the psychology literature, and is only suggestive. Elder and Zucker[14] and Kovacs and Julesz[29] provide evidence that closure plays a significant role in human perceptual organization.

A number of object recognition systems have in practice relied on finding convex groups of edges in images. Kalvin et al.[26] describes a 2-D recognition system that indexed into a library of objects. It began by finding unoccluded convex curves in the image, which it used for indexing. Jacobs[22, 23] describes a system that combined several convex portions of the image to find enough information to index into a library of objects. This system demonstrated that grouping using convex parts could speed up recognition by a factor of several hundred to a thousand.

Some systems have used image groups that are not only convex, but contain other special properties as well. Acronym[5] modeled objects using generalized cylinders that were convex. These convex parts projected to convex collections of edges in the image that were found by a bottom-up grouping system. Lowe's[30] system used bottom-up grouping to find groups containing a variety of non-accidental properties, including convex polygons with proximity and parallel relationships between their sides. For example, trapezoids are found that have nearby endpoints.

A variety of authors have proposed more general approaches to recognition that rely on finding the parts of objects. Hoffman and Richards[19], for example, suggest dividing objects into parts at concave discontinuities. And Biederman[2] suggests performing recognition using the invariant qualities of an object's parts and their relations. These parts tend to be convex. In fact, in implementing a version of Biederman's work, Bergevin and Levine[1] rely on convexity to find the parts of an object.

Convexity may be useful for other types of matching problems, such as motion analysis or stereo. Mohan and Nevatia[33], for example, perform stereo matching between groups of line segments that form partial parallelograms in each image. This reduces the combinatorics of matching. Sawhney[39] uses convexity and proximity as clues for forming groups to track in motion sequences. Such structure is likely to come from a single object face, which is necessary to the tracking method used.

Clearly, a number of researchers have recognized the potential value of convexity in perceptual organization. In some of the systems described above, such as Kalvin et al., Jacobs, Sawhney, and Wayner, convexity is used as a way of finding image features likely to come from a single object. The system we describe in this paper is most suitable for that purpose, since it robustly finds convex groups of edges, accounting for gaps, noise and occlusion. It is possible that our system could form the basis for robust grouping systems that find convex groups with special properties, such as those used by Brooks and Lowe, but this is less straight-forward. In other work, such as Hoffman and Richards, Biederman, and Bergevin and Levine, convexity is used as a method of finding object parts. In many cases the boundaries of convex parts form convex groups containing gaps, where each part joins others. Again our method may be suited to the detection of such groups, although this issue requires further exploration.

To summarize, the novelty of the work presented here lies in three areas. First, we present a concrete derivation that shows how to interpret convexity within the framework of non-accidental properties. At the same time that our approach is clearly derived from past efforts to interpret Gestalt grouping clues in terms of their inferential leverage, our work uses this type of analysis to understand new grouping clues. Second, we present a new approach to efficiently implementing grouping methods. The efficiency of our approach comes not from making local grouping decisions, but from the fact that our grouping criteria are unlikely to be met by random groups. Third, we present a robust tool that may be of value in a number of approaches to image understanding that require finding convex regions of images.

3. Precise Statement of the Problem

Here we precisely describe the convex sequences of line segments the algorithm will produce, and introduce some useful notation.

The system begins with line segments that we obtain by running a Canny edge detector[8] (in the experiments shown, $\sigma = 2$), and then using a split-and-merge algorithm based on Horowitz and Pavlidis[35] to approximate these edges with straight lines. This system approximates curves with lines whose end points are on the curves, such that the curves are no more than three pixels from the line segments.

We call a line segment "oriented" when one endpoint is distinguished as the first endpoint. If l_i is an oriented line segment, then $l_{i,1}$ is its first endpoint, and $l_{i,2}$ is its second. The image contains n line segments, and so it has $2n$ oriented line segments. The convexity of a set of oriented line segments depends on their having the appropriate orientation. That is, a set of oriented line segments is convex if for each oriented line segment, all the other line segments are on the same side of the oriented line segment as its normal, where we define the normal as lying to the right when we traverse the line segment from the first endpoint to the second.

Let S_n be a cycle of oriented line segments: $(l_1, l_2, ... l_n)$ (i.e., l_1 follows l_n). We define L_i to be the length of l_i, that is:

$$L_i = ||l_{i,1} l_{i,2}||$$

We define G_i to be the gap between L_i and L_{i+1}, that is:

$$\begin{aligned} G_i &= ||l_{i,2} l_{i+1,1}||, \quad \text{for } i < n \\ G_i &= ||l_{i,2} l_{1,1}||, \quad \text{for } i = n \end{aligned}$$

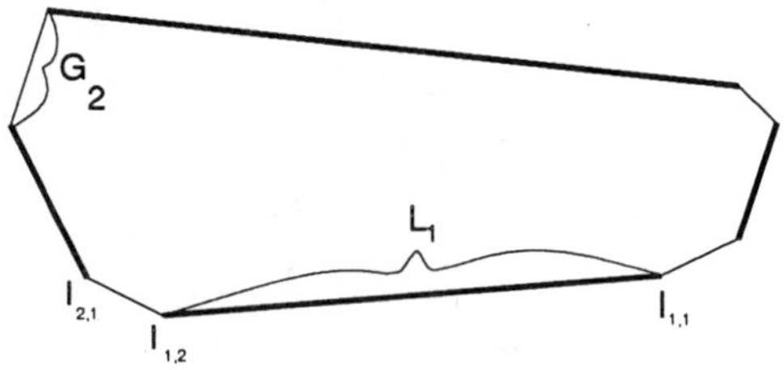

FIGURE 1. The thick lines represent the lines in a group. The thin lines show the gaps between them. The salience fraction is the sum of the length of the thick lines divided by the sum of the length of all the lines.

We then let:

$$L_{1,n} = \sum_{i=1}^{n} L_i$$
$$G_{1,n} = \sum_{i=1}^{n} G_i.$$

We say that S_n is *valid* if and only if connecting the line segments in sequence would create a convex polygon, and if:

$$\frac{L_{1,n}}{L_{1,n} + G_{1,n}} > k$$

for some fixed k. We define:

$$\frac{L_{1,n}}{L_{1,n} + G_{1,n}}$$

to be the *salience fraction* of the convex group (see Figure 1).

The output of this algorithm satisfies a simple, global criterion. One way to think of the salience fraction is that if the lines forming a group originally came from a closed convex curve, the salience fraction tells us the maximum fraction of the curve's boundary which has been detected as image edges. Previous approaches to convex grouping have either grouped one line with another if it is the nearest, mutually convex line ([23, 21]) or if it is within some preset distance ([39]). Because the salience threshold depends on the overall structure of the group, it can bridge gaps whose size is dependent on the overall group size, and hence is scale independent. Furthermore, the system is resistant to clutter; introducing new, spurious lines near a group will not prevent the algorithm from considering more distant lines that lead to groups with a strong overall salience.

4. The Grouping Algorithm and its Complexity

In this section we present an algorithm for finding these salient convex groups. We begin by describing a basic backtracking algorithm. We analyze this algorithm theoretically in order to bound its expected run time and the expected size of its output. We show that the actual results of running the algorithm match our theoretical predictions. Both our experiments and theoretical analysis indicate that the expected running time of the algorithm will be $O(n^2 \log(n) + mn)$ when we use a salience threshold designed to find m groups in an image having n lines. Furthermore, our analysis demonstrates that salient convexity can be a strong clue that a configuration of lines did not occur at random.

Finally, we make some modifications to the basic algorithm, which make it more robust, but which make complexity analysis more difficult. Accordingly, we use experiments to show that these modifications do not significantly affect the algorithm's performance.

4.1. The Basic Algorithm. Our problem definition takes the form of global constraints on the groups of line segments we seek. To perform an effective backtracking search, we must convert these into local constraints whenever possible. That is, we need constraints that determine whether a sequence of line segments could possibly lead to a valid sequence, so that we may prune our search.

We provide the following recursive definition of an *acceptable* sequence, assuming that S_i is acceptable.

(1) Any sequence of a singleton oriented line segment is acceptable.
(2) S_{i+1} is acceptable only if $l_{i+1} \notin S_i$ (i.e. no line may appear more than once in a group).
(3) S_{i+1} is acceptable only if the oriented line segments in it are mutually convex. This will be the case if the sum of the absolute values of the angles turned is 2π when one travels from the first endpoint of the first line to each additional endpoint in turn, returning finally to the first endpoint.
(4) S_{i+1} is acceptable only if: $G_i < \frac{L_{1,i}(1-k)}{k} - G_{1,i-1}$. This is equivalent to stating that $\frac{L_{1,i}}{L_{1,i}+G_{1,i}} > k$.

We prune all unacceptable sequences reached in our search. It is not hard to see that enforcing these constraints will not eliminate any correct groups, while guaranteeing that the search produces only correct groups. We prove this in [25].

To further reduce the run time of our algorithm we notice that some values are re-used many times in the course of such a search, so we precompute these results and save them in tables. In particular, we often wish to know whether two oriented line segments are mutually convex, and if they are we want to know the distance from the end of one segment to the beginning of the other. It is also convenient to keep, for each oriented line segment, a list of all other oriented line segments with which it is mutually convex, sorted by the distance that separates them. Finally, we precompute the angle that is turned when going from one oriented line segment to another. Calculating this information takes $O(n^2 \log(n))$ time, because we must sort $2n$ lists that can each contain up to $2n-2$ items.

We may now describe the backtracking search in more detail, noting how these results are used. The search begins by trying all oriented line segments in turn as singleton sequences. Given an S_i, we calculate $\frac{L_{1,i}(1-k)}{k} - G_{1,i-1}$. From constraint (4), we know that we only want to consider adding a line, l_{i+1}, when the distance from l_{i2} to $l_{i+1,1}$ is less than or equal to this quantity. Clearly we only want to add l_{i+1} if it is mutually convex with l_i. So we can find all candidates to add to the sequence by referencing our precomputed list of line segments that are convex with l_i. Since these lines are sorted by their distance from l_i we may loop through them, stopping once we reach line segments that are too far to consider. By limiting

FIGURE 2. A salient convex group may be formed by choosing any line from each of the four sides.

ourselves to these candidates, we have enforced constraint (4). In addition, we check that l_{i+1} is convex with l_1 using our precomputed results.

We can then enforce constraint (3) by keeping a running count of the angles turned as we traverse the line segments in S_i. A table lookup will tell us the angles added to go from l_i to l_{i+1} and from l_{i+1} to l_1. Therefore, we can ensure that the entire sequence is mutually convex by checking that the absolute values of the angles turned in traversing it sum to 2π. And constraint (2) is simply checked explicitly.

4.2. Complexity Analysis of the Basic Algorithm. We now analyze the performance of this algorithm on randomly generated line segments. This analysis has two goals. First, we wish to understand when salient convexity is a useful grouping clue. By determining the expected number of groups produced, as the salience threshold, k, and the number of lines, n, vary, we obtain a basis for determining the relative likelihood that a group was produced by a scene structure, as opposed to a random process, under various conditions. This analysis will lead us to a method for choosing k appropriate to the input size, so as to produce manageable and useful output. Second, we want to understand when the algorithm will be fast. We also determine the algorithm's expected run time, and show that if we choose a salience fraction that will produce a manageable sized output, that the system's expected run time will also be reasonable.

In the worst case, the algorithm will be exponential in both run time and in the size of its output, for any choice of k. As a simple example of this, in Figure 2 we show eight lines formed into a squarish shape. Even for fairly high values of k, we may form a salient convex group using either of the two lines on each side of the square, for a total of 2^4 groups. If we formed n-tuples of lines around an m sided convex polygon we would have an output of at least m^n groups. By making the sides' endpoints close together, we can ensure that these groups are judged salient for any value of k less than 1. And the work required by the system is at least equal to the size of the output.

However, since we are especially interested in the size of the output produced by random data, an expected case analysis of the algorithm on random data is called for. We also find that performance on real data is well predicted by this analysis. We need first to choose a model of random image generation. We have chosen a simple random model, for which we provide a conservative estimate of our algorithm's performance.

Perhaps the most generic possible model of random line generation would be to assume that lines have lengths, angles and positions chosen from independent uniform distributions. If we assume that camera position is random, there is no reason to expect any bias in line orientation or position so uniform distributions on these variables are particularly natural. Non-uniformity might, however, occur in line length in natural settings. For example, if three-dimensional lines were uniformly distributed in the world, one would tend to see more distant lines than nearby ones, and these distant lines would appear shorter. Furthermore, the actual distribution of line lengths we find in images will depend on our methods of detecting edges and approximating them with line segments. In the analysis that follows, we will none-the-less assume that the image contains n line segments with lengths uniformly distributed between 0 and M, the maximum allowed length. But it is straight-forward to repeat the analysis for other distributions, and to see that distributions that favor shorter lines will lead to results that indicate that less work is required to find salient convex groups than our present analysis indicates.

We will simplify our analysis by assuming that the beginning point of a new oriented line segment is uniformly distributed within a circle of fixed radius, $R = M$, centered at the second endpoint of the last oriented line segment in the current sequence. This case is more tractable to analyze than the assumption that lines are uniformly distributed throughout the image, because through this new assumption the position of each line becomes independent of the structure of the previous lines in a group. This assumption will cause us to conservatively estimate the work required to find convex groups, since randomly distributed lines are closest together when distributed in a circle, leading to more saliency.

We begin by examining the effectiveness of distance and angle constraints in pruning our search. Using these results, we compute the expected work and output size of the search, assuming a particular value of k. This allows us to see how we must adjust k as n grows to keep the output size reasonable, and to see how work depends on k, n, or the desired output size.

4.2.1. *The Distance Constraint.* We need to compute the probability that a new line will lie close enough to the current end of a group to meet the distance constraint. That is, from condition (4) of Section 4.1, we need the probability that: $G_i < \frac{L_{1,i}(1-k)}{k} - G_{1,i-1}$. We define:

$$r_i = \frac{L_{1,i}(1-k)}{k} - G_{1,i-1}$$

and let $h(r_i)$ denote a probability density function on r_i. We call r_i the *excess gap*, since it indicates the amount of allowable gap due to the length of the group that has not yet been "used up" by gaps in the group. The distance constraint requires l_{i1} to fall somewhere in a circle of radius r_i. We assume that the point is located somewhere in a circle of radius R according to a uniform distribution. Therefore, the probability of the distance constraint being met at any one step is the ratio of the area of these two circles, which we can compute once we know $h(r_i)$.

Next, we define:

$$s_i = r_i - G_i$$

That is, s_i is the gap left over after a new gap is added to the group. We let $g(s_i)$ denote its probability density function. Note that:

$$r_i = s_{i-1} + k' L_{i-1}$$

We may therefore recursively compute $h(r_i)$ and $g(s_i)$.

To begin, we have assumed that the lengths of the line segments are uniformly distributed between 0 and M. Therefore the density function of m_i is $\frac{1}{M}$, and the density:

$$h(r_2) = k'm_1 = \frac{1}{Mk'}.$$

Given a density for $h(r_i)$ we will want to compute two quantities. First, we will compute the probability that l_{i1} falls inside a circle of radius r_i. We indicate this as:

$$Pr(||l_{(i-1)2}l_{i1}|| \leq r_i) = \int_0^R h(r_i)\frac{r_i^2}{R^2}dr_i$$

Second, we want to use $h(r_i)$ to determine $g(s_i)$, given that l_{i1} does fall in a circle of radius r_i. That is, we need to find the density function of the distance from a point randomly chosen inside a circle of radius r_i to the boundary of the circle. First let $\overline{s_i}$ be some arbitrary possible value of s_i, (i.e. $\overline{s_i} \leq r_i$). Note that:

$$Pr(G_i \leq r_i - \overline{s_i}|r_i) = \frac{(r_i - \overline{s_i})^2}{r_i}$$

Since $s_i = r_i - G_i$:

$$Pr(s_i \leq \overline{s_i}|r_i) = 1 - \frac{(r_i - \overline{s_i})^2}{r_i^2} = \frac{2\overline{s_i}}{r_i} - \frac{\overline{s_i}^2}{r_i^2}$$

Taking the derivative, we find:

$$g(s_i|r_i) = \frac{2}{r_i} - \frac{2s_i}{r_i^2}$$

Note that this equation holds only when $s_i \leq r_i$, since the probability that $s_i > r_i$ is always 0.

To compute $g(s_i)$ we must consider all values of r_i greater than s_i, and for each, determine the likelihood of such an s_i occurring. That is:

$$g(s_i) = \int_{s_i}^{\max(r_i)} h(r_i)\left(\frac{2}{r_i} - \frac{2s_i}{r_i^2}\right)dr_i$$

where max(r_i) is the largest possible value of r_i. r_i is bounded by $(i-1)k'M$, which is the maximum length of a group of $(i-1)$ lines, times k'. r_i is also bounded by R.

Next, given $g(s_i)$, we want to determine $h(r_{i+1})$. For a given s_i, $r_{i+1} = s_i + k'm_i$. The density of the sum of two values is found by convolving the densities of each value. In this case, since $k'm_i$ is at most $k'M$ and no less than 0 we only consider values of s_i between $r_{i+1} - k'M$ and r_{i+1}, so:

$$h(r_{i+1}) = \int_{\max(0,(r_{i+1}-k'M))}^{r_{i+1}} \frac{g(s_i)}{Mk'}ds_i$$

In fact, this integral is a bit of a simplification, because r_i may never be bigger than R. If we ignore this effect, we are only exaggerating the likelihood of an additional line meeting the distance constraint, and hence overestimating the work that the system performs.

Given these relationships, we may compute any $h(r_i)$. This means that given that we have a group of i lines, we may compute the probability that another line

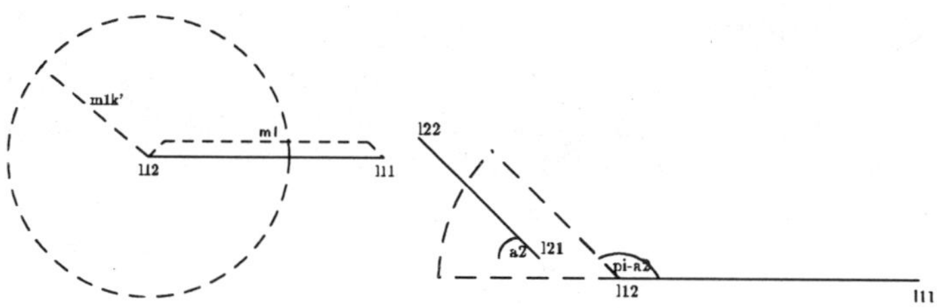

FIGURE 3. On the left l_{21} must appear within the dashed circle to satisfy the distance constraint derived from the requirement that groups be salient. On the right, the dashed circular wedges show where l_{21} must lie in order to satisfy the distance and convexity constraints.

will fulfill our distance constraints. While we could in principal find these values analytically, the integrals quickly become complicated, and so it is more convenient to compute them numerically.

4.2.2. *The Angle Constraint.* There are two parts to our treatment of the angle constraint. First, we consider the probability that a line that passes the distance constraint will be locally convex with just the previous line. When a line, l_i, passes the distance constraint, we know that l_{i1} will be uniformly distributed in a circle about $l_{(i-1)2}$. The location of l_{i1} in this circle is constrained to lie in a wedge (see Figure 3), and so the probability of this occurring depends only on the angle of the wedge, and is independent of the radius of the circle. The wedge's angle is $a_i - a_{i-1}$, the angle of l_i relative to l_{i-1}, provided that $a_i - a_{i-1} \leq \pi$. Otherwise, the wedge's angle is $2\pi - (a_i - a_{i-1})$. For a given angle of l_i, the likelihood of l_i being compatible with l_{i-1} is just the angle of this wedge divided by 2π. So integrating over all angles, which we have assumed are uniformly distributed, we find that there is a probability of $\frac{1}{4}$ that the lines will be compatible.

We must also consider the probability that a sequence of i lines will be mutually convex. We derive a density function on the sum of the angles that must be turned as we go from one line to the next, and use this to determine the probability that this sum is less than 2π. This is a necessary, though not a sufficient condition on convexity. The density function for each such angle is independent of the others and of the distance between the lines. So we need only consider the density on one of these changes of angle, and then convolve this density with itself i times to find the density of the sum of i such angles.

We know that the angle of one line relative to the previous one is uniformly distributed between 0 and 2π. We also know the relationship between the angle of the line and the probability that it is compatible with the previous line. So this probability of compatibility gives us a density on the relative angle of a new line, once we normalize it. If we let f be a probability density function on the change in angle of a new line, we have:

$$f(a_i - a_{i-1}) = \begin{cases} \frac{a_i - a_{i-1}}{\pi^2} & \text{if } a_i - a_{i-1} \leq \pi \\ \frac{2\pi - (a_i - a_{i-1})}{\pi^2} & \text{if } a_i - a_{i-1} > \pi \end{cases}$$

Convolving this density with itself is perhaps facilitated by noticing that this density function is the convolution with itself of a uniform density from $\frac{\pi}{2}$ to $\frac{3\pi}{2}$.

4.2.3. *Expected Work of the Algorithm.* We now determine the expected work our algorithm must do. As we have stated, there is a fixed overhead of $O(n^2 \log(n))$ work. We will determine now the expected number of steps that must be taken by the backtracking search, in addition to this preprocessing. It is important to note that one unit of preprocessing is not directly comparable to one node searched. The relative costs of these operations will depend on many implementation details. We will consider both the absolute number of expected nodes searched, and also the order of the run time. In this section, we discuss the case in which we hold the salience fraction, k, fixed and scale the number of line segments, n. After analyzing output size, we will later discuss the case in which we increase k as n increases so as to hold the size of the output constant.

To find the number of nodes searched, we sum over all i the amount of work that must be done when we consider extending all groups of i lines by an additional line. If our image contains n line segments, we must consider two orientations for each line segment, so there are $2n(2n-2)...(2n-2i+2)$ possible ordered sequences of i line segments. Let δ_i be the probability that the i'th line will pass the distance constraint, given that the previous lines have, and let λ_i be the probability that a group of otherwise compatible lines will have angles that sum to less than 2π. Then the expected number of groups of size i that we must consider, which we will call E_i, is:

$$E_i = \lambda_i 2n(2n-2)...(2n-2i+2)\delta_2...\delta_i \left(\frac{1}{4}\right)^{i-1}$$

where $E_1 = 2n$.

For each group with i lines that we reach, there are potentially $2n - 2i$ lines that we must consider adding to the group. However, our preprocessing has sorted these lines, so that we only need to explicitly consider the ones that meet the distance constraints and that are convex with the last line in our group. We call the expected number of possible extensions to a group of size i, X_i, and:

$$X_i = \frac{1}{4}(2n-2i)\delta_{i+1}$$

The total amount of work that we must perform in the course of our search is proportional to the number of times that we must consider extending a group. This work is given by:

$$W_n = \sum_{i=1}^{n} E_i X_i$$

This expression allows us to see that asymptotically, even the expected amount of work is exponential in n, because our run time is of the form: $c_1 n + c_2 n^2 + c_3 n^3 ...$ where the c_i values are constants. To predict at what point the computation will become unmanageable, we compute the expected amount of work for a variety of realistic situations. As described in [25], we are able to simplify the above expressions so that we need only compute the values for λ_i and σ_i once, through numerical integration. Then we can analytically compute the expected work for any values of R, M or k. We do not prove that the summation of $E_i X_i$ converges, but we find in practice that the terms become tiny for high values of i.

Table 1 shows the expected work of the system as n and k vary. Table 2 shows the amount of work of the system divided by $(2n)^2 \log(2n)$. This tells us roughly

Expected Work, for $M = R$							
Number	k						
of lines	.6	.65	.7	.75	.8	.85	.9
200	$1.99\ x10^7$	2,270,000	346,000	65,800	15,100	4,280	1,650
300	$5.82\ x10^8$	$4.14\ x10^7$	4,200,000	561,000	93,200	19,000	5,200
400	$8.73\ x10^9$	$4.23\ x10^8$	$3.04\ x10^7$	3,020,000	386,000	61,000	12,700
500	$8.53\ x10^{10}$	$3.05\ x10^9$	$1.62\ x10^8$	$1.24\ x10^7$	1,260,000	162,000	26,800
700	$3.45\ x10^{12}$	$8.09\ x10^{10}$	$2.65\ x10^9$	$1.29\ x10^8$	8,830,000	791,000	90,700
1000	$2.25\ x10^{14}$	$3.76\ x10^{12}$	$7.54\ x10^{10}$	$2.11\ x10^9$	$8.77\ x10^7$	5,060,000	378,000

TABLE 1. This table shows the expected work required to find convex groups. As the number of lines in the image and the salience fraction, k, vary, we show the number of nodes in the search tree that we expect to explore. The table does not show the number of steps spent in a preprocessing step. By $M = R$ we indicate that the length of lines are distributed according to a uniform distribution, with a maximum length equal to the radius of the image.

Ratio of search to preprocessing, for $M = R$							
Number	k						
of lines	.6	.65	.7	.75	.8	.85	.9
200	14.4	1.65	.25	.048	.011	.003	.001
300	175	12.5	1.26	.169	.028	.006	.002
400	1,410	68.6	4.92	.489	.063	.010	.002
500	8,560	306	16.3	1.25	.127	.016	.003
700	168,000	3,950	130	6.28	.431	.039	.004
1000	5,140,000	85,800	1,720	48.2	2.00	.115	.009

TABLE 2. This table is an adjunct to Table 1. It shows the expected number of nodes explored in the search tree divided by the number of steps in a preprocessing phase, for various image sizes and salience fractions.

the proportion of the system's work that is spent in search, as opposed to fixed overhead, although one step of overhead is not directly comparable to one step of search. When this ratio is low, the growth in run time is dominated by the growth in the expression $(2n)^2 \log(2n)$. This analysis tells us that when we hold k fixed as n grows, the asymptotic performance of the system is exponential, but that this exponential growth does not dominate the system's performance in many realistic situations.

Later, we compare this to the results of the full system on real data. For now, we compare this theoretically derived estimate of the system's work to simulations on random data, to determine the effects of various approximations that we have made. Our primary approximation has been to assume that the end point of one line will be uniformly distributed in a circle about the end point of a previous line. However, in simulation, lines will be uniformly distributed in a fixed image. Furthermore, we do not apply the full convexity constraints in our analysis, because it does not consider whether a newly added line is convex with the first line in our group. It is possible that connecting these two lines would cause a concavity. We expect that these approximations should make our analysis conservative, overestimating the work required.

In this test we first generate collections of random line segments in a square. To do this, we choose the length of the line segment from a uniform distribution between 0 and half the width of the square, and we choose the angle from a uniform distribution. We draw the location of the line from a uniform distribution, assuming that the line is completely inside the image. Table 3 shows the results of these experiments for various combinations of values of k and n.

It is useful to compare these results with Table 1. We expect first of all that our analysis will overestimate the amount of work required by the system. Second, since we are overestimating the constants in an exponential series, we expect this overestimate to become greater as the later expressions in the series come to dominate. That is, we overestimate the number of pairs of lines that our search will encounter by underestimating the effect of our constraints. When we consider the number of triples of lines encountered, this overestimate is compounded. We expect the overestimate to be more severe in situations where the higher order terms of our series come into play. This occurs as k shrinks and as n grows (i.e. when the number of larger groups considered becomes substantial).

This prediction is supported by the data. When our analysis predicts that only a few thousand nodes will be searched, we find that our analysis has overestimated the required work by about a factor of 10. This gap widens to the point where, when millions of search nodes are expected, this is actually an overestimate by about a factor of 40. Considering that we are predicting the behavior of an exponential process, this seems like a good, (albeit conservative) analysis. It is sufficiently accurate to provide us with a good idea of the circumstances under which our run time will be dominated by preprocessing.

In Section 4.3 we will discuss some additions to the algorithm that make it more suitable for real images, and then describe experiments run on real images before we discuss the significance of the system's run time. But first, we consider the size of its output.

Actual work for random lines, $M = R$							
Number of lines	k						
	.6	.65	.7	.75	.8	.85	.9
200	347,000	112,000	31,300	8,590	2,270	601	123
300	3,120,000	892,000	208,000	45,600	8,760	1,670	306
400	17,400,000	4,970,000	1,130,000	172,000	27,000	4,050	692

TABLE 3. This table shows the actual number of nodes in the search tree that were explored when finding convex groups in randomly generated images. The lengths of the lines were generated from a uniform distribution from zero to half the width of the image. The points were then randomly located in a square image.

4.2.4. *Expected Size of the Output.* It is also important to determine the size of the output of our algorithm. First it will help us to assess the significance of salient convexity as a grouping clue. Second, if we can predict the size of the output ahead of time, we can use this to decide how high to set our salience threshold based on the number of lines in the image to produce an output of the desired size. Third, it will help us to see the extent to which the system's run time depends on the size of its output.

The expressions above for E_i provide the expected number of groups of any particular length that we will encounter in our search. We could use this as a bound on the size of the output, but this would be an oversimplification. As previously noted, the values for E_i are somewhat exaggerated because they do not reflect all the constraint provided by the convexity requirement. In addition, because a group is reached in our search does not mean it will be accepted. When we reach a group of length i in our search, we have yet to take account of the length of the i'th line, or the gap between the i'th line and the first one. It is difficult to determine the probability distribution of this final gap, because it is dependent on the combination of i previous processes that built up the group. But we can approximate it very simply by assuming that the last end point in the group is randomly distributed with respect to the first end point. Using this approximation, the expected number of groups that the system will produce is:

$$O_n = \sum_{i=2}^{n} \frac{1}{4} \delta_{i+1} E_i$$

We sum for i from 2 up because groups of size one are never salient unless the salience fraction is less than or equal to .5.

Table 4 shows the number of groups that we expect to find using this method. Table 5 shows the number of groups that we found in experiments with random line segments. There is good agreement between these results. Our analysis typically overestimates the size of the output by about a factor of 2, except when the expected output is quite large. All of these results are consistent in predicting when the system will break down. Our analysis and experiments both agree on the values of k for which the system will run in a reasonable amount of time and produce a reasonable sized output.

This result has two implications. First, it allows us to identify those circumstances where salient convexity will provide a strong clue that a group did not emerge from a random process. As a simple example of this, suppose we expect an

Expected number of groups produced, for $M = R$							
Number	k						
of lines	.6	.65	.7	.75	.8	.85	.9
200	51,400	5,850	885	166	36.3	8.75	2.09
300	996,000	70,600	7,120	946	154	28.8	5.62
400	1.12×10^7	539,000	38,600	3,820	484	72.8	11.8
500	8.69×10^7	3,100,000	165,000	12,500	1,270	158	21.7
700	2.50×10^9	5.86×10^7	1,920,000	92,800	6,340	561	57.8
1000	1.14×10^{11}	1.90×10^9	3.80×10^7	1,060,000	44,100	2,530	179

TABLE 4. This table shows the expected number of salient convex groups that we expect our algorithm to produce, as the size of the image and the salience fraction vary.

Actual number groups for random lines, $M = R$							
Number	k						
of lines	.6	.65	.7	.75	.8	.85	.9
200	7,500	2,590	782	234	70	15	5
300	45,800	13,800	3,540	867	159	32	5
400	189,000	55,500	13,300	2,300	392	58	6

TABLE 5. The actual number of salient groups that were found in images of randomly generated line segments.

image of 300 lines to contain 20 convex groups that come from convex objects (or parts) and that have a salience threshold of .8 or more. Table 4 shows that we would expect a random process to produce 154 convex groups with this salience, implying that we would have to examine 174 groups output by our algorithm to find the 20 correct groups. Conversely, if the convex groups belonging to real objects contained four lines each, for example, we would have to search through about 300 million quadruples of lines to find the 20 genuine convex groups, and this assumes that we knew to only look at quadruples of lines. This implies that for this example, salient convexity indicates more than a million-fold increase in the likelihood that a group comes from a single object. It is obvious from this result that if we can choose k so that few random groups will be salient, but some of the real structure of the scene will produce salient convex groups, we will have a valuable grouping clue. To determine the general value of salient convexity, however, depends on the number of salient convex groups produced by real structure in scenes, and the distribution of values of k for these groups. It is beyond the scope of this paper to consider these issues further.

Our estimates of the output size of the algorithm not only allow us to determine the value of salient convexity as a grouping clue, but also allow us to determine how the asymptotic run time of the algorithm depends on the output size. To see this, we consider a strategy in which we adjust k as n grows to maintain a constant sized output. Note that:

$$\begin{aligned} W_n &= \sum_{i=1}^{n} E_i \frac{1}{4}(2n - 2i)\delta_{i+1} \\ O_n &= \sum_{i=2}^{n} \frac{1}{4}\delta_{i+1} E_i \end{aligned}$$

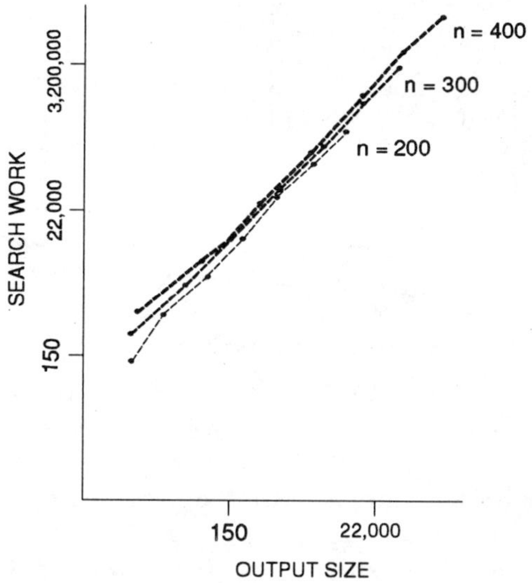

FIGURE 4. The number of search steps performed on simulated data, plotted against output size. The plot is on a log-log scale. The three graphs show different values of n.

And therefore:

$$W_n \leq 2nO_n + E_1X_1 \leq 2nO_n + n^2$$

This tells us that, if we denote output size by m, our search is $O(nm+n^2)$, and total run time of the algorithm is $O(n^2 \log(n) + nm)$. This confirms that the algorithm will be efficient whenever the output is kept small.

This conclusion is based on derivations of W_n and O_n that are bounds on the true values. It is conceivable that this conclusion might not hold if there were a very tight bound on W_n, and a very loose bound on O_n. However, our experiments indicate that this is not the case. Our data suggest that run time grows roughly linearly with the size of the output. In Figure 4 we plot the amount of work performed by the search, for fixed values of n, as k, and therefore the output size, varies. We can see that the search does grow linearly with the output size.

4.3. Applying the Algorithm to Real Images. Up until now we have presented our algorithm in its simplest form, to make it easier to understand its performance. We now describe some modifications that make the system more robust by finding groups that are only approximately convex, and that reduce the size of the system's output by eliminating groups that are similar or identical. Since these modifications make analysis difficult, we present experiments to show their effect on the algorithm's performance.

Error in sensing or feature detection can cause lines that are convex in the world to appear a bit non-convex in an image. To account for this, we allow groups to contain small concavities. Or, one line, when extended might intersect the end of a second line, forcing a concavity when the lines are joined (see Figure 5). To handle this possibility, we allow a portion of a line to participate in a convex group. We also may use just a portion of a line in a group even if using the whole line would

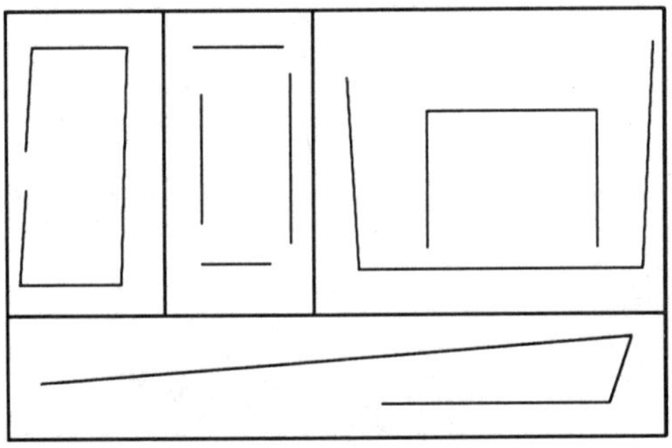

FIGURE 5. Four groups that our basic algorithm would not find. On the left, two of the lines in the group are nearly collinear, but not quite. In the middle, the upper line extends a little too far to the left to maintain convexity. On the right, the bottom line is not convex with the middle ones, but the central part of the bottom line can help form a strong convex group. In the bottom example, we can form a much more salient group by including only part of the top line.

still produce a convex group, if using just a portion of the line will make the group more salient by reducing the gaps between lines.

Our system will often find a number of similar convex groups. We therefore eliminate duplicate groups in a post-processing phase. Additionally, we eliminate supersets of groups that are less salient than the initial group; often spurious small lines may be added to groups that reduce their salience.

We repeated our experiments to determine the effect these modifications have on the run time of the system and on the size of its output. We ran both the basic algorithm and the modified algorithm on a set of real images, so that by comparing these results to our previous results we can tell how much of the change is due to the use of real images, and how much is due to the additional constraints.

Section 5 shows examples of the images we used. Table 6 shows the number of nodes explored in the search, for both the basic and full systems on these and similar images. Table 7 shows the number of groups produced by both variations of the algorithm on these images. Finally, Figure 6 graphically compares the results of our previous analysis and experiments to these new results.

We can see that our theoretical predictions almost always bound the performance of the system. Moreover, our analysis predicts the system's performance roughly to within an order of magnitude. For example, when fewer than one million search steps are expected, our analysis overestimates the work that will be required by the basic algorithm by roughly a factor of 2. This is quite good agreement, considering that the amount of search required by two different images of similar size may also vary by a factor of 2. Our predictions concerning the size of the output are similarly good.

It is perhaps surprising that the system does not spend significantly more time searching for groups in real images than in randomly generated ones. There are two significant ways in which real images differ from the random images that we have analyzed. First, the real images contain many more short than long line segments;

Actual work for real images								
No. lines	Alg. Type	k: .6	.65	.7	.75	.8	.85	.9
183	basic	1,800,000	548,000	133,000	28,900	7,030	2,000	613
	complete	284,000	166,000	75,000	37,000	15,400	6,710	2,470
265	basic	5,630,000	1,660,000	420,000	102,000	27,200	6,370	1,410
	complete	496,000	288,000	136,000	55,300	18,200	7,440	2,800
271	basic			816,000	193,000	47,000	9,740	1,590
	complete			106,000	59,400	24,200	9,810	3,840
296	basic	7,200,000	1,820,000	429,000	93,300	16,300	3,350	946
	complete	273,000	163,000	89,400	41,800	14,800	6,170	2,900
375	basic	689,000	226,000	78,600	27,100	9,620	3,410	1,390
	complete	201,000	104,000	54,600	31,300	18,500	9,610	3,610
450	basic	2,090,000	696,000	227,000	69,000	21,300	6,420	2,160
	complete	295,000	163,000	92,500	48,400	24,800	11,400	4,440
461	basic			72,000	26,700	9,970	3,560	1,250
	complete			105,000	37,500	19,300	8,200	3,130

TABLE 6. This is the number of nodes explored in the search tree for some real images. The second column indicates whether we used the modifications to the algorithm described in the text to make it more robust, or whether we used just the basic algorithm.

Actual number of groups found for real images								
No. lines	Alg. Type	k: .6	.65	.7	.75	.8	.85	.9
183	basic	16,300	4,760	1,170	248	110	60	35
	complete	2,160	1,190	494	315	134	69	31
265	basic	32,300	8,930	1,850	404	182	98	51
	complete	1,750	982	542	276	120	47	27
271	basic			5,600	598	148	64	36
	complete			540	291	157	65	34
296	basic	47,400	12,800	2,670	474	136	73	42
	complete	2,040	1,180	620	312	152	85	42
375	basic	23,100	11,100	5,250	2,390	1,020	406	163
	complete	1,680	1,020	536	331	188	122	48
450	basic	74,100	37,900	18,500	7,840	2,960	965	293
	complete	2,160	1,340	797	430	235	125	52
461	basic			754	376	194	96	49
	complete			863	368	178	84	34

TABLE 7. This table shows the number of salient convex groups produced by the two variations of our algorithm, when applied to a number of real images.

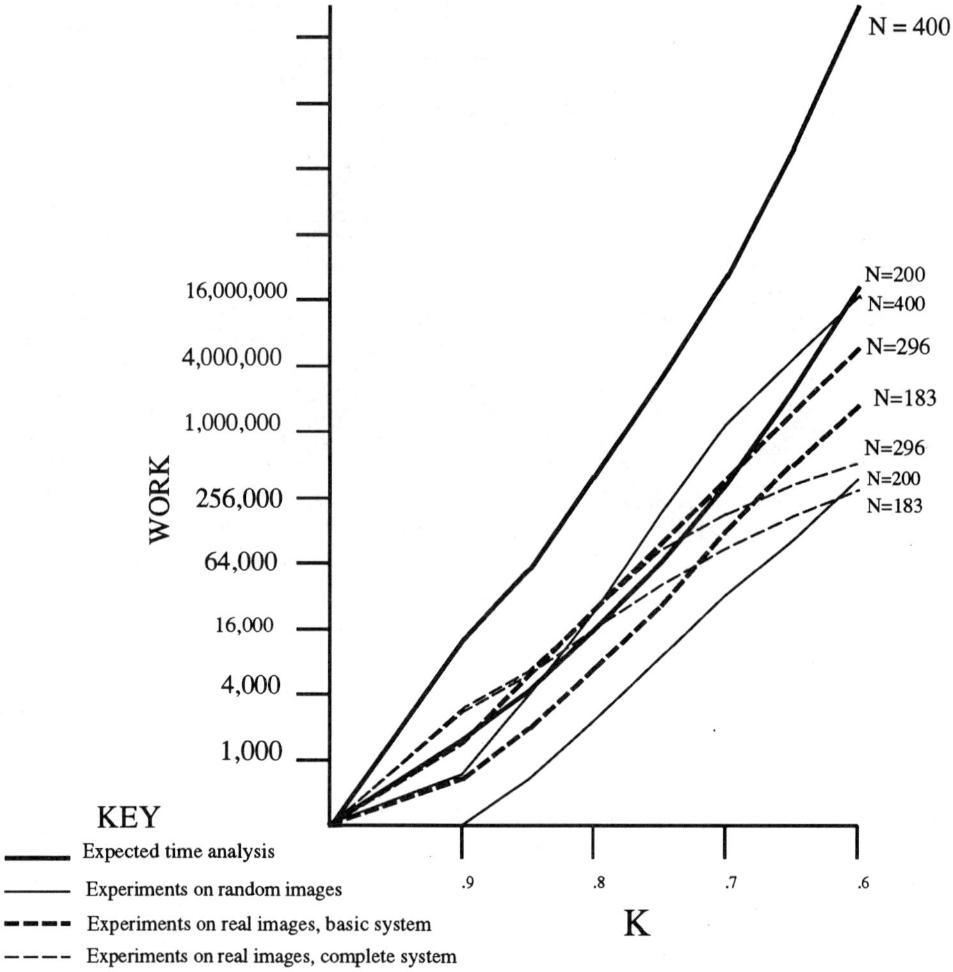

FIGURE 6. This graph compares the expected and actual number of nodes explored in the search for salient convex groups. The number of nodes are graphed on a log scale as k varies, for four cases: our theoretical predictions of expected work when $M = R$; the number of nodes actually searched with randomly generated images; the number of nodes searched on real images with the basic system; and the number of nodes searched on real images by the complete system. Only a sample of the results are graphed here, for clarity. Different cases are graphed with different brushes. On the right, N indicates the number of lines in the image. These are 200 and 400 for the analysis and random images, and 183 and 296 for the real images.

our assumption about the distribution of the lengths of lines was quite conservative for the methods of edge detection and line estimation that we used. Second, real images contain much more structure than random images. One might expect the structure of real images to cause considerable extra search. It appears, however, that even with real images, the system spends little time identifying structure, and most of its time in pruning combinations of lines that lack structure.

Since we are interested in practical, not just theoretical performance, we also measured the run time of our computer implementation. However, because these programs have not been carefully optimized, numbers concerning run times should be regarded with some skepticism. The system ran on a Symbolics 3640 Lisp Machine. On an image with 246 lines, the basic algorithm spent 48 seconds on preprocessing overhead. The search tree was explored at a rate of between 450 and 2,300 nodes per second. Our implementation of the complete algorithm was approximately a factor of 20 slower. Preprocessing was particularly slow in the complete algorithm. However, our implementation of the complete algorithm was simple and inefficient, and we believe that most of the additional time it required could be eliminated in a more careful implementation. These numbers indicate that the overall system could be expected to run in a few minutes or less in a practical implementation, and that the difference in cost between a step of preprocessing and a step of search is about one order of magnitude.

4.4. Evaluation of Grouping Complexity. We have considered our algorithm at three different levels. We have determined its asymptotic performance, its performance in the range of values for n and k that are of interest in real images, and its actual run time. Although the worst-case asymptotic performance of the algorithm is exponential, we have shown that its expected performance is $O(n^2 \log(n) + nm)$, when k is chosen to produce an output of size m from an image of n lines. This translates into an algorithm that requires approximately one million simple steps, and that runs in several minutes in realistic situations. With careful coding and modern hardware, much faster performance should be possible.

The efficiency of the algorithm stems from the fact that precisely when salient convexity is a sufficient constraint to select a small number of groups of lines, it will also be a powerful constraint that we can use to prune our search for objects. Consequently, the key limiting factor in the algorithm's performance is that we must increase k as n increases, in order to keep the run time and the size of the output reasonable. For images containing a few hundred lines, we have shown analytically and experimentally that we must use a value for k ranging between .70 and .85. These values of k allow the system to find convex groups even in the presence of a considerable amount of occlusion or noise. At the same time, the algorithm is still able to exhaustively search for groups that satisfy a global criterion. We therefore know the exact nature of the groups we find as a function of group and gap size. The algorithm will not be distracted by clutter into making good local decisions that turn out to be poor decisions when one considers the entire image.

Although the algorithm stands or falls with its performance on real images of interest, we believe there is much value to a theoretical analysis of the algorithm's behavior. Our analysis has predictive value, it allows us to see when the system will break down. We conclude that as images become more cluttered, our system will continue to work if we demand greater salience in the groups that it produces. This allows us to set the salience threshold dynamically, if we wish, to ensure that our system will run quickly and produce a reasonably small output. Furthermore, our analysis shows that the use of a higher salience threshold is necessary in cluttered scenes if the groups that we find are to be truly salient, in the sense that they are unlikely to arise randomly.

5. Grouping Performance

So far, we have evaluated our grouping system in terms of its efficiency, and the probability that the groups it produces will reflect real scene structure, rather than a random process. We now present some examples of the system's performance on real images. In particular, we have integrated our grouping system into a complete recognition system, to demonstrate its potential value.

Our system, described in detail in [25], first forms convex groups, as we have described here. Additionally, it only uses a group if none of its oriented line segments appear in a more salient convex group. This prevents us from having to consider several, similar groups. For each group, we find 2-D point features at places where the lines in the group have an intersection point that is stable with respect to error. The recognition system then considers all pairs of salient convex groups found. The 2-D points in a pair of salient groups are used to index into a lookup table, where we have represented comparable groups of 3-D model points, at compile time. This indexing phase, which matches groups of 2-D image points to geometrically consistent 3-D model points is also described in [24]. We then explore the matches hypothesized by indexing, beginning with those generated by pairs of image groups that matched the fewest pairs of model groups. A verification step uses matches between 2-D points and 3-D points to determine a hypothetical location of a model in the image. It then searches for additional evidence of the model's presence at the location, to confirm or reject a hypothesis.

We first show the most salient groups found in several images, along with their associated point features. Figure 7 shows the most salient groups found by the grouping system in an image of an isolated telephone. Many of the groups found here show up reliably in other pictures of the phone, taken from different viewpoints.

In Figure 8 we see some groups found in a scene that includes the telephone. Almost all the telephone's convex groups are at least partially occluded in this picture. However, we find unoccluded portions of these groups, many useful groups from the stapler, and some of the salient structure of the mugs. Figure 9 shows the groups found in another scene containing the occluded telephone.

Figure 10 shows the results on a different scene, which was taken at the CMU calibrated image lab. Although the edges are noisy and hard even for a person to interpret, we can see that the system finds much of the rectangular structure inherent in the buildings in the scene.

In each of the pictures shown, many of the most salient groups come entirely from the convex structure of a single object, making them useful for recognition. We also see many remaining challenges to grouping, because many of the groups found either do not appear perceptually salient, or appear to either combine lines from two different objects, or to combine strong lines from one object with noisy or unstable lines. For example, in the fourth set of groups shown in Figure 8, we can see that a strong group from the face of the telephone includes an external edge from the rim of an occluding coffee mug.

These groups, however, do enable our system to efficiently recognize objects. Although [25] details the performance of the system, Figure 11 shows two samples of our results. Pairs of groups of points from the telephone have been represented in the indexing table ahead of time. The figure shows that the system correctly locates the telephone. Circles indicate the location of image points used for indexing, which come from two salient convex groups. In each example, one group is from the side of the telephone. In the example on top, the other group comes from the rectangular face of the phone; in the example on the bottom, the second group comes from the small plate containing the phone number. We can see from these examples that grouping produces a relatively small set of convex groups, some of which lead to correct recognition of the object sought. See [25] for further details on the overall efficiency of the recognition system, using grouping.

6. Conclusions

Convexity is just one potential grouping clue, but it is an important one to understand thoroughly. Objects often contain at least some convex parts, especially in two important application areas, recognition of buildings and of manufactured objects. Something similar to the salient convexity measure that we consider is an essential part of the grouping methods used by Jacobs[22], Wayner[42], and Sawhney[39]. And related grouping methods based on convexity have been widely used in various approaches to recognition. But convexity has usually been handled in ad-hoc ways that are sensitive to local perturbations of the image.

In this paper we use a simple, global measure of a convex group's salience. A global definition of our output has the strong advantage of allowing us to anticipate our output, independent of unrelated context. We show here that much of the global constraint provided by salient convexity can be converted into a local form in which it can be applied at each step of the search, and that this allows us to build an efficient system.

Three key questions arise if we wish to understand the performance of salient convexity. First, can we compute it efficiently? Second, how likely is it that salient convex groups reflect the underlying structure of the scene, as opposed to simply resulting from a random process? Third, how robust can salient convex groups be to possible occlusions or clutter? We have performed a theoretical analysis, backed up by experiments, that allow us to answer these questions.

We show that in realistic situations, we can efficiently find the most salient convex groups of an image. Our algorithm takes $O(n^2 \log(n) + mn)$ expected time to find the m most salient groups in an image with n line segments. This theoretical result is also practical. In real images, our system runs sufficiently fast to be useful as an experimental system.

The systems runs quickly precisely because salient convex groups do not often arise by chance. This allows us to effectively prune our search for groups. It also allows us to show that the groups that we do find are likely to reflect some non-random scene structure. Previously, Gestalt grouping phenomena have been explained by showing that they are unlikely to occur at random. We show how to apply this criterion to identify grouping laws not described by the Gestaltists. This

FIGURE 7. This shows the most salient groups in an image of a telephone. These are all the groups that have a salience fraction of at least .75, and that consist entirely of oriented line segments that do not appear in more salient convex groups. The dotted lines show the edges of the image. Within the image there is a box around each separate group. Solid lines show the lines that form the group. Circles show the corners found in the group.

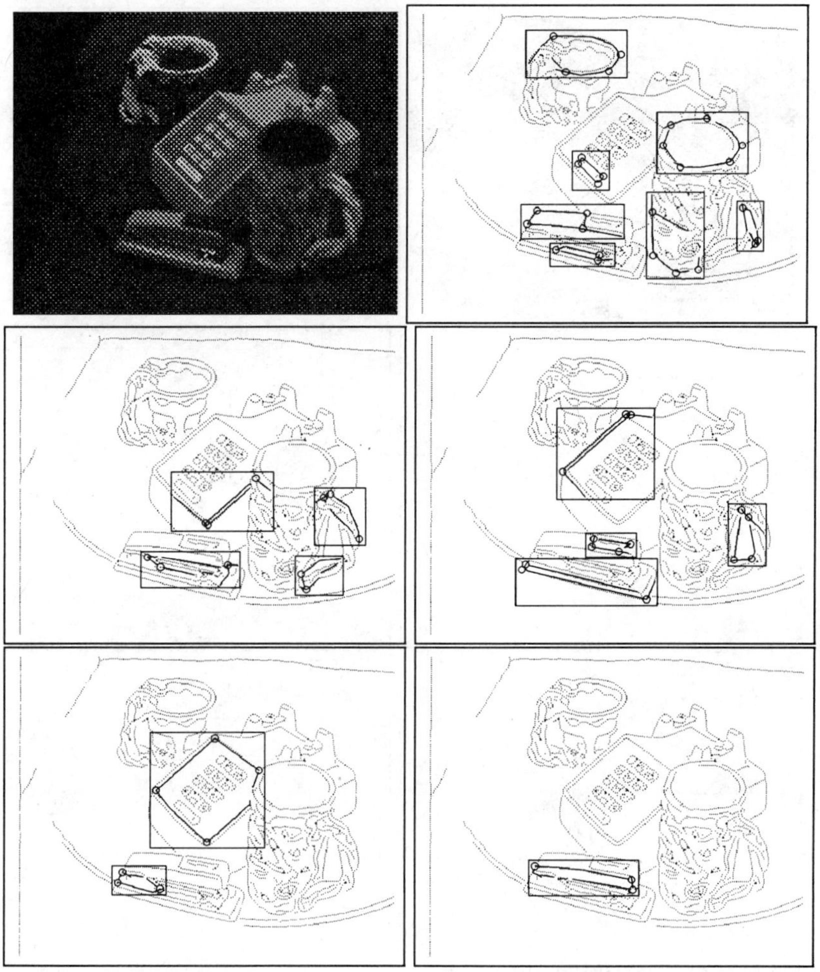

FIGURE 8. Similarly, a scene is shown in the upper left corner, and the remaining pictures show all the most salient groups found.

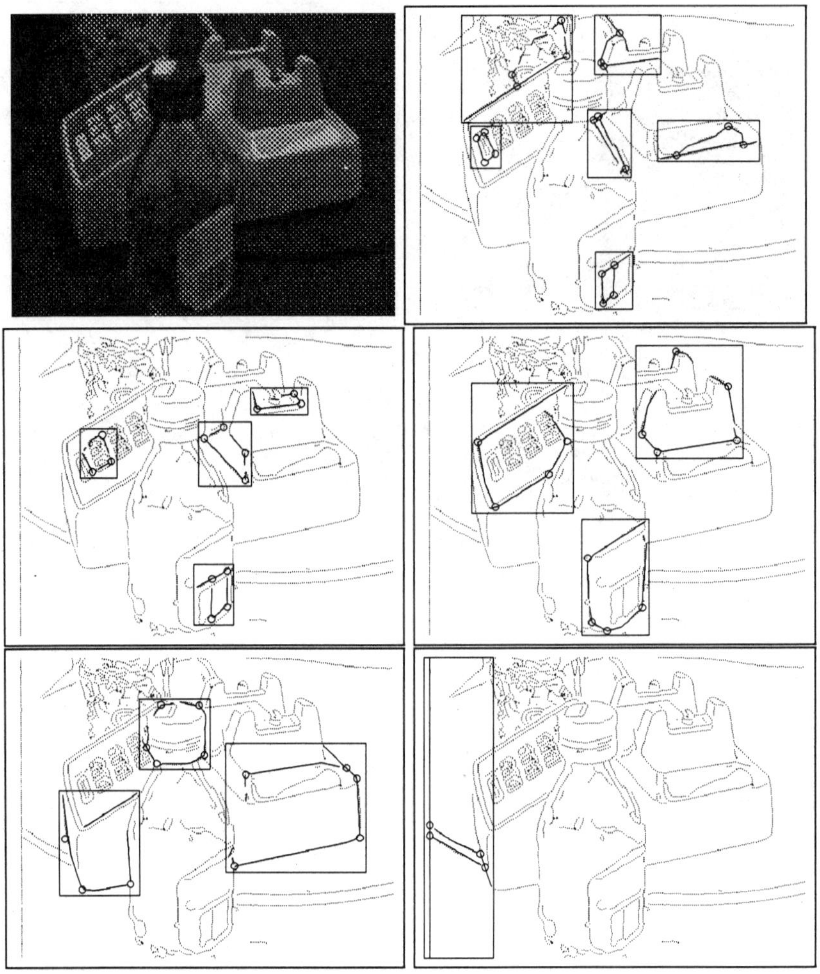

FIGURE 9. The salient convex groups found in another scene.

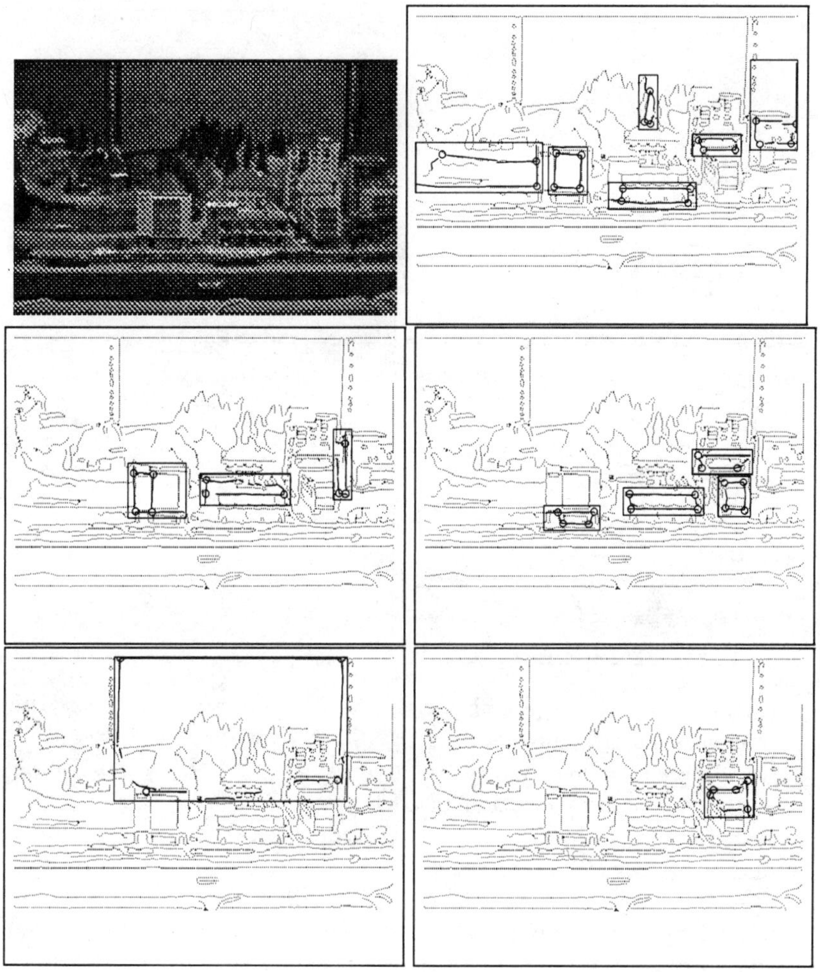

FIGURE 10. An arbitrarily chosen picture from the CMU calibration lab, courtesy of David Michael, along with the most salient groups found.

FIGURE 11. This figure shows the results of a recognition system that uses salient convex grouping. On the left are edges found in two different scenes. On the right we show the correct hypothesis of the telephone's location in each scene, which is found by the system. Lines, which indicate the hypothetical location of model lines, are shown superimposed over a dotted edge map of the image. Circles indicate the location of image points that were used for indexing. Squares show the hypothesized location of the corresponding model points.

is necessary if we wish to perform grouping reliably in a wide range of situations, where compelling Gestalt clues are not always present. To this end our work is complementary to, not competitive with, Gestalt grouping. We believe that reliable grouping systems will need to make use of as many possible sources of information as possible.

Finally, our analysis shows how to choose a measure of convex salience appropriate to the size of the image, and of the desired output. For realistic sized images, we show that a salience threshold ranging from about .70 to .85 will produce a fast algorithm, with a manageable output size. This tells us that our grouping system can find convex groups that have between 30% and 15% of their boundary missing due to occlusion or noise. The presence of clutter also affects the robustness of the algorithm; in an uncluttered scene with fewer line segments we could use a lower salience threshold, as we add clutter, our output will grow unless we raise this threshold. Since our output is characterized by a simple salience criteria, we can predict precisely how robust our algorithm will be.

Finally, we have demonstrated the potential value of salient convex grouping by incorporating our system into a complete recognition system. Grouping reduces the combinatorics of recognition by focusing the search for an object on subsets of the image that are likely to come from a single object. It is not necessary that every group of lines that we find in the image actually comes from the object for which we are looking. It is sufficient if we can locate enough image groups to allow us to recognize an object without having to examine too many irrelevant image groups, that is, our groups need to provide points that are more likely to come from the object for which we are looking than are randomly selected groups. The greater this likelihood is, the greater is the advantage provided by grouping. The grouping system presented here supports a very focused recognition system that requires little search.

REFERENCES

1. Bergevin, R. and Levine, M., 1993, "Generic Object Recognition: Building and Matching Coarse Descriptions from Line Drawings," *IEEE Transactions on Patern Analysis and Machine Intelligence*, **15**(1):19-36.
2. Biederman, I., 1985, "Human Image Understanding: Recent Research and a Theory," *Computer Graphics, Vision, and Image Processing*, (32):29-73.
3. Binford, T., 1981, "Inferring Surfaces from Images," *Artificial Intelligence*, **17**:205-244.
4. Boldt, M. R. Weiss, and E. Riseman, 1989, "Token-Based Extraction of Straight Lines," *IEEE Transactions on Systems, Man and Cybernetics*, **19**(6):1581-1594.
5. Brooks, R., 1981, "Symbolic Reasoning Among 3-D Models and 2-D Images," *Artificial Intelligence*, **17**:285-348.
6. Burns, J. and E. Riseman, 1992, "Matching Complex Images to Multiple 3D Objects using View Description Networks," *IEEE Conference on Computer Vision and Pattern Recognition*, pp. 328–334.
7. Califano, A. and R. Mohan, 1991, "Multidimensional Indexing for Recognizing Visual Shapes," *IEEE Conference Computer Vision and Pattern Recognition*, pp. 28–34.
8. Canny, J., 1986, "A Computational Approach to Edge Detection," *IEEE Transactions on Patern Analysis and Machine Intelligence*, **8**(6):679-698.
9. Clemens, D., 1991, *Region-Based Feature Interpretation for Recognizing 3D Models in 2D Images*, MIT AI TR–1307.
10. Clemens, D. and Jacobs, D., 1991, "Space and Time Bounds on Model Indexing," *IEEE Transactions on Patern Analysis and Machine Intelligence*, **13**(10):1007-1018.

11. Cox, I., J. Rehg, and S. Hingorani, 1992, "A Bayesian Multiple Hypothesis Approach to Contour Grouping," *European Conference on Computer Vision*, pp. 72–77.
12. Denasi, S., G. Quaglia, and D. Rinaudi, 1991, "The use of Perceptual Organization in the Prediction of Geometric Structures," *Pattern Recognition Letters*, **13**(7):529-539.
13. Dolan, J. and E. Riseman, 1992, "Computing Curvilinear Structure by Token-based Grouping," *IEEE Conference Computer Vision and Pattern Recognition*, pp. 264–270.
14. Elder, J. and S. Zucker, 1993, "The Integration of Figure Fragments into Representations of Planar Shape," McGill Research Centre for Intelligent Machines TR-93-2.
15. Finkel, L. and P. Sajda, 1992, "Object Discrimination Based on Depth-from-Occlusion," *Neural Computation*, **4**(6):901-921.
16. Grimson, W.E.L., 1990, *Object Recognition by Computer: The role of geometric constraints*, MIT Press, Cambridge.
17. Guy, G. and G. Medioni, 1993, "Inferring Global Perceptual Contours from Local Features," IRIS-USC Technical Report.
18. Hérault, L. and R. Houraud, 1993, "Figure-Ground Discrimination: A Combinatorial Optimization Approach," *IEEE Transactions on Patern Analysis and Machine Intelligence*, **15**(9):899-914.
19. Hoffman, D. and Richards, W., 1984, "Parts of Recognition," *Visual Cognition*, edited by S. Pinker, MIT Press, Cambridge.
20. Huang, T., 1991, "Reply Computer Vision Needs More Experiments and Applications," *Computer Vision, Graphics and Image Processing*, **53**(1):125-126.
21. Huttenlocher, D. and Wayner, P., 1992, "Finding Convex Edge Groupings in an Image," *International Journal of Computer Vision*, **8**(1):7-29.
22. Jacobs, D., 1987, "GROPER: A Grouping Based Object Recognition System for Two-Dimensional Objects," *IEEE Workshop on Computer Vision*, pp. 164-169.
23. Jacobs, D., 1989, "Grouping for Recognition," MIT AI Memo 1177.
24. Jacobs, D., 1992, "Space Efficient 3D Model Indexing," *IEEE Conference Computer Vision and Pattern Recognition*, pp. 439–444.
25. Jacobs, D., 1992, "Recognizing 3-D Objects Using 2-D Images," MIT AI TR–1416.
26. Kalvin, A., E. Schonberg, J. Schwartz, and M. Sharir, 1986, "Two-Dimensional, Model-Based, Boundary Matching Using Footprints." *The International Journal of Robotics Research*, **5**(4):38-55.
27. Kanade, T., 1981, "Recovery of the Three-Dimensional Shape of an Object from a Single View," *Artificial Intelligence*, **17**:409-460.
28. Kohler, W., 1959, *Gestalt Psychology*, Mentor Books, New York.
29. Kovacs, I. and B. Julesz, 1993, "A Closed Curve is Much More than an Incomplete One: Effect of Closure in Figure-Ground Segmentation."]it Proc. Nat. Acad. Sci., USA, **90**:7495-7497.
30. Lowe, D., 1985, *Perceptual Organization and Visual Recognition*, Kluwer Academic Publishers, The Netherlands.
31. Mahoney, J., 1987. *Image Chunking: Defining Spatial Building Blocks for Scene Analysis*, MIT AI TR–980.
32. Marr, D., 1977, "Artificial Intelligence - A Personal View," *Artificial Intelligence*, **9**:37-48.
33. Mohan, R. and Nevatia, R., 1989, "Using Perceptual Organization to Extract 3-D Structures," *IEEE Transactions on Patern Analysis and Machine Intelligence*, **11**(11):1121-1139.
34. Nitzburg, M. and D. Mumford, 1990, "The 2.1-D Sketch," *IEEE ICCV*:138-144.
35. Pavlidis, T. and S. Horowitz, 1974, "Segmentation of Plane Curves," *IEEE Transactions on Computers*, **C**(23):860-870.
36. Richards, W. and A. Jepson, 1992, "What Makes a Good Feature?" MIT AI Memo 1356.
37. Sarkar, S. and K. Boyer, 1993, "Integration, Inference, and Management of Spatial Information Using Bayesian Networks: Perceptual Organization," *IEEE Transactions on Patern Analysis and Machine Intelligence*, **15**(3):256-274.
38. Saund, E., 1992, "Labeling of Curvilinear Structure Across Scales by Token Grouping," *IEEE Conference Computer Vision and Pattern Recognition*, pp. 257–263.
39. Sawhney, H., 1992, *Spatial and Temporal Grouping in the Interpretation of Image Motion*, University of Massachusetts, COINS TR92-05.
40. Sha'ashua, A. and Ullman, S., 1988, "Structural Saliency: The Detection of Globally Salient Structures Using a Locally Connected Network," *IEEE ICCV*:321-327.

41. Syeda-Mahmood, T., 1992, "Data and Model-Driven Selection using Color Regions," *Second European Conference on Computer Vision*:115-123.
42. Wayner, P.C., 1991, "Efficiently Using Invariant Theory for Model-based Matching," *IEEE Conference Computer Vision and Pattern Recognition*:473-478.
43. Williams, L. and A. Hanson, 1994, "Perceptual Completion of Occluded Surfaces," *IEEE Conference Computer Vision and Pattern Recognition.*
44. Witkin, A. and J. Tenenbaum, 1983, "On the Role of Structure in Vision," In *Human and Machine Vision*, edited by Beck, Hope, and Rosenfeld. Academic Press, New York.
45. Zucker, S., 1983, "Cooperative Grouping and Early Orientation Selection," In *Physical and Biological Processing of Images*, edited by Braddick and Sleigh. Springer-Verlag, Berlin.

NEC RESEARCH INSTITUTE, 4 INDEPENDENCE WAY,, PRINCETON, NJ 08540,
E-mail address: dwj@research.nj.nec.com

DIMACS Series in Discrete Mathematics
and Theoretical Computer Science
Volume 19, 1995

MIXTURE MODELS FOR OPTICAL FLOW COMPUTATION

ALLAN JEPSON AND MICHAEL J. BLACK

ABSTRACT. The computation of optical flow relies on merging information available over an image patch to form an estimate of 2D image velocity at a point. This merging process raises a host of issues, which include the treatment of outliers in component velocity measurements and the modeling of multiple motions within a patch which arise from occlusion boundaries or transparency. We present a new approach which allows us to deal with these issues within a common framework. Our approach is based on the use of a probabilistic *mixture model* to explicitly represent multiple motions within a patch. We use a simple extension of the *EM-algorithm* to compute a maximum likelihood estimate for the various motion parameters. Preliminary experiments indicate that this approach is computationally efficient and can provide robust estimates of the optical flow values in the presence of outliers and multiple motions. The basic approach can also be applied to other problems in computational vision, such as the computation of 3D relative motion, which require the integration of several partial constraints to obtain a desired quantity.

Motion analysis
Optical flow, probabilistic mixture models, motion boundaries.

1. INTRODUCTION

The computation of optical flow relies on merging information available over an image patch to form an estimate of 2D image velocity at a point. The well known aperture problem for optical flow computations [11] states that, given information available from only a small spatial aperture, we can expect to derive only a partial constraint on the image motion. In order to fully constrain the optical flow we need to integrate several such constraints obtained over a larger spatial neighborhood. As the size of this neighborhood grows there is an increased likelihood that it will span an object boundary in the scene which will result in multiple motions within the region. Multiple motions can also be the result of transparency, highlights and shadows. In these situations, the assumption of a single motion within the region results in inaccurate estimates of the optical flow. We relax the single-motion assumption and, instead, assume that the motions within any particular region can be described by a probabilistic *mixture* of distributions.

1991 *Mathematics Subject Classification.* Primary 68U10.

We thank J. Heel for providing the Pepsi-can image sequence and J. Bergen for providing the transparency sequence. Funding for this work was provided, in part, by the ARK (Autonomous Robot for a Known environment) Project, which receives its funding from PRECARN Associates Inc., Industry Canada, the National Research Council of Canada, Technology Ontario, Ontario Hydro Technologies, and Atomic Energy of Canada Limited. The authors also gratefully acknowledge the financial support of NSERC Canada.

1052-1798/95 $1.00 + $.25 per page

We observe that, when multiple motions are present, the motion estimates within a region form distinct *clusters.* We employ a simple extension of the *EM-algorithm* [14] to isolate these clusters and estimate their likelihood. This approach has a number of benefits. Like robust regression techniques [3], the approach allows us to robustly estimate the dominant motion within a region. Morevoer, by assuming the motion is due to a mixture of distributions we are able to recover multiple coherent motions and identify *outlying* measurements which do not correspond to a coherent motion. The recovered information about the presence of multiple motions may prove useful for the early detection of surface boundaries.

In this paper we describe the problems caused by multiple motions and briefly review previous approaches for dealing with them. We then introduce the theory of *mixture models* and describe the EM-algorithm. This basic approach has more general applicability than motion estimation and can be applied to other problems in computational vision, such as the computation of 3D relative motion, which require the integration of several partial constraints to obtain a desired quantity. Here we illustrate the theory with a series of experiments with natural image sequences containing motion boundaries, noise, and transparency. These preliminary experiments indicate that this approach is computationally efficient and can provide robust estimates of the optical flow values in the presence of outliers and multiple motions.

2. Integration of Partial Constraints

We consider the problem of estimating the optical flow from constraints available within a particular image region. Let $S(\vec{x},t)$ denote an image sequence formed, possibly, by some preprocessing of the original sequence $I(\vec{x},t)$. For example, $S(\vec{x},t)$ might be obtained from a smoothed or filtered version of $I(\vec{x},t)$. To extract motion information we apply the *data conservation constraint*, which states that S is preserved locally in space and time in the direction of image motion. That is, upon differentiating the conservation constraint

$$S(\vec{x}(t),t) = constant,$$

we obtain the "motion constraint equation"

$$\vec{c}(\vec{x},t)\cdot\vec{v}(\vec{x},t) = 0. \tag{1}$$

Here the "motion constraint vector", $\vec{c}(\vec{x},t)$, is the spatiotemporal gradient of S, namely $\vec{\nabla}S(\vec{x},t)$, and $\vec{v}$ is a 3-vector representing the local image velocity. Usually $\vec{v}$ is taken to be $(v_1, v_2, 1)$, where v_1 and v_2 are the components of the 2-D image velocity in the image directions x_1 and x_2, respectively.

The motion constraint equation (1) provides a single (linear) constraint on the two unknowns v_1 and v_2, and is therefore insufficient to determine a unique 2D image velocity. This is commonly referred to as the *aperture problem* [11]. As a result, we are faced with collecting several such constraints, say from a spatiotemporal neighborhood of the point $(\vec{x},t)$, in an attempt to infer a particular 2D image velocity. Within this neighborhood one typically assumes that the motion can be described by a single parametric model which is commonly taken to be constant, affine, or quadratic. With this approach, the neighborhood must be taken to be sufficiently large to include several constraints having different orientations.

There are other constraints, however, on the choice of the "aperture" size which require that the aperture be kept small. For example, our model of the motion typically will only provide a good approximation to the true image motion over small neighborhoods. Additionally, as the region size grows, it is more likely to contain multiple surfaces with different motions whose constraints will contaminate the single-motion estimate. We refer to this dilemma surrounding the choice of aperture as the *generalized aperture problem.*

A second issue that must be faced is that the constraint (1) arises from an assumption about the conservation of the image structure $S(\vec{x}, t)$, and this assumption may sometimes be inappropriate. In the simplest example, with S just taken to be the original image I, we have the assumption that $I(\vec{x}(t), t)$ is (locally) constant for paths $(\vec{x}(t), t)$ moving with the correct image motion. This constraint is clearly violated in many natural situations, such as cases with shading variation, highlights, transparency, or occlusion boundaries. While more effort can be spent on developing more sophisticated structure assumptions to deal with some of these cases (as has been done, for example, by Fleet [6]), we cannot avoid the need to make some assumption in connecting image structure to the motion of points in the scene. When this assumption is violated we can expect our motion constraint vector, $\vec{c}(\vec{x}, t)$, to be meaningless. Thus we should expect any method for the measurement of motion constraint vectors to produce at least the occasional outlier, and therefore the method used for the integration of these constraints must be robust to such outliers.

In summary, optical flow computation relies on a somewhat tenuous link to the motion of the scene provided by assumptions about data conservation and, moreover, these assumptions are insufficient on their own to determine a unique image velocity. However, on the positive side, in many situations of interest the desired flow field is spatially coherent over relatively large regions of the image. For example, it has been shown that an affine flow model is a reasonable approximation in many cases [1], such as for a smooth surface having a sufficiently small variation in relative depth. If such a surface is textured then it can be expected to give rise to a large number of motion constraint vectors, and thus the equations for the six parameters of the affine flow can be massively redundant. In this sort of situation we should expect to be able to compute an accurate representation of the flow field for the region.

Consider, for example, the situation that occurs when the spatial neighborhood is centered at a motion boundary. In this case, approximately half of the constraints will correspond to one side of the boundary and half to the other. Such an example is shown in Figure 1 where the bottom edge of the figure is the v_1 axis, the left edge is the v_2 axis, and the dark lines are the constraints lines, $\vec{c}(\vec{x}, t)$. In the figure two "clusters" of constraint intersections can be observed. If we assume a single translational motion over the region the "optimal" motion estimate will lie somewhere between the two actual motions and will result in a *blurring* of the flow field at the motion boundary. A similar situation results in cases of multiple transparent motions and fragmented occlusion where no clear surface boundary exists. These situations result in exactly the same type of clusters of constraints.

To cope with situations such as this we relax the single motion assumption in two ways. First, we assume that a region may contain multiple coherent motions. We can think of these multiple motions as corresponding *layers* [15] whose spatial

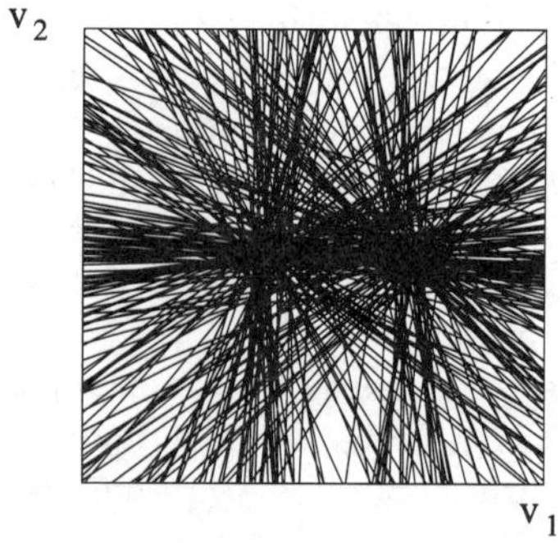

FIGURE 1. Constraint lines from differently moving surfaces within an aperture give rise to distinct "clusters" of constraint line intersections.

extent may be the entire region. Each layer contains a single consistent motion and each layer may be described by different parametric motion models. This is an important point since, in the case of transparency, there is more than one motion at each point in the image corresponding to different surfaces in the scene. Second, we assume that multiple motions and noise will occasionally result in constraint vectors which are outliers and, these should be identified and rejected.

2.1. Previous Approaches. A host of techniques have been developed to deal with the generalized aperture problem and outliers.[1] Example techniques include the use of "adaptive windows" which adjust their size and shape in an attempt to capture constraints from a single smooth surface patch [16]. While such an approach cannot cope with transparency and fragmented occlusion, the area-based regression approach of Bergen et al. [2] provides an iterative method for recovering two transparent motions from three frames. For coping with motion boundaries, an alternative approach is provided by regularization, in which the effective integration domain for constraints on a single smooth surface is dictated implicitly through a smoothness model and iteration. Smoothing over motion discontinuities is avoided either by explicitly introducing boundaries (as line processes) [9] or by using weak continuity constraints [4]. To cope with cases of fragmented occlusion, Darrell and Pentland [5] have proposed a method for segmenting the motions into distinct layers. The use of multiple layers within a robust regularization framework is discussed further in the chapter by Madarasmi and Kersten in this volume [13].

A host of other techniques fall under the category of *robust estimation* [10] in which the goal is to recover the dominant motion while treating the inconsistent constraints as outliers and reducing their influence on the solution. The robust estimation framework introduced by Black [3] has been applied to area-based regression techniques as well as correlation. The framework also generalizes the regularization techniques by applying outlier rejection to both the data conservation and spatial smoothness assumptions.

One problem with the robust statistical techniques is that, in cases of multiple motions, they treat "secondary" structure as noise. In the previous figure which contained multiple motions, the robust techniques would accurately recover one

[1]See [3] for a review.

of the two motions. An additional procedure would be needed to recognize that the "outliers" have a coherent structure and that two motions are present. Our approach, as opposed to making the single-motion assumption, explicitly models multiple motions and outliers, and hence is able to capture the more complex structure present in the data.

A somewhat different approach, referred to as "constraint line clustering", has been proposed by Schunck [17]. The basic idea is that the redundancy in the motion constraint vectors $\vec{c}(\vec{x}, t)$ arising from a smooth surface patch should be recognizable from the constraint vectors themselves. Indeed, for a patch moving with a nearly constant velocity, the "constraint lines" will all nearly pass through the same point in the (v_1, v_2)-plane. Therefore, given the constraints from an image patch, the idea is to seek such clusters of constraint lines. If the cluster detection process could be designed to be insensitive to outliers, and if the location of each identified cluster could be made insensitive to other clusters in the data set, then the approach should be able to provide accurate flow estimates without the need for a detailed knowledge of the appropriate integration regions. We follow this general strategy here, although our cluster detection process is quite different than the one-dimensional technique proposed by Schunck. In particular, Schunck does not model multiple motions within a patch and therefore cannot detect and exploit information about multiple motions when it exists.

3. Mixture Models of Flow

For a given image region we attempt to model the flow in terms of a handful of smoothly varying layers. For example, $\vec{v}(\vec{x}; \vec{a})$ may represent a constant velocity field for one layer, or it could denote an affine flow where the components v_1 and v_2 are given by linear functions of the image position $\vec{x}$. In the first case the parameter vector $\vec{a}$ is 2-dimensional, while it is 6-dimensional in the affine case.[2] Multiple motions within a particular patch are represented by selecting more than one set of parameters $\vec{a}$. However, note that at this stage of analysis we have not modeled *where* in the image patch each of the various models are appropriate. Thus transparent motion, with two different velocity fields realized over the whole patch, will be initially modeled in the same way as an occlusion boundary. A subsequent level of analysis is needed to determine which of these two interpretations is appropriate for a particular patch.

We wish to consider fitting a layered flow model to the set of motion constraint vectors measured within an image patch. In particular, we seek the parameter values $\vec{a}_n$, $n = 1, \ldots, N$ for N possibly distinct smooth fields, one for each layer. For the nth layer, the probability of observing a constraint vector $\vec{c}_k$, given that the observation is at the spatial location $\vec{x}_k$, is modeled by the "component probability" distribution $p_n(\vec{c}_k | \vec{x}_k, \vec{a}_n)$. In addition we also have a model for outlier processes given by $p_0(\vec{c}_k)$. Finally, the probability of selecting layer n is given by the "mixture probabilities" m_n, which are treated as further parameters we need to fit. Together these pieces provide the overall probability of observing the constraint $\vec{c}_k$, namely

$$p(\vec{c}_k | \vec{x}_k, \vec{m}, \vec{a}_1, \ldots, \vec{a}_N) = \sum_{n=0}^{N} m_n p_n(\vec{c}_k | \vec{x}_k, \vec{a}_n). \tag{2}$$

[2]In practice, it is often useful to add parameters representing the uncertainty of $\vec{a}$.

Here the mixture probabilities m_n, for $n = 0, 1, \ldots, N$ must sum to one.

Given a set of motion constraint vectors obtained within a patch at time $t = t_0$, say $\{\vec{c}_k(\vec{x}_k, t_0)\}_{k=1}^K$, we seek parameter values $\{\vec{a}_n\}_{n=1}^N$ and mixture probabilities $\{m_n\}_{n=0}^N$ which provide a *maximum likelihood* fit to the data set. In particular, the log likelihood of generating this set of observations from a specific model is

$$\log L(\vec{m}, \vec{a}_1, \ldots, \vec{a}_N) = \sum_{k=1}^{K} \log p(\vec{c}_k | \vec{x}_k, \vec{m}, \vec{a}_1, \ldots, \vec{a}_N). \tag{3}$$

At a local extrema, it can be shown that the parameters $\vec{m}$ and $\vec{a}_n$ for $n = 0, \ldots, N$ must satisfy

$$\sum_{k=1}^{K} q_{nk} = \lambda m_n, \tag{4a}$$

$$\sum_{k=1}^{K} q_{nk} \frac{\partial}{\partial \vec{a}_n} \log p_n(\vec{c}_k | \vec{x}_k, \vec{a}_n) = 0. \tag{4b}$$

Here the quantities q_{nk} represent the "ownership probabilities", that is, the probability that the kth constraint belongs to the nth layer. These ownership probabilities are defined by

$$q_{nk} = \frac{m_n p_n(\vec{c}_k | \vec{x}_k, \vec{a}_n)}{\sum_{j=0}^{N} m_j p_j(\vec{c}_k | \vec{x}_k, \vec{a}_j)}. \tag{5}$$

These equations (4) for a maximum likelihood fit have been derived by a number of authors; for further details see [14]. The first equation (4a) comes from the condition that the partial derivative of $\log L$ with respect to the mixture proportion m_n must be equal to the Lagrange multiplier λ. This Lagrange multiplier arises by imposing the constraint that the mixture proportions must sum to one. The second equation is obtained simply by requiring that the partial derivative of $\log L$ with respect to the parameters $\vec{a}_n$ must vanish.

These equations suggests an iterative algorithm, known as the EM-algorithm [14], for obtaining a maximum likelihood fit for the parameters m_n and $\vec{a}_n$, for $n = 0, \ldots, N$. Given an initial guess for these parameters we first estimate the ownership probabilities q_{nk} for each constraint belonging to each component. This is the expectation, or "E"-step, and simply involves the evaluation of the right hand side of (5). Given these ownership probabilities q_{nk}, we need to find parameter values $\vec{a}_n$ which satisfy (4b). This is equivalent to performing a maximization step, that is the "M"-step, on the expected value of the log probabilities $\log p_n(\vec{c}_k)$. As we see below, for Gaussian distributions this maximization step can be easily solved. The result is a simple iterative algorithm which is guaranteed to increase the log likelihood of its fit each iteration.

3.1. Mixtures of Constant Velocity Models. Our purpose in this paper is simply to demonstrate the utility of considering mixture models of optical flow. As such we restrict our attention to the simplest case, in which the flow field is decomposed into patches with the flow in each patch treated as constant velocity plus noise. To deal with simple occlusion boundaries and transparency, we allow two different constant velocity layers to be extracted for each patch. As a result, we take N to be 2, and take the parameters $\vec{a}_n$ to be simply $\vec{v}_n$, the 3-vector representing the constant velocity of the model (recall that an image velocity is represented by the 3-vector $(v_1, v_2, 1)$). In addition, we also attempt to identify outliers, corresponding to the 0th component of the mixture. This component is, roughly speaking, modeled by a uniform distribution and does not require any parameters to be fit other than its mixing proportion m_0. Details of the outlier model are given further below.

Given the choice of using noisy uniform flow in each patch, the next step is to define the component densities p_n of the mixture distribution. For $n > 0$ (i.e. other than the outlier process), $p_n(\vec{c}_k|\vec{v}_n)$ is meant to represent the likelihood of measuring the motion constraint vector $\vec{c}_k$ within a patch which has mean image velocity $\vec{v}_n$.

For the moment assume that the actual velocity is given by $\vec{v}_n$, then an exact constraint vector $\vec{c}_k$ would lie on the plane perpendicular to $\vec{v}_n$. In [6] it is shown that a reasonable approximation for the distribution of errors in component velocity measurements is given by a roughly Gaussian distribution for the *angular* error between $\vec{c}_k$ and the plane perpendicular to $\vec{v}_n$. The appearance of this angle should not be too surprising since, after all, we are simply measuring the orientation of a surface in space and time. We will make an additional assumption that this angular error distribution is independent of the actual image velocity $\vec{v}_n$ and is roughly isotropic.[3] Given these assumptions, the probability of observing $\vec{c}_k$, given that the actual image velocity is $\vec{v}_n$, is modeled by a Gaussian distribution in $d(\vec{c}_k, \vec{v}_n)$ defined by

$$d(\vec{c}_k, \vec{v}_n) = \frac{\vec{c}_k \cdot \vec{v}_n}{||\vec{c}_k||\,||\vec{v}_n||}.$$

Here $d(\vec{c}_k, \vec{v}_n)$ is simply the sine of the angular error which, for the small angles we are concerned with here, is roughly equal to the angular error itself. An important point about this error distribution is that it is only meant to model the measurements that are reasonably accurate. While the actual error distributions have longer tails than one might expect from such a Gaussian model (see [6]), this is not critical for our current situation since we also incorporate a model for outliers as a separate component in the mixture model. In effect, the longer tails are modeled by this outlier process, rather than by the above Gaussian distribution.

In addition to measurement noise within each patch, we also wish to accommodate the deviations of the actual flow from our constant velocity approximation. This additional variability is also modeled using angular errors from the mean velocity $\vec{v}_n$. This is chosen simply for convenience, since the variance of this error can simply be added to the variance of the measurement error, to obtain the following Gaussian model for the nth component distribution

$$p_n(\vec{c}|\vec{v}) = \frac{1}{\sqrt{2\pi}\sigma_v} \exp(-\frac{d^2(\vec{c}, \vec{v})}{2\sigma_v^2}). \tag{6}$$

[3]The measurement scheme we use in Section 4 uses only two frames while the spatial support of the component velocity measurements is considerably larger. Thus we would expect the spatial orientation error to be smaller than the speed error, that is, the noise should not be isotropic. This could be taken into account in a more detailed model.

Here σ_v^2 is an estimate for the combined variance of the component velocity measurement errors and the modeling error in assuming a uniform velocity within the patch.

Given this specification of p_n we can solve the maximization step in closed form. Recall that this involves finding a solution of (4b), for a fixed set of mixture probabilities m_n and ownership probabilities q_{nk}. We omit the derivation, and simply state that the solution is given by choosing the new approximation for $\vec{v}_n$ as the eigenvector corresponding to the minimum eigenvalue of the 3×3 matrix

$$D_n \equiv \sum_{k=1}^{K} \frac{q_{nk}}{\sigma_v^2 ||\vec{c}_k||^2} \vec{c}_k \, \vec{c}_k^{\,T}. \tag{7}$$

This result is easy to justify intuitively. Consider the quadratic form

$$\begin{aligned} \vec{v}_n^{\,T} D_n \vec{v}_n &= \sum_{k=1}^{K} \frac{q_{nk}}{2\sigma_v^2 ||\vec{c}_k||^2} (\vec{c}_k \cdot \vec{v}_n)^2, \\ &= \sum_{k=1}^{K} \frac{q_{nk}}{2\sigma_v^2} d^2(\vec{c}_k, \vec{v}_n), \end{aligned}$$

which we recognize as minus one times the expected value of the exponent in the probability distribution p_n. By choosing the eigenvector $\vec{v}_n$ associated with the minimum eigenvalue of D_n we are simply maximizing the expected value of this exponent, as is standard in maximum likelihood estimates.

3.2. Modeling Outlier Processes. The distribution $p_0(\vec{c}_k)$ in the mixture is meant to model outliers. One common approach is to choose p_0 to be a Gaussian distribution with a large variance. The large variance attempts to model the long tails in the typical error distributions for component velocity measurements. Unfortunately, a single Gaussian outlier model is often insufficient. For example, an important case for motion estimation is the situation in which a patch overlaps an occlusion boundary. In such a case we may obtain two "signal" distributions, with one peak at the velocity of the foreground and a second peak at the velocity of the background. Each of these peaks has the typical long tails, and moreover, measurements straddling the boundary can provide constraints far from either peak. In order to model this behaviour we would need at least two broad Gaussians to approximate the tails around the peaks, and a third distribution to capture the yet more widely distributed responses scattered by the occlusion boundary. The main point being that it is very unlikely that we can find a reasonable model of the outlier processes in terms of a single Gaussian distribution.

It is worthwhile to consider the opposite extreme, where we use many Gaussian distributions to model the outlier distribution. As we show below, this leads to a computationally simple implementation. Imagine that we covered the sphere of possible image velocities (represented by unit vectors) using Gaussians having a significantly larger standard deviation than σ_v, the standard deviation of the signals we are seeking. At any point on the sphere, the sum of the values of all the Gaussians in this cover should be roughly given by a constant, as is depicted in Figure 2. By using a larger standard deviation for this cover we are constraining any fit of an outlier distribution to be a smoothly varying function over the sphere.

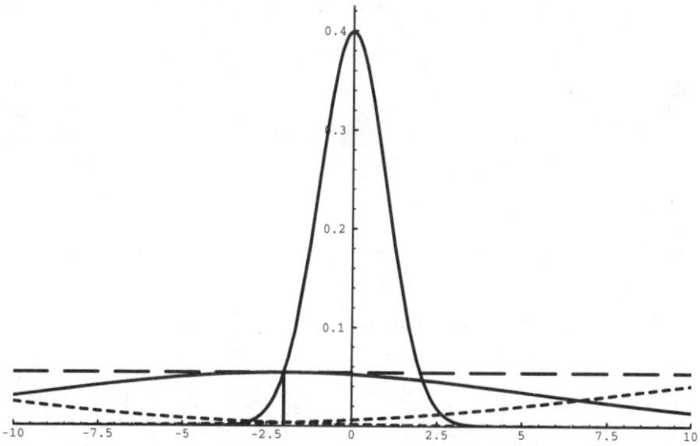

FIGURE 2. A coherent motion is represented by a Gaussian with small variance while outliers are represented by a tiling of broad Gaussian distributions. The dashed line represents an outlier threshold which is an upper bound on the sum of the outlier distributions; in this case, the intersection with the coherent motion distribution is set to be two standard deviations.

Appropriate mixture coefficients for the various Gaussians, holding the means and variances fixed, is provided by the ownership probabilities, q_{nk}, for each broad Gaussian. As we saw in the discussion of the E-step above, these quantities can easily be computed from an equation of the form (5).

In fact, it can be shown that an *upper bound* for the outlier probability is provided by the case in which all of the mixture probability for the members of the cover is concentrated on a single element of the cover, and this element is such that the constraint agrees with its mean value. (We are free to play with the positioning of the elements in the cover, since we are just seeking the value of this upper bound.) This upper bound can also be obtained by simply using p_0 to be the constant value provided by the cover, and treating the mixture probability m_0 as the probability of getting an outlier, integrated over all possible positions. Using p_0 and m_0 in equation (5) gives an estimate q_{0k} which is an upper bound for the probability that constraint k belongs to the outlier distribution. The corresponding algorithm for identifying outliers is thus trivial and it is conservative with respect to which constraints are treated as signal rather than outliers.

All that remains is to discuss the choice of the constant p_0. We find it convenient to consider a situation in which the outlier probability should be about 1/2, and use this situation to set p_0. For example, consider a single Gaussian having a standard deviation of σ_v, with a mixture probability of m'_1, while outliers account for the remaining data. Moreover, assume that for this choice of mixture proportions, data a distance $\rho\sigma_v$ from the mean is to be assigned an outlier probability of 1/2. These parameters m'_1 and ρ are then used to set the value of p_0, which then remains fixed during the execution of the algorithm. In particular, using (5) we find

$$p_0 = \frac{m'_1}{(1 - m'_1)\sqrt{2\pi}\sigma_v} \exp(-\rho^2/2). \tag{8}$$

In our experiments we use $m'_1 = 0.9$ and $\rho = 2.5$. In this case, when we have 10% outliers, the quantity ρ dictates some trust in constraints coming within 2.5 standard deviations of our estimated mean. However, as the mixture proportions change during the execution of the algorithm, so will this region of trust; the region decreases with an increase in the percentage of outliers even though p_0 remains fixed.

3.3. Summary of the EM-algorithm. Given a choice of the constants p_0 and σ_v, the EM-algorithm proceeds as follows. First, using (5) the ownership probabilities are computed according to the current values of the mixture probabilities, $\vec{m}$, and the current estimates for the mean velocities $\vec{v}_n$, $n = 1, 2$. The mean velocity $\vec{v}_n$ can then be updated by constructing the 3×3 matrices D_n, and finding the eigenvector associated with the minimum eigenvalue. This needs to be done for $n = 1$ and 2, but there is no need to fit any such parameters for the outlier process. Also, a new set of mixture probabilities is obtained from (4a), followed by a renormalization to ensure the sum of the mixture probabilities is one. Note that here we also update the mixture probability m_0 for the outlier process. This entire EM-iteration is repeated until the change in the parameters is sufficiently small. For our experiments here we simply used 10 iterations, which turned out to be more than sufficient.

4. Computational Examples

In order to demonstrate the feasibility of the approach we consider two real image sequences, one of which involves a simple occlusion boundary, while the other involves transparency. We examine the behaviour of our approach in areas of the images which contain multiple motions as well as noisy regions and regions which do not conform to the simple uniform-motion assumption.

From the wide range of different measurement strategies for component velocities or, equivalently, for the motion constraint vectors $\vec{c}_k$, we chose a phase-based approach. In previous work a similar method has been shown to provide reasonably accurate component velocities, with a low outlier rate [6]. The particular approach we use is based on only two consecutive frames, and the actual component velocity measurement method is similar to the phase-based stereo disparity measurement scheme discussed in [12]. Briefly, the basic steps in the component velocity measurement are to first convolve each frame with the complex band-pass filter $G_2 + iH_2$, for each of four different spatial orientations of the filter kernel [8]. This kernel is chosen because it is compact (eg. the fine-scale version is 9×9), has a simple analytical form, and the real and imaginary parts nearly form a quadrature pair. Two different spatial scales were used, one tuned to a spatial wavelength of 4 pixels while the other is tuned to a wavelength of 8. The complex responses of these convolutions were sampled at the rate of 1/4 of the wavelength and quantized to 8 bits. The spatiotemporal phase derivatives were calculated using these subsampled and quantized results. The phase derivatives in the x_1, x_2 and t directions supplied the coefficients of the motion constraint vector $\vec{c}_k$. Points at which the complex convolution response was below 4% of the maximum possible value were discarded (since responses at this low level have been crudely quantized to only a few different gray levels), as were points where the phase and amplitude derivatives failed the test for a singularity neighborhood [7]. Finally, in order to bring the two frames into rough alignment, the frames were sometimes spatially shifted by an integral multiple of the filter spacing. The appropriate shift was easily obtained by applying

FIGURE 3. The **Pepsi Can** image sequence. The mixture approach is applied to the boxed regions of the scene (numbered 1 to 4 from left to right).

the mixture model approach first to the wavelength 8 case, and using the results to choose the appropriate shift for the wavelength 4 case.

4.1. An Occlusion Boundary. Component velocities were obtained for the Pepsi can sequence in Figure 3. As suitable test cases for our mixture model approach we chose various 32×32 image patches, as depicted in Figure 3. The camera motion is purely translational and the image motion is to the left with speeds ranging roughly between 1.6 and 0.7 pixels/frame (for the can and the background, respectively). In all cases we take the standard deviation of the mixture model within a patch to be $\sigma_v = 0.2$ pixels/frame.

The second patch from the left in Figure 3 is roughly centered on an occlusion boundary, and we begin our discussion with this patch. A third of all the motion constraints obtained for this patch are depicted in Figures 4*d*. Note the presence of the two clusters associated with the motions on either side of the boundary which are made clearer in Figures 4*a* and *b* where the two clusters are shown separately. The two white "X"'s mark the peaks of the extracted mixture model for this example, while the convergence to these peak values is illustrated by black in "X"'s in Figure 4*d*. This rapid convergence behaviour was typical in all our tests and moreover the convergence appeared to be rather insensitive to the initial guess. The recovered velocities were $(-1.53, -0.02)$ for the portion of the can, and $(-0.70, 0.01)$ for the background. The mixture probabilities were $(m_0, m_1, m_2) = (0.03, 0.67, 0.30)$ for the outliers, the can, and the background, respectively. The method has clearly recovered the velocities of both sides of the occlusion boundary without difficulty.

A couple of additional points can be made using this same patch. Figure 4*a* shows all the constraints deemed to have an ownership probability for the first motion larger than 0.4 (i.e. $q_{1k} > 0.4$), with the darkness of the constraint lines increasing with the probability that the constraint belongs to the first motion. A similar plot is given in Figure 4*b* for the constraints having an ownership probability

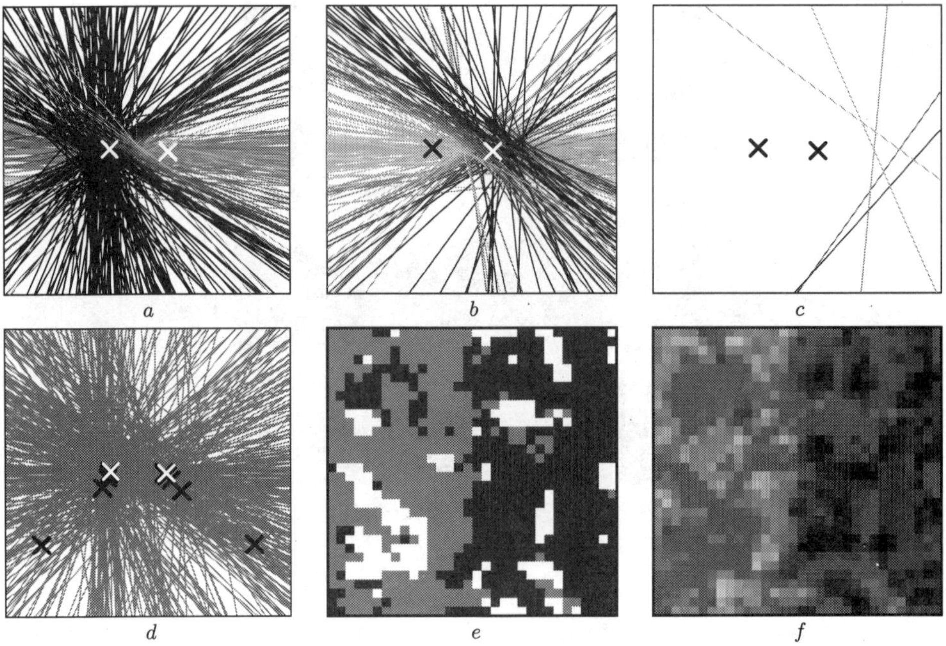

FIGURE 4. Region 2 (see text).

larger than 0.4 for the second motion. Clearly the mixture components have picked out appropriate clusters of constraints. Note that constraint lines that are roughly horizontal, and pass close to both peaks, have low ownership probabilities. This is appropriate since these constraints are roughly equally likely to arise from either motion, and thus their probabilities for any particular motion is close to 1/2. This illustrates the competition between the various components in the mixture model for the ownership of each constraint. There are only a handful of constraint lines that are outliers and are depicted in Figure 4*c*. Finally, in Figures 4*e* and 4*f* we show the spatial distribution of responses for the horizontal velocity $v_1(\vec{x})$ and the ownership probabilities $q_{2k}(\vec{x})$, respectively, for $\vec{x}$ varying over the patch. The general spatial distribution of the ownership probabilities reflects the structure and location of the occlusion boundary within the patch. To show the area of support for each motion, we have depicted the maximally probable horizontal velocity in Figure 4*e* (the white areas are regions where there were no component velocities, due to low amplitude or removal by the singularity neighborhood test). The majority of incorrectly classified pixels arise in areas where the ownership probabilities are near 1/2 (seen as neutral gray areas in Figure 4*f*).

The results for the other patches in Figure 3 are shown in Figure 5. Consider the results for Region 3 (the third patch from the left in the image) which contains only the single motion of the can. In this case, the mixture model collapses both peaks onto the same point, given by the velocity $(-1.61, -0.006)$ pixels/frame. This value is in excellent agreement with the velocity obtained for the can in the Region 2, indicating that the presence of the second motion there did not significantly perturb the responses.

In general, given a single motion within a patch, the situation is not this simple with both mixture components collapsing to a point. For example, in the patches on

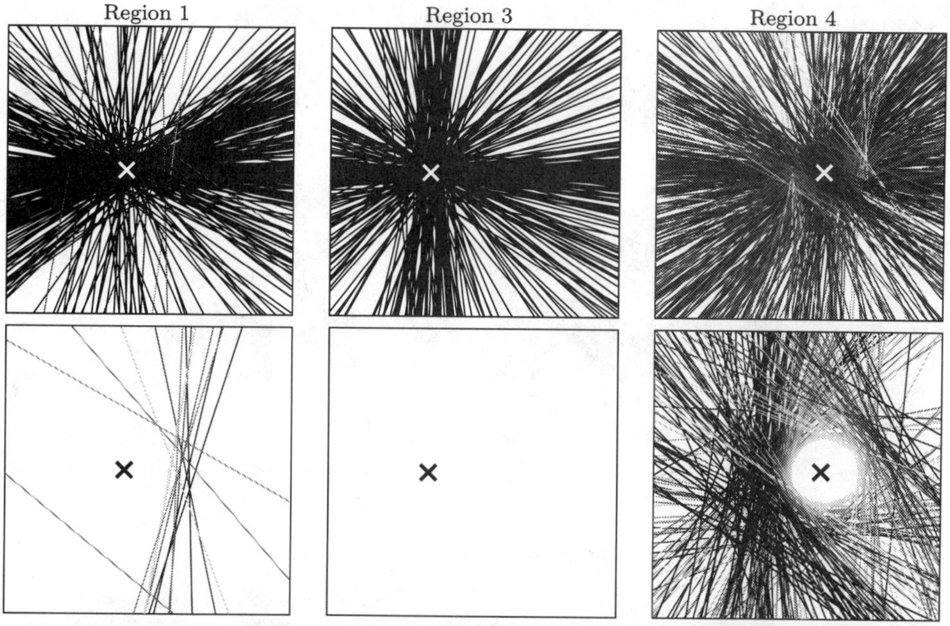

FIGURE 5. Constraint lines for Regions 1, 3, and 4. The top row shows the constraints corresponding to the dominant motion and the "X" marks the recovered motion (note that there is only a single motion in these regions). The bottom row shows the outliers (none for Region 3 and many in the noisy Region 4).

the extreme left and right (Regions 1 and 4) which both contain a single motion, the mixture model initially converges to descriptions having two different peak locations. Region 1 contains a planar surface slanted away from the camera. In this case, our simple assumption of uniform motion over the region is not a good approximation to the true motion and the resulting constraint lines do not form a tight cluster. When we assume that two uniform motions are present the method recovers horizontal component velocities of -1.4 and -1.0 pixels/frame, while the vertical components are essentially zero. The distance between these two motions, however, is only $2\sigma_v$, which is not a sufficient spread in order for the sum of the two Gaussian distributions to have more than a single peak. A similar situation occurs in Region 4 on the far right of the image. In this region there is a considerable amount of noise in the measurements, with 17% of the constraints labeled as outliers. In this noisy situation the mixture model also chooses a pair of velocities roughly separated by $2\sigma_v$ and, thus, a unimodal distribution.

These observations suggest a simple decision criterion for whether the velocities within a patch belong to a single layer. In particular, we consider the criteria that the minimum probability on the line connecting the two peaks must be at least half the height of the peaks in order to be merged. This criterion can also be based on the Mahalanobis-distance between the two peaks. When we recognize that the motions within a region should be merged we rerun the EM-algorithm assuming a single motion and outliers. The results for Regions 1 and 4 are shown in Figure 5.

Figure 6 shows the result of applying the mixture model approach over the entire image in 32×32 patches which are separated from each other by 8 pixels in both

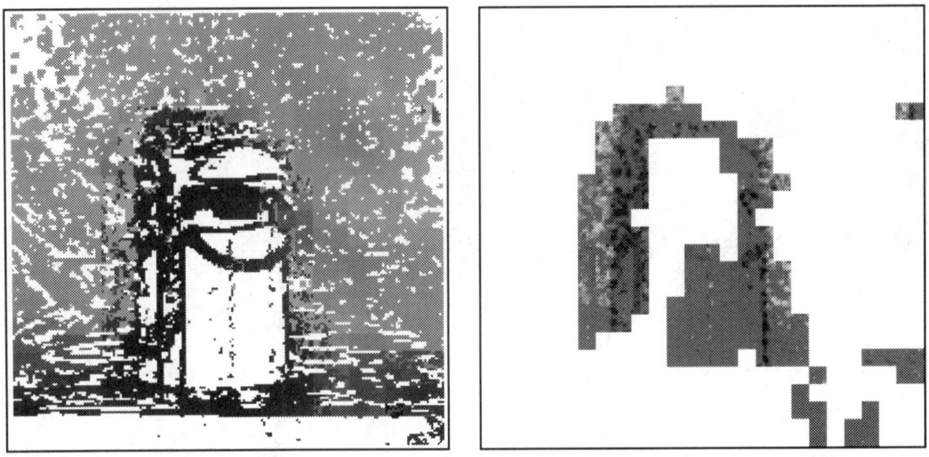

(a) Horizontal Flow *(b) Mulitple Motions*

FIGURE 6. Mixture models applied to the entire Pepsi-can image (see text).

a *b* *c*

FIGURE 7. **Where's Jim?** Transparency sequence containing the reflection of a mystery vision researcher (see text).

directions. We use a simple uniform motion model within the patches and begin by assuming two coherent motions and outliers. After convergence, we apply the above decision criteria to determine if there is one motion or two within each patch. The horizontal component of the flow is shown in Figure 6*a* (in the same format as Figure 4*e* where white indicates no information, and the magnitude of the dominant motion is displayed in shades of gray). The regions containing two well-separated motions are shown in Figure 6*b* (in the same format as Figure 4*f*). Note that most of the boundary of the can has been identified, along with a few incorrectly classified regions in which the flow is quite noisy.

4.2. Transparency. We next consider a case of additive transparent motion in which a face is reflected in the glass covering an Escher print (Figure 7*a*). The entire image was treated as a single region, two uniform translational motions were assumed, and a noise estimate of $\sigma_v = 0.1$ was used. The initial motion constraints were computed with wavelength 8.[4] The recovered motion parameters were $(-3.31, 0.02)$ for the Escher print and $(-0.79, 0.01)$ for the reflection of the face.

The accuracy of these motion parameters can be evaluated by performing a simple computation. We compute the difference between the second image and the first image shifted by one set of motion parameters. This has the effect of canceling the intensity structure which is consistent with the motion and revealing the structure of the other surface. Figures 7*b* and *c* show the results obtained by canceling the effect of the moving print and reflection respectively. These results obtained from two frames compare favorably with those of [2] which required three frames to recover the two motions.

5. Conclusion

We have examined the problems posed by multiple motions and outliers in the intergration of motion information over a spatial neighborhood. We relax the assumption of a single motion and, instead, view image regions as containing multiple layers corresponding to surfaces with different image motions. We also cope with outliers which can decease the accuracy of the recovered flow. To achieve this we introduced the idea of using mixture models for integrating noisy constraints when there are multiple interpretations and we provided details of the EM-algorithm for computing the maximum-likelihood estimate of the motion parameters. Our experiments demonstrate the feasibility of the approach and indicate that it is computationally efficient and can provide robust estimates of the optical flow values in the presence of outliers and multiple motions.

References

1. J. R. Bergen, P. Anandan, K. J. Hanna, and R. Hingorani. Hierarchical model-based motion estimation. In G. Sandini, editor, *Proc. of Second European Conference on Computer Vision, ECCV-92*, volume 588 of *LNCS-Series*, pages 237–252. Springer-Verlag, May 1992.
2. J. R. Bergen, P. J. Burt, R. Hingorani, and S. Peleg. A three-frame algorithm for estimating two-component image motion. *IEEE Transactions on Pattern Analysis and Machine Intelligence*, 14(9):886–896, September 1992.
3. M. J. Black. *Robust Incremental Optical Flow*. PhD thesis, Yale Univeristy, New Haven, CT, 1992. Research Report YALEU/DCS/RR-923.
4. A. Blake and A. Zisserman. *Visual Reconstruction*. The MIT Press, Cambridge, Massachusetts, 1987.
5. T. Darrell and A. Pentland. Robust estimation of a multi-layer motion representation. In *Proc. IEEE Workshop on Visual Motion*, pages 173–178, Princeton, NJ, October 1991.
6. D.J. Fleet. *Measurement of Image Velocity*. Kluwer, Boston, 1992.
7. D.J. Fleet, A.D. Jepson, and M. Jenkin. Phase-based disparity measurement. *CVGIP: Image Understanding*, 53(2):198–210, March 1991.
8. W.T. Freeman and E.H. Adelson. The design and use of steerable filters. *IEEE Transactions on Pattern Analysis and Machine Intelligence*, 13(9):891–906, 1991.
9. S. Geman and D. Geman. Stochastic relaxation, Gibbs distributions and Bayesian restoration of images. *IEEE Transactions on Pattern Analysis and Machine Intelligence*, PAMI-6(6):721–741, November 1984.
10. F. R. Hampel, E. M. Ronchetti, P. J. Rousseeuw, and W. A. Stahel. *Robust Statistics: The Approach Based on Influence Functions*. John Wiley and Sons, New York, NY, 1986.
11. B. K. P. Horn. *Robot Vision*. The MIT Press, Cambridge, Massachusetts, 1986.

[4] A large wavelength was necessary to simultaneously compute constraints for both motions since they differ by approximately 2.5 pixels.

12. A.D. Jepson and M. Jenkin. Fast computation of disparity from phase differences. In *Proc. Computer Vision and Pattern Recognition, CVPR-89*, pages 398–403, San Diego, 1989.
13. S. Madarasmi and D. Kersten. The visual perception of surfaces, their properties, and relationships. this volume.
14. G.J. McLachlan and K.E. Basford. *Mixture Models: Inference and Applications to Clustering.* Marcel Dekker Inc., N.Y., 1988.
15. M. Nitzberg and D. Mumford. The 2.1-D sketch. In *Proc. Int. Conf. on Computer Vision, ICCV-90*, pages 138–144, Osaka, Japan, December 1990.
16. M. Okutomi and T. Kanade. A locally adaptive window for signal matching. *International Journal of Computer Vision*, 7(2):143–162, January 1992.
17. B. G. Schunck. Image flow segmentation and estimation by constraint line clustering. *IEEE Transactions on Pattern Analysis and Machine Intelligence*, 11(10):1010–1027, October 1989.

Canadian Institute for Advanced Research, and, Department of Computer Science, University of Toronto, 6 King's College Rd, Rm 283C, Toronto, Ontario, Canada M5S 1A4

Xerox Palo Alto Research Center, 3333 Coyote Hill Rd., Palo Alto, CA 94304

DIMACS Series in Discrete Mathematics
and Theoretical Computer Science
Volume 19, 1995

MULTILEVEL DETECTION OF STEREO DISPARITY SURFACES

YIBING YANG AND ALAN L. YUILLE

ABSTRACT. We present a multilevel algorithm for matching a pair of stereo images. In contrast to previous work, we formulate stereo matching in terms of surface detection in a three-dimensional spatio-disparity space. This can be thought of as using clustering to perform stereo matching. Our method involves a fine-to-coarse sweep, in which the surface is enhanced using a non-linear surface-sensitive filter, and a coarse-to-fine sweep, in which the surface is detected. The basic operations include convolution and local winner-take-all. The algorithm is non-iterative, simple, efficient, and gives good results for complex scenes.

1. INTRODUCTION

Stereo matching is an important problem in computer vision and photogrammetry [4, 10, 16, 20, 25]. Given a pair of stereo images, the problem is to establish the correspondence between left image points and right image points. Once correspondence is established, we are able to determine the 3D structure of the surface in the scene by means of triangulation.

For a point in the left image and a point in the right image to originate from the same point on an observed surface, they must appear "locally similar". This suggests that we can determine a match by conducting a two-dimensional search. In fact, by taking advantage of the epipolar geometry, we can reduce the 2D search space to a 1D search along the appropriate epipolar line. Usually the epipolar lines are just the raster lines. Let the image position be specified by (x, y), where the x-axis is parallel to the raster lines. Then a point (x, y) in one image should be matched to (x', y) in the other image with the same y-coordinate. The difference in the x-coordinate, $U = x' - x$, is called the *stereo disparity*. Stereo matching amounts to detecting disparities.

The local similarity between a point (x, y) in the left image and a point (x', y) in the right image can be measured by using window correlation [13, 29]. The window centered at (x, y) specifies an image patch and that centered at (x', y), another image patch. These two image patches are to be compared using a metric such as the sum of squared differences (SSD) or normalized correlation. This essentially assumes that the disparity is constant within the window.

A question that naturally arises is how large ought the window to be? On the one hand, the window is should be large so that enough image information is available to resolve ambiguous matches. On the other hand, the window should be small so that the disparity estimate is accurate. These conflicting requirements

1991 *Mathematics Subject Classification.* Primary 68U10.

lead to the classical *multiresolution paradigm* for stereo matching which is well-documented in the literature (see, e.g., [1, 12, 14, 15, 17, 18, 22, 24, 27, 32, 35, 36]). In essence, the idea is first to use a large window to obtain a rough initial disparity estimate and then to refine the estimate by decreasing the window size. (In practice, multiresolution representations of the left and right images are frequently used.)

Although the disparity may be assumed constant in a small window, this constant disparity assumption can be misleading if the window is large. The disparity variation gives rise to geometric distortions [19, 28] – a window of points in one image appear deformed in the other image. As a result, a roughly correct match may have poor local similarity, and be overridden by a completely false match [37, 38]. Therefore using a large window to obtain a rough initial disparity estimate is impractical.

Motivated by the distortion problem, we present a new approach, called the *multilevel paradigm,* for computing stereo disparities. Key ideas involved in this paradigm include the following:

(1) Computing the local similarity for *every* match, using a small window for which the constant disparity assumption should apply.
(2) Integrating neighborhood support for every match, taking into account the spatial variation of the disparity. We introduce a simple nonlinear filter to accomplish this task.
(3) Making use of a pyramid-like data structure to integrate support from increasingly larger neighborhoods and then detecting the disparity in a coarse-to-fine manner.

The multilevel algorithm has the following attributes:

(1) The algorithm is simple and modularized. The modules include "local matching", "surface enhancement", and "surface detection".
(2) The basic operations are one-dimensional convolution and local maximum-picking (winner-take-all).
(3) The algorithm has a level-recursive, non-iterative control structure and is computationally efficient.

2. Overview

Our approach to stereo matching makes use of the divide-and-conquer principle. As illustrated in Fig. 1, three self-contained modules are involved in the matching process: local matching, surface enhancement, and surface detection.

By local matching each and every possible match is assigned a real value measuring the goodness of the match. A match between point $(x+u/2, y)$ in the left image and point $(x-u/2, y)$ in the right image can be represented by the triple (x, y, u). Thus local matching produces a 3D image $\Xi = \Xi(x, y, u)$, where the image intensity $\Xi(x, y, u)$ reflects the goodness of the match (x, y, u). We call the xyu-space the spatio-disparity space (SDS) and the 3D image $\Xi = \Xi(x, y, u)$, the SDS image. For example, SDS image can be defined as $\Xi(x, y, u) = -\big(I^l(x+u/2, y) - I^r(x-u/2, y)\big)^2$ where I^l, I^r are the left and right images respectively.

The local matching module takes the stereo images as input and outputs the SDS image. Suppose at (x,y) the disparity is $U(x, y)$. Then $u = U(x, y)$ specifies a surface in the SDS. Each surface point $(x, y, U(x, y))$ is associated with a high

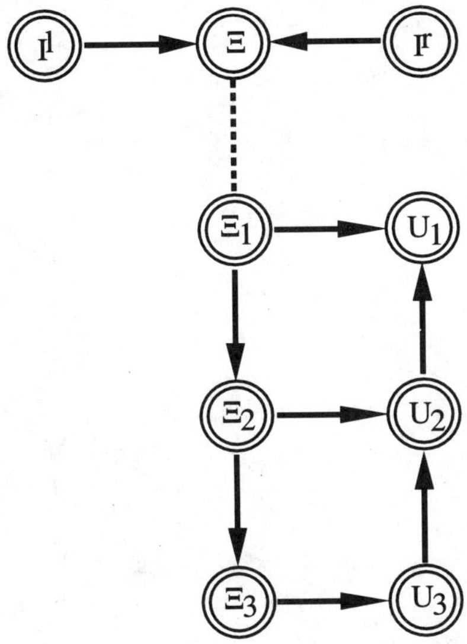

FIGURE 1. The multilevel paradigm for disparity detection. Local matching produces an SDS image $\Xi(x, y, u)$ representing the goodness of the match (x, y, u). A nonlinear filter, integrating local support, is applied to the SDS image to obtain a multiresolution representation Ξ_1, Ξ_2, Ξ_3. At each level a disparity estimate can be made by winner-take-all. A better procedure CfT, described in section 6.2, allows the disparity estimates at the coarse levels to guide the estimates at the fine level.

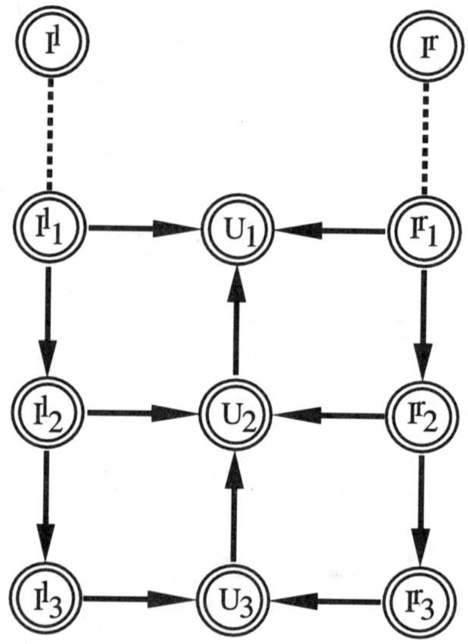

FIGURE 2. The multiresolution paradigm for disparity detection. In this approach multiresolution representations of the images are constructed, typically be Gaussian filtering, then matching is performed in a coarse-to-fine manner starting from the coarsest scale.

intensity $\Xi(x, y, U(x, y))$. Our task is to detect the "bright" disparity surface $u = U(x, y)$ in $\Xi = \Xi(x, y, u)$. This can be thought of as clustering in the SDS image.

The surface enhancement module produces a multilevel representation of the SDS image. More specifically, a hyperpyramid $\{\Xi_m\}$ is constructed from $\Xi = \Xi(x, y, u)$, where m is the level index. At the first or the finest level, $\Xi_1 \equiv \Xi$. At level m, Ξ_m is computed from Ξ_{m-1} by applying a filter to the latter and then downsampling the filtered image. The filter serves to to enhance the disparity surface in the SDS image and to reduce ambiguities. It plays a critical role in this approach.

While the disparity surface is enhanced in the fine-to-coarse sweep, it is detected in the coarse-to-fine sweep. Disparity surface detection is conducted first for the SDS image at the coarsest level and then, for the SDS images at successively finer levels. The disparity surface computed at one level is used to prune the search space in detecting the surface at the next, finer level.

The traditional *multiresolution paradigm* is illustrated in Fig. 2 for comparison. In this paradigm, multiresolution representations [7] are first constructed for the two input images. This involves successive Gaussian filtering and downsampling. Then matching is conducted in a coarse-to-fine manner starting from the coarsest scale.

The remaining of this paper is organized as follows. The next section discusses local matching or computing the SDS image. Section 4 introduces the surface-sensitive filter for constructing the SDS image hyperpyramid. Sections 5 and 6 are devoted to global matching, concerning surface enhancement and surface detection, respectively. Section 7 demonstrates experimental results. This is followed by a discussion and concluding remarks.

3. Local Matching

3.1. Measuring Local Similarity. A pair of corresponding points in the stereo images must be locally similar, since they are the projections of the same point on the surface being viewed. This is the so-called *compatibility constraint*, without which matching would be impossible.

A SDS image is produced from the stereo pair by local matching (in fact almost any stereo algorithm could be used to assign an intensity value to each and every SDS point). Although, unlike many existing algorithms, the SDS image is explicitly constructed as an intermediate result to be used for future processing.

The SDS image intensity has the property of being upper-bounded:

$$\Xi(x, y, u) \leq \Xi_{\max}$$

Due to the compatibility constraint, a correct match is expected to have an intensity value close to, if not equal to, the theoretical maximum $\Xi_{\max}$. Two interrelated issues arise in computing the SDS image: (i) what image primitives should be matched? and (ii) how to evaluate the similarity?

Squared Difference

The matching primitive can simply be the individual pixel values in the stereo pair. Assuming that a world point is expected to have the same intensity value in the two images, we may compute the SDS image by the negative of the squared difference:

$$\Xi(x, y, u) = -\left(I^l(x + u/2, y) - I^r(x - u/2, y)\right)^2 \tag{1}$$

In this case $\Xi(x, y, u)$ is close to

$$\Xi_{\max} = 0$$

for a correct match. A well-known problem with this method is that, due to photometric distortions, corresponding points may have quite different intensity values.

Normalized-correlation

An alternative matching primitive is a local image region. The similarity between a pair of image regions can be measured by using the variance-normalized cross correlation of windowed image patches:

$$\Xi(x, y, u) = \frac{\int\int d\xi d\eta \theta_l \theta_r \omega(\xi, \eta)}{\sqrt{\int\int d\xi d\eta \theta_l^2 \omega(\xi, \eta) \int\int d\xi d\eta \theta_r^2 \omega(\xi, \eta)}} \tag{2}$$

where

$$\theta_l = I^l(x + u/2 + \xi, y + \eta)$$

$$\theta_r = I^r(x - u/2 + \xi, y + \eta)$$

and $\omega(x, y) = \omega(x)\omega(y)$ is a separable windowing function. In this case

$$0 \leq \Xi(x, y, u) \leq \Xi_{\max} = 1$$

This measure has the advantage of being robust against photometric distortions, where the images differ by locally constant multiplicative factors. All the experimental results in this paper were obtained using this measure.

3.2. Computation. We now present an efficient technique for constructing the SDS image defined in terms of variance-normalized correlation, using one-dimensional convolution as the basic operation. The key point is to use convolutions to compute local cross-correlation and auto-correlations of the stereo images.

Local auto-correlation of the left and right images are respectively defined as

$$\Phi_{ll}(x,y) = \left(I^l(x,y)\right)^2 * \omega(x) * \omega(y)$$

$$\Phi_{rr}(x,y) = \left(I^r(x,y)\right)^2 * \omega(x) * \omega(y)$$

where $*$ denotes the convolution operation. Local cross-correlation of the left and right images is defined as

$$\Phi_{lr}(x,y) = \left(I^l(x+u/2,y)I^r(x-u/2,y)\right) * \omega(x) * \omega(y)$$

¿From eqn. (2) we derive

$$\Xi(x,y,u) = \frac{\Phi_{lr}(x,y)}{\sqrt{\Phi_{ll}(x,y)\Phi_{rr}(x,y)}} \tag{3}$$

where the SDS image is defined in terms of variance-normalized correlation. Computing the SDS image involves the following steps (in this procedure "black" and "white" are arbitrary array names).

Procedure LoM: $I^l, I^r \rightarrow \Xi$

1. Compute $\Phi_l(x,y)$:
   $black(x,y) = \left(I^l(x,y)\right)^2$; $white(x,y) = black(x,y) * \omega(x)$;
   $black(x,y) = white(x,y) * \omega(y)$; $\Phi_{ll}^{1/2}(x,y) = \sqrt{black(x,y)}$;
2. Compute $\Phi_r(x,y)$:
   $black(x,y) = \left(I^r(x,y)\right)^2$; $white(x,y) = black(x,y) * \omega(x)$;
   $black(x,y) = white(x,y) * \omega(y)$; $\Phi_{rr}^{1/2}(x,y) = \sqrt{black(x,y)}$;
3. Compute $\Xi(x,y,u)$:
   $black(x,y,u) = \Phi_l(x+u/2,y)\Phi_r(x-u/2,y)$;
   $white(x,y,u) = black(x,y,u) * \omega(x)$;
   $black(x,y,u) = white(x,y,u) * \omega(y)$.
   $\Xi(x,y,u) = black(x,y,u)/\Phi_{ll}^{1/2}(x+u/2,y)\Phi_{rr}^{1/2}(x-u/2,y)$

A couple of remarks are in order. First of all, the above procedure has a nice parallel structure; each step involves only simple, local, identical operations. Hence implementations using SIMD computers are especially suitable. Secondly, the procedure is computationally efficient; computing each SDS image intensity involves about two one-dimensional convolution operations.

3.3. The Problem of Local Ambiguity. Although a SDS point on the disparity surface appears bright in the SDS image, the reverse is not necessarily true. In other words, local matching is intrinsically ambiguous.

To detect the disparity surface from the SDS image, the simplest way to pick the maximum along the line of sight. Using maximum-picking along the disparity direction, we can obtain a disparity surface

$$U(x,y) = \arg\max_u \Xi(x,y,u) \tag{4}$$

which is represented in cyclopean cordinates.

In fact, this area–correlation technique is the oldest technique for stereo matching. A major problem with this method is that the intensity profile along a line of sight often has multiple peaks and spurious peaks may override the true peaks. This ambiguity problem may be caused by local repetitive texture patterns in the stereo images, as well as by image noise. As a result, the disparity estimates may contain many large errors.

It is sometimes argued that the ambiguities could be reduced by employing large windows for correlation, so that enough information is available to determine a unique, though rough, estimate of the disparity. Then the rough estimate could be refined by reducing the window size. This is the idea on which multiresolution methods are based.

By increasing the window size, however, the problem of geometric distortion is also exaggerated. Even for a planar surface the distortion in a large window can be serious if the surface is heavily slanted. Consequently, a correct match may yield a correlation value much smaller than $\Xi_{\max}$ and a false match may have a bigger correlation value. In other words, increasing the window size does not lead to disambiguation, unless the distortions are negligible.

4. A Surface-Sensitive Filter

4.1. The Disparity Surface. Due to the cohesive nature of objects in the physical world, neighboring images points usually have similar depth values. In other words, the disparity $U = U(x, y)$ depicts a coherent surface in the SDS. By taking advantage of this property, we will be able to identify the true match despite the local ambiguities.

In the SDS image, points on the disparity surface appear to be "bright". Thus stereo matching amounts to detecting the brightest surface in the SDS image. Surface detection is reminiscent of the problem of curve detection which is popular in low-level vision. Although, at first sight, the former is merely an extension of the latter to three-dimensional images there exist special properties of the geometry and the photometry that simplify the detection task.

We first consider the geometric properties. Firstly, the disparity surface can be written with u as an explicit function of x and y. In contrast, in curve detection in general a curve can usually only be specified as a implicitly function of the two image coordinates. Moreover, the slope of the disparity surface is strictly constrained.

By convention the SDS is defined with respect to the cyclopean view, so that a point (x, y, u) represents the match between the point $(x + u/2, y)$ in the left image and the point $(x - u/2, y)$ in the right image. As a result, in the xu-space the left view lines of sight have a slope -2 and the right view lines of sight have a slope $+2$. For a surface patch of an object to be binocularly visible, the partial derivative of the disparity surface along the epipolar direction has to be bounded:

$$|\partial U/\partial x| < 2$$

This constraint is illustrated in Fig. 3, where three segments of the disparity curve are depicted in bold.

In practice the disparity surface gradient $\nabla U(x, y)$ is often close to zero. In fact small disparity gradients are favored by the human visual system. As psychophys-

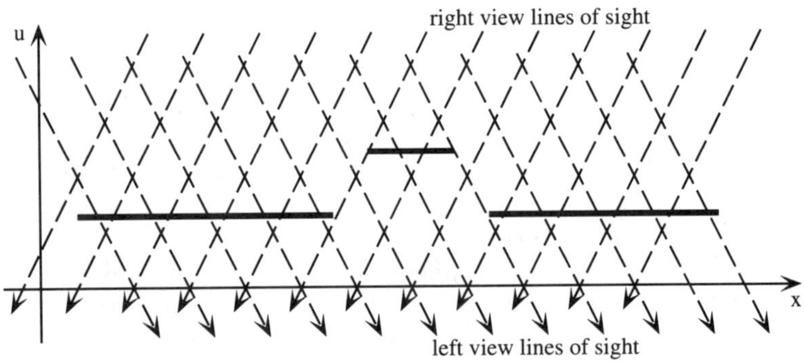

FIGURE 3. Left view lines of sight and right view lines of sight.

ical studies [6] demonstrated, stereoscopic fusion is hard to achieve if the disparity gradient is greater than one. Thus is it reasonable to assume that, in a small region, the disparity variations are usually small.

As far as the photometric aspects are concerned, an important property is that, on the disparity surface, each SDS point has an intensity which is close to $\Xi_{\max}$. In other words, in the SDS image the disparity surface points are not only "bright" but also approximately have the same equal brightness. In contrast, in conventional curve detection the curve is characterized only as being "bright"; curve points are, in general, not equally bright.

To summarize, stereo matching can be formulated as a surface detection problem subject to the following two constraints:

(1) (**Weak continuity**) In a local spatial neighborhood most points have similar disparities.
(2) (**Equal brightness**) In the SDS image, the intensity is close to $\Xi_{\max}$ for a point on the disparity surface.

As a remark, the individual disparity curves can be computed from $\Xi(x, u) = \Xi(x, y, u)$, where y is a dummy variable, by using dynamic programming [2, 5, 8, 11, 26].

4.2. A Surface-Sensitive Nonlinear Filter. The last section explored the geometric and photometric properties of the disparity surface. We will now introduce a nonlinear filter which will enhance the sought-after disparity surfaces in the SDS image and simultaneously suppress ambiguities. This filter exploits the weak continuity constraint and the equal brightness constraint discussed in the last subsection.

One might consider using a feature detector to enhance the disparity surface. In curve detection, it is a common practice to use a set of bar-sensitive linear filters to extract the curve and to get rid of image noise. In the same spirit, one might design a linear surface-sensitive filter.

However, such a filter is not appropriate for enhancing the disparity surface because it fails to take into account the equal brightness constraint. As a result, the true peak along the profile $\Xi(u) = \Xi(x, y, u)$ may have a small response if the peak is not "sharp". Although the absolute brightness is not informative in identifying a curve point it is important in detecting disparities.

Our surface enhancing filter ϕ is based on the idea of integrating neighborhood support. It is defined in terms of the response it generates when applied to a SDS

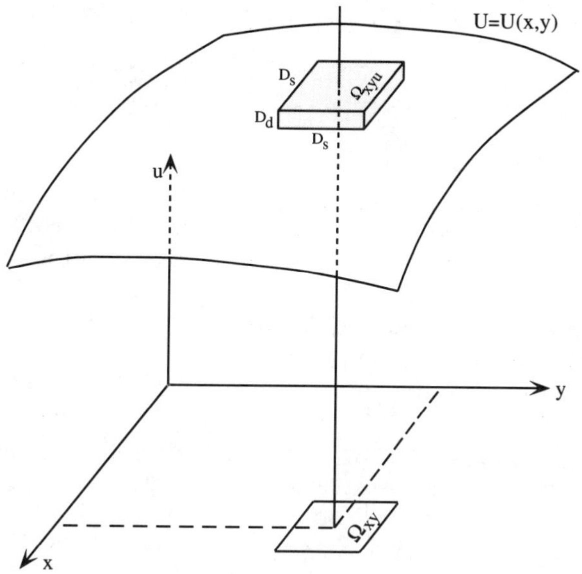

FIGURE 4. The disparity surface and a "receptive field".

image. The response associated with a SDS point (x, y, u) is computed from its "receptive field":

(5) $$\Lambda_{xyu} = [x - D_s, x + D_s] \times [y - D_s, y + D_s] \times [u - D_d, u + D_d]$$

which is a translation of the solid

(6) $$\Lambda = [D_s, D_s] \times [D_s, D_s] \times [D_d, D_d]$$

from the origin to (x, y, u). The size of the receptive field is $2D_s \times 2D_s \times 2D_d$. Fig. 4 illustrates the receptive field Λ_{xyu} located at (x, y, u).

A SDS point is supposed to have a large response if its receptive field happens to capture a patch of the disparity surface $u = U(x, y)$. Consider the surface patch defined on

(7) $$\Omega_{xy} = [x - D_s, x + D_s] \times [y - D_s, y + D_s]$$

The response at (x, y, u) is expected to be weak unless u is close to $U(x, y)$. This suggests that we can compute the response at P in two steps:

(1) Detect a candidate surface patch defined on Ω_{xy} that is totally confined within the receptive field Λ_{xyu}. This involves the "winner-take-all" operation in the disparity direction.
(2) Integrate the SDS image intensities for the SDS points residing on the candidate surface.

More formally, the filter ϕ maps one SDS image $\Xi = \Xi(x, y, u)$ to another SDS image $\Xi' = (\phi \circ \Xi)(x, y, u)$ given by

(8) $$\Xi'(x, y, u) = \sum_{\xi,\eta} g(\xi - x, \eta - y) \max_{\mu} \Xi(\xi, \eta, \mu), \quad (\xi, \eta, \mu) \in \Lambda_{xyu}$$

where $g(\cdot,\cdot)$ is a normalized weighting function:

$$\sum_{-D_s \leq \xi,\eta) \leq D_s} g(\xi,\eta) \equiv 1 \tag{9}$$

Due to the equal brightness constraint, the SDS points on the disparity surface have intensities close to $\Xi_{\max}$. If a SDS point (x, y, u) has a receptive field that captures a patch of the disparity surface, then its response $(\phi \circ \Xi)(x, y, u)$ will be close to $\Xi_{\max}$; otherwise it will have a weak response, even though $\Xi(x, y, u)$ may be close to $\Xi_{\max}$.

The filter defined above is nonlinear; the response cannot be expressed as a linear combination of the intensity values in the receptive field. Obviously, the nonlinearity comes from the inhibitive interaction (winner-take-all) along the disparity direction.

As shown in Fig. 4, the receptive field is a "flat" solid; its height D_d is small relative to D_s. In fact, the receptive field implies that most disparities in a neighborhood $\Omega_{x,y}$ are within the range $[U(x,y) - D_d, U(x,y) + D_d]$. Thanks to the weak continuity constraint, this range is small.

The receptive field should be extended in the u-direction so as to tolerate the undulations of the disparity surface. If it contains only one SDS point in the u-direction, then ϕ degenerates into a linear filter: $(\phi \circ \Xi)(x, y, u) = \Xi(x, y, u) * g(x, y)$. In this case, a SDS point (x, y, u) on the disparity surface may not have a strong response if the surface is not flat in the neighborhood Ω_{xy}.

As a remark, the weak continuity constraint is a new contribution, which contrasts the existing smoothness assumptions [3, 9, 23, 30, 31] about the disparity surfaces.

4.3. Two Subfilters. The surface-sensitive filter ϕ defined in the last subsection can be decomposed into two subfilters:

$$\phi = \phi_s \circ \phi_d$$

where the first subfilter ϕ_d takes effect in the disparity domain and the second subfilter ϕ_s, in the spatial domain. Although ϕ is new, the two subfilters are well-known to the computer vision community.

The First Subfilter

The first subfilter ϕ_d is a morphological filter. Let $\mathcal{F} = \mathcal{F}(x, y, u)$ be a SDS image. For each fixed (x, y) we have a 1D image $\mathcal{F}(u) = \mathcal{F}(x, y, u)$. The morphological filter performs the dilation operation, denoted as $\oplus$, on the 1D image. More formally, the response can be expressed as

$$(\phi_d \circ \mathcal{F})(x, y, u) = \mathcal{F}(x, y, u) \oplus e(u) \tag{10}$$

where e is the structuring element

$$e(u) = \begin{cases} 1 & \text{if } -D_d \leq u \leq D_d \\ 0 & \text{otherwise} \end{cases}$$

The structuring element does not extend in the spatial directions.

The Second Subfilter

The second subfilter ϕ_s is simply a linear filter. For each fixed u we have a 2D image $\mathcal{F}(x,y) = \mathcal{F}(x,y,u)$. The linear filter "smoothes" this 2D image. More formally, the response can be expressed as

$$(\phi_s \circ \mathcal{F})(x,y,u) = \mathcal{F}(x,y,u) * g(x,y) \tag{11}$$

where $g = g(x,y)$ serves as the convolution kernel.

Thus applying ϕ to Ξ involves first a morphological dilation operation along the disparity direction and then, a linear convolution operation in the spatial directions. This decomposition provides new insight into ϕ. Besides, it leads to a computationally efficient method of applying the filter, as to be shown in the next section.

5. Surface Enhancement

5.1. Pyramids. "Pyramid-like" representations [33, 34] are useful in computer vision and image processing. Pyramids, in general, are data structures that provide successively condensed representations of the information in the two-dimensional input image. The information being condensed might simply be image intensity, so that the successive levels of the pyramid are reduced-resolution versions of the input image; but it can also be descriptive information about features in the image, so that successive levels represent increasingly coarse approximations to these features.

Let us review the problem of pyramid construction, using the Gaussian pyramid as an example. Let the input image be $I = I(x,y)$, and the Gaussian pyramid be $\{I_m\}$, where m is the index of levels. Constructing the pyramid involves scale-recursive procedure. At the first level, $I_1 \equiv I$. At the level $m > 1$,

$$I_m(x,y) = (G \circ I_{m-1})(2x, 2y) \tag{12}$$

where $G = G(x,y)$ is a Gaussian and $G \circ I$ is in fact the convolution of G with I. Note that conceptually each intra-level recursion that computes I_m from I_{m-1} involves two operations:

(1) Applying a compact, level-independent operator to I_{m-1}, producing I^-_{m-1}. Since G is a low-pass filter, it removes the high frequency components in I_{m-1}.
(2) Downsampling I^-_{m-1} by a factor 2 in each image direction. If $I = I(x,y)$ is $N \times N$, then $I_2 = I_2(x,y)$ is $N/2 \times N/2$, and the pyramid has no more than $M = \log N + 1$ levels.

In fact, a Gaussian pyramid is a multiresolution representation of the input image. The first level, where the input image resides, corresponds to the highest resolution. Resolution decreases, by a factor of two in each image direction, from one level to the next level. The multiresolution representation is the basis for the multiresolution paradigm for image registration and in particular, stereo matching. In this paradigm, the Gaussian pyramids are first constructed for the two images respectively. Then matching is conducted using a coarse-to-fine control structure.

5.2. SDS Image Hyperpyramid. The SDS image hyperpyramid is constructed in an analogous way to the Gaussian pyramid. Given a SDS image $\Xi = \Xi(x, y, u)$, computing the hyperpyramid $\{\Xi_m\}$ involves successive filtering and downsampling. At the first level, $\Xi_1 \equiv \Xi$. At level $m > 1$, the SDS image is defined as

$$\Xi_m(x, y, u) = (\phi \circ \Xi_{m-1})(2x, 2y, 2u) \tag{13}$$

where ϕ is the surface-sensitive filter defined in the last section. In other words, Ξ_m is derived by first applying ϕ and then downsampling by a factor of 2 in every direction of the SDS. Note that the SDS image hyperpyramid is defined recursively, with the same filter being used at each level.

Let the disparity surface represented at level m be $U_m = U_m(x, y)$, where $U_1 \equiv U$. The disparity surface at level m is an isotropically shrunk version of that at level 1. More specifically,

$$U_m(x, y) = 2^{-(m-1)} U(2^{-(m-1)}x, 2^{-(m-1)}y) \tag{14}$$

In addition to the x- and y-axis, shrinking also occurs along the u-axis, by the same factor.

Consider the problem of detecting the disparity surface in $\Xi_m = \Xi_m(x, y)$. Observe that

(1) The equal brightness constraint for $\Xi = \Xi(x, y)$, as elaborated in Section 3, also apples to $\Xi_m = \Xi_m(x, y)$, $\forall m$. That is, in the SDS image Ξ_m a point on the disparity surface U_m has an intensity close to $\Xi_{\max}$.
(2) The weak continuity constraint for the disparity surface $U = U(x, y)$ also applies to $U_m = U_m(x, y)$, $\forall m$. If $U = U(x, y)$ is assumed to be a fractal surface, then it is justified to keep the size of the receptive field level-independent. Since a fractal surface is scale-invariant, we may use the same enhancement filter at different levels.

While our goal is to detect U from Ξ, the problem at level m is to detect U_m from Ξ_m. As we know, it is difficult to determine the disparity surface from $\Xi_1 = \Xi$ because of the ambiguities. As we go to level 2, however, the problem is simplified; the ambiguities are reduced and the disparity surface enhanced. At level 3, the ambiguities are further reduced. In a word, the surface detection problem gets simpler at coarser levels.

In essence, we circumvent the problem of ambiguity by accumulating neighborhood support. Decimating in the spatial directions allows us to accumulate neighborhood support over successively larger neighborhoods. Decimating in u enables the receptive field to remain level-dependent. In fact, applying ϕ blurs SDS image not only in the spatial directions, but also in the disparity direction as well.

5.3. The Size of the Receptive Field. A problem that arises in computing the hyperpyramid in concerned with the size of the receptive field. The receptive field is a $2D_s \times 2D_s \times 2D_d$ solid, as shown in Fig. 4. It is reasonable to let D_s be proportional to D_d, due to the weak continuity constraint. Thus the problem reduces to determining the height of the receptive field.

For the sake of computational efficiency we want the receptive field to be as small as possible. In other words, D_d should be small. However, the receptive field has to cover 2 or more SDS points in the u-direction, as indicated in the last section. Thus the minimal receptive field is

$$2D_s \times 2D_s \times 2$$

in size. By virtues of the minimal receptive field principle, we can get

$$(15)\ \Xi_m(x,y,u) = \sum_{\xi,\eta} g(\xi - x, \eta - y) \max_{\mu=\{0,1\}} \Xi_{m-1}(2x, 2y, 2u+\mu), \quad \xi, \eta \in \Omega_{xy}$$

from eqn. (8) and eqn. (13). The minimal receptive field principle plays an important part in the multilevel algorithm. The performance of the algorithm will degenerate if a non-minimal receptive field is used.

At level 2, each SDS point is associated with a region of size $2D_s \times 2D_s \times 2$ at level 1. Similarly, at level 3, each SDS point is associated with a region of size $2D_s \times 2D_s \times 2$ at level 2. As a result, each SDS point at level 3 is associated with a region of size $4D_s \times 4D_s \times 4$ at level 1. More generically, each SDS point at level m is associated with a region of size

$$2^{m-1}D_s \times 2^{m-1}D_s \times 2^{m-1}.$$

That is, the size of the region grows geometrically with respect to the level index.

At level 3, Ξ_3 is defined in terms of Ξ_2 which in turn is defined in terms of Ξ_1. Notice that Ξ_3 cannot be directly defined in terms of Ξ_1, nor can Ξ_m, $m > 3$. In other words, although a point (x, y, u) at level $m \geq 3$ corresponds to a region of size $2^{m-1}D_s \times 2^{m-1}D_s \times 2^{m-1}$ at level 1, $\Xi_m(x,y,u)$ cannot be directly computed from Ξ_1 in the region; $\Xi_m(x,y,u)$ can only be recursively defined in terms of $\Xi = \Xi(x,y,u)$.

5.4. Computation. As indicated before, ϕ can be realized by cascading two subfilters ϕ_d, ϕ_s. To derive Ξ_m from Ξ_{m-1}, we first apply the subfilter ϕ_d and downsample in the u-direction. Then we apply the subfilter ϕ_s and downsample in the spatial directions. Applying the subfilter ϕ_s amounts to convolving with the kernel $g(x,y)$. Since $g(x,y) = g(x)g(y)$ is separable, the 2D convolution can be computed in an efficient way, using 1D convolution operations.

Let $-$ denote pre-decimation and $+$ denote post-decimation. Then Ξ_m is computed as follows:

$$\Xi_{m-1} \equiv \Xi^{---} \rightarrow \Xi^{--+} \rightarrow \Xi^{-++} \rightarrow \Xi^{+++} \equiv \Xi_m$$

where Ξ^{--+} and Ξ^{-++} are intermediate results. Suppose Ξ_{m-1} is $A \times A \times B$, then Ξ^{--+} is $A \times A \times \frac{B}{2}$, Ξ^{-++} is $A \times \frac{A}{2} \times \frac{B}{2}$, and finally Ξ_m is $\frac{A}{2} \times \frac{A}{2} \times \frac{B}{2}$. An explanation of the notations is given below.

- $\Xi^{--+}(x,y,u)$ is nothing but the maximum of $\Xi_{m-1}(x,y,2u)$ and $\Xi_{m-1}(x,y,2u+1)$. If u is kept constant, then Ξ^{--+} is a 2D image whose nodes are denoted by $\circ$ in Fig. 5.
- Ξ^{-++} is the convolution of $\Xi^{--+}(x,y,u)$ and $g(y)$, computed at the locations marked by $|$ in Fig. 5. It is worth emphasizing that, instead of first convolving and then downsampling, we compute the convolution at predefined location.
- Ξ^{+++}, or Ξ_m, is the convolution of $\Xi^{-++}(x,y,u)$ and $g(x)$, computed at the locations marked by $-$ in Fig. 5.

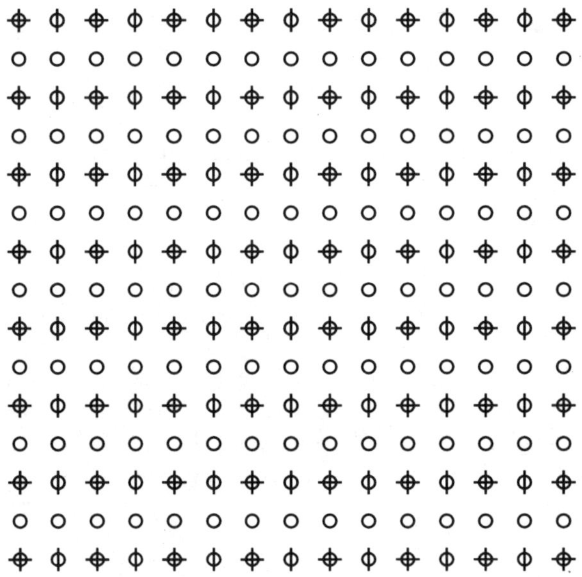

FIGURE 5. Sampling the spatial domain.

To summarize, the SDS image hyperpyramid is constructed in the following steps.

Procedure FtC: $\Xi \to \{\Xi_m\}$

1. Initialize $m = 1$; $\Xi_m = \Xi$.
2. Increment $m \leftarrow m + 1$.
3. Compute $\Xi_m = \Xi_m(x, y, u)$:
$$\Xi^{--+}(x, y, u) = \max_{\mu=\{0,1\}} \Xi_{m-1}(x, y, 2u + \mu)$$
$$\Xi^{-++}(x, y, u) = \big(\Xi^{--+}(x, y, u) * g(y)\big)(x, 2y, u)$$
$$\Xi_m(x, y, u) = \big(\Xi^{-++}(x, y, u) * g(x)\big)(x, 2y, u)$$
4. If $m < M$ go to Step 2; otherwise stop.

The above procedure is simple and computationally efficient. Computing an intensity $\Xi_m(x, y, u)$, $m > 1$ involves only 3 one-dimensional convolution operations. Moreover, at different SDS locations, Ξ^{--+} can be computed in parallel from Ξ_{m-1}, so can Ξ^{-++} from Ξ^{--+} and Ξ_m from Ξ^{-++}. If the SDS image at the finest level is $N_s \times N_s \times N_d$ in size, then constructing the hyperpyramid involves a total of

$$3N_s^2 N_d\Big(\frac{1}{8} + \frac{1}{64} + \cdots + \frac{1}{8^{(m-1)}}\Big) < \frac{3}{7} N_s^2 N_d$$

one-dimensional convolution operations.

6. Surface Detection

6.1. Ambiguity and Accuracy. Recall that the output of local matching, i.e., the SDS image $\Xi = \Xi(x, y, u)$, is intrinsically ambiguous. For fixed x and y, the 1D profile $\Xi(u) = \Xi(x, y, u)$ may have multiple peaks; the true peak may be overridden by the spurious ones. Also recall that in the SDS image pyramid, Ξ_m is

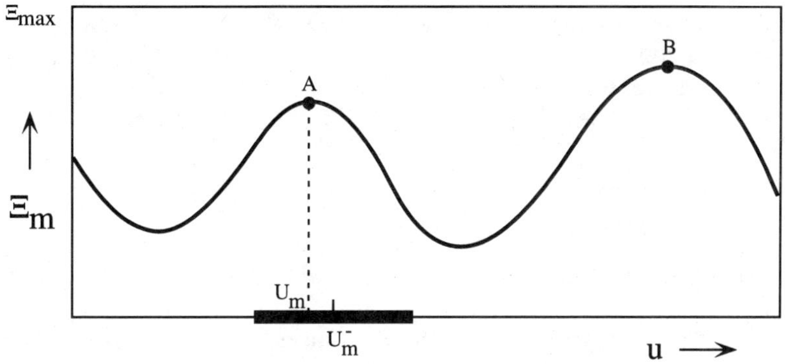

FIGURE 6. Searching in a pruned space.

a reduced-size version of Ξ_{m-1} where ambiguities are suppressed and the disparity surface enhanced. Therefore, to avoid ambiguities we prefer to compute the disparity surface at a coarse level.

The problem is that the result obtained at a coarse level, albeit robust, can hardly be accurate. At level m, information over a neighborhood of size $2^{m-1}D_s \times 2^{m-1}D_s$ is integrated in computing the SDS image intensity Ξ_m. Consequently, the disparity detected at a coarse level is a somewhat averaged version of the true disparities over a large neighborhood.

Summarizing, robustness can be achieved at coarse levels and accuracy, at fine levels. This suggests that a coarse-to-fine control structure can be used to first obtain a rough estimate of the disparity surface at the coarsest level, and then to refine the estimate at finer levels, as shown in Fig. 1. At the coarsest level, the disparity $U_M(x, y)$ is straightforward to detect, by picking the maximum of $\Xi_M(u) = \Xi_M(x, y, u)$. Once U_{m+1} is obtained, it is used to derive U_m. This process is repeated until U_1 is computed.

There are numerous ways to refine an estimate of the disparity surface, among which the simplest way is local searching. Suppose we have a rough disparity estimate $U_{m+1} = U_{m+1}(x, y)$ at level m. Deriving U_m from U_{m+1} involves two steps.

(1) (**Prolongation**) Expand the surface U_{m+1} and obtain its representation

$$U_m^-(x, y) = 2U_{m+1}(x/2, y/2)$$

at level m.

(2) (**Local Maximum-picking**) Use $U_m^-(x, y)$ to prune the space in which to search for $U_m(x, y)$.

The second step is illustrated in Fig. 6, where the curve depicts the SDS image intensity as a function of the disparity $\Xi_m(u) = \Xi_m(x, y, u)$ (x, y are fixed), and the black bar represents an *active internal* specified by the rough estimate $U_m^-(x, y, u)$. Instead of searching along the whole disparity range, we focus on the active internal in search for the intensity maximum. The peak at A gives the refined disparity estimate $U_m(x, y, u)$, although it is no higher than the peak at B.

By using a coarse disparity estimate to prune the search space, we are able to resolve the ambiguities in the SDS image. The disparity surface detected at level $m+1$ specifies an *active layer*

$$\{(x,y,u) : u \in [2U_{m+1}(x/2,y/2) - D_d, 2U_{m+1}(x/2,y/2) + D_d]$$

in the SDS image at level m, where D_d is a constant. Searching for U_m is confined within the active layer.

6.2. Coarse-to-Fine Computation. Two computational issues are involved in the coarse-to-fine sweep. The first issue is concerned with interpolating a disparity surface. Linear interpolation may be used to map U_{m+1} to U_m^-. The second issue is concerned with the size of the pruned search space. What we propose is a 4-ary tree search.

Linear Interpolation

Suppose U_{m+1} is $N \times N$, then U_m is $2N \times 2N$. A simple way to compute U_m^- is to use linear interpolation:

$$\begin{aligned} U_m^-(x,y) &= \frac{1}{2}(U_{m+1}(\lceil x/2 \rceil, \lceil y/2 \rceil) + U_{m+1}(\lceil x/2 \rceil, \lfloor y/2 \rfloor) \\ &\quad + U_{m+1}(\lfloor x/2 \rfloor, \lceil y/2 \rceil) + U_{m+1}(\lfloor x/2 \rfloor, \lfloor y/2 \rfloor)) \end{aligned} \tag{16}$$

where $\lceil\ \rceil$ and $\lfloor\ \rfloor$ denote "ceil" and "floor" respectively. In the above equation "=" means that U_m equals *the integer* that is closest to the r.h.s. As such $(x, y, U_m(x,y))$ is a grid point in the SDS at level m. Of course more sophisticated interpolation methods can be used in place of linear interpolation. But they do not seem to be necessary.

4-ary Tree Search

Fig. 7 illustrates the nodes at different levels at different positions along the disparity direction. In the fine-to-coarse sweep, each node at level m is associated with a pair of nodes at level $m-1$, as shown by the downward arrows. In fact, if the convolution kernel is a Kronecker delta function:

$$g(x,y) = \begin{cases} 1 & \text{if } x = y = 0 \\ 0 & \text{otherwise} \end{cases}$$

then $\Xi_m(x,y,u)$ is computed as:

$$\Xi_m(x,y,u) = \max\{\Xi_{m-1}(2x,2y,u), \Xi_{m-1}(2x,2y,u+1)\}$$

In the coarse-to-fine sweep, each node is connected to its four children. If a node at level $m+1$ corresponds to a disparity estimate, then its four children at level m are activated to compete with one another. The child node with maximal SDS image intensity represents the disparity estimate at level m. More formally, local maximum-picking computes

$$U_m(x,y) = \arg\max_{u \in \mathcal{U}} \Xi_m(x,y,u) \tag{17}$$

where

$$\mathcal{U} = \{U_m^-(x,y) - 1, U_m^-(x,y), U_m^-(x,y) + 1, U_m^-(x,y) + 2\}$$

and $U_m^-(x,y)$ a prolongated version of U_{m+1}.

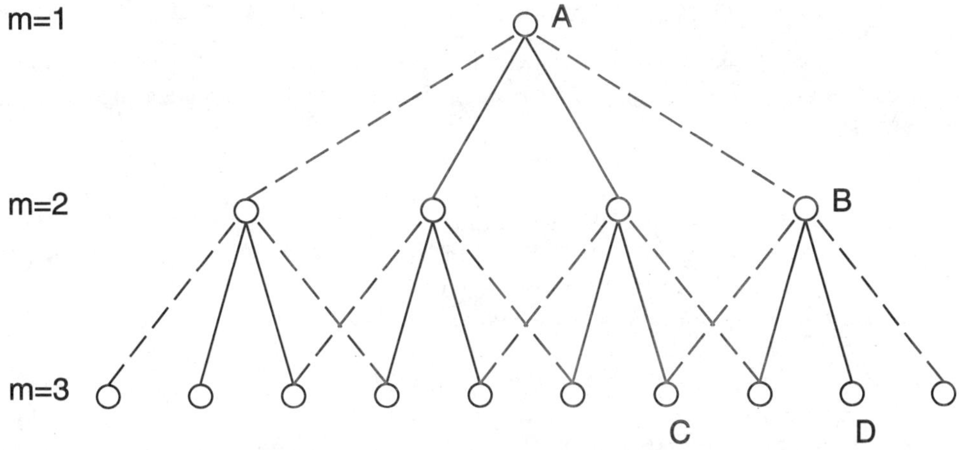

FIGURE 7. Tree search in the coarse-to-fine sweep.

If we know $U_{m+1}(x, y)$ which corresponds to a node at level $m + 1$, then its children at level m are activated in computing $U_{m+1}(2x, 2y)$. The parent-children relationship is described by a n-ary tree. We choose $n = 4$ for a number of reasons. First of all, n should be even, since the coarse-to-fine sweep generates a binary tree. Secondly, if $n = 2$, then the accuracy at the finest level is determined by that at the coarsest level. Suppose that in Fig. 7, node A is the initial disparity estimate. If the true disparity at level 1 is at node B, then it cannot be reached by searching the binary tree rooted at A. In fact all the nodes to the right of node C cannot be reached. Finally, n should be small to enable the search to be localized.

To summarize, the disparity surface is computed from the SDS image pyramid in the following steps.

Procedure CtF: $\{\Xi_m\} \to U$

1. Let m=M and compute $U_m(x, y) = \arg\max_u \Xi_m(x, y, u)$;
2. Decrement $m \leftarrow m - 1$;
3. Compute U_m^- from U_{m+1} using (16);
4. Compute U_m using (17);
5. If $m > 1$ go to Step 2; otherwise let $U = U_1$ and stop.

In the discrete SDS which is isotropically sampled from its continuous counterpart, the disparity $U(x, y)$ produced by the above procedure is accurate to one pixel. In fact, subpixel accuracy can be achieved by interpolating the SDS image profile $\Xi_1(u) = \Xi_1(x, y, u)$. For example, we may use a quadratic function as the interpolant.

Only at the finest is the SDS image profile to be interpolated. At level m, $m > 1$, it not necessary to keep subpixel accuracy, since $U_m(x, y)$ is supposed to be a rough estimate. In short, to achieve subpixel accuracy, only an extra step to refine Ξ_1 is needed. As an alternative way, we can use a Bayesian approach to compute the disparity surface representation from the output of the above procedure, incorporating a prior model about the disparity surface.

7. Experimental Results

We have tested this algorithm on different types of stereo pairs including outdoor scenes and aerial images. The results are encouraging; a typical example is presented as follows.

Shown in the upper of Fig. 8 is the "parking meters" stereo pair (obtained by T. Kanade at CMU) which can be fused using a stereoscope. The image size is 256 by 240. A three-level hyperpyramid was constructed for the SDSI. The disparity surfaces $\{\hat{U}_m\}$ computed at the 3 levels, Ξ_1, Ξ_2 & Ξ_3, using global WTA (i.e., eq. (4)) are shown in the middle-left, lower-left, and lower-middle respectively as intensity images, with closeness encoded by brightness. As we can see, the ambiguities are reduced level by level and Ξ_3 is basically unambiguous.

The disparity surfaces $\{U_m\}$ computed using CtF, the procedure defined in section 6.2, are shown in the middle-right, lower-middle, and lower-right, respectively. Notice that $\{U_3\}$ is the same as $\{\hat{U}_3\}$. All the results shown here are accurate to one pixel. As we can see from $\{U_1\}$, the algorithm has captured the 3 parking meters in front of the shrubberies, the side view of the sign board which is between the second and the third parking meters, and the large depth gap between the front and the building. This result is comparable to that in [21] which was obtained using the multiple-baseline method.

A further example is the classic "Pentagon image". The results are shown in Fig. 9.

8. A Degenerate Case

It is illuminating to explore the relationship between the multilevel paradigm and the multiresolution paradigm. In the first section we indicated that the multilevel method is motivated by the multiresolution method. Here we will show the latter is a degenerate case of the former.

Proposition. *In the degenerate case that the receptive field of ϕ is a flat sheet or $\Lambda_{xyu} = [x - D_s, x + D_s] \times [y - D_s, y + D_s] \times [u, u]$, the multilevel method is in spirit the same as the multiresolution method.*

To make this proposition clear, we will represent the SDS image at level m in terms of the input stereo images. Suppose that the SDS image is defined as

$$\Xi(x, y, u) = -\left(I^l(x + u/2, y) - I^r(x - u/2, y)\right)^2 \tag{18}$$

By definition the SDS image at level m is expressed as

$$\Xi_m(x, y, u) = (\phi \circ \Xi_{m-1})(2x, 2y, 2u) \tag{19}$$

If the receptive field is a flat sheet, then ϕ reduces to its linear subfilter ϕ_s:

$$(\phi \circ \Xi)(x, y, u) = \sum_{\xi, \eta \in \Omega_{x,y}} g(\xi - x, \eta - y)\ \Xi(\xi, \eta, u) \tag{20}$$

where

$$\Omega_{x,y} = [x - D_s, x + D_s] \times [y - D_s, y + D_s]$$

FIGURE 8. "Parking meters" and results. See text.

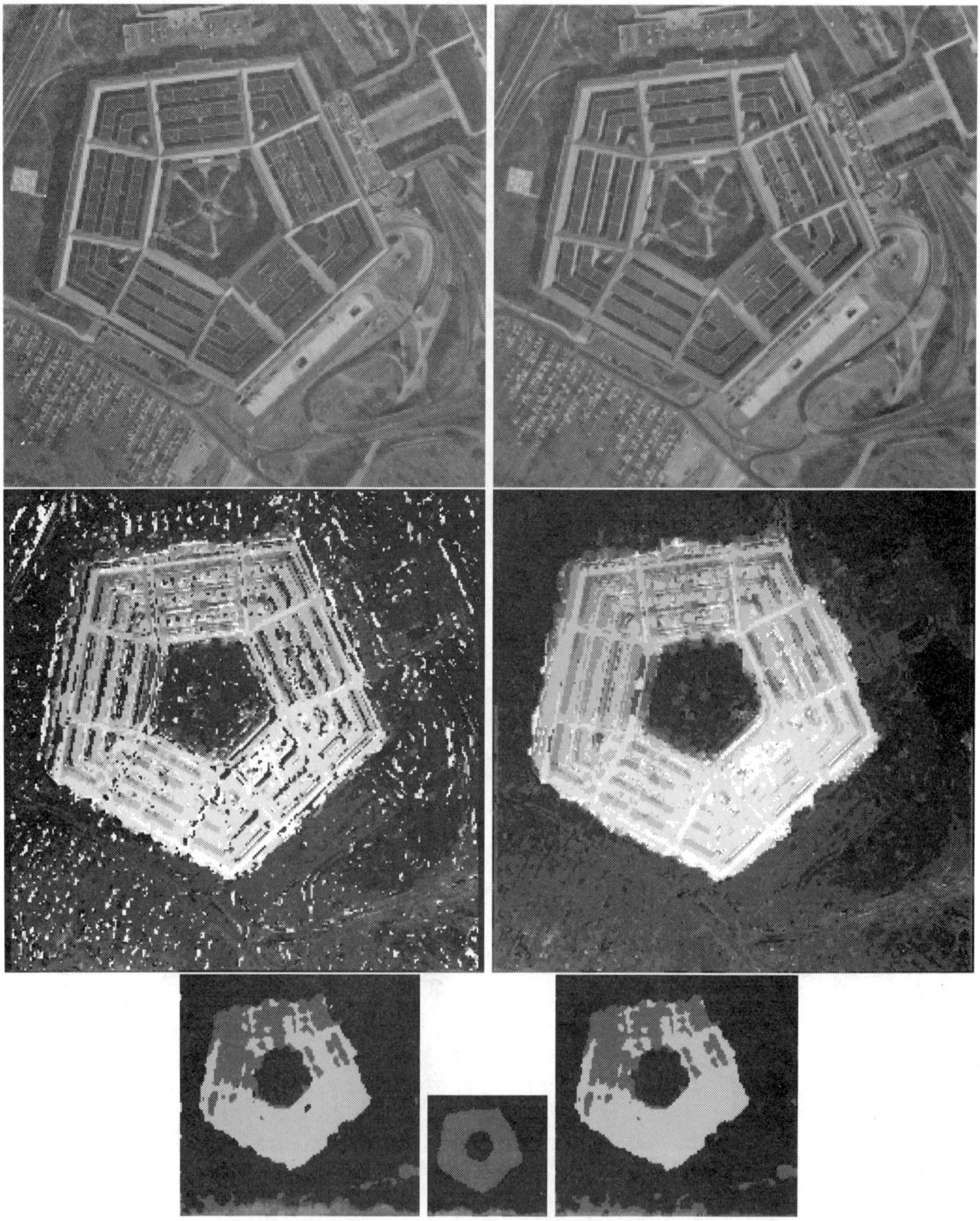

FIGURE 9. "Pentagon image" and results. Same format as Fig. 8, see text.

By simple algebraic manipulations we yield

(21)

$$\Xi_m(2^{-(m-1)}x, 2^{-(m-1)}y, 2^{-(m-1)}u) = - \sum_{\xi,\eta \in \Omega^{*(m-1)}_{x,y}} g^{*(m-1)}(\xi - x, \eta - y) \left(I^l(\xi + u/2, \eta) - I^r(\xi - u/2, \eta)\right)^2$$

where

$$\Omega^{*m}_{x,y} = [x - 2^m D_s, x + 2^m D_s] \times [y - 2^m D_s, y + 2^m D_s]$$

$$g^{*m}(x,y) = \underbrace{g(x,y) * g(x,y) * \cdots * g(x,y)}_{m}$$

The above equation shows that $\Xi_m(x,y,u)$ is the local similarity between point $(x, y+u/2)$ in I^l and point $(x, y-u/2)$ in I^r, where the the window size is proportional to m, the level index. At a coarse (fine) level, a large (small) window is used to compute the sum-of-squared-difference.

In the degenerate case the multilevel paradigm is characterized by the following points:

(1) Compute area-correlation, using a window that is sufficiently large, to derive a rough disparity estimate.
(2) The disparity within the window is assumed to be constant.
(3) Refine the disparity estimate by using successively smaller windows.

It turns out that the above properties are characteristic of the multiresolution method. So it is evident that the multiresolution paradigm is equivalent to the multilevel paradigm provided that the receptive field does not extend along the disparity direction.

9. Concluding Remarks

Stereo matching amounts to surface detection in the SDS image. In some sense the SDS image contains all the information available in the stereo pair. In the case that intensity difference is used as the local similarity measure, the left and right images can be recovered (up to a constant) from the SDS image. The SDS image enable us to cope with a group of matches whose disparities are not necessarily the same.

The surface-sensitive filter ϕ plays an important role in the multilevel paradigm. Surface enhancement using ϕ is effective and efficient. ϕ is a novel nonlinear filter which can be regarded as the cascade of a morphological filter and a linear filter; it adaptive nature enables us to accommodate surface variations in a simple way.

The algorithm employs a level-recursive control structure, with no infra-level iterations. The basic operations are 1D convolution and local maximum-picking. SIMD computers are especially suitable for implementation.

Acknowledgments. The first author is indebted to David Mumford and James Clark for their insights and advice. This research was supported in part by the Brown/Harvard/MIT Center for Intelligent Control Systems with U.S. Army Research Office grant number DAAL03-86-K-0171.

References

1. Anandan, P. (1989) A computational framework and an algorithm for the measurement of visual motion. *Int. Journal of Computer Vision*, 2:283-310.
2. Baker, H.H. & T.O. Binford (1981) Depth from edge- and intensity-based stereo. *Proc. 7th IJCAI*, pp. 631-636.
3. Barnard, S.T. (1989) Stochastic stereo matching over scale. *Int'l J. Computer Vision.* 17-32.
4. Barnard, S.T. and M.A. Fischler, (1987) Stereo vision. In *Encyclopedia of Artificial Intelligence*, S.C. Shapiro ed., pp. 1083-1090.
5. Belhumeur, P.N. & D. Mumford (1992) A Bayesian treatment of the stereo correspondence problem using half-occluded regions. *Proc. IEEE CVPR.*
6. Burt, P. & B. Julesz, (1980) A disparity gradient limit for binocular fusion. *Science*, 208:651-657.
7. Burt, P.J. (1981) Fast filter transformations for image processing. *Computer Vision, Graphics and Image Processing*, 16:20-51.
8. Cox, I.J., et al. (1992) Stereo Without Disparity Gradient Smoothing: a Bayesian Sensor Fusion Solution. *British Machine Vision Conference,* D. Hogg and R. Boyle ed., Springer-Verlag, pp. 337-346.
9. Drumheller, M. & T. Poggio (1986) On parallel stereo. *Proc. IEEE Robotics and Automation*, pp. 1439-1448.
10. Forstner, W. & A. Pertl (1986) *Photogrammetric Standard Methods and Digital Image Matching Techniques for High Precision Surface Measurements*, Elsevier Science Publishers B.V.
11. Geiger, D., B. Ladendorf & A.L. Yuille (1993) *Occlusion and Binocular Stereo.* Tech. Rep. no. 93-1. Harvard Robotics Lab, Cambridge, MA.
12. Grimson, W.E.L. (1981) *From Images to Surfaces.* M.I.T. Press, Cambridge, Mass.
13. Hannah, M.J. (1974) *Computer Matching of Areas in Stereo Images.* Ph.D. dissertation, Stanford University.
14. Hannah, M.J. (1989) A system for digital stereo image matching. *Photogrammetric Engineering and Remote Sensing,* 55(12):1765-1770.
15. Hoff, W. & N. Ahuja, (1989) Surface from stereo: Integrating feature matching, disparity estimation, and contour detection. *IEEE Trans. Patt. Anal. Machine Intell.,* 11(2):121-136.
16. Horn, B.K.P. (1985) *Robot Vision.* Cambridge, MA: MIT Press.
17. Hsieh, Y.C., D.M. McKeown, Jr. & F.P. Perlant, (1992) Performance evaluation of scene registration and stereo matching for cartographic feature extraction, *IEEE Trans. Patt. Anal. Machine Intell.,* PAMI-14(2):214-238.
18. Jones, D. G. (1991) *Computational models of binocular vision.* Ph.D. thesis, Stanford University.
19. Jones, D.G. & J. Malik (1992) Determining three-dimensional shape from orientation and spatial frequency disparities. *Proc. European Conf. on Computer Vision.*
20. Julesz, B. (1971) *Foundations of Cyclopean Perception.* The University of Chicago Press.
21. Kanade, T. & M. Okutomi, & T. Nakahara (1992) A multiple-baseline stereo method. *Proc. DARPA Image Understanding Workshop*, pp. 409-426.
22. Lucas, B.D. (1984) *Generalized image matching by the method of differences.* Ph.D dissertation, Carnegie-Mellon University.
23. Marr, D. & T. Poggio (1976) Cooperative computation of stereo disparity, *Science*, 194:283-287.
24. Marr, D. & T. Poggio (1979) A theory for human stereo vision. *Proc. Roy. Soc. Lond.* B 204:301-328.
25. Marr, D. (1982) *Vision.* San Francisco: W.H. Freeman and Co.
26. Matthies, L. (1992) Stereo vision for planetary rovers. *Int'l J. Computer Vision,* 8(1):71-91.
27. Moravec, H. P. (1980) *Obstacle avoidance and navigation in the real world by a seeing robot rover.* Ph.D. dissertation, Standford Univ, Standford, CA.
28. Okutomi, Masatoshi & Kanade, Takeo (1990) A locally adaptive window for signal matching. *Proc. of Int'l Conf. on Computer Vision.*
29. Panton, D.J. (1978) A flexible approach to digital stereo mapping. *Photogrammetric Eng. Remote Sens.,* 44(12):1499-1512.
30. Pollard, S.B., J.E.W. Mayhew & J.P. Frisby (1985) PMF: a stereo correspondence algorithm

using a disparity gradient limit. *Perception*, 14:449-470.
31. Prazdny, K. (1985) Detection of binocular disparities. *Biological Cybernetics*, 52:93-99.
32. Quam, L.H. (1984) Hierarchical warp stereo. *Proc. DARPA Image Understanding Workshop*, pp.149-156.
33. Rosenfeld, A. (ed.) (1984) *Multiresolution Image Processing and Analysis*, New York: Springer-Verlag.
34. Terzopoulos, D. (1986) Image analysis using multigrid relaxation methods. *IEEE Trans. Patt. Anal. Machine Intell.*, PAMI-8:129-139.
35. Weng, J., N. Ahuja, & T.S. Huang (1992) Matching two perspective views. *IEEE Trans. Patt. Anal. Machine Intell.*, PAMI-14(8):806-825.
36. Witkin, A., D. Terzopoulos & M. Kass (1987) Signal matching through scale space. *Int'l J. Computer Vision*, pp. 133-144.
37. Wood, G.A. (1983) Realities of automatic correlation problem. *Photogrammetric Engineering and Remote Sensing*, 49:537-538, April 1983.
38. Yang, Y., A.L. Yuille, & J. Lu (1993) Local, global, and multilevel stereo matching. *Proc. IEEE Conf. on Computer Vision and Pattern Recognition*, pp. 274-279.

DIVISION OF APPLIED SCIENCES, HARVARD UNIVERSITY, CAMBRIDGE, MA 02138
E-mail address: yibing@@hrl.harvard.edu

Part IV

Human Vision

DIMACS Series in Discrete Mathematics
and Theoretical Computer Science
Volume 19, 1995

Some Problems of Visual Shape Recognition to Which the Application of Clustering Mathematics Might Yield Some Potential Benefits

Irving Biederman

ABSTRACT. An overview of recent theory and research into human shape recognition is presented. Shape allows an extraordinarily large number of descriptive possibilities. Do one of these descriptions characterize the representations that mediate humans object recognition? Can the mathematics of clustering suggest others or how others can be evaluated?

1. Overview of Shape Recognition

In a fraction of a second humans are able to comprehend unanticipated novel images of real world scenes, as when we "channel surf" while watching television. Approximately half of the primate cortex is devoted to this activity. On closer analysis we see that this extraordinary feat is executed not by a single undifferentiated mass of tissue, but by upwards of 30 different cortical areas or modules. Why the high degree of modularity? Recent work on artificial neural networks suggests that problems with independent dimensions are best learned by separate networks, each devoted to solving a single aspect of the problem. For example, given a set of shapes that can appear in a variety of locations, separate networks for processing location and shape each converged on a solution much more rapidly than a single network that tried to learn both (Rueckl, Cave, & Kosslyn, 1989). Similarly, separate cortical modules may be involved in the representations of faces and objects. Even for faces, separate subareas

---

1991 Mathematics Subject Classification. Primary 22A05, 73K12

Sections of this paper are slightly modified from an abstract that appeared in the British Machine Vision Conference Proceedings, 1993 and from a revision of a chapter to appear in Invitation in Cognitive Science, 2nd edition, MIT Press. This research was supported by Grants from the Air Force Office of Scientific Research (90-0274) and the McDonnell-Pew Foundation Program in Cognitive Neuroscience (T89-01245-029).

may be involved in the perception of facial identity and facial emotion. Currently, the identification of the functions of the various modules has enjoyed some success because the classes reflect, more or less, aspects of the investigators' own subjective experience. Do the mathematics of clustering allow us to gain a deeper understanding as to why the brain evolved to form these particular modules for the representation of information in the visual signal and not others? Although I pose this question most specifically in the domain of shape, the problem is actually quite general and relates not only to all of vision but to the entirety of the modularity of the brain.

In this paper I will elaborate on the potential organization that is apparent in the visual cortex in general but I will then provide some overview to the specific problems of shape recognition as that domain may provide the most fruitful for the application of clustering mathematics.

2. What are the Classes of Shape Recognition Problems?

Let us simplify (perhaps) the answer to the question that heads this section by considering only the information that we might obtain from a brief, unfamiliar, unanticipated snapshot of a scene to a single, stationary eye. We thus ignore contributions to vision from binocularity, motion, eye movements, the nulling of the effects of head and eye movements, context, anticipation, etc. One can readily experience this domain by a personal experiment performed at home with a remote television control. Turn on the television set. With the sound turned off and with your eyes closed change channels. Now blink briefly with one eye. You will find that you will readily *comprehend* the scene; not just as an array of colors, textures, and shapes, but as an *interpreted* interacting set of creatures and objects about which one might be able to tell a five minute story, specifying the tone and nuances of the scene. If movement is present, that would be integrated into the account. Relations among entities are described with prepositions, verbs, and comparatives, such as "on," "larger than," or "holding." Often the prepositions are understood, as in " a man writing with a pen."

Up to a few years ago, if we looked at a typical text book in visual perception, we would have found virtually no exposition of scene comprehension, which is overwhelmingly the most important visual entreé to cognition. After the early chapters on transduction, optics, early neural coding, and visual pathways to the brain, we would find chapter headings with titles like "motion," "color," "texture," and "monocular and binocular cues for depth." Until recently, discussions of "shape," or "object perception" were left as afterthoughts for a few sentences at the end of a chapter on contrast sensitivity, perhaps combined with boxes on the Gestalt organizational principles, illusions, and other curiosities.

But how are scenes and objects actually *recognized*? By recognition, I mean not that the exact same scene is classified as exactly matching one viewed earlier, but how any scene--novel or familiar, but especially novel--can be understood in terms of our world knowledge. And how can it be done so quickly and effortlessly? Note that in the brief glimpse at the television screen, you did not have to try to interpret the scene. You couldn't help it. The interptetation was automatic.

I will focus next on what appear to be, at least on the surface, problems of object perception for purposes of recognition (rather than, say, purposes of grasping) that might be amenable to clustering analyses. The issue here is why the particular organization that we see in the representation of object shape, rather than others. The extension of this theory to the representation of inputs of multiple objects when they may form a scene is considered in section 4.

In the remaining sections I review the current statement of a theory of human object recognition, its empirical status, and some of the challenges that remain. The study of human shape recognition is of particular importance because the human represents the only existence proof that a general shape recognizer is even possible. First we present a brief discussion of the goals of this research.

3. What Should be Modeled in a Theory of Human Object Recognition?

There is likely no single answer to this question in that humans can activate an apparently unbounded set of classes for any given object image and achieve this activation in a variety of ways. For example, if the goal is to distinguish among the contents of a bin of parts, model-based matching, in which the image is matched against an exact, metrically specified, object representation (the model) can be successful (e.g., Lowe, 1987; Ullman, 1989). The major problem is estimating the pose of the object. No concern need be paid to the extent to which the theory's performance resembles that of a human.

We have concentrated on modeling *primal access* (Biederman, 1987): The initial activation in a human brain of a basic-level representation of an image from an object exemplar, even a novel one, in the absence of any context that might reduce the set of possible objects. This commits us to take seriously the data (especially reaction times) obtained during real-time performance. By concentrating on *basic-* (actually *entry-*) level classification, we account for the kinds of classification by which humans gain most of their knowledge about their world, that something is a sofa or an elephant, for example. "Basic level" refers to the level of abstraction of visual concepts that maximizes between-category distinctiveness and within-category informativeness. Specifying the *subordinate* level class, for example that something is an African (vs. Indian) elephant or is a particular style of sofa, provides only a slight increase in informativeness at an enormous loss of distinctiveness. That is, the differences among sofas are much smaller (and less significant) that the difference between a sofa and an elephant. Similarly, the *superordinate* level, that something might be an animal or an article of furniture, sacrifices informativeness with only a slight gain, if any, in distinctiveness. The difference between the classes of animals and furniture may be slightly greater than the differences, say, between a sofa and a lamp or an elephant and a rabbit, but that slight gain in distinctiveness comes at an enormous loss in the additional information that we obtain in knowing that something is a lamp, and not a sofa; or a rabbit and not an elephant. By modeling entry-level rather than basic level, we can treat an exemplar of a class that differs greatly in shape from others of that class. Thus *penguin* is considered a separate entry-level class from the class *birds* . Entry level terms for an object are the first to enter a child's vocabulary and are used at least ten times more frequently than other level terms to refer to the same entity (Biederman, 1987).

4. Geon Theory

Three striking and fundamental characteristics of human object recognition are its invariance with changes in viewpoint, its ability to operate on unfamiliar objects, its robustness in the face of occlusion or noise, and its speed, subjective ease and automaticity. Geon theory (Biederman, 1987) offers an account of this extraordinary capacity.

Geon theory assumes that objects are represented as an arrangement of simple, viewpoint-invariant, volumetric primitives, termed *geons*, such as bricks, cylinders, wedges, cones, and squashes, and their curved axis counterparts, as illustrated in figure 1. The arrangement matters, as illustrated with the cup and the bucket. In the cup, the curved cylinder is connected in (two) END-TO-MIDDLE joins and is SIDE-OF the cylinder. In the bucket, the curved cylinder, is ON-TOP-OF and is connected in (two) END-TO-END joins.

The viewpoint invariance derives from a classification of the edges corresponding to the orientation and depth discontinuities of the object's surfaces, according to viewpoint invariant contrasts (VICs). VICs are differences in nonaccidental properties, i.e., properties of edges that are unaffected (or largely unaffected) by rotation in depth, such as straight vs. curved, vertices formed at the cotermination of edges (L, Y, arrow, and tangent Y), and whether edges are parallel or nonparallel. In the current version of the theory there are 24 geons generated by these contrasts which serve as a partition of the set of generalized cylinders into convex or singly concave volumes. Double concavities are taken as candidates for parsing points by which a complex object is decomposed into its geons.

In a current theoretical effort, we are exploring a redefinition of the geons in terms of those volumes that minimize viewpoint uncertainty over the aspect graph (Koenderink & Van Doorn, 1979). For example, a cylinder can be identified as such from almost any viewpoint, except for

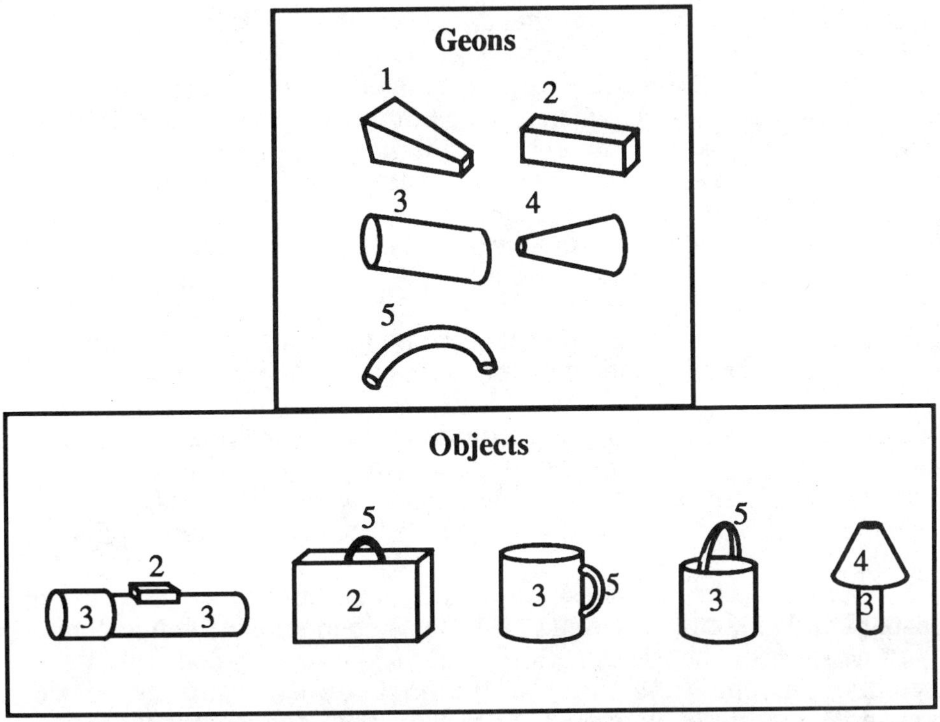

Figure 1. Five geons and five objects. Note that the pail and the bucket are composed of the same geons but in different relations. TOP-OF is a relation. If the page is rotated 180° the pail will resemble a cap and the lamp a trowel or shovel.

those cases where it projects as an ellipse or quadrilateral. If the areas of the viewing sphere corresponding to each projection of a volume, ignoring aspect ratio and degree of curvature, are converted to probabilities, say .80 for the cylinder, and .10 for the ellipse, and .10 for the quadrilateral, then geons have the property of low uncertainty in the Shannon sense, i.e., they yield low values for $-\sum p(i)\log_2 p(i)$, where each of the p(i)s refers to the proportion of the surface of the viewing sphere occupied by each interpretation. A complex volume or shape would change its interpretation many times over the viewing sphere and thus have high uncertainty. The lowest uncertainty, zero, would be for a sphere. Obviously, low viewpoint uncertainty is equivalent to high viewpoint invariance. Although it is likely that the volumes that would pass the criterion of low viewpoint uncertainty would largely correspond to the current set of geons, this redefinition allows a more principled basis for the derivation of the geons. In particular, low viewpoint uncertainty, by minimizing change as different views of an objects are encountered, may yield optimal conditions for a self-organizing neural network to develop hidden units that function as geon detectors.

4.1 Relation of geon theory to other current models of biological object recognition. All models of biological object recognition assume an input layer that can be approximated by a lattice of filters that cover the visual field. Each node in the lattice is occupied by a number of simple filters (typically modeled as a Gaussian damped sinusoid, termed a Gabor filter), each "tuned" to a particular orientation and spatial frequency at that position in the visual field, though there is considerable overlap in the coverage of the filters at nearby nodes (e.g., Buhmann, Lange, von der Malsburg, Vorbrüggen, & Würtz, 1991). Each of the Gabor filters corresponds to a simple cell and each node corresponds to a single hypercolumn in the initial region of the cortex that receives visual inputs (area V1). There are approximately 2,100 hypercolumns in the V1 area of each hemisphere. At the object layer are units corresponding to the various object categories that the network is supposed to differentiate.

Most current theories of object recognition can be distinguished along two dimensions: a) the degree to which they assume intermediate representations between input and output, and b) whether they assume the representation is defined in a coordinate space or is a structural description.

4.1.1 *Intermediate Representations.* A recent review article by Dickinson, Pentland, and Rosenfeld (1992) provided an excellent account of this dimension of theorizing. Some theories assume no intermediate representation, mapping the output of the filters directly onto the object layer, as in the theories of Poggio & Edelman, 1990, and Buhmann, et al., 1991). At the other extreme, are those theories such as geon theory that assume volumetric primitives. Between these two extremes, are theories that specify lines (as in Lowe's, 1987, model) and surfaces, for example. Dickinson et al observed a tradeoff between the ease at which the representation could be determined from the image and the ease at which object representations in the data base could be successfully activated. Thus, filter values are easy to compute but are of very little help in selecting among objects given that viewing conditions can change. Geons are extremely difficult to extract but can be quite powerful in accessing object representations.

4.1.2 *Coordinate Space.* Another important dimension by which theories can be classified is whether they assume a coordinate space that preserves retinal proximities for the matching of input against stored representation (as in Poggio & Edelman, 1990) or whether a *structural description*, consisting of elements (such as parts) and *explicit* relations among these elements, as in Hummel and Biederman (1992).

4.2 Scenes. How we can gain such quick access to a semantic representation of a visual input consisting of not one object but an array of objects, that is, a scene? It is possible that an extension of geon theory (Biederman, 1989) might offer some explanation of this feat. It is not the case that scenes are recognized as the result of the recognition of the individual objects, although may occur on occasion. Rather the arrangement of interacting objects shapes in the scene, particularly when one object occludes another as when a desk occludes all but the back of a chair, may produce a familiar arrangement of geons in specified relations, such as a large horizontal brick (the desk) occluding a smaller vertical slab (the chair back). Such *geon clusters*, even though they do not correspond to entry level objects, may well function as visual objects. If an array of objects is not arranged in such a cluster (or clusters), the perception of that array will proceed serially, object-by-object.

5. The Psychophysics of Entry Level Classification

5.1 Picture Priming Experiments. What are the data that we wish to model? We have recently completed a number of picture-priming experiments in which subjects name, as quickly and as accurately as possible, briefly (100-500 msec) presented line drawing of objects. The image is followed by a mask consisting of a random-appearing arrangement of contours. The pictures are then shown a second time, several minutes later (in a different order). There is marked facilitation (or *priming*) in the speed and accuracy of naming on this second block of trials. A part of the facilitation is visual (and not just verbal or conceptual) in that an image of the same basic level class but of another shape, such as another type of chair, is named more slowly than the original object.

5.1.1 *Strong Invariance.* What happens if the image, on its second presentation, is projected to another part of the retina (an equal distance from fixation as on its first presentation), or at another size, or at another orientation in depth than what it was when first presented, or reflected? Would there be any priming? A weak form of invariance would be supported if there was some priming, but less than if the object was at its original position, size, or orientation. Remarkably, the results clearly supported strong invariance--there was *no* effect of changing position, size, reflection, or orientation in depth (up to parts occlusion) (Biederman & Cooper, 1991a,b; 1992; Biederman & Gerhardstein, 1993). Biederman & Gerhardstein also showed that depth invariance could readily be achieved with nonsense objects (in a same-different matching task), indicating that invariance could occur in the absence of a familiar object model.

On computational grounds, the invariances seem entirely reasonable in that the alternative, a separate representation of an object for each of its image manifestations, would require a prohibitively large number of representations. The invariance in recognition *speed*, i.e., the strong invariance, moreover, is inconsistent with the hypothesis (such as that advanced by Ullman, 1989) that recognition is achieved through template *transformations* for translating, scaling, or rotating an image or template so as to place the two in correspondence, as such transformations would (presumably) require time for their execution, not to mention the formidable initial problem of selecting the appropriate transformation to apply to an unknown image. Transformational models can achieve weak invariance, but not strong invariance (unless they assume transformations with no time cost).

The phenomenon of strong invariance for position, reflection, size, and orientation in depth, may not be just a psychophysical curiosity but may reflect a fundamental partitioning in the way in which the brain handles shape. In recent years it has become apparent that there are least two extrastriate cortical visual systems. Both start at the striate cortex (V1), the primary projection area in the occipital cortex, which receives direct inputs from the retina, by way of the lateral geniculate body. One system, termed the "where" system, extends dorsally from V1, to the posterior parietal (PP) cortex, and has been implicated in memory for location. The other, termed the "what" system, extends ventrally, from V1 to V2, V4, and then to the inferior temporal (IT) cortex and appears to determine object recognition.

Why should these particular visual systems have evolved separately? A possible clue lies in the realization that "where" may be too narrow a characterization of the function of the dorsal system. Instead, there is ample evidence that the dorsal system mediates *motor interaction*, in general. To be sure, location ("where") is a critical component of successful motor interaction: To pick up a coffee cup requires that one reach for the cup in a given location. If the cup is on the left side, one cannot reach for it on the right. Similarly, the metrics and dynamics of the grasp are closely tuned to the actual size and orientation of the handle. The motor interaction is not limited to reaching and grasping but also includes other motor functions such as navigation toward some location and avoiding obstacles along the way.

We can now appreciate a possible computational basis for why two separate visual systems might have evolved. The one for recognition must be able to activate the same representation despite variation in stimulus parameters that are critical for motor interaction, viz., position, size, and orientation in depth. Similarly, motor interaction does not require knowing what the object is. That is, we can reach for or navigate to or around an object without identifying it. It is our conjecture that the motor interaction system employs metric information in a coordinate space but the recognition system employs qualitative (viewpoint-invariant) differences in a structural description.

5.1.2 *Parts*. We thus have ample evidence for invariances in recognition, but how can we describe the representation itself? Should it be the particular edges and vertices presented in the image? Or a specific object model, such as a grand piano? Or of the object's parts? Or all three? Somewhat surprisingly, there is a single answer to this question. The magnitude of the perceptual priming is *completely* determined by the capacity to activate representations of the *parts* of an object; there is no contribution from the features (vertices and edges) actually present in the image or the global shape or an object model.

To assess what information is affected by priming, Biederman and Cooper (1991a) measured naming speed and accuracy with briefly presented stimuli by deleting every other image feature (edge and vertex) from each geon to create two *complementary images* of each object, as shown in Figure 2. That is, the two images for each object, when superimposed, would form an intact picture with no overlap in contour. The complementary images were created in such a way that each part (or geon) of the object could be recovered (or fail to be recovered) from each of the images. Although complementary images shared no edges and vertices, they presumably would activate the same components. Because the amount of contour deleted from each image was substantial and included vertices, it is unlikely that a local process of filling-in could have completed the contour of these images (see Biederman & Cooper, 1991a for a more complete discussion).

These results are supportive of an earlier demonstration by Biederman (1987) who showed that pictures of common objects were unrecognizable when the contour was deleted in such a manner that the geons could not be recovered. When the same amount of contour deletion allowed recovery of the geons, recognition could be perfect.

On a first block of trials, subjects viewed a number of brief presentations of one member from each complementary pair which they named as quickly and as accurately as possible. On the second block, they would see either the *identical* image, its *complement*, or a *same name-different -exemplar* image (also contour-deleted) from a category with the same name and basic level concept but with a different shape. Mean correct naming reaction times and error rates were markedly lower to the identical image than the different exemplars, indicating that a portion of the priming was indeed visual. The critical comparison, however, concerned the relative performance of the complementary condition. If priming was a function of repetition of the specific vertices and edges in the image, then the complementary condition would have been equivalent to the different exemplar condition, as neither shared any features with the original image. Remarkably, there was no difference in performance in naming complementary and identical images, indicating that *none* of the priming could be attributed to the specific vertices and lines actually present in the image.

What then caused the priming? One possibility is that it was the parts (or geons), as they were common in the two conditions. Another possibility is that a semantic model of a subordinate category, e.g., a grand piano, rather than the basic level category, e.g., piano, was activated in the initial presentation. To test this possibility, an experiment was run in which complementary images were created by deleting half the *parts* of the objects.

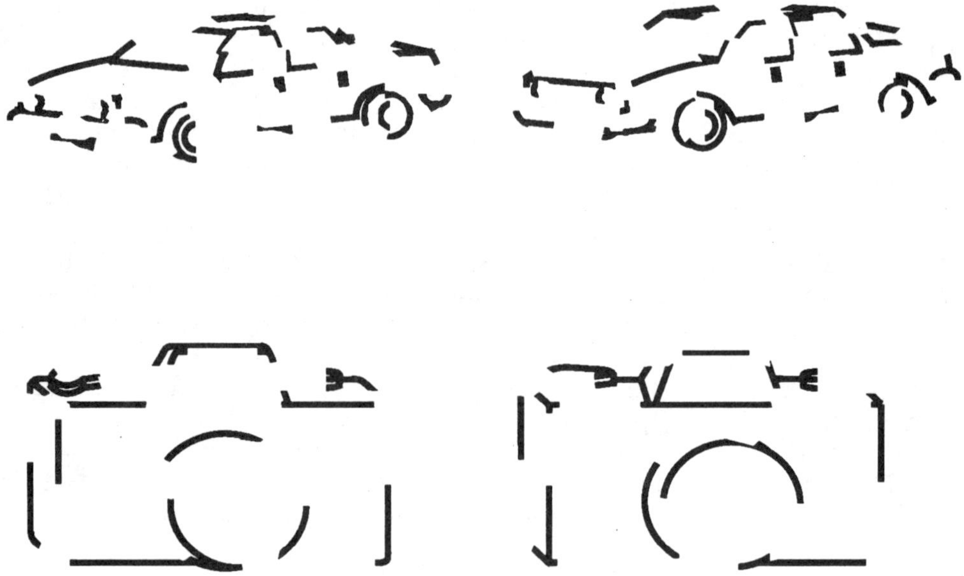

Figure 2. Sample complementary images produced by deleting alternate vertices and edges from each geon. From Cooper, Biederman, & Hummel (1992).

With these stimuli, presumably, the same subordinate category would be activated from either members of a complementary pair, but through different parts. (This experiment required the use of objects that would require at least six parts to look complete.) The design was otherwise identical to that of the previous study. As with the first experiment, performance with the identical images was better than with the different exemplars. Now, however, performance with the complements was equivalent to that with the different exemplars, indicating that *none* of the priming could be attributed to a subordinate semantic model. By elimination, the two experiments, taken together, suggest that *all* of the priming can be attributed to a representation of the parts (and their interrelations) of the object.

5.2 Viewpoint-Invariant vs. Metric Properties? But are these parts geons? Cooper & Biederman (1993) recently reported a series of experiments in which they compared the relative importance of differences in aspect ratio of a part (a metric property) with a difference in a viewpoint invariant characterization of the part. For example, a lamp that had a cylinder as its base could have been changed to one of a different aspect ratio or the cylinder could have been changed to a brick of the same aspect ratio. The results clearly supported the greater importance of the geon changes. Similarly, Biederman & Gerhardstein (1993) found that there was no effect of rotation in depth on recognition speed in a priming task as long as the object could be readily described as an arrangement of distinctive geons and the original geons remained in view. The rotation, of course, altered the aspect ratio and degree of curvature of the objects.

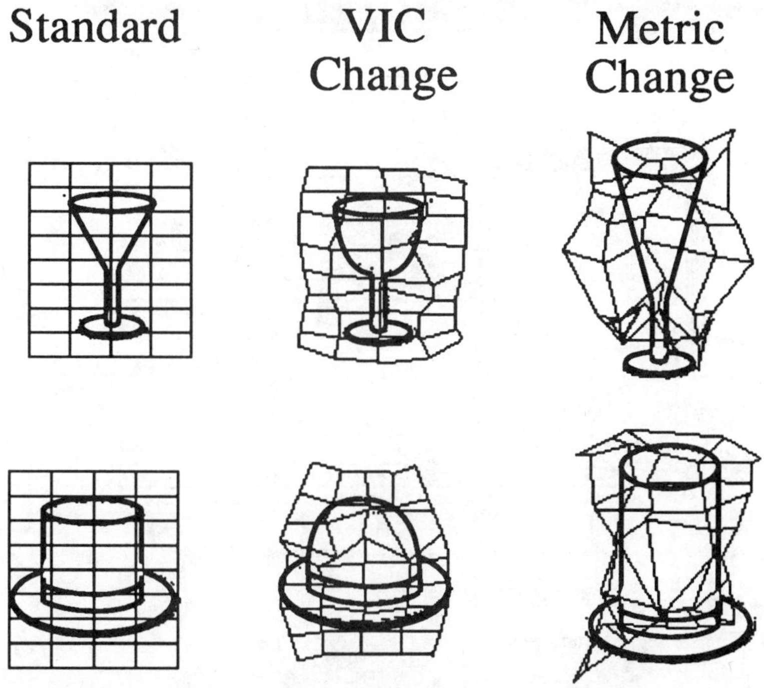

Figure 3. Sample stimuli used in the Cooper & Biederman (1983) Same-Different Name Matching Experiment. If he object in the left column might have been presented first on a "Same" trial (the *Standard* shown in the left column), the second object would either be one in which one of the parts differed in a viewpoint-invariant contrast (*VIC Change*, as shown in the middle column) or in which one of the parts differed in aspect ratio (Metric Change, as shown in the right column). The distortion of the grids superimposed on the stimuli in the middle and right columns shows how closely the images resembled the standard according to a model of V1 by Buhmann et al (1991) based on von der Malsburg's (1987) dynamic links architecture (described below): The greater the distortion, the less the resemblance.

6. A Neural Net Implementation of Geon Theory

A neural net implementation of geon theory proposed by Hummel and Biederman (1992) is a seven layer network whose architecture is shown in figure 4. The network takes as input a line drawing representing the orientation and depth discontinuities of an object and activates units representing a viewpoint-invariant structural description of the object specifying its geons and their relations. This description is activated regardless of whether the model has previously been exposed to the object.

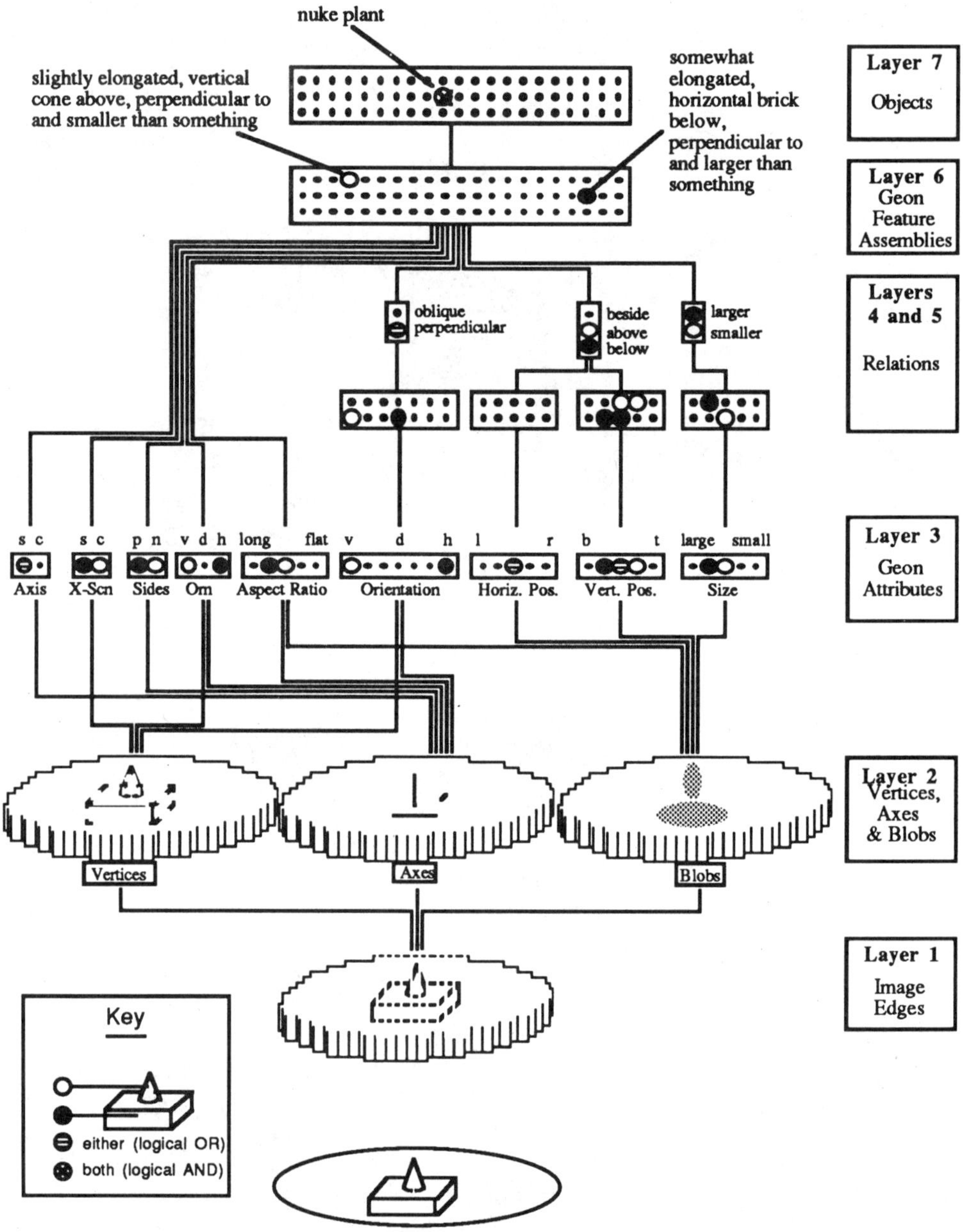

Figure 4. Neural net model for object recognition. From Hummel & Biederman (1992). With permission of the American Psychological Association.

The model's capacity for structural description derives from its solution to the *dynamic binding problem* of neural networks (specifying what goes with what): Independent units representing an object's parts (in terms of their shape attributes and relations) are bound *temporarily* when those attributes occur in conjunction in the system's input. Binding is initially achieved through "fast enabling links" (FELs) that phase lock the oscillatory activity of cells that are tuned to oriented image edges (in the first layer, which is a toy V1). In particular, the FELs cause synchrony in the firing of activated units that are collinear, parallel, or coterminate.

In the third layer (L3) of figure 4, for example, the units marked by filled circles represent the attributes of a brick. These all fire together and out of phase with the units marked by the open circles, representing the cone. (s = straight; c = curved for the Axis and Cross-Section. p = parallel; n = nonparallel for Sides. v,d, and h refer to vertical, diagonal, and horizontal for the Orientation of the axis [fine and coarse]. b = bottom; t = top for Vertical Position.) Only 36 cells in that layer are required to specify the information for each part. Because the binding is temporary, these same cells can be used to code the other parts of the object as well as the parts of other objects, no matter where they are in the visual field. The binding is thus achieved without positing additional units for "anding."

L4 and L5 derive invariant relations (so that the same "above" cell fires in phase to the cone independent of where it is next located).The outputs of L3 that represent the distributed values of a geon and its orientation and aspect ratio, along with the outputs of L5 representing the relations, provide an input vector that self-organizes a unit in L6, termed a *geon feature assembly*. Units in L7 are object cells that self organize to an integration over successive outputs from L6. These operations produce a parts-based structural description that is subsequently used directly as a basis for viewpoint-invariant recognition. The model's recognition performance conforms well to the results from the shape priming experiments. Moreover, the manner in which the model's performance degrades due to accidental synchrony produced by an excess of phase sets suggests a basis for a theory of visual attention.

6.1 Binding via Fast Enabling Links (FELs)

A major contribution of the model is its proposal of a solution to the binding problem--determining what goes with what. Each cell has two kinds of connections to other cells: a) the standard connections that excite or inhibit the firing of a target cell, and b) *fast enabling links* (FELs) that cause (enable) cells that are simultaneously active to fire together if their receptive fields are: a) cocircular (or collinear), b) closely parallel, or c) coterminate. If a cell fires, it passes activation (i.e., excitation and inhibition) in a standard manner and an enabling signal over its FELs. In general, the activation and inhibition will not be to the same units that share FELs. The FELs produce synchronous firing of all the cells that are activated by a given geon while allowing cells activated by different geons to fire out of phase with each other. By not having FELs between the segments comprising a T vertex, as where the sides of the cone in the image in Figure 3 occlude the back edge of the brick on which it rests, the model causes all the features activated by each geon to fire in phase, but out of phase with the features activated by other geons.

7. Geon Structural Descriptions and Clustering

An important distinction between a structural description theory, such as geon theory, and template models is that the former assumes separate units (or networks) for the representation of shape primitives (the geons), geon attributes, and relations. No such distinctions are made in template theories, which do not even assume shape primitives. In a sense the allocation of separate units to distinguish relations from shape primitives renders this distinction explicit, rather than implicit. It is a simple matter in geon theory to know if the relation ON-TOP-OF was activated by a particular pattern. It becomes a significant undertaking to determine whether such a relation is activated with a template model.

The advantages of a structural description model that assumes a high level shape primitive are that such shape primitives are very efficient for accessing a data base of objects (Dickinson, et al, 1993). Moreover, precious few units are required to represent any input. The disadvantages are that high level descriptions are extremely difficult to determine from the input. Part of this problem may be a consequence of the current state of early visual preprocessing algorithms. Few models can derive a clean line drawing of an unknown object, representing only the orientation and depth discontinuities, under most viewing conditions. Yet people can perform this operation with ease. Perhaps a more challenging problem is that many objects do not readily allow a geonic description. These problems are discussed in the next section.

The advantages of most template theories are that they do not require the difficult stage of determining shape primitives, relations, or attributes. For the most part, they perform operations directly off the gray level image. Their potential shortcomings are that they fail to capture the invariances of vision and have difficulty in representing novel objects.

At the core of this issue, when approached as a problem in developmental neurobiology, is the issue addressed by the Rueckl et al. (1989) simulation. Is it the case that shape primitives, attributes, and relations represent "different tasks" so that separate networks are best allocated to them? Or do they represent distinctions that are best left to cognitive operations that follow object recognition, rather than preceding it? It would be interesting to hear the thoughts of clustering mathematicians on this issue.

As an immediate subgoal, however, I wonder if there can be any contribution from the mathematics of clustering to the problem of which relations are specified. In the current neural net implementation of geon theory, relative size (larger-than, small-than, equal-to), position (above, below, side-of) and relative orientation (parallel-to, orthogonal-to, oblique-to) are specified. To these three, the type of join (end-to-end; end-to-middle) can be specified and whether the join is centered (centered; off-centered) for end-to-middle joins. But why these relations and not others? Is there some way of conceptualizing relations from the standpoint of clustering that the degree of their viewpoint invariance is captured?

8. Nongeonic Objects

We need to consider those classes of objects that ultimately achieve recognition, such as those that do not readily decompose into parts or are highly irregular. RBC predicts that these will, in general, be recognized more slowly than their more regular counterparts. Moreover, where the differences are not geonic, viewpoint invariance over depth rotation is lost (Biederman & Gerhardstein, 1993) and it is unlikely that differences in entry level classes in any culture would distinguish among such entities. The difficulty that people have with such objects serve to confirm geon theory.

In some cases, however, recognition does proceed quickly, although any currently implemented geon finder would have a terribly difficult time. Often these are with objects that have a great deal of decoration or detail. Our best guess is that in such instances activation of geons at a coarse scale has occurred, despite the detail. But it remains a challenge as to how to achieve such recognition in an implemented model.

Faces may represent a special case, in which the two-layer networks are more appropriate than an invariant-parts model. A face-recognition system of Buhmann et al. (1991) does an excellent job in recognizing faces, even with changes in expression and modest changes in orientation. The neural net implementation of geon theory would fail at this--it would know that something was a face but not whose face it was. But Fiser, Biederman, & Cooper (1993) showed that the Buhmann et al system was insensitive to the effects of potent psychophysical variables in entry-level object recognition. For example, it recognized nonrecoverable as well as recoverable images but recognized a stored member of a complementary pair (as in figure 2) better than the complement. People cannot recognize nonrecoverable images and their recognition of a complement is as good as the originally presented image of a complementary pair.

Evidence that face recognition may require a different solution can be seen in the near impossibility of recognition in the presence contrast reversal (as with a photographic negative) or rotation in the plane. Entry-level object recognition is hardly affected by the former and only modestly by the latter. Perhaps it is not surprising that a neurological condition exists, prosopagnosia, which is results in impairment of face recognition but not entry-level object recognition.

REFERENCES

Biederman, I. (1987). Recognition-by-components: A theory of human image understanding. *Psychological Review*, 94, 115-147.

Biederman, I. (1988). Aspects and Extensions of a Theory of Human Image Understanding. In Z. Pylyshyn, (Ed.) *Computational Processes in Human Vision*. Pp. 370-428. New York: Ablex.

Biederman, I., & Cooper, E. E. (1991a). Priming contour-deleted images: Evidence for intermediate representations in visual object recognition. *Cognitive Psychology*, 23, 393-419.

Biederman, I., & Cooper, E. E. (1991b). Evidence for complete translational and reflectional invariance in visual object priming. *Perception*, 20, 585-593.

Biederman, I., & Cooper, E. E. (1992). Size invariance in visual object priming. *Journal of Experimental Psychology: Human Perception and Performance*, 18, 121-133.

Biederman, I., & Gerhardstein, P. C. (1993). Recognizing depth-rotated objects: Evidence for 3D viewpoint invariance. *Journal of Experimental Psychology: Human Perception and Performance*, 19, 1162-1182.

Buhmann, J., Lange, J., von der Malsburg, C., Vorbrüggen, J. C., & Würtz, R. P. (1991). Object recognition in the dynamic link architecture: Parallel implementation of a transputer network. In B. Kosko (Eds.), *Neural Networks for Signal Processing*. (pp. 121--159). Englewood Cliffs, NJ: Prentice Hall.

Cooper, E. E., & Biederman, I. (1993). Metric versus viewpoint-invariant shape differences in visual object recognition. Poster presented at the Annual Meeting of The Association for Research in Vision and Ophthalmology, Sarasota, Fl. May.

Cooper, E. E., Biederman, I., & Hummel, J. E. (1992). Metric invariance in object recognition: A review and further evidence. *Canadian Journal of Psychology*, 46, 191-214.

Dickinson, S. J., Pentland, A. P., & Rosenfeld, A. (1992). From volumes to views: An approach to 3-D object recognition. *Computer Vision, Graphics, and Image Processing: Image Understanding*, 55, 130-154.

Fiser, J., Biederman, I., & Cooper, E. E. (1993). To what extent can matching algorithms based on direct outputs of spatial filters account for human shape recognition? Submitted for publication.

Hummel, J. E., & Biederman, I. (1992). Dynamic binding in a neural network for shape recognition. Psychological Review, 99, 480-517.

Koenderink, J. J., & van Doorn, A. J. (1979). The internal representation of solid shape with respect to vision. *Biological Cybernetics*, 32, 211-216.

Lowe, D. G. (1987). The viewpoint consistency constraint. *International Journal of Computer Vision*, 1, 57-72.

O'Reilly, R. C., Kosslyn, S. F., Marsolek, C. J. (1990). Receptive field characteristics that allow parietal love neurons t encode spatial properties of visual input: A computational analysis. *Journal of Cognitive Neuroscience,* 2, 141-155.

Poggio, T., & Edelman, S. (1990). A network that learns to recognize 3D objects. *Nature*, 343, 263-266.

Rueckl, J. G., Cave, K. R., & Kosslyn, S. M. (1989). Why are "what" and "where" processed by separate cortical visual systems? A computational investigation. *Journal of Cognitive Neuroscience*, 1, 171-186.

Ullman, S. (1989). Aligning pictorial descriptions: An approach to object recognition. *Cognition*, **32**, 193-254.

von der Malsburg, C. (1987). Synaptic plasticity as a basis of brain organization. In J. P. Chaneaux & M. Konishi (Eds.), *The Neural and Molecular Bases of Learning* (pp. 4111-432). John Wiley & Sons Limited

Department of Psychology, University of Southern California
Hedco Neuroscience Bldg., University Park, CA 90089-2850, USA
E-mail address: ib@rana.usc.edu.

DIMACS Series in Discrete Mathematics
and Theoretical Computer Science
Volume 19, 1995

PERCEPTUAL MODELS OF SMALL DOT CLUSTERS

JACOB FELDMAN

ABSTRACT. This paper investigates how human observers draw perceptually coherent clusters out of fields of dots, a process analogous to but different from statistical clustering. We focus on the way perceptually minimal ("molecular") clusters—dot triplets—are categorized by human observers as having arisen from either a curvilinear generating process (a chain) or an unordered random process (a cluster). In order to draw perceptual decompositions from larger fields of dots, observers must attach class-conditional probability densities to these minimal clusters; that is, they must have a model of just how straight is straight "enough" for a dot triplet to be reasonably classified as a chain rather than a cluster. We first consider the normative statistical facts concerning how three completely random dots can be expected to be distributed. Human observers lack this knowledge, however, and must use internal models of each generating process to construct prior distributions for the shape of both random and "ordered" triplets. Here, we work out a theory of such models that enables the observer to construct usable (though not actually "correct") class densities. In accordance with the theory, human subjects interpret a configuration of three equidistant dots (counter-normatively) as the most typical ("generic") configuration of three randomly positioned dots—apparently because this case is maximally distant in the dot triplet space from the non-generic case of three collinear dots. Subjects' category judgments precisely resemble the orthodox Bayesian likelihood inference they would draw if their implicit probabilistic models took the regular categorical form described by the theory. In this sense, subjects' prior probabilistic beliefs, and hence their qualitative interpretations of observed dot patterns, are built around *regularities* that they tacitly believe exist in the dots world.

1. PERCEPTUAL CLUSTERING OF DOTS

The parallel between clustering a set of points in a data space by statistical techniques (on the one hand) and decomposing a field of dots in the visual field by direction inspection (on the other) is an appealing one—and yet the two process are peculiarly different. Consider, for example, Fig. 1 (below). To a human observer (such as the reader), it is obvious that there are two natural groups or clusters of dots. This inference comes with no conscious effort or thought, and seems both stable and compelling. Notoriously, however, statistical techniques find this kind of inference a difficult one. Commonly, the number and general form of clusters

1991 *Mathematics Subject Classification.* Primary 92J30

I am grateful to Whitman Richards and Allan Jepson for many helpful discussions of the ideas in this paper, and to an anonymous reviewer for comments on an earlier draft; and to Janice Taylor for running the subjects and coding their responses. A final version of this paper will be submitted for publication elsewhere. This research was supported by the Rutgers University Center for Cognitive Science.

must be supplied to the algorithm (see [4] or [15] for introductory surveys). Though techniques exist to cope with clusters both of unknown number (e.g. hierarchical techniques) and of unknown form (e.g. nonparametric techniques such as that of [13], their behavior differs from that of human observers in intriguing and systematic ways. Most techniques have particular trouble, for instance, distinguishing among clusters of different apparent dimensionality; or, putting it slightly differently, in determining the exact dimensionality at which a given subset of the data is best modeled (a problem so pervasive it is often termed "the curse of dimensionality"). A normative definition of the "correct" dimension for a given set of data is elusive, if not impossible—and consequently, so is a definition of the "correct", or even the "most reasonable" decomposition of a given data set. Human observers, however, have strong intuitions on the question. This paper attempts to characterize these intuitions formally, focusing on a critical sub-problem: the classification of minimal groups consisting of three dots.

The three dots case can be seen as the minimal one in the sense that three dots constitute the smallest dot group that has *shape* (cf [10]):that is, structure after the similarity groups (translation, orientation, and scale) have been removed. Hence three dots form the smallest cluster that can exhibit any tendency to cohere due to shape—in particular, as we shall focus on in this paper, due to alignment or *collinearity.* This perceptual tendency, which will be investigated empirically as well as theoretically below, can be seen as a minimal kind of categorical inference, in that (as shall be argued below) three nearly-collinear dots appear to be classified as having been produced by a qualitatively different kind of generating process than three far-from-collinear dots (that is, a curvilinear one rather than an unordered random one). In a more abstract sense, we will propose, three collinear dots, but not three unordered dots, are sensibly interpreted as having been generated by a *causally coherent* process: what Jepson and Richards [9] (cf [7]) call a "regularity".

Describing the human inference in this domain turns turns critically on the exact nature of human perceivers' *prior probabilities*—that is, their primitive probabilistic assumptions about exactly how likely various configurations of dots are to occur in the dots world. (Here by "priors" we mean both the prior probability of a each class of dot configurations occurring in the world, as well as the class-conditional densities that dictate how likely each configuration is within each class.) We start by assuming that subjects assign different prior probabilities to different dot triplets, the minimal case; and then draw clustering inferences by combining decisions about triplets. The bulk of this paper is occupied with asking what the human intuitive priors for dot triplets are, and why. At the end we return only briefly to the question of how the decisions about triplets are combined to form overall grouping interpretations, deferring a more complete theory of this overall inference to a longer future paper. For the three dots case, it will turn out that the human subjects draw inferences in a recognizably Bayesian (though unorthodox) way. It will be argue that the exact assumptions prior they make in order to do so, as suggested above, reflect an attempt to recover regularities—the observable evidence of underlying causal forces—in the dots world.

The central issue of the paper, then, concerns human observers' tacit beliefs about the probability of various configurations of dots in the plane, in the absence of any explicit domain-specific knowledge. To provide some kind of a standard against which these mental constructs can be compared, we first present some quasi-

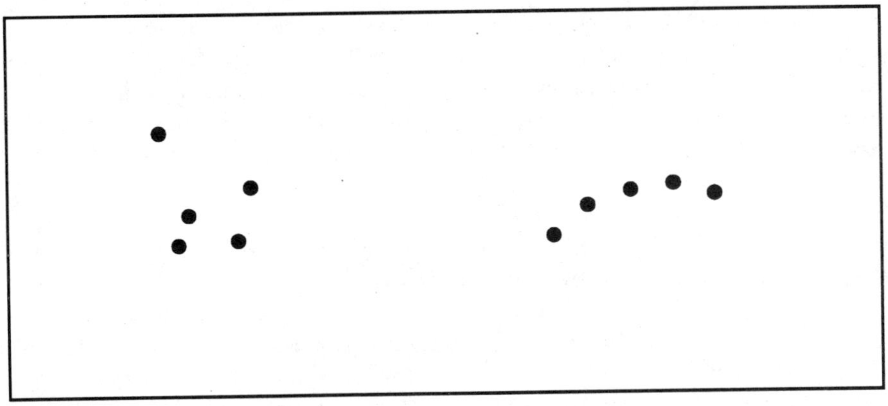

FIGURE 1. Two clusters, of different perceptual "types".

normative facts about dot densities due to the statisticians Kendall and Kendall [10] These densities, of which subjects are presumably completely unaware (they are in fact counterintuitive in several respects) serve as kind of neutral backdrop for observers' naive hypotheses. Section 2.2 lays out these basic statistical "facts." In grouping dots, as will be discussed below, observers need values need to know how just how typical the observed configuration is of each of several hypotheses, in order to judge whether a given grouping hypothesis explains the observation to a substantial degree. But since the true normative priors are unavailable to them, they are forced to substitute "reasonable" guesses as to how just how the configurations generated by each grouping hypothesis are distributed. Just how these imaginary distributions are constructed will be the topic of the next section.

Preview of the theory. We will first propose that minimal dot groups (triplets of dots) are divided into two qualitatively distinguishable categories. Then, each category will have a separate "modal" class-dependent density, centered on a "most typical" case. (We call such distributions "modal" because the density peaks correspond to qualitative types.) Within each mode, the highest (most typical) point is stipulated (heuristically and non-normatively) to occur at a point that is maximally distant, and thus maximally distinguishable, from the most typical point adjacent modes. The observer thus implicitly imputes a certain convenient, regular, structure to the world—a structure that is highly speculative and not necessarily "true," but which nevertheless aids the observer in drawing robust and useful inferences about regularities in the world (like tendencies of dots to cohere to one another in particular ways). Thus the clustering scheme described here can be thought of as depending on a kind of "kind world" assumption (cf [1]).

The next section begins by situating the three dots case, and the detection of collinearity in fields of dots, in both perceptual psychology as well as in a clustering framework.

1.1. Two kinds of dot clusters. Going back to Fig. 1, notice that the two clusters in the figure seem to be of two perceptually distinct types: the cluster on the left seems to scatter randomly in two dimensions, while the cluster on the right seems to be an ordered chain of dots falling along some sort of curvilinear path. The Gestalt psychologists first observed the two apparently distinct grouping principles at work here, calling them the principle of *proximity* and *good continuation*, respectively. Gestalt grouping principles have more of a descriptive than an explanatory flavor, however, in that they seem not to provide any justification or reasoning under which these types of grouping are superior to any other. Moreover Gestalt principles provide no prediction of the exact magnitude of the tendency for dots to cohere in various ways.

The extraction of curvilinear structure from fields of dots, in particular Glass patterns, has been approached from a number of directions since. Some (e.g. [18]) have modeled it using low-level orientation-tuned operators similar to the type believed to operate in visual cortex. On this view the perceptual salience of a group of nearly collinear dots depends on its literal resemblance (shared low spatial frequency component) to an oriented segment, the ideal form preferred by some cortical cells. Other work [16] has asked how curvilinear groups of dots can be efficiently computed, arguing that perceptual phenomena can be seen as emerging from the heuristics the visual system implicitly employs. Witkin and Tenenbaum [17] have argued that this sort of low-level structure and regularity in the image serves as an underpinning for a "meaningful" representation of causal forces at work in the world, a view more in sympathy with the model developed below.

There has generally not been a motivated way to predict the exact magnitude of the off-angle response—the function governing how quickly the cohesiveness of dot clusters degrade as they range from perfect collinearity to extreme non-collinearity. The analysis below will provide exact (and apparently empirically correct) predictions for the magnitude of the off-angle response, which emerge essentially from an analysis of the structure of the space of grouping hypotheses.

The goal, then, is to construct a concrete model of the inference scheme that underlies the grouping reflex. Ideally, such a scheme would attach to each configuration of dots an exact numerical estimate of the probability that the configuration merits a particular category label—that is, that it was produced by a certain type of dot-generating process.

First of all, we assume that exactly two kinds of dot-generating process inhabit the dots world: 2-D patches in which dots can be generated uniformly (like a Poisson process); and 1-D curves along which dots are generated at random intervals of arclength[1] (see Fig. 4). The curvilinear process may be thought of as a kind of "regularity" in this domain (see [6] for further discussion of this notion), while the 2-D patch process may be thought of as a region of non-regularity, i.e. a region in which dot locations are all generated independently of one another, rather than being causally co-related. These two types of probabilistic process underlie all the discussion below, though the two types will be represented in several other ways, all of which are in the cause of understanding what observers perceive as "typical" outcomes of each generating process.

We denote individual dots in the plane by $\mathbf{x} \in R^2$, and an observed set of N dots $\{\mathbf{x}_1 \ldots \mathbf{x}_N\}$ by $\mathbf{X}^N$, omitting the N whenever unambiguous. Thus to each set of dots $\mathbf{X}$ we would like to attach two inferred quantities: the probability of a "chain" category, $p(C_c|\mathbf{X})$, and the probability of a "random cluster" category, $p(C_r|\mathbf{X})$.

[1] The fact that we restrict our attention in this paper to small dot clusters (triplets) relieves us of worrying about global properties of these two generating patches, such as the border of the 2-D patches and the connectivity and complexity of the 1-D curves.

FIGURE 2. "Typical" examples of (a) a 3-CHAIN and (b) a 3-CLUSTER

Notice that it is only because of our assumptions about the dots world—that it contains two types of regular process, the dot-producing curve and the internally unstructured cluster—that these posterior probabilities represent meaningful inferences about something going on in the world.

1.2. Molecular clusters. In order to construct these functions we begin with the smallest observation from which anything about curvilinear processes can be inferred. We would like to ignore the effects on $p(C_i|\mathbf{X})$ of translation, rotation, and scale of $\mathbf{X}$. Since all pairs of points are equivalent under these operations, we take triplets as a minimal set. Since, in this sense, a triplet is the smallest set of points that can still meaningfully exhibit "chain"-ness or "cluster"-ness, we call it a "molecular" chain or cluster respectively. A larger chain would then be made up of consecutive interlocking 3-CHAINs, e.g. $\{\mathbf{x}_1, \mathbf{x}_2, \mathbf{x}_3\}$, $\{\mathbf{x}_2, \mathbf{x}_3, \mathbf{x}_4\}$, etc. Note that for the moment postpone choosing a convenient parameterization for each triplet—such as by its main angle and the lengths of its two virtual legs—until one can be motivated more carefully in Sec. 2.1.

A minimal chain, which we will refer to as a "3-CHAIN", is a set of three dots generated by the curvilinear process; a minimal cluster (a 3-CLUSTER) is a set of three dots produced by the patch process. We assume that these two processes occur in the world with probabilities p(3-CHAIN) and p(3-CLUSTER) respectively.

Naturally we do not expect the two generating processes to produce identical-looking dot triplets. Rather, we expect 3-CHAINs to "typically" be straighter than 3-CLUSTERs. "Typical" examples are depicted in Fig. 2). To be more precise, for any three dots, we pick the straightest of the three possible orderings of the dots; and then measure how far even this straightest ordering is from perfect collinearity. We then expect that the distribution of this angle among 3-CHAINs would be centered relatively near zero (collinearity), and that of 3-CLUSTERs relatively far from zero. (The parameterization we will introduce later is designed to help make idea explicit.) The two kinds of generating process and their expected outcomes are depicted schematically in Fig. 3.

For most of the remainder of the paper we will be concerned with the two quantities p(3-CLUSTER$|\mathbf{X}$) and p(3-CHAIN$|\mathbf{X}$). We assume that these two interpretations are the only two categories into which a triplet of dots can be put: this assumption will be spelled out in more detail and defended below. Intuitively, the question then is: in order for a triplet of dots to be considered a 3-CHAIN, how straight is straight enough; and why?

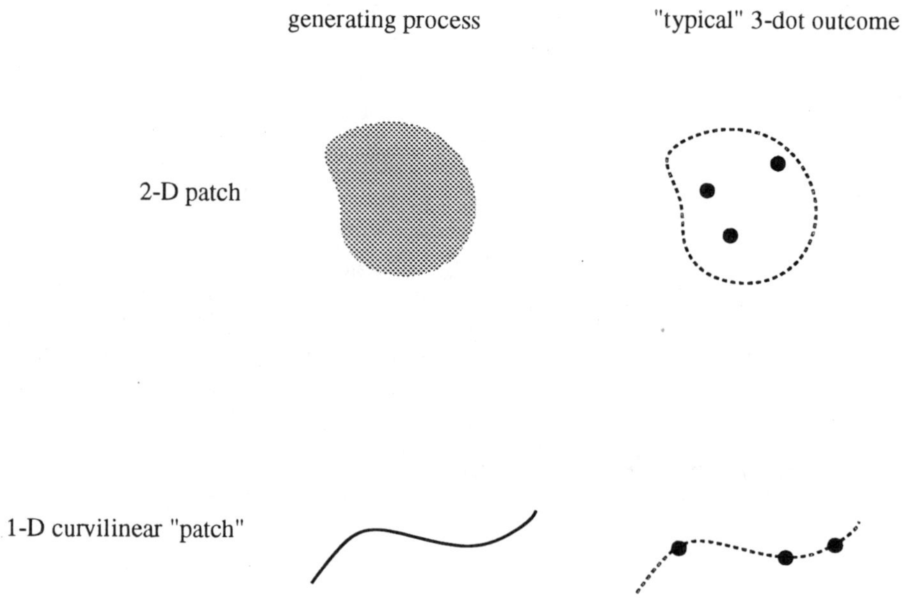

FIGURE 3. A schematic depiction of the two generating processes or "regularities" under consideration, showing the form of the underlying generating process, and a "typical" outcome.

Notice that we are implicitly defining a 3-CLUSTER only by default with respect to a 3-CHAIN: a 3-CLUSTER is a triplet of dots *not* straight enough to be considered a 3-CHAIN. That is, $p(\textsc{3-cluster}|\mathbf{X}) = 1 - p(\textsc{3-chain}|\mathbf{X})$ by stipulation. The structure in the observer's category decision will come instead from the two class densities $p(\mathbf{X}|\textsc{3-cluster})$ and $p(\mathbf{X}|\textsc{3-chain})$. Various theories about the proper structure of these densities will be take up in detail in the next section. Once priors have been determined, and assuming the two categories themselves are mutually exclusive[2] and jointly exhaustive, Bayes dictates that the posterior probability of the 3-chain category should be calculated as

$$(1.1) \qquad p(\textsc{3-chain}|\mathbf{X}) = \frac{p(\textsc{3-chain})p(\mathbf{X}|\textsc{3-chain})}{p(\mathbf{X})}$$

where the denominator $p(\mathbf{X})$ is composed of contributions from the two generating processes:

$$(1.2) \qquad p(\mathbf{X}) = p(\textsc{3-chain})p(\mathbf{X}|\textsc{3-chain}) + p(\textsc{3-cluster})p(\mathbf{X}|\textsc{3-cluster})$$

where again $p(\textsc{3-chain})$ and $p(\textsc{3-cluster})$ are the prior probabilities with which these two categories occur in the world. Whether human observers draw their category judgments in this normative manner, though, and what values they give

[2] Care must taken on this point, since the two distributions will in fact overlap. What we mean by "mutually exclusive" is that both category labels cannot be applied *correctly* to the same observed triplet.

to the two scalar priors, are largely empirical questions, which will be taken up as such in Section 3.

2. Statistical and perceptual models of dot-producing processes

In probabilistic theory, and by extension in the theory of clustering, it is natural to think in terms of a "generating process," some agent that gives rise to a particular set of observed outcome events with particular probabilities. Such an idea has, perhaps peculiarly, been slower to gain prominence in the theory of perceptual inference. Recently, however, Leyton ([11],[12]) has drawn attention to the central role of generative models in perceptual description, in particular in the theory of shape representation. Each shape, in Leyton's model, is conceptualized as having been produced by the deformation of some simpler shape. The resulting representation of a complex object is as a nested series of generative operations from which the object is interpreted as having been generated. This kind of representation thus corresponds to a kind of "causal history" by which the observed object is interpreted as having been generated. This process imparts to the interpretation more of a quality of explanation rather than mere description (see also [17]).

In earlier work ([6],[7]) it was observed that the hierarchical organization of such generative processes can be usefully captured by a certain discrete structure called the *category lattice* (see also [8] and [9]). In such a lattice, each node represents a class of objects in the overall space. Moving down the lattice, one moves from general and inclusive classes to more structurally constrained and specific classes. The generative history of an object corresponds, in this scheme, to the exact route by which the object's history has transformed it from the completely structurally fixed class at the bottom of the lattice (the simple primitive object). A critical idea in this scheme is the idea of *genericity*: loosely, that each observed object should be associated with a node on the lattice in which it is generic, that is in which all the generative operations in its description have been carried out to some non-zero degree. Our use of this scheme in representing the two categories 3-CHAIN and 3-CLUSTER will be relatively simple and self-contained, but it is worth noting that it is consistent with this more fully articulated and complex scheme which covers a wider range of perceptual categories in other types of object spaces.

A few remarks are in order to place the category apparatus into the context of existing perceptual theory. The lattice scheme is predicated on the search for "regularities" in the observed world, a world which exhibits more consistently regular structure than is in principle necessary (see also [14]). An observer detects this structure in order to help recover the causal forces at work in the environment, and thereby to infer reliable and useful facts about the world. In [6] the notion of regularity was cast, broadly and abstractly, as a lower-dimensional manifold through some higher-dimensional embedding space of objects. In the dots world this notion reappears concretely as our 1-D curvilinear process. As will be detailed below, this curvilinear process is our model of how chain clusters come about; and, critically, is the process whose presence in the environment means that the inference of a "chain" is actually correct. Thus while the theory below is intended to have direct application to the problem of psychological clustering, it may also be seen as an attempt to focus probabilistic machinery on concepts of regularity and categorization that were originally conceived to apply to a wider range of perceptual inferences.

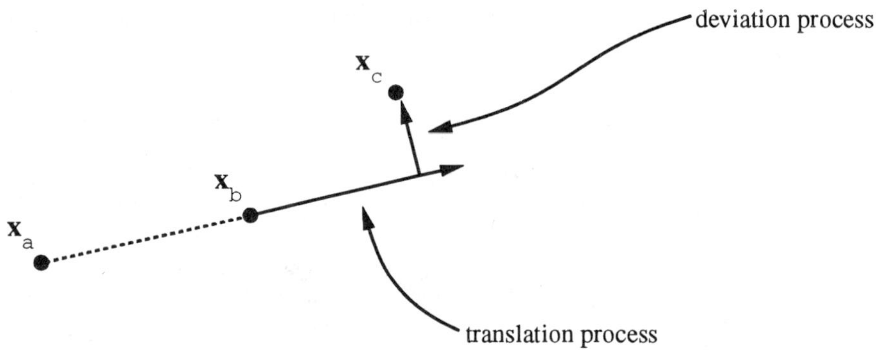

FIGURE 4. The "generative model" for dot triplets. A "translation process" carries $\mathbf{x}_c$ out from $\mathbf{x}_b$, along the line joining $\mathbf{x}_a$ to $\mathbf{x}_b$. A transverse "deviation" process carries $\mathbf{x}_c$ away from this line.

For our purposes here, the category structure and genericity notions are necessary to provide an underpinning for a more focused set of probabilistic questions: questions concerning the class-conditional densities associated with each of the modal categories, and about the decision procedure by which human observers place a set of observed dots into one of the categories.

2.1. A generative model of dot triplets. Following [7], we now construct a generative model for dot triplets, which will serve to determine a parameterization for dot triplets. This in turn will determine a category lattice formally expressing the admittedly simple relationship between the categories 3-CLUSTER and 3-CHAIN.

We begin with a reference dot, $\mathbf{x}_b$ (Fig. 4). We imagine a "translation process" that positions a dot $\mathbf{x}_c$ along some translation vector $\vec{t}$ away from $\mathbf{x}_b$, $\mathbf{x}_c = \mathbf{x}_b + \beta_1\vec{t}$ for some scalar β_1. A third dot, $\mathbf{x}_a$, is required to determine a reference direction for $\vec{t}$, i.e. $\vec{t} \sim \mathbf{x}_b - \mathbf{x}_a$. Thus we can think of an entire space of dot triplets that are spanned by the this translation process alone, consisting of the exactly collinear triplets in which $\mathbf{x}_c - \mathbf{x}_b = \beta_1\vec{t}$.

To express dot triplets outside this special subspace, we need a second process, which we call the "deviation" process $\vec{d}$; this process carries the third dot in a direction normal to the tangent $\mathbf{x}_a\mathbf{x}_b$ (so that $\vec{d} \perp \vec{t}$). Critically, all dot triplets can be expressed as a weighted sum of these two processes, i.e. any triplet can be written as $\{\mathbf{x}_a, \mathbf{x}_b, \mathbf{x}_b + \beta_1\vec{t} + \beta_2\vec{d}\}$ for some scalars β_1 and β_2. The two operations are of course isomorphic to the familiar tangent and curvature vectors of a curve respectively, except reconceived constructively as producing discrete dots at intervals of non-vanishing arclength.[3]

In combination, these two processes can produce any triplet. Triplets produced by the combined operation of some non-zero amount of *both* operations ($\beta_1 \neq 0$, $\beta_2 \neq 0$) are called *generic* in the space defined by the two operations, i.e. in the triplet space. Triplets produced by the translation process *only* are by definition non-generic in the overall space, but can be thought of as generic in a subspace.

[3]Readers may also note the relevance of the "fundamental theorem of curves" (see [3]) which tells us that in a certain sense these two vectors specify all there is to know about the shape of the underlying curvilinear process near the point $\mathbf{x}_b$.

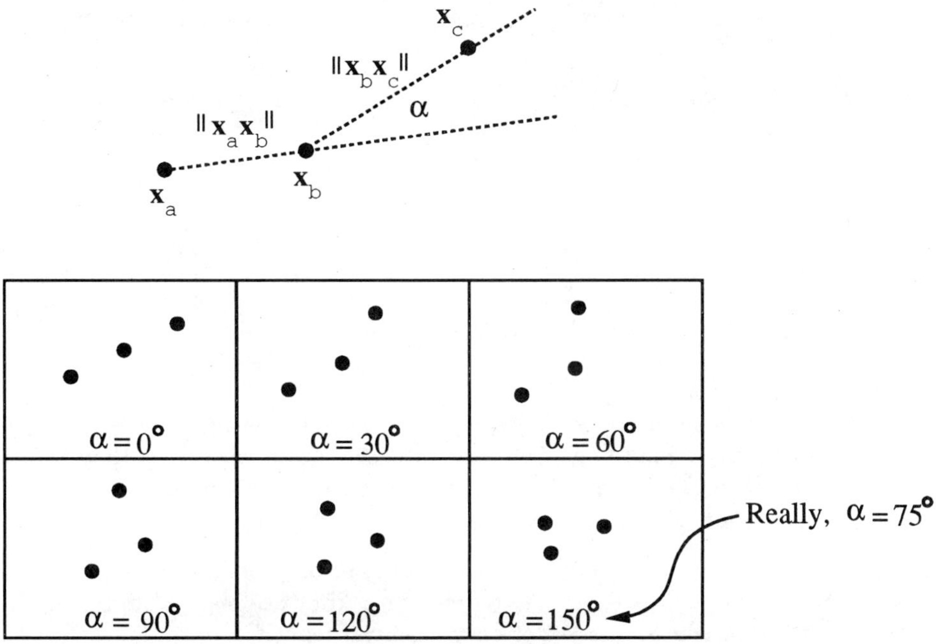

FIGURE 5. We can parameterize any triplet, up to similarity, by α (the angular deviation from straight) and $\frac{\|\mathbf{x}_b\mathbf{x}_c\|}{\|\mathbf{x}_a\mathbf{x}_b\|}$, the ratio of the length of the second "leg" to the length of the first leg. Below, examples are given of dot triplets with α at various values (and $\frac{\|\mathbf{x}_b\mathbf{x}_c\|}{\|\mathbf{x}_a\mathbf{x}_b\|} = 1$). Notice that α cannot exceed 120°, since it is always measured as the angle nearest to straight (illustrated in the bottom right example).

The space spanned by the two processes, i.e. the set of all dot triplets, can of course be conveniently parameterized by magnitudes to which the two processes were carried out, β_1 and β_2, for which we substitute the more meaningful (but diffeomorphic) angle α and line-length ratio $(\frac{\|\mathbf{x}_b\mathbf{x}_c\|}{\|\mathbf{x}_a\mathbf{x}_b\|})$. This natural parameterization of the space by leg-length ratio $(\sim \beta_1)$ and angle $\alpha(\sim \beta_2)$ is depicted in Fig. 5. The idea is that this parameterization is not simply an arbitrary choice, but rather corresponds straightforwardly to the particular generative model of triplets we have adopted.

For the remainder of the paper we will restrict attention to the angle α, the parameter on which the regularities turn out to be defined; the line-length ratio is devoid of special values and hence of special subcategories. Therefore we will usually write $p(3\text{-CHAIN}|\alpha)$ rather than $p(3\text{-CHAIN}|\mathbf{X}^3)$, etc.

The close relationship between the chosen parameterization and the chosen generative model provides a clean formalism for distinguishing among subcategories of triplets that have been produced by qualitatively different generative histories—namely, either by the full model $\vec{t} + \vec{d}$ (i.e., generic triplets) or by the submodel $\vec{t}$, at which the angle α is 0 and the deviation process completely drops out. At this point, the curvature of the inferred underlying generating process vanishes. This

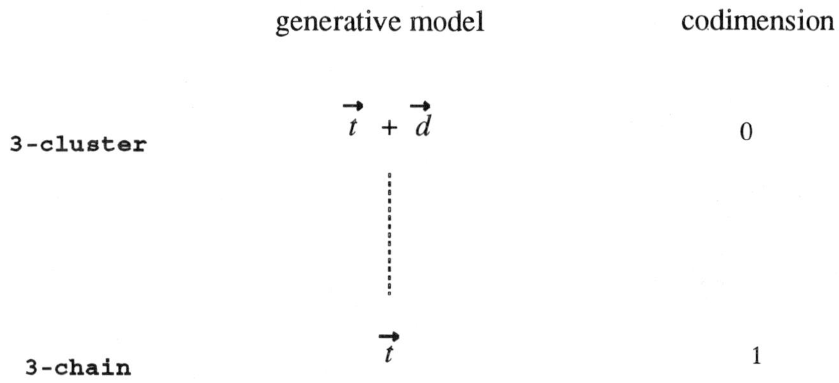

FIGURE 6. The category lattice depicting the relationship between 3-CHAIN and 3-CLUSTER. For each category, the lattice gives the symbolic representation of its generative model, making explicit the dimensional relationships among distinct categories.

nongeneric subspace of the overall space, containing only collinear triplets, is of course a very special subcategory. It represents a kind of idealized form for the output of the curvilinear process—not only a central tendency but a kind of "best case", a case from which detection of the curvilinear process is maximally (though not perfectly) reliable (see [5] for some subtleties on this point).

The category lattice depicting this structure is shown in Fig. 6. The parametric collapse from $\vec{t}+\vec{d}$ to $\vec{t}$ (we can omit the weights because we are not distinguishing among generic members of the same category) is realized as a transition from the *generic* dot triplet category at the top of the lattice to the lone special subcategory at the bottom. This category transition carries a change in dimension from the 2-D space spanned by both parameters to the 1-D subspace spanned by the translation parameter only. (The change in dimension from 2 to 1 is more conveniently expressed as change in codimension[4] from 0 to 1, for reasons outside the scope of the current discussion.)

Now that we have a model of how members of each category 3-CLUSTER and 3-chain are generated, we can ask what "typical" products of each generative process would look like. This is a question to be addressed by statistical techniques. That is, given a randomly chosen member of the generic triplet category, how straight do we expect the angle between the dots to be; and what is the distribution of angle over many such randomly chosen triplets? (Equivalently, how much of the "deviation" process do we typically expect to have been applied in the construction of a random generic triplet?) The true, statistical "fact of the matter" can be seen as a kind of normative backdrop against which various perceptual and cognitive beliefs about the same question can be compared. As it happens, this question was answered for the generic category by Kendall and Kendall [10], whose results are summarized in the next section. (Section 2.2 also discusses the corresponding statistical "facts" for the curvilinear case (codimension 1 case).) Before looking at the normative question ("what do random members of each generative category look

[4] Codimension is the difference in dimension between a given category and the overall embedding space (the generic category).

like?"), though, we start with psychological question ("what do subjects think they look like?"). To answer this question for the generic (codim-0) category, we begin by asking where in dot-triplet space would this generic category be centered: that is, where in the observer's mental construction of this space would the maximally generic point be located.

2.1.1. *The maximally generic point in triplet space.* In [7], evidence was presented (in a different domain) that subject construct prior probability distributions—that is, they create their own beliefs about the statistics of the world— in such a way that there are peaks (modes) at each distinct category in their model of it. In the dot triplets domain, this would mean simply that there would be two peaks in their mental density function: one at the nongeneric (collinear) subcategory, i.e. at $\alpha = 0$, and one at a point construed to be *maximally* generic. Putatively, following [7], maximally generic simply means maximally *distant* from nearby non-generic points. In the dot-triplet domain, the most non-generic angle is simply the angle most distant from the non-generic point $\alpha = 0°$: namely 120°, the largest value the angle can take, in accordance with the symmetry argument presented above. Though it is easy to see that 120° is the angle most distant from straight, it is worth expanding on this point a bit in order to clarify the internal structure of the triplet space.

Consider the space of dot triplets as itself a geometric object. We can display its structure in a convenient form by the contrivance of placing the first and second at dots, $\mathbf{x}_a$ and $\mathbf{x}_b$ at points (0,-1) and (0,0), respectively, so that the position of the third dot $\mathbf{x}_c$ is parameterized by the ordinary Cartesian coordinates of the space (Fig. 7). In order to focus only on angle and ignore leg-length ratio, consider the part of this space containing just dot triplets with equal interdot distances, $\frac{||\mathbf{x}_b\mathbf{x}_c||}{||\mathbf{x}_a\mathbf{x}_b||} = 1$. This part of the space falls naturally on the circle $||\mathbf{x}_c||^2 = 1$ (shown in the figure).

Now consider how our requirement that we always measure the straightest angle in any dot triplet imposes a kind of *symmetry structure* on this circular space. Imagine moving the dot $\mathbf{x}_c$ along the periphery of this circle starting from the right most point. In this region, labeled **A** in the figure, the angle of the corresponding dot triplet changes from 0° at the rightmost point to 90° at the topmost point, etc. If one were to continue past the point $(\frac{-1}{2}, \frac{\sqrt{3}}{2})$ (where $\mathbf{x}_c$ is actually drawn in the figure), though, one would begin to form triplets whose straightest vertex was not equidistant from the other two dots. So instead we can continue moving $\mathbf{x}_c$ straight *down*, along the segment marked **B**, until we meet the x-axis. Notice that in segment **B** we encounter each and every triplet from **A** again, only scaled, rotated and translated. From the point of view of shape alone, that is, the two sections **A** and **B** contain exactly the same triplets. In a similar though less subtle way, the parts labeled **A'** and **B'**, mirror images of **A** and **B** respectively, also contain exactly the same set of triplets, except reflected. Thus one can move continuously from the rightmost point on the circle all the way around and back to the rightmost point, encountering each unique triplet exactly four times.

In order to take "shape" only, then, we can map a single one of these iterations (the figure indicates section **A**, but the choice is an arbitrary one) to a linear space containing each triplet just once (shown on the right in the figure). This space can now be taken to represent the triplet space in its most intrinsic form. Notice

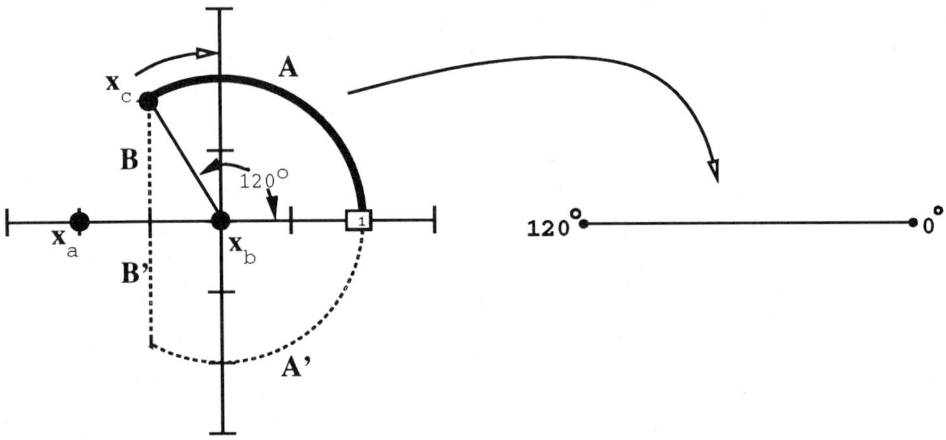

FIGURE 7. Schematic depiction of the space of dot triplets, making clear how $\alpha = 0$ and $\alpha = 120°$ are maximally distant points. The space of equal-leg-length dot triplets can be depicted (left) as a part of a circle embedded in the 2-D space of general dot triplets; the highlighted part (labeled A in the figure) contains each distinct triplet exactly once. This space can be rewritten (right) as a linear space containing two definite endpoints, 0° and 120°. These endpoints, which are maximally distant from each other in a metric-invariant sense, become the prototypes of the two opposed categories: one the non-generic point, the other the maximally generic point.

that it has two definite endpoints: 0° and 120°, respectively. Notice further that these two points are maximally distant from each other in a sense that is completely insensitive to the choice of metric. Now since 0° has been identified as a non-generic point in the space, we can confidently identify 120°, the equilateral triplet, as the maximally non-generic point.

Intuitively, this point in the space is the triplet that is (psychologically) the least straight, in the sense of being maximally remote from the nearest possible attribution of straightness. That is, the ordering of the three dots is as ambiguous as possible, and the triplet's probability of assignment to the category 3-CHAIN ought to be as low as possible. These two maximally remote cases thus become the ideals or prototypes of the two qualitatively opposed generating processes. The idea, then, is that human observers would construct their probabilistic beliefs so that the complementary category, 3-CLUSTER, would be maximally probable at this point, so that the two categories could be optimally distinguished. Section 3 will present empirical evidence that subjects do in fact construct their class density functions in exactly this manner.

2.2. Statistical models of dot triplets. We now consider ways in which probability distributions for first 3-CLUSTER and then 3-CHAIN may be obtained. For each type of molecular cluster, there is a quasi-normative distribution due to [10], which we present first. Then we consider alternative distributions constructed around the category theory presented in the previous section. The latter probabilities, which

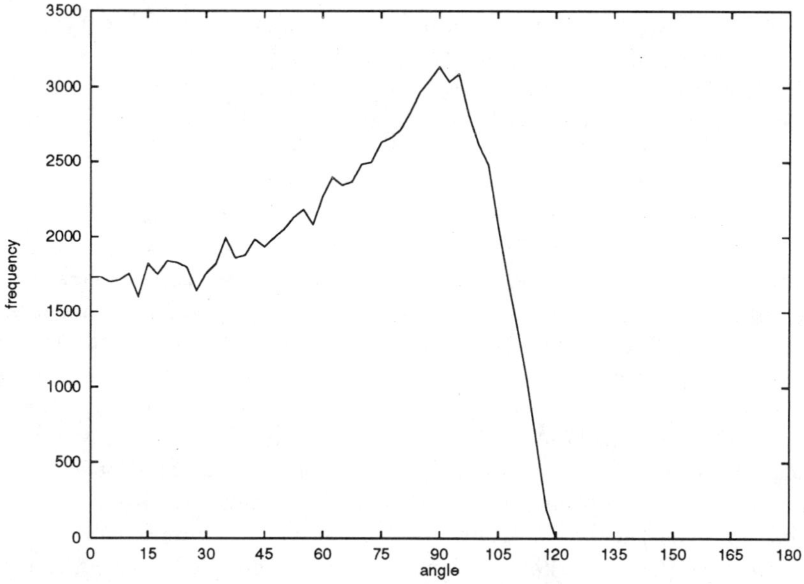

FIGURE 8. Simulation of the distribution (derived in [10]) of the straightest angle of random dot triplets. Here we generated 100,000 triplets (drawn from a Gaussian distribution), and measured their smallest deviation from straight (0°). (Thus this value is naturally bounded by 0° and 120°.) The theoretical peak is at 90°.

the experiments in Section 3 suggest subjects actually adopt, seem to reflect their tacit beliefs about the regularities extant in the dots world.

2.2.1. *Random clusters of three dots.* Kendall and Kendall [10] (see also [2] considered the question: when triangles are generated at random (i.e. with dots drawn from a 2-D Gaussian distribution in the plane), what distribution of largest internal angle should we expect? This is equivalent, of course, to asking what random members of 3-CLUSTER look like. Though the problem is simple to state, the answer turns out to be somewhat counterintuitive. A simulation of the resulting curve is presented in Fig. 8. (Notice that in the figure the angle measures deviation from straight, and is thus the complement of the largest interior angle in the random triangle.) The distribution is naturally bounded by 0° and 120°, since we are selecting the largest interior angle, and all triangles have at least one interior angle at least as large as 60°.

The peak of the distribution is at 90°—perhaps counterintuitively. One intuition about why the peak is not larger is that values larger than 90° come from triangles close to equilateral, requiring in effect a close accidental alignment between the two independent angles.

2.2.2. *Random chains of three dots.* The problem of finding a density for 3-CHAIN is somewhat different from the 3-CLUSTER case. Here, the natural way to obtain a distribution of triplets is to impose a distribution on the expected curvature of the underlying curvilinear process. For example, we might stipulate a distribution on the family of curves; then choose a curve randomly from this distribution; then generate each triplet by choosing three points at random along this curve; and then compute the density of triplets generated in this manner. This scheme amounts to assuming something about the world, though—that is, we get out a distribution that reflects, in effect, how much bend per arclength we expect curve processes in the dots world to contain. Thus we will not get a single normative answer as in the 3-CLUSTER CASE, but rather a family of answers keyed to assumptions.

A loose but simple argument suggests that this distribution can be expected to be approximately normal in form. Since in a completely arbitrary curve, nearby points can be thought of as independent (i.e. the curve can bend an arbitrary amount between adjacent sampled points), we can expect the density of the deviations from tangency at various sampled points along the curve to act as the sum of a number of independent, identically distributed random variables—and thus, by the central limit theorem, normal in the limit. The mean is presumably zero by symmetry (strictly on the contingent assumption that curve processes in our world are unbiased with respect to handedness). But the standard deviation is arbitrary—it depends on the assumptions about the distribution of curves and inter-sample arclengths, and thus on the distribution of the random variable.

Hence we can conclude that randomly chosen members of 3-CHAIN will have angles chosen from a normal distribution with mean 0° and *arbitrary* width. Like the 3-CLUSTER case, this is in some respects counterintuitive. Imagining random members of this category, one imagines only relatively straight triplets (as do subjects: see Experiment 2 below). That is, the only triplets one imagines generated are those that one would actually classify as belonging to this category—that is, as having been generated by 3-CHAIN. But under certain assumptions about the curve, the tails of this distribution will actually produce triplets that human observers would not put into this class, but would rather classify as "random"—i.e., as members of 3-CLUSTER.

In both cases, 3-CLUSTER and 3-CHAIN, literally random members of each category turn out to be distributed in ways that the intuition finds "incorrect"—reflecting, presumably, that our intuitions are based on category models such as the ones describe in Section 2.1, rather than on tutored statistical beliefs. *Psychologically* typical members of each category, by contrast, are generated in such a way as to respect the structure of the psychologically qualitative categories. The next section describes one scheme by which such densities might be generated.

2.3. Constructing probability densities for the category models. As noted above, ordinary human observers are presumably ignorant of Kendall and Kendall's distribution of random triplets, and substitute for it some kind of heuristic expectation. The critical intuition concerns what human observers, as opposed to statisticians, mean by "random". In this context, a statistician means by the word random "chosen without bias on the space," like Kendall and Kendall's triplets. Under the category theory presented above, though, what the perceptual system means by "random" is: *not regular*, or, more specifically, *not non-generic*. That is, the non-genericity $\alpha = 0$ represents a class of triplets that have some regular structure, and which trigger an inference of an underlying curve. Psychologically random triplets, putatively, should be biased towards triplets that do *not* trigger such an inference (cf [1] on the need for bias on the space of perceptual hypotheses).

Under this reasoning, it makes sense (as alluded to above) to construct the class

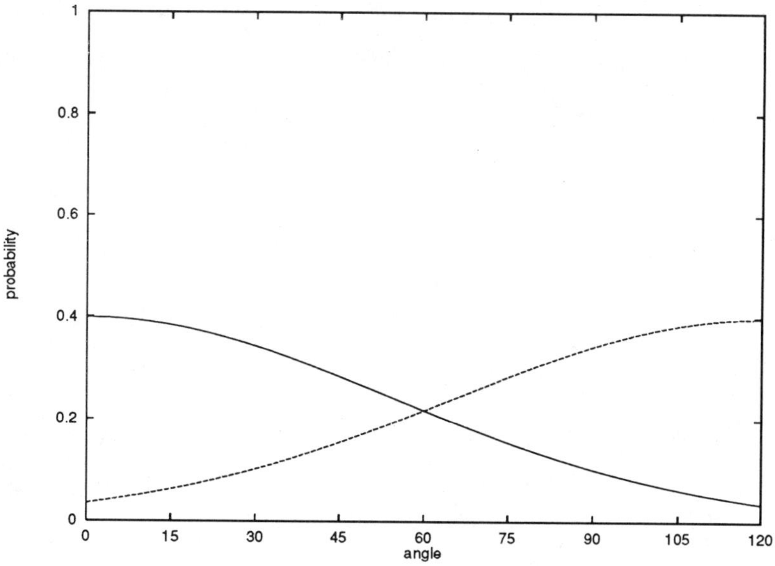

FIGURE 9. The two hypothetical prior "modes", one at 0° (straight), the other at 120° (generic). Standard deviations are each 53.059°, the fitted value drawn from Experiment 1.

density $p(\mathbf{X}|\text{3-CLUSTER})$ by placing a peak at the angle furthest in the space from straight, a point established in Sec. 2.1.1 to be at $\alpha = 120°$. Similarly, we construct $p(\mathbf{X}|\text{3-CHAIN})$ with a peak at $\alpha = 0°$. We assume, arbitrarily but reasonably, that these two densities are each normal in shape.

If these two peaks were "real" densities, determined by the world, then the world could also give them different heights (corresponding to the scalar priors, $p(\text{3-CHAIN})$ and $p(\text{3-CLUSTER})$), and different widths (standard deviations). Moreover, in the world there might be some other regularities that occur—other factors contributing to the density $p(\alpha)$. Here, however, the densities are being constructed heuristically by the observer rather than dictated by the environment—that is, they are "subjective" densities, constructed internally as part of an inferential machine. Hence the observer can unilaterally disallow the complicating factors of varying heights, varying widths, and additional regularities, about which it has no knowledge. The resulting rather severe set of regularizing assumptions—that the peaks (1) have equal height (2) have equal width, and (3) mutually exhaust the set of events that occur in this domain—enforce the observer's desire to impose on (or impute to) the world as little structure as possible, and no more than is necesary to distinguish the categories of interest. The resulting class densities consist entirely of two normal "humps," with the same height and width, one centered on each regularity (depicted in Fig. 9).

How is the standard deviation of each "hump" determined? Loosely speaking, the idea is that the curve descends from a maximum at its central mode down to some "significance" level, near zero, at the location of the other node. (This value is literally a significance level in that it denotes the probability of the observed triplet having been generated by the alternative "null" hypothesis of no structure). A theory whereby these widths can be derived will be presented in the discussion

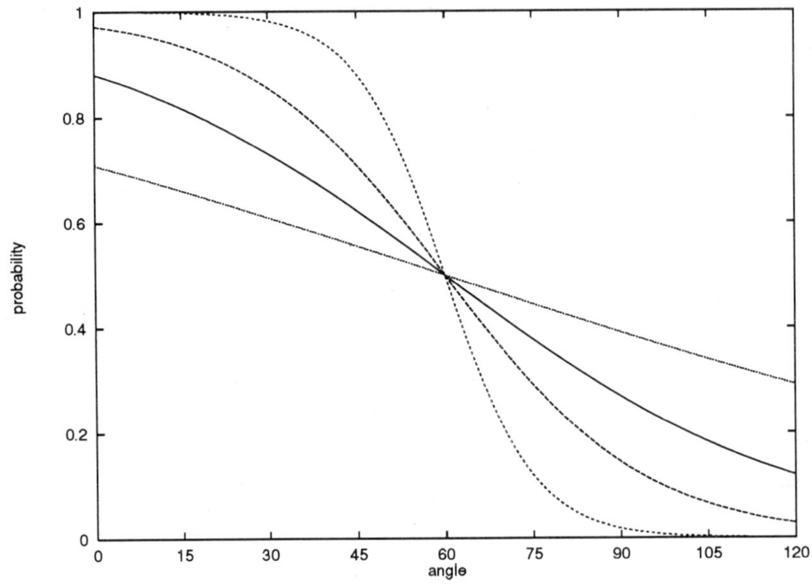

FIGURE 10. The hypothetical a posteriori curve, shown at four different standard deviations ($\sigma = 15°, 45°, 75°$, and $105°$).

of Experiment 1, below. Loosely, the idea will be that variances are selected that capture the categorical distinction between the two classes as cleanly as possible, and support an optimally transparent posterior classification. The widths used to draw Fig. 9 are drawn empirically from Experiment 1, but match those predicted by this scheme almost exactly.

With the two conditional densities established as described above, we are now in a position to construct the a posteriori functions $p(3\text{-CHAIN}|\alpha)$ and $p(3\text{-CLUSTER}|\alpha)$, by which observers actually assign a category label to an observed triplet. First, by stipulation that these two categories are the only ones under consideration, $p(3\text{-CHAIN}|\alpha) = 1 - p(3\text{-CLUSTER}|\alpha)$, so we can look only at the chain case. Substituting in the assumption of normal class densities, we have

$$p(3\text{-CHAIN}|\alpha) = \frac{N(0°,\sigma)}{N(0°,\sigma) + N(120°,\sigma)} \tag{2.3}$$

where $N(\alpha, \sigma)$ is the normal distribution with mean α and standard deviation σ.

This curve (plotted for four values of σ in Fig. 10) has the familiar sigmoid shape characteristic of many psychometric functions, with a transition from one category to the other whose sharpness depends on σ. Thus the category theory predicts that human observers will draw category distinctions between 3-CHAIN and 3-CLUSTER according to a curve drawn from somewhere in this family. The next section presents an experiment that tests this empirically.

3. Experiments on human observers

We now report an experiment designed to directly probe human subjects' intuitions about membership in the categories 3-CHAIN and 3-CLUSTER. In the main part of the experiment (labeled Experiment 1), subjects were presented with triplets of dots, and asked to judge whether they should be classified as having been produced by a curved 1-D generating process or a 2-D patch process (via suitably comprehensible instructions, described below). In a follow-up to Experiment 1 (labeled Experiment 2), the same subjects were asked to generate a number of "typical" members of each of the two categories, thus probing their class-conditional densities in a more directly productive manner. It should be noted that subjects were instructed that two categories of dot triplets were under consideration, so neither experiment tests our assumption that perceivers automatically assign prior probability to only these two classes; rather, the experiments are a way to reveal the exact form of the class density functions subjects unconsciously assign to the two classes under consideration.

3.1. Experiment 1: Perceptual classification of dot triplets. This experiment investigated subjects' beliefs about the dots world by directly asking them to classify dot triplets as the products of either a 2-D patch process or a 1-D curvilinear process.

Method

Subjects. Sixteen subjects drawn from the university community were paid for their participation.

Procedure. Subjects were instructed that they were to be presented with triplets of dots, to be interpreted as the markings on two equinumerous species of bottom-dwelling ocean creatures: a kind of *worm*, whose three markings were randomly placed along its curvilinear body; and an approximately round *flatfish*, whose three marking were randomly placed on its flat surface. The instructions explicitly drew subjects' attention to the fact that any triplet could be either type of creature: worms can curl up arbitrarily, while the dot markings on flatfish might just "happen" to fall nearly in a line. As illustration, subjects were shown two worms and two flatfishes, arranged so that exactly the same two dot triplets were given as an example of both categories.

Thus subjects understood that their task was in essence to determine how likely each dot configuration was to be a worm based on how nearly collinear the three dots were. For each triplet, subjects responded by choosing a number from 1 to 5, meaning (1) "almost definitely a flatfish", (2) "probably a flatfish", (3) "either one, with about equal probability", (4) "probably a worm", and (5) "almost definitely a worm".

After practice trials, subjects were presented with 396 such trials, crossing 33 angles (from $-120°$ to $120°$ in increments of $7.5°$) with 12 levels of line-length ratio (from 1.00 to 2.85 in integral powers of 1.1).

Results

The results of Exp. 1 are plotted in Fig. 11. The data can be taken as representing subjects' posterior classification function, $p(\text{3-CHAIN}|\alpha)$. In the plot, the Y-axis is subjects' numeric response, normalized to the interval [0,1] (i.e. converted into a probability judgment). The X-axis collapses over line-length ratio and over sign of the angle. Thus each plotted point corresponds to 24 observations per subject, by 16 subjects (except for $\alpha = 0$ which is only 12 observations, since there is no sign).

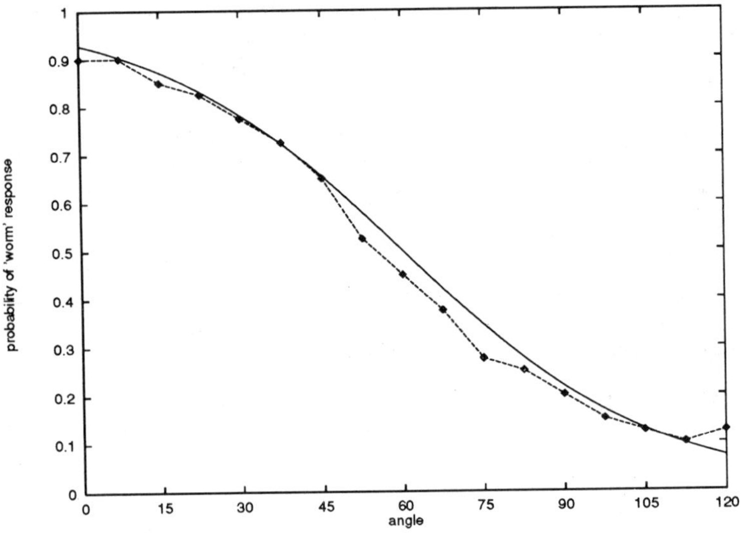

FIGURE 11. Data from Experiment 1. On the Y-axis are subjects' numeric responses, converted to a judgment of probability that the observed triplet was a "worm". On the X-axis is the angle of the dot triplet, collapsing over line-length ratio and sign of angle. Also plotted is the best-fit curve from the likelihood family described in the text, in which the two prior peaks each have standard deviation $= 53.059°$.

We can use these data to estimate the means of the two (implicit) prior peaks from which the posterior function is drawn. Using a conservative model in which the two standard deviations and the two heights are allowed to vary independently, the two means are estimated at $1.439°$ (99% confidence interval from $-0.168°$ to $3.046°$) and $118.539°$ (99% confidence interval from $116.909°$ to $120.169°$). Using the more theory-dependent model in which the two peaks have the same height and width, the two means are estimated at $-2.566°$ (99% confidence interval from $-5.061°$ to $-0.071°$) and $116.372°$ (99% confidence interval from $113.924°$ to $118.820°$). The peaks clearly come close to the $0°$ and $120°$ predicted by the category model, though the more theory-dependent estimates leave some room for doubt as to the precise locations of their centers, in each missing the predicted values by a small amount. This slight deviation may well be due to some unknown additional factor, though it might just as well be due to a small degree of kurtosis in the subjects' densities. In any case the subjects' "generic" peak is quite distant from the normative value of $90°$.

The figure shows the best fit (least-squared error) curve from the family of posterior functions described in Eq. 2.3, $R^2 = 0.988$, $F_{1,16} = 1215.8, p \ll 0.01$. Here we use the theoretical model in which both peaks have the same height and width, so that the only free parameter is their common standard deviation, σ. This value has been estimated at 53.059°. Note that because this posterior curve has no separate height parameter, the close fits at the point-predictions for $\alpha = 0°$ and $\alpha = 120°$ are particularly impressive.

Discussion

It is clear from the results that subjects classified dot triplets according to Eq. 2.3, to an extremely close approximation, using standard deviations of about 53° for each mode. Subjects were in effect placed into a novel, imaginary domain of bottom-dwelling creatures of which they had no prior knowledge whatsoever. They were thus left to create class-conditional density functions wholly out of their own intuitions—i.e., putatively, wholly out of the sort of regularity-driven category-generating apparatus described above.

Accounting for the distribution widths. To fully characterize these data, we would like to be able to say something about the sole remaining fitted parameter, σ, the standard deviation of each of the two prior modes. What value would we have predicted for this parameter?

One prediction for the value of σ can be derived from the premise that the observer is attempting to construct priors so that an optimally simple maximum-likelihood scheme can be employed in order to calculate posterior classifications. That is, given an observed triplet with angle α, the observer would like to be able to read off the posterior probability of membership in the class 3-CHAIN, $p(\text{3-CHAIN}|\alpha)$, directly from the prior density $p(\alpha|\text{3-CHAIN})$. Since these two quantities are actually related by Eqs. 1.1 and 1.2 (in which we can ignore the constant terms $p(\text{3-CHAIN})$ and $p(\text{3-CLUSTER})$), equating them simply requires that the denominator (Eq. 1.2) sum identically to unity. Pictorially, the prior and posterior probabilities become identical when the two "humps" sum together to produce a flat horizontal line of height unity (Fig. 12). Putting it another way, this condition requires that between the two generating processes, all shapes of triplets are equally probable in the world.

Hence in order to derive a prediction for the exact shape of the two humps, we seek the value of σ for which the function $f(x) = N(0°, \sigma) + N(120°, \sigma)$, in the interval $[0°, 120°]$, deviates as little as possible (in a least-squares sense) from the line segment $f(x) = 1$. That is, we seek a value of σ that as nearly as possible satisfies the equation

$$\text{(3.4)} \qquad \frac{d}{d\sigma}\left[\int_0^1 [e^{-(\frac{x^2}{2\sigma^2})} + e^{-(\frac{(x-1)^2}{2\sigma^2})} - 1]^2 dx\right] = 0.$$

An analytic solution to this minimization is problematic because of the Gaussian terms, but numeric estimation provide a figure of $\sigma \approx 0.4435(120° - 0°) = 53.225°$, almost exactly the empirical figure of 53.059° (only about 0.31% high). Anther way of expressing the closeness of this fit is to plug this value back into a model of the subjects' data: the model now has 0 degrees of freedom (a point prediction) but accounts for 98.7% of the variance (and has infinite F-ratio).

It is worth remarking on the fact that subjects' induced density functions (Fig. 9) bear a plain resemblance to the canonical signal detection set-up. In both cases, a single discriminant parameter is inhabited by two Gaussian peaks of equal height and equal variance. In effect the subjects here have fabricated two class densities exactly as if to simulate signal and noise, with the triplets generated by the generic

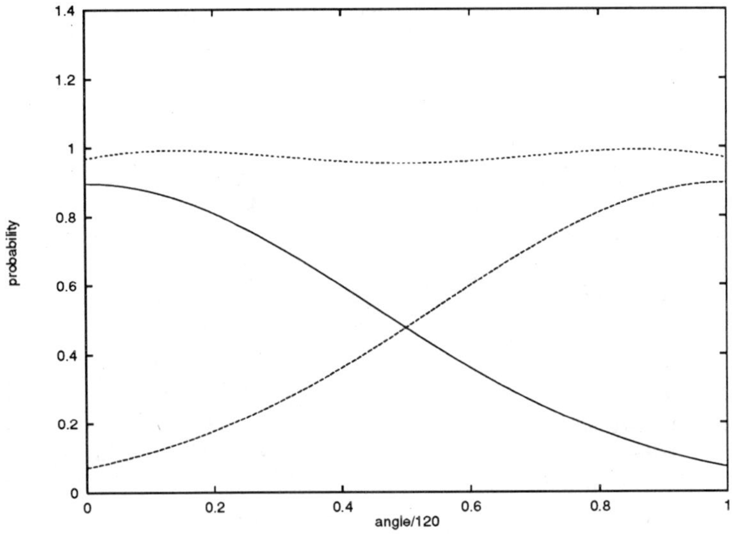

FIGURE 12. The two class densities sum identically to about unity, meaning that all triplet angles are about equally likely, and prior and posterior probabilities are as nearly identical as possible.

category, 3-CLUSTER serving as noise, and $d' = 2.2599(= 1/0.4435)$. In a sense this is consistent with the idea that the generic category serves as a random backdrop or distractor to the regular (codimension-1) target category 3-CHAIN. The difference is an epistemic rather than a technical one. There is no extrinsic noise source in the dot triplet space; the variance in both categorical densities is a subjective invention on the part of subjects, rather than anything like literal uncertainty in the physical detection of a signal. The probabilities, that is, are subjective ones; the "signal" is a concept of order or regularity in an abstract hypothesis space, and the "noise" is a concept of disorder or irregularity; or, more precisely, of causal independence in the generation of the dots.

Because the subjects' responses are at odds with the normative distributions within each class, it is tempting to conclude from this they are simply *inaccurate* in their attempts to guess what typical members of each class, selected randomly, would look like. It makes more sense, though, to point out that observers are following a useful strategy. Their model of collinearity constitutes a kind of theory about a basic form that regularity in this unknown domain might take. They are making the quite plausible assumption, in effect, that observed objects in the new domain are *not* generated completely randomly; but rather that they are subject to causal forces that, while mysterious in exact nature, are coarsely predictable in form.

3.2. Experiment 2: Generation of dot triplets. Experiment 1 allowed us to infer the form of subjects class conditional probability densities indirectly. Experiment 2 takes a different tack by probing these densities directly (though less sensitively), by asking the same subjects from Experiment 1 to generate "random" members of each set.

Method

Immediately after completing Experiment 1, the subjects were given questionnaires, asking them to draw 10 different examples of each type of creature, "worms" and "flatfish". It was feared that if subjects were asked to produce a large number of examples, they would tend to increasinly produce deliberately (an unrepresentatively) atypical ones On the other hand, asking for only a very small number would provide an overly impoverished picture of their probabilistic beliefs. The number chosen (10) is a somewhat arbitrary compromise between these concerns. It should be noted though that any progression in the subjects' drawings from more to less prototypical has been ignored in the analysis below. This fact may well be taken to cast a certain amount of doubt on the idea that the procedure directly elicits the subjects' priors. Due to the coarseness and origins of the data, then, the plots below are to be viewed more impressionistically than analytically, and more for insight into subject's priors than for final adjudication of theoretical questions.

Results

Each of the resulting dot triplets were coded by hand as an angle (the positive deviation from straight, as above) and a line-length ratio. Again, we collapse over line-length ratio, and plot frequency as a function of angle. Fig. 13 and Fig. 14 show the results for "worms" and "flatfish" respectively. The peaks at 0° and 120° are clearly visible[5], indicating the subjects produced many exactly straight "worms" and many almost exactly equilateral "flatfish". Intriguingly, though, both plots seem to have one additional peak: a second "worm" peak at about 45°, and a second "flatfish" peak at about 90°. Unfortunately, the limited amount of data available (constrained by the factors discussed above) prevents these data from being analyzed other than impressionistically.

Discussion

An argument from [7] suggests an explanation for the additional peaks. Each generic triplet, putatively, is generated by starting from some regularity and moving some distance off through the space, a distance far enough to make it distinguishable from objects at the regularity, but not so far as to encroach into the next categorical region of the space. Thus additional qualitative categories spin off in the region near a regularity, exactly one per "qualitative" region. Hence the four spikes in the two frequency plot can perhaps best be described (in order of increasing angle), as $0, 0 + \epsilon_1, 120 - \epsilon_2$, and 120. Again, the categorical scheme provides a formalism in which each qualitative category is assigned a unique symbolic description. The origin of the exact values of the ϵ's is unknown. Rather, the point is simply that the qualitative symbolic description generated by the category theory is isomorphic to the modal peaks evident in the subjects' data.

3.3. General discussion of experiments. Taken together, the two experiments paint a hybrid picture of human observers' clustering interpretations, in which posterior inferences are drawn using distinctly Bayesian machinery, and yet in which critical antecedents such as the class-conditional densities are fabricated virtually

[5]Note that the slight migration of the 120° peak towards smaller angles is attributable to subjects' inaccuracy in drawing the figures. That is, *any* deviation from an exact equliteral triangle will be *decrease* the measured angle, since we always measure at the largest interior angle.

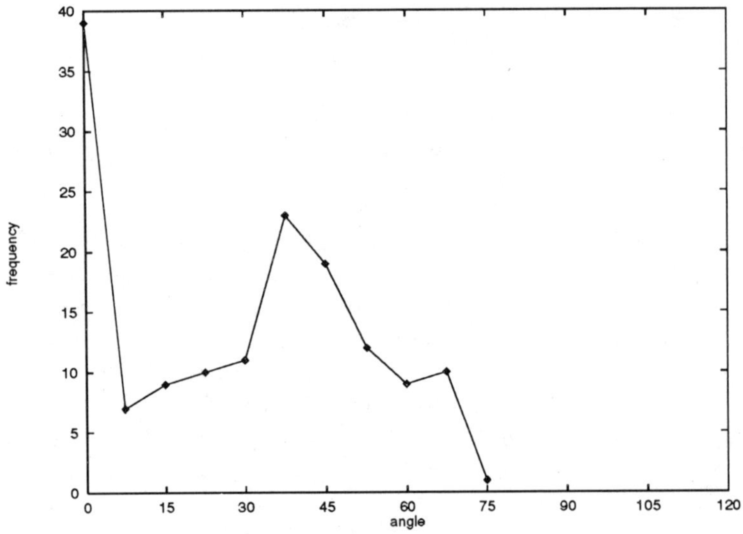

FIGURE 13. Data from Experiment 2 ("worms"). Frequency is plotted as a function of the angle of the dot triplet. Two modes are visible, one at straight, and one at slightly off from straight.

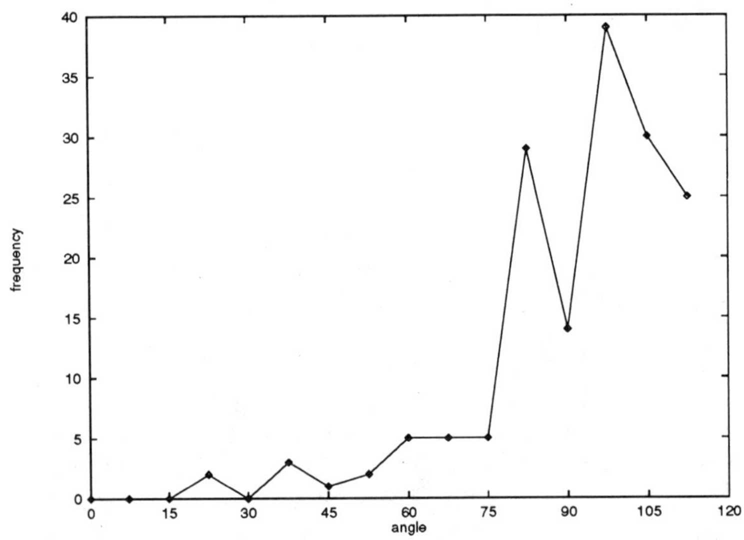

FIGURE 14. Data from Experiment 2 ("flatfishes"). Frequency is plotted as a function of the angle of the dot triplet. Two modes are visible, one at about equilateral (120°) and one at about 90°.

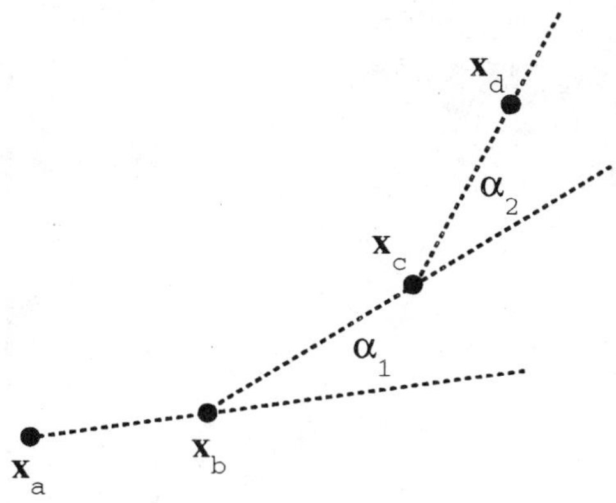

FIGURE 15. Parameterization for four dots—a multi-molecular cluster.

out of whole cloth. While the relationship between the densities drawn from Experiment 2 and those implied by Experiment 1 is unclear, subjects' responses in Experiment 1 clearly demonstrate that they classify dot triplets in an approximately normative, *given* that they draw category densities according to the category lattice scheme. In the simplest terms what this means is that subjects seem to make a particular set of tacit assumptions about the regularities governing the dots world. They then use these regularity notions to structure their beliefs about what is likely and what unlikely given various hypotheses about the state of the world. From these beliefs in turn they can infer which hypotheses about the state of the world are more and less likely to be true.

4. Composing molecular clusters into larger groups

With the exact class-conditional densities for molecular clusters in hand, an observer is theoretically in a position to determine likelihoods for arbitrarily large sets of dots, by composing molecular decisions into larger groups. In this manner a psychologically correct decomposition of dot configurations such as that in Fig. 1 could be computed in a Bayesian way. We now briefly consider how this might be done, and show that the normative way of doing so does not work: the human intuitions are, again, somewhat more complicated.

Consider a chain of four dots $\{\mathbf{x}_a, \mathbf{x}_b, \mathbf{x}_c, \mathbf{x}_d\}$ defined by two angles α_1 and α_2 (Fig. 15), again ignoring relative inter-dot distances. We ought to be able to treat the two triplets $\{\mathbf{x}_a, \mathbf{x}_b, \mathbf{x}_c\}$ and $\{\mathbf{x}_b, \mathbf{x}_c, \mathbf{x}_d\}$ as separate trials, computing a likelihood $p(3\text{-CHAIN}|\alpha_i)$ for each triplet separately. The key question concerns the independence of these two "trials". If the two triplets did not originate from a common process, the probabilities that each one is actually a 3-CHAIN ought to be independent. On the other hand, if the four in a row were all generated by a common curvilinear process, then we would not expect their class labels to be independent. Rather, in a world in which curvilinear regularities exist, very straight triplets might be expected to be commonly followed by additional straight triplets. Critically, such an expectation is justified *only* in such a world; the expectation is not justified in a world lacking such a regularity (see [14] on this point).

If the two constituent triplets in the ordered four-tuplet are treated as independent events, the likelihood of the overall configuration under a chain interpretation

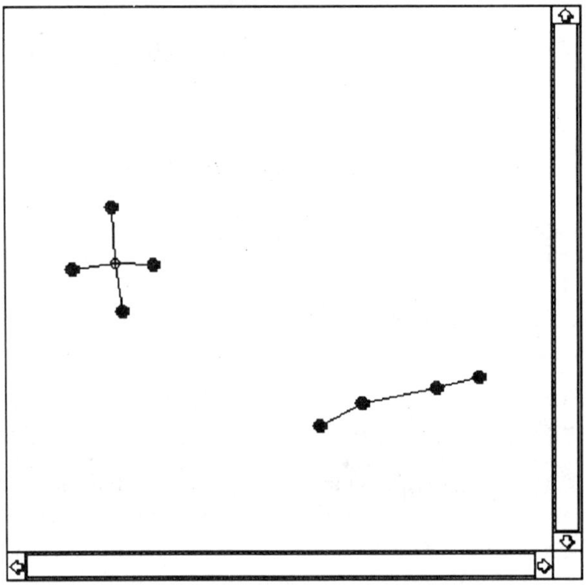

FIGURE 16. Using a maximum a posteriori grouping procedure as described in the text, the computer finds two groups: an unordered cluster of four dots on the left (drawn linked to their centroid) and a chain of four dots on the right (drawn with links from dot to dot).

would be

$$p(\mathbf{X}^4|\text{curvilinear process}) = p(\alpha_1|3\text{-CHAIN})p(\alpha_2|3\text{-CHAIN}). \tag{4.5}$$

Substituting Gaussian densities for each triplet density, this become approximately

$$p(\mathbf{X}^4|\text{curvilinear process}) \approx \frac{1}{\sqrt{2\pi}}e^{[\frac{1}{2\sigma^2}(\alpha_1^2+\alpha_2^2)]}. \tag{4.6}$$

This equation ought to tell us the degree to which a given chain extraction can be taken to explain a given set of dots. Assuming completely neutral priors (chains and clusters equally likely, and groups of all sizes equally likely: cf the celebrated "Principle of Indifference"), we can compute a maximum a posteriori interpretation for each field of dots. This interpretation in principle represents the inferred generating process for the observed dots that, among all models under consideration, is actually the most likely to be the correct.

For many dot configurations, the a maximum a posteriori interpretation does in fact result in the clustering that human observers extract. For example, consider the field of eight dots shown in Fig. 16. The groups discovered by the computer are indicated: dots belonging to a chain are linked by line segments in sequence, and dots belonging to a cluster are each linked to a virtual dot (drawn as an unfilled circle) at the centroid of the cluster.

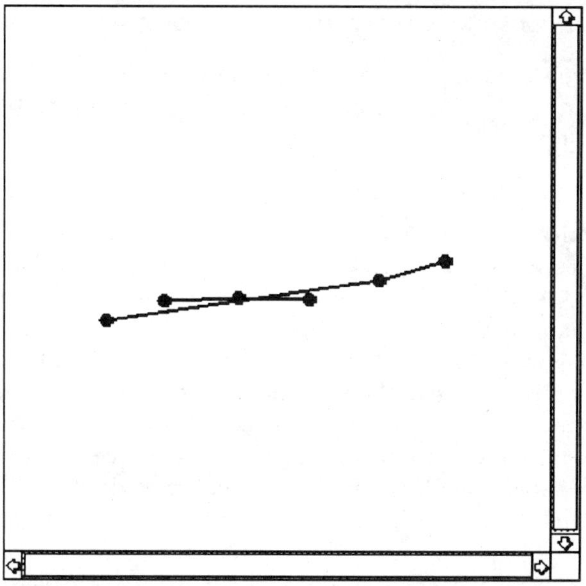

FIGURE 17. The maximum a posteriori clustering of these six dots (given the assumptions in the text) is as two distinct but intertwined chains, not the single long chain that human observers prefer.

Here, the computer's interpretation is the same as the intuitive one (compare Fig. 1). It is not difficult to find examples, however, in which human intuitions conflict with the computed solution, for interesting reasons. An example with six dots illustrates the problem (Fig. 17). Again, the chains found by the program are linked up by line segments. Most human observers probably see the configuration as one long but very coherent chain of six dots. The decomposition discovered by the program, however, separates the dots into two distinct intertwined triplets. That is, each of these two triplets is internally so nearly collinear, and thus accounts for its subset of the data so completely, that the combination beats out the more curvaceous but also more intuitive six-dot chain.

What strikes the observer as implausible about this decomposition, perhaps, is that the two triplets are not in a generic configuration with respect to each other—they are nearly parallel and nearly coincident. This configuration is highly special, and highly unlikely if the two triplets are independent; but of course it is actually to be expected if all six dots actually originated from one long curvilinear process. In order to render the cluster composition fully competent to parse fields of dots, then, the genericity idea must be extended up to slightly more abstract level: the *relative* configuration among sub-decompositions. Just as three dots within a perceived cluster must be in general position with respect to each other (given the inferred category), the various clusters (cells) in in a hypothetical dot decomposition must all be in general pose with respect to one another, in some suitably defined sense. A full dot decomposition theory, correctly accounting for observers' decompositions of arbitrary fields of dots, requires such a definition, which we defer to another paper.

5. Conclusion: causal models of dot groupings?

The inferential machinery presented in this paper focuses a great deal of apparatus on what is, after all, a very simple domain. The data presented in Section 3, however, make it quite clear that human observers' intuitions about this domain contain a considerable amount of structure not given by the problem. The triplet classifications, for example, reveal a nest of tacit assumptions about what ought to be seen as "typical" in the dots world: the centers and widths of the class-conditional densities are drawn from the head, not from the world, and do seem to be built around the categorical structure described here. Even the reader's perception of a dot chain in Fig. 1 reveals the reader's tacit belief in the existence of "regularities" in the dots world. Perhaps counterintuitively, if there were no curvilinear dot-generating processes in the domain, then *any* perceived chain, no matter how compelling, would be a false target—an accidental alignment. The fact that one does, automatically, perceive such chain groupings suggests that our perceptual system in some sense believes that causal forces in the environment will, with some non-zero prior probability, cause dots to be generated along curves (again see [14] on this point).

Theorists of clustering per se, in evaluating the utility of novel clustering techniques, are often necessarily guided by their intuitions about what clustering solutions make "sense" and thus ought to be recovered by the technique. At the most abstract level, the kind of formalism described in this paper can be seen as shedding light on just in what respect some clustering intuitions make more sense than others, and on the logical structure of the inference mechanism that underlies such intuitions.

References

1. Bennett, B., Hoffman, D. and Prakash, C. *Observer mechanics: A formal theory of perception,* Academic Press, London, 1989.
2. Broadbent, S.R. *Simulating the ley hunter,* J. Royal Statistical Society A **143(2)** (1980), 109–140.
3. Do Carmo, M. *Differential geometry of curves and surfaces,* Prentice-Hall, Englewood Cliffs, N.J., 1976.
4. Duda, R. and Hart, P. *Pattern classification and scene analysis,* John Wiley, New York, 1973.
5. Feldman, J. *Perceptual decomposition as inference: continuous curvilinear processes,* Occasional Paper #42, M.I.T. Cognitive Science Center, 1991.
6. ______ *Constructing perceptual categories,* Proc. IEEE Conf. Computer Vision and Pattern Recognition (Champagne-Urbana, Illinois, 1992), 1992, 244–250.
7. ______ *Perceptual categories and world regularities,* Ph.D. thesis, Department of Brain & Cognitive Sciences, M.I.T., 1992.
8. Jepson, A. and Richards, W. *What is a percept?* Tech. Report RBCV-TR-93-43, Dept. Computer Sci., Univ. of Toronto, 1993. (Also Occasional Paper #43, M.I.T. Cognitive Science Center, 1991.)
9. ______ *What makes a good feature?* Spatial Vision in Humans and Robots, (L.Harris and M. Jenkin, eds.), Cambridge University Press, Cambridge (to appear). See also A.I. Memo #1356, M.I.T., 1992.
10. Kendall, D.G. and Kendall, W.S. *Alignments in two-dimensional random sets of points,* Adv. Applied Probability **12** (1980), 380–424.
11. Leyton, M. *Perceptual organization as nested control,* Biological Cybernetics **51** (1984), 141–153.
12. ______ *Inferring causal history from shape,* Cognitive Science **13** (1989), 357–387.

13. Matthews, G. and Hearne, J. *Clustering without a metric,* I.E.E.E. Trans. Pattern Analysis and Machine Intelligence **13(2)** (1991), 175–184.
14. Richards, W., Jepson, A., and Feldman, J. *Priors, preferences, and categorical percepts,* Proc. Workshop on Psychophysics and Computation (Chatham, MA, 1993), to appear.
15. Späth, H. *Cluster analysis algorithms for data reduction and classification of objects,* John Wiley, New York, 1980.
16. Stevens, K.A. *Computation of locally parallel structure,* Biological Cybernetics **29** (1978), 19–28.
17. Witkin, A. and Tenenbaum, J. *On the role of structure in vision,* Human and Machine Vision (J. Beck, B. Hope, and A. Rosenfeld, eds.), Academic Press, New York, 1983.
18. Zucker, S.W. *Early orientation selection: tangent fields and the dimensionality of their support,* Computer Vision, Graphics, and Image Processing **32** (1985), 74–103.

Rutgers University Center for Cognitive Science, Psychology Building New Wing, Busch Campus, Rutgers University, New Brunswick, New Jersey, 08855

E-mail address: jacob@ruccs.rutgers.edu

DIMACS Series in Discrete Mathematics
and Theoretical Computer Science
Volume **19**, 1995

SUBJECTIVE CONTOURS IN EARLY VISION AND BEYOND

BELA JULESZ

ABSTRACT. This is a review of research on illusory contours by the author and his collaborators starting with cyclopean contours using random- dot stereograms (Julesz,1960), and continuing with their suggestion of how such contours can explain the asymmetry effect of preattentive texture discrimination (Williams and Julesz, 1991,1992). Some recent results are discussed that show that beyond local interactions illusory contours depend on global parameters such as closure of stacked lines (Kovacs and Julesz, 1993). The coherence distance between incomplete line segments is increased manyfold when they belong to a long line or closed loop. Much of these phenomena can be regarded as belonging to early vision that might be related to neurophysiological findings in V2 in the monkey cortex von der Heydt, R. Peterhans, E. and G. Baumgartner (1988). A model will be discussed based on local exitation and global inhibition of orthogonal line segments. However, our research might go beyond early vision discussing a case (Gorea and Julesz, 1990) in which cognitive contours of familiar objects (a schematic human face) might play a role.

1. INTRODUCTION

With this talk I planned to commemorate Gaentano Kanizsa's 80th birthday last month, but alas he died a few days before he could have received my manuscript. So, I decided to read this talk to you in somewhat modified form to be of interest to colleagues in machine vision too, and by doing so give a tribute to Prof. Kanizsa who both as a psychologist and painter had great impact on visual perception. A shortened version of this paper - written for psychologists - will appear in the Giornale Italiano di Psicologia: Special Issue.

After F. Schumann (1904) drew attention to the concept of illusory (or subjective) contours this important perceptual phenomenon had been practically forgotten for a generation. Perhaps his induced white square (by partially enclosing a square area by half circles on the left and right sides and by horizontal and vertical lines as shown in Fig. 1A) was not vivid enough to make a lasting impression. A few decades later Walter Ehrenstein produced his illusory circles (see Fig. 1B) that were rather impressive, but still were largely ignored [as reviewed by T.E.Parks (1987) with these original demonstrations reprinted]. It was Gaetano Kanizsa who finally succeeded to draw attention to this concept in 1955 [Margini Quasi-percettivi in Campi con Stimulazione Omagenea, Rivista di Psicologica, 49:7-30] with his many vivid demonstrations that inspired decades of research that contains even to this date. [Here I demonstrate only a few illusory contours from the Parks (1987) review in Fig. 1 and the interested reader is referred to his article]

1991 *Mathematics Subject Classification.* Primary 92J30.

I was impressed with the Kanizsa - contours from the very first moment I ventured into visual perception. Indeed, my interest in stereopsis started with trying to extract contours from real-life half-tone images. I knew from Ernst Mach's work that the second-gradient of the luminance distribution (now called the Laplacean) was a good way to extract contours. However, when I looked at an illusory contour there was no trace of any luminance gradient, yet vivid contours were perceived. It appeared that some more global configuration caused the formation of these contours, and globality in my mind was strongly linked to higher semantic and Gestalt-like cues - that I regarded enigmatic and wanted to avoid. Furthermore, I was convinced that several of these illusory contours were formed at an early stage in spite of the opinion of some influential thinkers, such as Richard Gregory, who regarded optical illusions and subjective contours as the result of some cognitive hypothesis testing. I do not doubt that such cognitive contours also exist, but first I discuss subjective contours of early vision as follows: In order to study subjective contours in early vision I turned my attention to stereopsis. As a former radar engineer, I knew that there was no camouflage in three-dimensions and that subjective contours could be created by binocular disparity gradients in the total absence of any monocular cues. In order to prove this point to psychologists I created the computer-generated random-dot-stereograms (RDS) that by their very definition were just an array of randomly chosen black and white dots, yet the hidden binocular correlation could be extracted by the visual system of primates and humans, and surfaces with their monocularly invisible contours would jump out in vivid depth (Julesz,1960). Such a RDS is depicted by Fig. 2 that portrays a cyclopean diamond hovering over a randomly dotted background when binocularly fused. [For a moment the reader should ignore the horizontal white stripe in the middle.] These cyclopean contours between the diamond and background can be regarded as illusory ones on their own right, since when monocularly viewed they do not exist.

As a matter of fact I even produced some Kanizsa-kind of illusory contours with RDS, as shown in Fig. 2. I cut the cyclopean diamond and the background with a horizontal blank strip in half. I wanted to know whether a subjective contour would be created as a perceptual continuation of the diamond shape and the diamond's contour would be seen with corners. However, as the reader can attest, instead, just two vertical contours are perceived, being the shortest path through the blank gap.

This demonstration convinced me that subjective contours do not obey complex laws of good Gestalt, but might be rather based on some co-linearity of adjacent contours that close gaps. Therefore, I regarded the neurophysiological findings of von der Heydt, R. Peterhans, E. and G. Baumgartner (1988) of great importance who discovered neurons in area V2 in the monkey cortex that fired for co-linear line segments separated by gaps better than for long uninterrupted lines. So, here was the first suggestion that the perception of illusory contours was done in early vision, since I regard areas V1, V2, V3, and V4 the neurophysiological sites of early visual perception (Julesz, 1991). [Even more importantly, these units in V2 are not only firing for adjacent co-linear structures, but also for neighboring structures that are orthogonal to each other! This duality of neural illusory contour detectors tuned both to co-linear and perpendicular relations is of utmost importance to perception, and I will return to this topic later.]

Fortified with this insight we were able to show that the asymmetry effect of preattentive texture discrimination - an other early visual process - is most likely based on subjective contour phenomena in V2 (Williams and Julesz, 1991, 1992). The asymmetry effect, using closed and opened circles as elements, is shown in Fig. 3.

The reader can see that the open circle(s) among closed circles pop-out effortlessly, while the closed circle(s) embedded in open circles can be only detected after some time-consuming scrutiny. Several investigators tried to explain this highly nonlinear phenomenon by complex ways from Weber's-law (Treisman and Paterson, 1984) or asymmetric effects of increased internal noise on texture discrimination (Rubenstein and Sagi, 1990). However, with Doug Williams we assumed that the phenomenon of illusory contours often does close the gaps. Indeed, if the open circles are closed by the illusory contour phenomenon with some probability (P), then to see the hidden closed circle in the open circles is very difficult since many of the open circles appear also closed. On the other hand, the open circle appears open with probability (1-P) among the array of closed circles. Since we showed (Williams and Julesz, 1991, 1992) that $P << 0.5$, therefore the asymmetry effect is quite dramatic. We also demonstrated that if the open circles in Fig. 3 had their gaps in parallel (as shown in Fig. 4), the asymmetry effect remained as strong as with the orientation of the gaps being jittered. This demonstration argues that it is not the increased noise of the background that causes the textural asymmetry effect (after all the background in Fig. 3 appears less noisy), but merely the closure of the gaps by the subjective contours.

Here we exploited the illusory contour effect again for explaining a complex textural asymmetry phenomenon without having to invoke some highly cognitive segmentation phenomena. Nevertheless, there are several perceptual closure effects that appear more global than those mentioned so far. One of them is the Glass effect shown in Fig. 5 as first reported by Glass (1969) . He took two identical transparencies of the same random array of black and white dots, and overlaid them after shifting, expanding, or rotating one with respect to the other. The resulting pattern contained global structures of gratings, radial lines, or concentric circles, respectively.

In Fig. 5 the pattern is rotated with respect to its identical dual, and the resulting pattern shows a global structure of concentric circular lines. It is easy to show that this is not merely a Moire effect, but has some global component, since restricting the field with a small aperture affects the visibility of the circular structure. There is also a perceptual limit to the maximal displacement for all three kind of operations, I will call the coherence limit (Δ), and the reader may consult Sagi and Kovacs (1993) for details. With Ilona Kovacs we became interested in the global extent of this coherence limit Δ. We embedded in an array of randomly oriented line segments - actually Gabor-patches, as shown in Fig. 5 - some target of quasi co-linear Gabor patches with gaps that appeared as "snakes" as shown in the inset.

The snake cannot be detected in a brief flash ($T \sim 180$ msec) provided the gap between its constituent Gabor-patches is above a critical size. However, if the snake "bites its tail", that is the target suddenly becomes closed, this closed loop pops-out from the clutter (Kovacs and Julesz, 1993). This remarkable phenomenon is depicted by Fig. 7.

These closed stacked curves cannot posses kinks, i.e., adjacent line segments (Gabor-patches) must be quasi-co-linear (within 30 degree orientation) for this robust pop-out effect. Also this perception of closure is the best for brief presentations. For $T >> 180$ msec the observer starts to scrutinize some other global structures at the expense of the primordial closed structure. So the reader will see a more enhanced effect when looking very briefly at Fig. 7. Of course, among these zig-zag shaped snakes a long really co-linear line pops out the best.

Here I refrain from speculations, yet note that our results with Ilona Kovacs suggest a local co-linear excitatory mechanism that perceptually connects incomplete segments, and an orthogonal inhibitory mechanism that acts over a longer distance and favors long lines and closed loops. Perhaps this dualism can be related to the neuronal mechanisms described by von der Heydt, R. Peterhans, E. and G. Baumgartner (1988). Whatever might be the case, the fact that tiny gaps in lines result in strong perceptual effects cannot be explained by some linear phenomena, such as some small missing components in the global Fourier spectra. Instead, a global nonlinear mechanism is required that causes "strong perceptual fields flow through these gaps" to use an old Gestalt terminology with some new possible meaning.

It seems that long lines, large circles, and relatively smooth closed loops are easily extracted by the early visual system when embedded in a noisy background. It seem logical that this type of early segmentation might have served well our arboreal ancestors in a dense foliage. Without the need of recognizing a leopard, a snake, or some prey lurking behind the leaves and branches, a line or closed contour detector can quickly segment the environment into figures and the ground.

Gestalt psychologists, particularly Kurt Koffka (1935), were fascinated with closure phenomena, and talked much about "good-Gestalt" and "Prĕgnanz". These concepts, however, were difficult to quantify with classical line drawings. It is only in texture perception of random elements - for which the geometry and statistics of the texture and targets are under computer control - when these Gestalt concepts gain some concrete meaning.

I am quite sure that in addition to these early global contour detectors some higher-order cognitive contour detectors do exist, since one can detect certain very familiar objects quicker and better than unfamiliar ones. All of us can detect our native tongue in the clutter of many foreign languages spoken simultaneously. The well known "Dalmatian dog hidden in clutter" emerges much slower when the dog is shown up-side-down. After all we are not accustomed to dogs walking up-side-down. This led to the following experiment:

With Andrei Gorea we showed four horizontal and vertical line segments embedded in an array of diagonal line segments as portrayed in Fig. 8. As shown, some of these non-diagonal line segments formed a texture gradient in the diagonally oriented array and observers were asked to detect when the diagonal array contained such horizontal and/or vertical line segments. Half of the time the non-diagonal line segments were a random cluster of three horizontal and one vertical element, but in the other half of presentations these same clusters depicted a schematic human face, where the three horizontal lines constituted the two eyes and the mouth, and the vertical line was the nose.

Although observers were unaware of the presence of the face, they detected the deviation in orientation from the diagonal somewhat better if the non-diagonal elements belonged to the face (Gorea and Julesz, 1990). This superiority of face versus nonsense clusters was present even when the cluster was symmetrical, but did not depict a face. This face superiority effect was rather weak, but was statistically significant. It certainly points to the possibility of some cognitive subjective contours. Biederman (1988), and Biederman and Cooper (1991) studied such cognitive contours using familiar objects with incomplete contours. According to the geon theory of Biederman incomplete drawings of objects might cue the combination of convex and concave stored geons (some kind of templates) of familiar objects. I do not wish to discuss the geon theory, and also omit my own texton theory of preattentive texture discrimination (Julesz, 1981), because in this article I concentrated mainly on observable robust perceptual phenomena in the spirit of the work of Kanizsa and omitted speculations.

Here I only summarized my own efforts with my collaborators that were based on illusory contours and by doing so I paid tribute to Gaentano Kanizsa without whom we might have missed this exciting field of research. In addition, I tried to emphasize the importance of subjective contours for experts in machine vision and demonstrate that contours go much beyond the concept of a Laplacean.

References

1. Biederman, I. (1988) *Aspects and extensions of a theory of human image understanding*, in Computational Processes in Human Vision, Z. Pylyshyn, ed., Ablex, Norwood, NJ, 370-427.
2. Biederman, I. and E.E. Cooper, (1991) *Priming contour-deleted images: evidence for intermediate representations in visual object recognition*, Cognitive Psychology, **23**, 393-419.
3. Glass, L. (1969) *Moiré effects from random dots*, Nature, **243**: 578-580.
4. Gorea, A. and Julesz, B. (1990) *Context superiority in a detection task with line-element stimuli: a low-level effect*, Perception **19**, 5-16, .
5. Julesz, B. (1960) *Binocular depth perception of computer- generated patterns*, Bell System Technical Journal, **39**, 1125-62 . Julesz, B. (1971) Foundations of cyclopean perception. Chicago, IL: Univ. of Chicago Press.
6. Julesz, B. (1981) *Textons, the elements of texture perception and their interactions*, Nature **290**, 91-97.
7. Julesz, B. (1991) *Early vision and focal attention*, Reviews of Modern Physics, **63**, 3, July 1991, 735-772.
8. Kanizsa G. (1955) *Margini Quasi-percettivi in Campi con Stimulazione Omagenea*, Rivista di Psicologica, **49**, 7-30.
9. Koffka, K. (1935) Principles of Gestalt Psychology, (Harcourt, Brace and Co. New York), 150.
10. Kovacs, I. and Julesz, B. (1993) Kovacs, I. and Julesz, B., (1993) *A closed curve is much more than an incomplete one: Effect of closure in figure-ground segmentation*, Proc.Natl.Acad.Sci.USA, **90**, 7495-7497.
11. Mach, E., 1886, The analysis of sensations and the relation of the physical to the psychical. First German edition in 1886; republished by Dover, NY, 1959 using the 5th German edition revised and supplemented by S. Waterlow.
12. Parks, T.E. (1987) *Illusory figures*, in The Oxford Companion to the Mind. ed. Gregory,R.L., 344-347. Oxford U. Press, Oxford, New York.
13. Rubenstein, B. and D. Sagi, (1990) *Spatial variability as a limiting factor in texture- discrimination tasks: implications for performance asymmetries*, J.Opt.Soc.Am.A., **9**, 1632-1643.
14. Sagi,D. and I. Kovacs (1993) *Long range processes involved in the perception of Glass patterns*, Investigative Ophthalmology and Visual Science, **34**, 4, Proc. of ARVO Meeting, May 1993. (published abstract)
15. Schumann, F. (1904) *Einige Beobachtungen über die Zusammenfassung von Gesichtseindrucken zu Einheiten*, Psychologische Studien, **1**, 1-32.
16. Treisman, A. and R. Paterson, (1984) *Emergent features, attention and object perception*, Journal of Experimental Psychology: Human Perception and Performance, **10**, 12-31.

17. von der Heydt, R. Peterhans, E. and G. Baumgartner (1988), *Illusory contours and cortical neuron responses*, Science, **224**, 1260-1262.
18. Williams, D. W. and Julesz B. (1991) *Filters vs. textons in human and machine texture discrimination*, in Neural Networks for Perception, **1** - Human and Machine Perception, Wechsler, H. (Ed.), Academic Press, Boston.
19. Williams, D.W. and Julesz, B. (1992), *Perceptual asymmetry in texture perception*, Proceedings of the National Academy of Sciences USA, **89**, 6531-6534.

LABORATORY OF VISION RESEARCH, RUTGERS UNIVERSITY, PSYCHOLOGY BLDG. - BUSCH CAMPUS, PISCATAWAY, NJ 08854

FIGURE 1. Several kind of subjective contours. [From Parks, T.E. (1987) "Illusory figures", in The Oxford Companion to the Mind. ed. Gregory,R.L. pp. 344-347. Oxford U. Press, Oxford, New York.]

FIGURE 2. Cyclopean illusory contour generated by binocular disparity gradient. This cyclopean contour can be regarded as a novel illusory contour per se, since monocularly no such contour exists. Furthermore, it demonstrates that the cyclopean diamond does not generate a classical illusory contour with corners according to good Gestalt continuation, but merely the shortest vertical line contours are seen. [From Julesz (1971).]

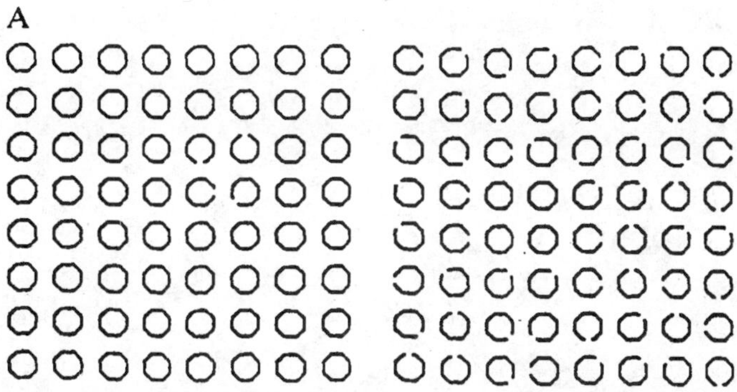

FIGURE 3. The asymmetry effect of texture discrimination showing that texture A embedded in B is perceived differently than vice versa. Left: The open circle among closed ones pops-out. Right: The detection of the closed circle among open ones requires scrutiny. [From Williams and Julesz (1991).]

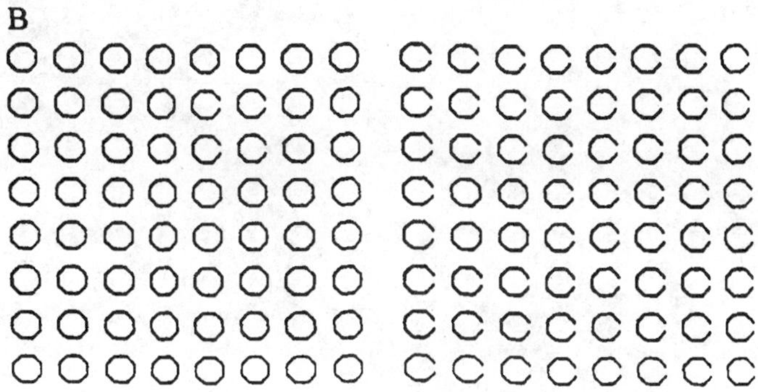

FIGURE 4. Same as Figure 3, except the array of open circles has their gap= s in parallel. Although this array is less noisy, than the one with jittered gap orientation, the asymmetry effect appears the same. [From Williams and Julesz (1991).]

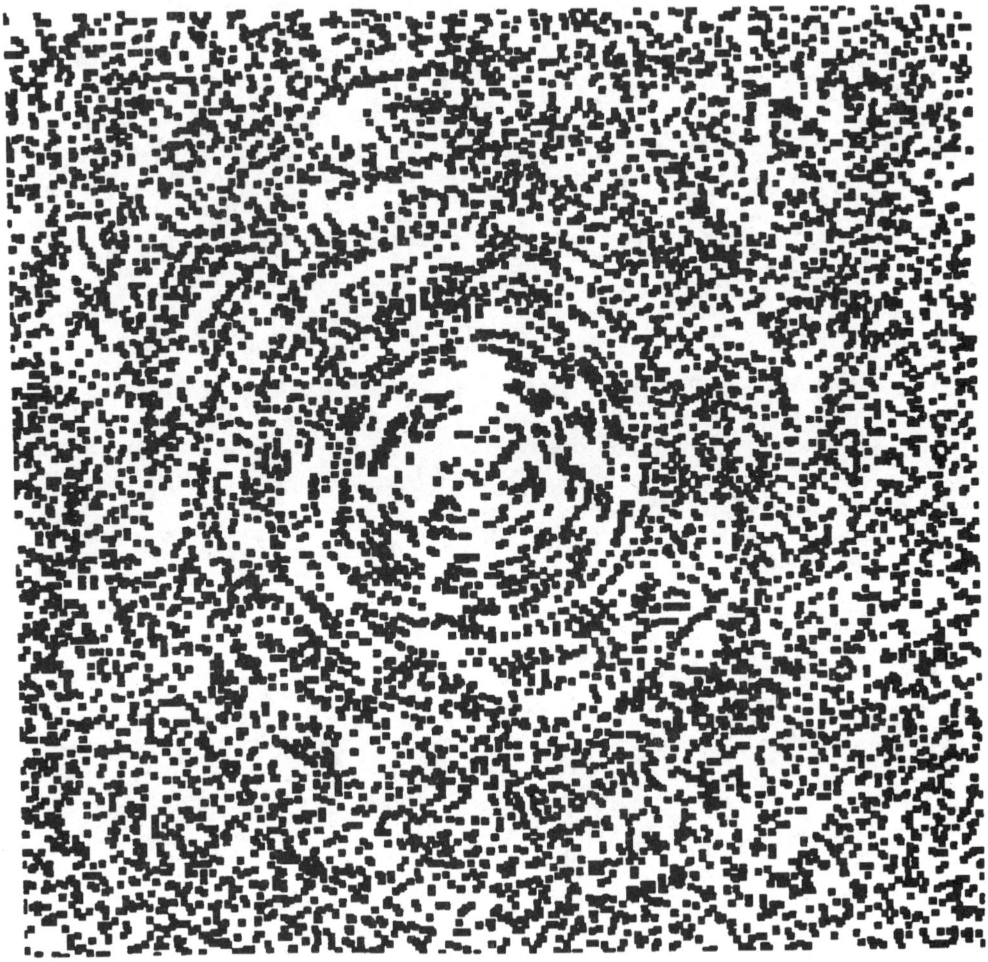

FIGURE 5. Demonstration of a Glass-pattern obtained by rotation. The circular structure is less visible when the size of the pattern is truncated. [After Glass (1969).]

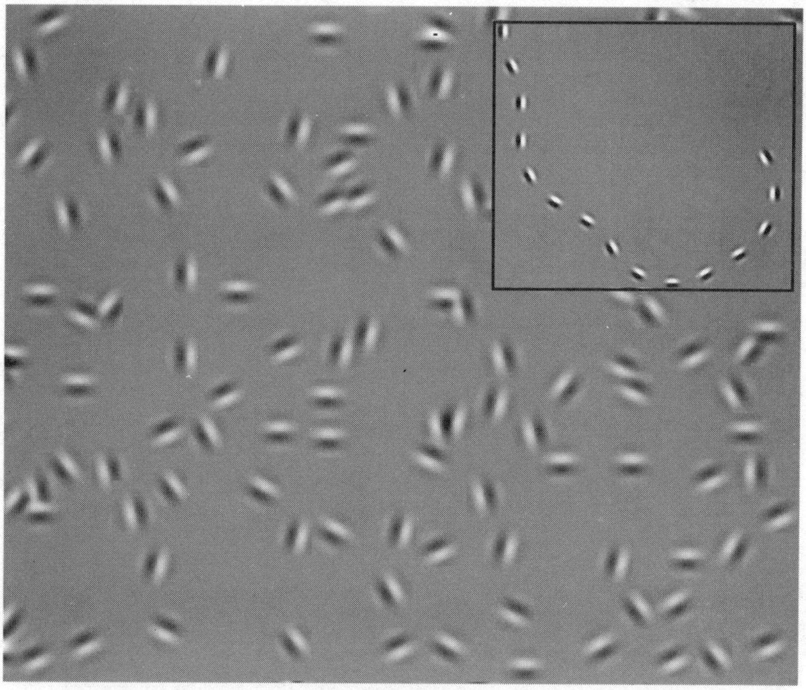

FIGURE 6. An array of randomly placed Gabor-patches in which a "snake"-shaped target (shown in the inset) appears hidden. [From Kovacs and Julesz (1993).]

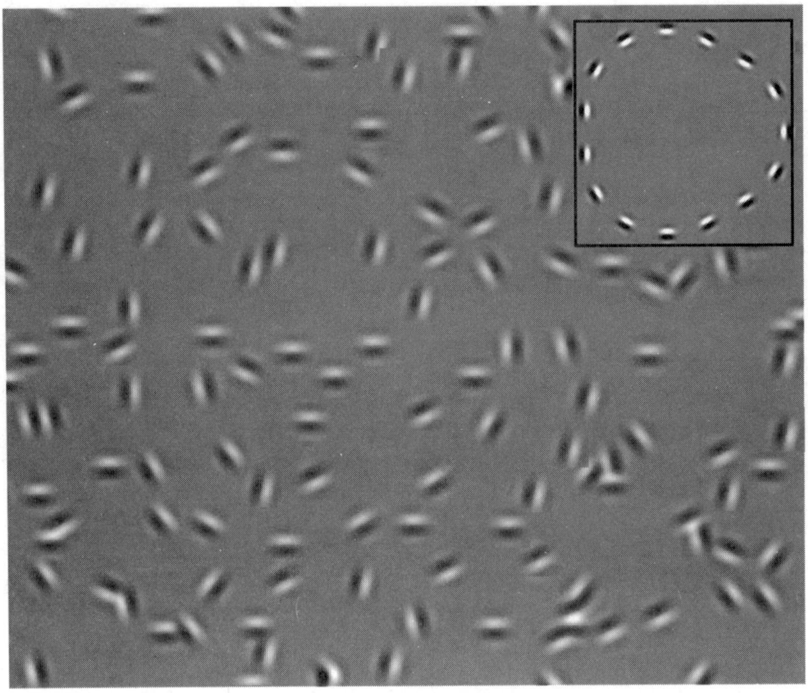

FIGURE 7. An array of randomly placed Gabor-patches in which a closed target (shown in the inset) appears to pop-out when briefly presented. [From Kovacs and Julesz (1993).]

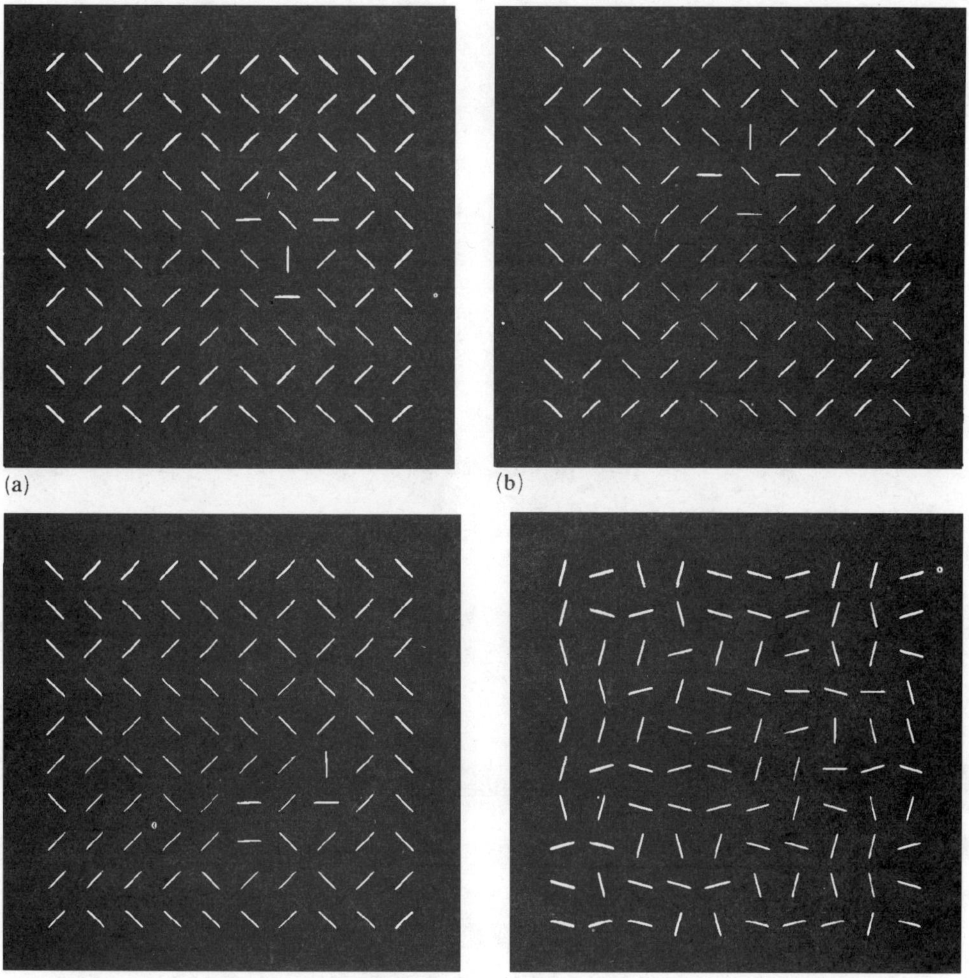

(a) (b)

(c) (d)

FIGURE 8. Context superiority effect of detecting line segments with orientation gradients. The cluster of horizontal and vertical line segments among diagonal ones is better detected when the cluster forms a schematic human face (left top), than for the other three cases in which the three horizontal and one vertical line segments are scrambled, or the line segments are jittered in their orientations (right bottom). [From Gorea and Julesz (1990).]

DIMACS Series in Discrete Mathematics
and Theoretical Computer Science
Volume 19, 1995

The Visual Perception of Surfaces, their Properties, and Relationships

Daniel Kersten and Suthep Madarasmi

ABSTRACT. Human perception of surfaces, their properties and relationships present formal problems that share similarities and differences with traditional studies of data clustering. We will describe three problems: occlusion, transparency, and lightness. For each we will show how the human visual system partitions or "explains" the image in terms of its causes--namely the shapes, materials, lighting, and depth relations. In occlusion the visual system segments an image into surfaces that are overlapping in the image. We will outline a Bayesian computational scheme called multi-layer segmentation that solves occlusion. Transparency perception demonstrates the partitioning of overlapping data based on material property. In transparency, a single image data point can belong to more than one grouping or surface. The third example illustrates human perception unraveling the hidden causes of shape, material and illumination implicit in an image.

The human visual system is a natural contrivance that is continually grouping and categorizing data that arrives in the form of image intensities. Much physiological research has gone into showing how receptive fields of single neurons can be understood in terms of the kind of information that is brought together. For example, it is well-known that image data along specific orientations funnels into "simple" and "complex" cells of the primary cortex of the brain. More recent neurophysiological research has shown that grouping of information in primary cortex and by single cells is just the beginning of the story. Information is transmitted from the primary visual area to a multitude of regions, the primary purpose of which may be getting information together that belongs together (Barlow, 1986). But we immediately ask ourselves: What belongs together? What are the criteria used for grouping? We believe that a careful study of the phenomena of perception can provide answers to guide both biological and computational studies of vision. In this paper, we'd like to describe some of our own experimental and theoretical work that pertain to these questions in a way that will hopefully interest both the mathematician and the psychologist.

The first step, not immediately apparent to a beginning student of perception, is to realize that an image is an encrypted form of useful information about object shape, object relations, material property, illumination, and viewpoint. It isn't phenomenally apparent

1991 *Mathematics Subject Classification.* Primary 68T10, 68U10, 92J30.
Supported by Air Force Office of Scientific Research grant 90-0274, National Science Foundation grant BNS-9109514, and the University of Minnesota/Army High Performance Computing Research Center.

that it is encrypted, because through the enormous processing power of our own visual systems, we experience the decrypted version of the image. Figure 1 shows a 3D surface plot of intensity data as a function of position x and y, taken from a natural image. It has been rendered in such a way as to make apparent the qualitative geometric features of the intensity pattern of the original image. It is non-trivial to spot the meaningful groupings of the data. The problem is that image intensities, depicted here as hills and valleys, do not uniquely specify the surface properties which the visual system is interested in. At the end of this paper, Figure 10 shows the same data represented as intensities in x and y.

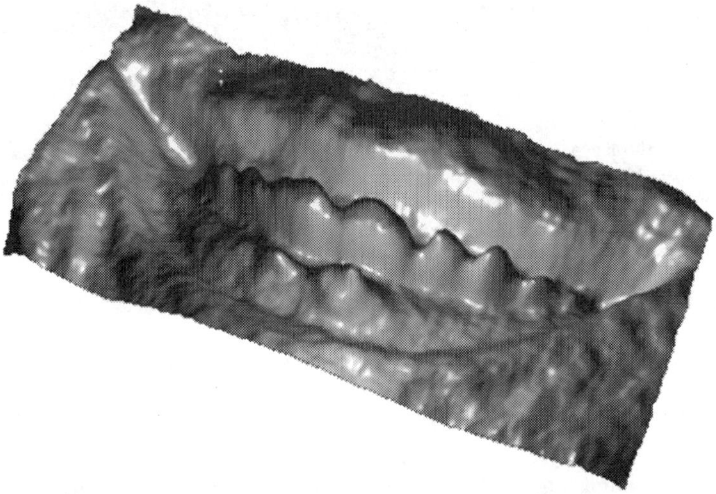

Figure 1. The intensities of an image of some familiar surfaces are represented as the height of a surface at spatial locations x and y, where the height is proportional to intensity. The data have been rendered as a smoothed plastic surface to highlight the form and structure the depth relations in the original image intensities.

There the data are in a form that our visual systems are built to decrypt and you can note that your perceptual system handles the required inferences automatically with little or no thought. So part of the visual decryption process is really an automatic inference of the causes (e.g. shape, material and illumination) of the image data. One needs to know how to compute these inferences and how to compare computational models with human perception.

This example shows that the grouping of image information must be based on a background theory of what constitutes the scene and how image data is generated from the scene being viewed. Grouping must take into account the fact that the intensity value of an image pixel depends on more than one cause in the scene. That is, partitioning should be based on scene causes such as shape and material, rather than image intensities. Because scenes consist of surfaces that may overlap, grouping must also be able to deal with surface occlusion--that is grouping processes must be able connect surface regions whose data support is not connected in the image. The perceptual problem of occlusion has clear relations to classical problems of data clustering based on similarity measures. Here similarity measures are based on depth, texture, or color of feature elements. However, even for occlusion, we will see that one cannot simply group image pixels based on the values of features, but one has to consider prior probabilities on the geometric regularities of the underlying surfaces and the form of their bounding contours.

In the sections below, we will first outline the Bayesian framework for making perceptual inferences. Then we will describe the problem of occlusion, and a Bayesian algorithm for grouping pixels likely to belong to the same surface that handles occlusion. Then we will describe some perceptual phenomena involving transparency and lightness that point to the challenges that future theoretical work faces.

1. Perception as Bayesian inference

The theory of Bayesian inference provides an abstract starting point for describing the perceptual process of image decryption (Clark & Yuille, 1990; Kersten, 1990; Knill & Kersten, 1991b). Visual systems estimate the causes, c, of image data d. This can be done by maximizing the posterior probability of c, given d. Or using Bayes' rule,

$$p(c \mid d) = \frac{p(d \mid c)p(c)}{p(d)},$$

maximizing the posterior probability is equivalent to maximizing the product of the prior, $p(c)$ and the likelihood, $p(d|c)$. These two probabilities represent two types of constraints that serve to determine the best bet of c. The likelihood, $p(d|c)$ says how well the data fit a particular model of the causes. The prior narrows the choices when there is more than one possible set of causes satisfying the likelihood constraint. The prior term, $p(c)$, does not depend on the data. It can be very specific, in the sense of the probability being concentrated (as delta functions) on a subspaces thereby effectively limiting the parameter space, c. Then there may be little ambiguity regarding the causes of the image data and the prior is incorporated into the choice of likelihood function. When $p(c)$ highly constrains the solution, the posterior is a *specific model* of the causes of the image. At the other extreme, if c is of higher dimension than the data, and p(c) is spread out, the data underdetermine the solution. Of course, expressing perception as a Bayesian inference is just the starting point. Some basic problems in applying the Bayesian framework are: the choice of which causes should be inferred out of myriad possibilities; modeling the likelihood and priors for this choice; finding a method to maximize the posterior probability; and experimentally testing the resulting scientific model. Vision is complex, and the problem has been subdivided into intermediate stages and modules by both computer and biological studies of vision. The resulting algorithms or theories can be primarily data-driven--that is, they emphasize processing the input data in ways that may be generally applicable to subsequent visual tasks. An algorithm or theory of vision may be model-driven--that is, it may focus on a specific application or task.

1.1. The data-driven approach. It seems clear that in order to get image information in a useful form, it has to be brought together in an organized fashion so that we can assign labels to regions that mean something--pencil, water or George. A reasonable intermediate stage would be to group pixels that correspond to single surfaces based on their properties, relationships and identities. At first blush, one might think that most of the work has been done by the very nature of the world and the optics of images--after all, natural images appear so organized. Objects themselves tend to consist of matter that clumps together and is bounded by a surface. Images to some extent preserve the connectivity of the source data. But grouping information based on intensity or color doesn't work because of the enormous fluctuation in lighting, and viewpoints. Using contrast or edges is certainly better. But there is a lot of fluctuation in image intensities for other reasons including varying surface texture and photon noise. The grouping of regions based on texture has been studied in both psychological and computational domains (Julesz, 1975; Malik & Perona, 1990). Useful information about surfaces (e.g. slant and tilt) can be extracted from image data measurements such as texture statistics, but general purpose applications have been difficult to develop.

1.2. The model-driven approach. At the other extreme from a primarily data-driven process of interpreting image data, is a "top-down" strategy in which information from a large dimensional input is brought together based on a specific hypothesized model of what the information source is. The idea of "deformable templates" is one such example that has been applied to detecting eyes (Yuille, Cohen & Hallinan, 1988). As another example, if the change in image intensities is due to the movement of lips or facial muscles, then a physics-based model of how facial muscles move and deform the image of a face can strongly limit the interpretation of the image data (Terzopoulos & Waters, 1993). This strategy limits the parameter space and in effect funnels information useful for the task of estimating facial expressions.

1.3. Middle-level constraints. It seems likely that biological visual systems combine both data- and model-driven strategies. Invertebrates with a limited repertoire of required behaviors can sacrifice some adaptability with a small discrete set of mechanisms each of which embodies a highly specific model for interpreting the perceptual data. Human perception is an example of a highly flexible system which may achieve its adaptability by using a large number of specialized modules. However, human vision also has considerable versatility. We can efficiently detect symmetry patterns in otherwise random dots and extract shape from novel synthetic images, such as random fractals. We can infer the form of the Martian landscape, albeit with some inaccuracies and well-known mis-interpretations, despite our lack of models for specific Martian object classes.

One way to combine the data- and model-driven approaches is to select data for possible further processing by comparing the discrepancy between the input data and the model's estimate of it. Given the enormous range of possible object models and tasks, it is difficult to see how to do this computation with the flexibility and speed of human vision without some intermediate organizational principles. Starting early in this century, the Gestalt psychologists studied how human vision organizes image information based on global principles (Rock & Palmer, 1990). What they lacked, however, was an appreciation for how this information could be used to solve problems related to natural visual function--that is an ecological perspective (Gibson, 1950). It is with the advent of computer vision that Gestalt and ecological studies of perception have promise of becoming integrated.

Marr (1982) was instrumental in establishing a bridge between several disciplines studying vision. He emphasized the importance of considering the functional goals of vision when searching for features and principles of organization for primarily data-driven vision. Functional considerations also motivate the search for middle-level constraints. The argument is that vision cares about objects and surfaces, so image data should be partitioned in an intermediate stage as representations of the surfaces and their properties that gave rise to the data. This intermediate stage would provide an input for recognition that is robust with respect to illumination variation and noise. This idea is not new (Barrow & Tenenbaum, 1978; Marr, 1982), but understanding how human vision might achieve this goal requires advances from computational and empirical studies. Intermediate level constraints can be used at various levels; for example, to group edge segments into contours (Cox, Rehg & Hingorani, 1993).

In the next section (Section 2) we describe perceptual phenomena of occlusion and transparency that illustrate middle-level representations and constraints used by human vision. We show how these constraints can be employed in an algorithm for solving occlusion. In Section 3, we will show evidence that the perception of object properties such as shape and material is mediated through representations of surface shape, material and lighting.

2. Occlusion and transparency

2.1. Perception of occlusion. One of the perceptual problems that shares similarities to formal problems of data partitioning is segmenting an image into regions with similar features (Duda & Hart, 1973). But what makes segmentation particularly hard is surface occlusion. In Figure 2, although the dark vertical bar divides the gray horizontal wedge into two sections, one infers that these two parts belong together. In visual perception, our ability to solve occlusion is closely connected to the figure-ground problem and involves the cooperation of numerous constraints from detecting low-level features such as collinearity of the bounding edges and similar textures, to the use of specific high-level models for familiar figures as is illustrated in the well-known "Dalmatian dog" picture (e.g. page 14, Gregory, 1970). Part of the process of detecting occluding surfaces relies on our ability to perform this grouping based on depth of surfaces, including stereoscopic depth (Nakayama, Shimojo & Ramachandran, 1990). One might want to group features in specific depth planes. However, there are a number of reasons why it may be better to organize information from various sources by attaching it to specific "layers" likely to belong to the same surface, rather than based on depth. As the wedge in Figure 2 shows, the distance from the eye to a single surface can change drastically. It is here where we see the utility of an intermediate-level organizing principle--grouping based on surface properties. Here the grouping is not data clustering according based simply on commonalty of image features, but also on likely surface membership. The next section illustrates an approach we call "multi-layer" segmentation to do this grouping (Madarasmi, Kersten & Pong, 1993; Madarasmi, Pong & Kersten, 1994). Closely related work can be found by Finkel and Sajda (1992) and Darrell and Pentland (1991).

2.2. An algorithm for grouping data, given occluding surfaces. The central idea behind multi-layer segmentation can most easily be illustrated by the problem of grouping input data that consists of depth, rather than image intensity data. The approach is easily extended to image data. For an example applied to stereo image pairs, see (Madarasmi et al., 1993). We consider "range data" which is a possibly sparse and noisy set of depth measurements, *d*, at each pixel site of the input. The goal is to assign these measurements to layers with each layer determining a label. Measurements with the same label should belong to the same surface regardless of whether these measurements are at the same depth or not. One of the key ideas is that sparse and noisy data give each other mutual support when they are fit well by a "smoothing layer" which, in effect, tries to fit an elastic membrane to the data (Figure 2). The smoothing layers allow one to interpolate across areas where there is little or noisy data, or occlusion.

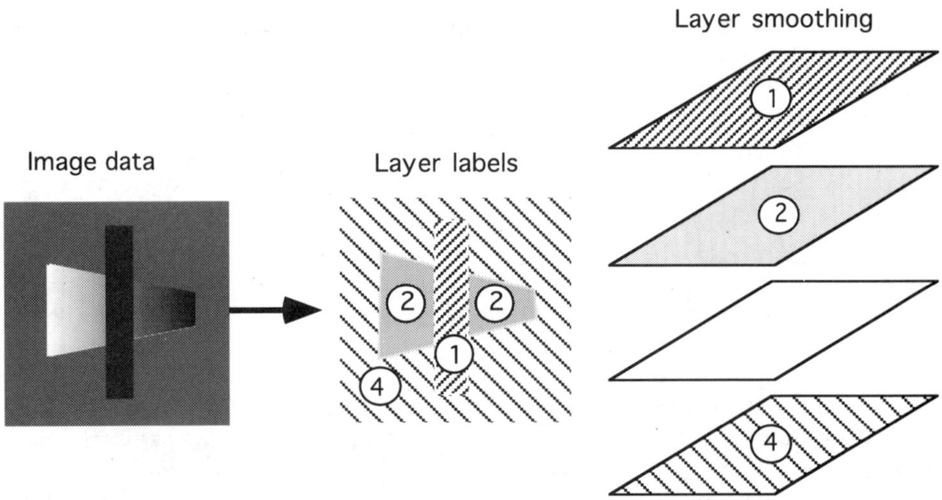

Figure 2. Pixels that have some feature in common (e.g. intensity, texture, depth, depth gradient) are linked based on layers whose data support can come from regions of the image separated because of occlusion. The middle-level constraint that surfaces have relatively smooth interiors is implemented by local propagation of smoothness constraints in a process called "layer smoothing". Layers compete to explain the data. The white layer "lost" by having no pixels assigned to it.

Bayes' rule expresses the posterior probability of a set of depths, u, and a set of layer assignments, C conditional on image data, d in terms of the likelihood and prior:

$$p(u,C|d)=\frac{p(d|u,C)p(u,C)}{p(d)}$$

We use probability models that are expressible as Gibbs distributions so that:

$$p(d|u,C)p(u,C)\propto e^{-E/T}$$

where E is a cost function. E is also called an energy function, because of the formal relation to statistical physics (Geman & Geman, 1984). T is a parameter (temperature) that controls the spread of the distribution. We would like to find u and C that maximize the posterior probability, or equivalently minimize E. Assuming statistical independence, we can write E as a sum of energy potentials:

$$E=\lambda_D V_D(u,C,d)+\lambda_S V_S(u,C,a)+\lambda_G V_G(u,C,a)+\lambda_E V_E(C)$$

where a is the activation level for each layer described below. The energy potentials include the data matching term V_D which corresponds to the Bayes likelihood, and the prior terms: the intra-layer smoothness V_S, the inter-layer "gap" term V_G, and the edge (or region configuration) potential V_E. Each of the above potentials can be expressed as a mathematical function to capture the underlying costs.

The data matching term V_D is a measure of the fidelity of the estimated depth value

$u_{C_i i}$ in the appropriate, classified layer C_i to the datum, d_i at each site i. Each pixel i is assigned to C_i, one of L layers. C_i is an element of {1,...,L}. There are no unassigned pixels, and each pixel is assigned to only one layer.[1] The data fit constraint is equivalent to assuming Gaussian measurement noise and is given by:

$$V_D = \sum_{i=1}^{M} (u_{C_i i} - d_i)^2$$

$u_{C_i i}$ represents the depth at the corresponding site of the layer where pixel i is classified.[2] There are M pixels in the image and M sites in each layer.

The equation for the smoothness term is given by:

$$V_S = \sum_{i}^{M} \sum_{l}^{L} \sum_{j \in N_i} V_s(u_{li}, u_{lj}) a_l$$

where N_i are the neighbors of i, V_S is the local smoothness potential, and a_l is the activation level for layer l defined by the proportion of pixels in the image belonging to the layer. This activation level is zero when a layer has no assigned data and therefore effectively turns off the contribution of that layer to the energy function.

The local smoothness potential is given by:

$$V_s = \sigma_1(u_{li}, u_{lj}) + \tau \sum_k \sigma_2(\Delta_k u_{li}, \Delta_k u_{lj})$$

where τ is the weighting term between first and second order differences. The first order model penalizes large first derivatives and favors constant solutions. The second order term penalizes large second derivatives and encourages planar surfaces as solutions. The Δ_k is

---

[1] Both of these restrictions can be relaxed. The restriction on unassigned pixels could be removed to allow for outliers. Transparency would require that pixels have multiple layer assignments.

[2] A data constraint for a pair of stereo images could be modeled to encourage matching similar intensities, (d^R, d^L) and contrasts (i.e. differences) in these intensities in the left and right eyes at pixel location (x_i, y_i):

$$V_D = \sum_{i=1}^{M} \left| d^R(x_i, y_i) - d^L\left(x_i + u_{C_i i}, y_i\right) \right|$$

$$+\lambda \sum_{i=1}^{M} \sum_{j \varepsilon N_i} \left| \left(d^R\left(x_j, y_j\right) - d^R\left(x_i, y_i\right)\right) - \left(d^L\left(x_j + u_{C_i i}, y_j\right) - d^L\left(x_i + u_{C_i i}, y_i\right)\right) \right|$$

M is the number of pixels in the image, C_i is the layer classification for site i, and u_{li} is the estimated disparity at layer $l{=}C_i$. λ controls the relative contribution of these two data constraint terms. See Madarasmi et al., (1993a).

the difference operator in direction k. The sigma function is designed to encourage piece-wise smooth surfaces:

$$\sigma_n = \begin{cases} \dfrac{(a-b)^2}{\beta_n} & \text{if } (a-b)^2 < T_n \\ 1 & \text{otherwise} \end{cases}$$

As a consequence of energy minimization, missing and hidden values are interpolated assuming small first and second order derivatives.

The gap constraint, V_G, helps to minimize the number of layers required to account for the data. It also prevents nearby pixels that are in different layers from being assigned the same values of u:

$$V_G = \sum_{i}^{M} \sum_{l \neq C_i}^{L} \sum_{j \in N_i} V_g(u_{C_i i}, u_{lj}) a_l a_{C_i}$$

$$V_g(u_{C_i i}, u_{lj}) = \begin{cases} 0 & \text{if } \left| u_{C_i i} - u_{lj} \right| \geq T_n \\ 1 & \text{otherwise} \end{cases}$$

The a's are the activation levels defined above. The gap constraint ensures that if a site i belongs to layer C_i, then neighboring points j for layers not equal to C_i must have different depth values u_{lj} than $u_{C_i i}$. The product of activation levels in the above equation plays a crucial role in determining which of the two layers involved in the gap constraint should be assigned the data. When two layers have an associated gap cost, the system is biased towards assigning the pixel to a layer with more data support (i.e. higher activation level).

Boundaries are encouraged to be smooth by giving appropriate weights to V_E which specifies preferences for particular 3x3 configurations of the attributes (Figure 3).

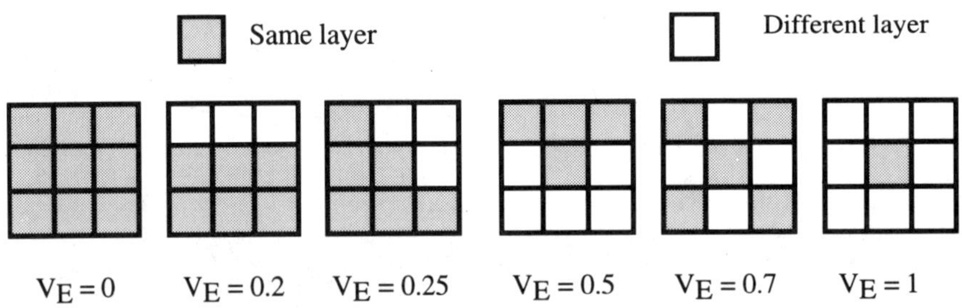

Figure 3. The local probability of various configurations are reflected in the costs V_E.

The energy function is typically non-convex and convergence to a solution can be a problem. There exist a number of methods for finding minimum energy, including graduated non-convexity (Blake & Zisserman, 1987), and applications of mean-field theory to convexify the energy function (Geiger & Yuille, 1991). We ran simulations on a

Connection Machine (CM-2) using the Gibb's Sampler with simulated annealing (Geman & Geman, 1984) together with intra-layer smoothing which is described in more detail in (Madarasmi et al., 1994). Figures 4 and 5 show some results from images of occluded surfaces.

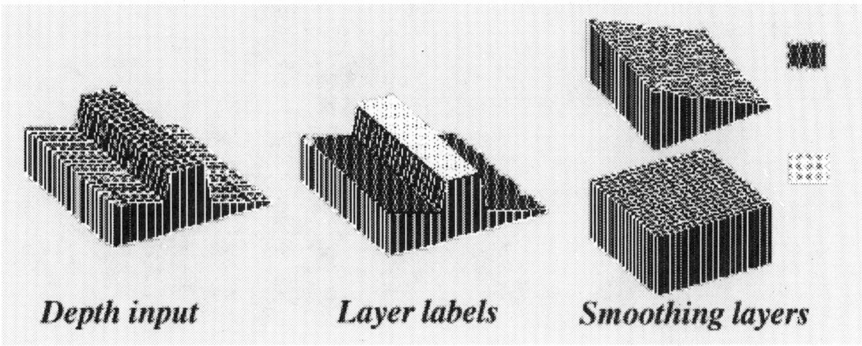

Figure 4. The layer assignment produced using the multi-layer algorithm for depth data in which a horizontal plane covers a slanted one. There are two labels.

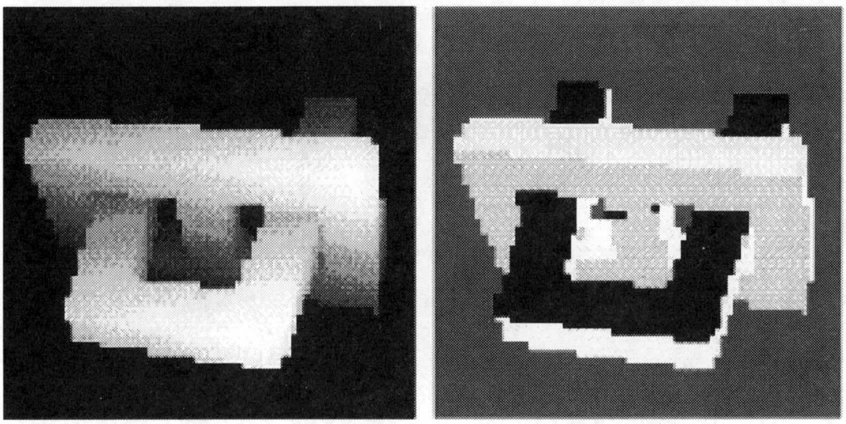

Figure 5. The segmentation produced using the multi-layer algorithm for a set of depth data. The left panel shows the input data where depth is coded as image intensity--the lighter regions are closer to the sensor. The right panel shows the labels assigned by the multi-layer algorithm. There are five dominant layers whose labels are represented by various gray-levels.

2.3. Perception of transparency: A one-to-many mapping. In the previous section (2.2), we saw how the multi-layer algorithm can be used to assign pixels that are likely to belong to a given occluded surface to different layers. But occlusion is, in a formal sense, a special case of another visual problem that human perception solves. Figure 6 shows overlapping diamond-shaped rings. Look at one of the overlapping pairs (e.g. upper left panel of Figure 6a), and ignore for the moment the fact that the image is a stereo pair. You should see one transparent ring covering another ring. The conditions for

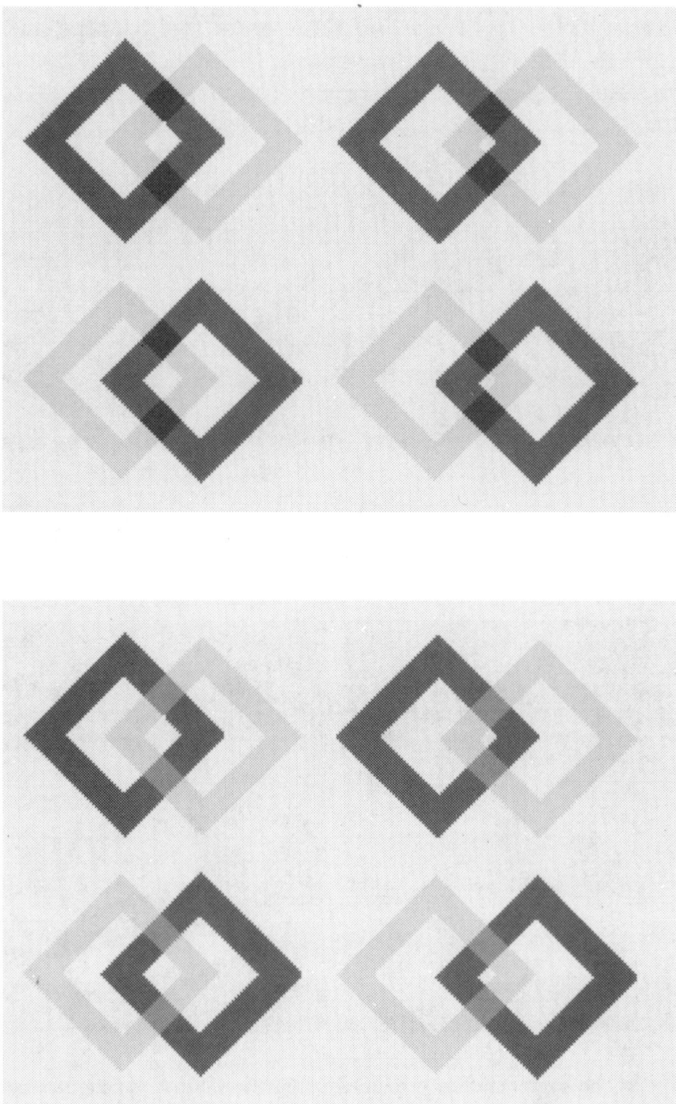

Figures 6a and b. When the transparency information is consistent with either the dark or light rings to be in front, then there is no conflict with stereo (Panel a). When the transparency information is consistent with just the light ring to be in front, then there is conflict when stereo information puts the dark one in front (lower panel of 6b). To see the stereoscopic effect, the two halves should be fused by crossing your eyes.

transparency to be seen have been well-studied. For example, if the intensities at the edge intersection fail certain conditions, transparency is not seen (Metelli, 1974). When viewed as a problem in grouping pixels likely to belong to the same surface, we see that the mapping is no longer one-to-one. The layer assignment, C, is no longer a function. Instead, C_i can be thought of as a non-empty subset of {1,...,L}. Pixels share membership in partitions. This is a consequence of the fact that the causes map many-to-one: a given image intensity is the combination (multiplicative and/or additive) of two sources. Further there is a reasonably a large set of constraints that govern the grouping. These constraints include contour continuation, and stereoscopic depth relationships (Kersten, 1991). If the left and right sides of Figure 6a are fused by crossing the eyes, one sees a light transparent ring floating in front for the top pair, and a dark ring in front for the bottom pair. If one changes the intensity of the intersection patches (Figure 6b), one can still see the light ring floating in front (top pair of 6b). But for the bottom pair of Figure 6b there is a now a conflict between the stereo and the transparency information and it is difficult to see the dark ring floating in front. Other recent studies are consistent with the conclusion that transparency is intimately tied to the nature of surfaces and their depth relationships (Nakayama et al., 1990; Trueswell & Hayhoe, 1993).

Transparencies are often ambiguous--the transparent and opaque regions can alternate perceptually. But, transparency is an example of a problem that cannot be easily solved by local resolution of ambiguity. Figure 7 shows that any local process that decides which contour is in front must agree with the other decisions at distant intersections. This agreement is straightforward to apply for noise-free occluding surfaces, but is not so for transparency. One percept that is never or rarely seen is both horizontal edges of a single transparent square to appear in front. This process of disambiguation is fast and sophisticated enough to interact with the perception of 3D motion (Kersten, Bülthoff, Schwartz & Kurtz, 1992).

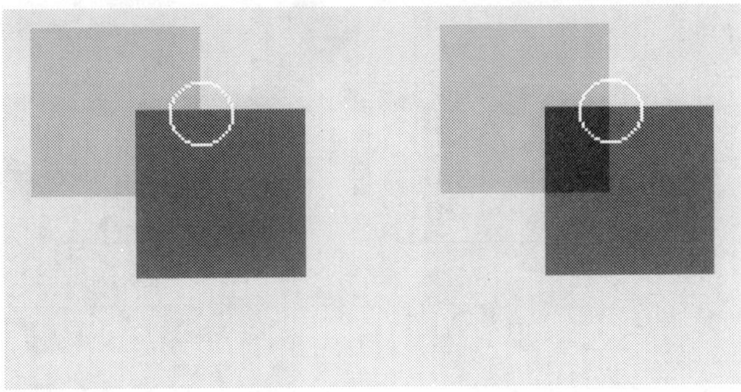

Figure 7. Look at the contour intersection inside the white circle. Is the vertical or horizontal edge in front? For the case of occlusion (left pair of squares), the "T" junction indicates that the horizontal edge is in front. For transparency (right pair of squares), the assignment can be ambiguous. If local processes make the decision, then global processing must be used to ensure consistency.

3. Shape, material, and illumination

In this section, we describe some phenomena involving the perception of lightness that further underscore the fact that visual perception seeks out the encrypted causes of the data. Because there are multiple causes of any given image intensity or intensity change, decrypting this information seems to require new ways of thinking about how we see even basic surface attributes. For example, how do we tell whether a surface is painted with a highly reflective white paint, or with a low-reflectivity dark gray paint? Reflectance is a surface property--it refers to the fraction of light returned by a surface. How human perception infers reflectance is a problem of lightness perception (Beck, 1972; Gilchrist, 1977).

One reason it is difficult to measure surface reflectance from image data is because the illumination level falling on a surface can give rise to enormous differences in intensities for a given surface reflectance. Even when the mean level of illumination on a surface remains fixed over time, spatial variations can also compound the problem. One of the first algorithms to filter out spatial changes in image intensity due to illumination was described by Land and McCann (Land & McCann, 1971). They pointed out how one could account for illusions of brightness noted earlier (e.g. O'Brien, 1958), by assuming that brightness, in part, was determined by the visual system's attempt to extract reflectance as a surface invariant. Figure 8 shows a stimulus similar to Land and McCann's "two-squares-and-a-happening" illusion. There are two apparently rectangular patches, one of which appears darker than the other. In actuality, the measurement of the intensity along an interior horizontal axis would show the left and right patches to have identical forms. There are two types of explanations offered for this illusion: mechanistic and functional.

The mechanistic explanation has been the basis of a number of computer vision algorithms (see (Hurlbert, 1986) for a review) and goes as follows. Sharp spatial changes in illumination are amplified relative to gradual changes. Then the small variations get filtered out by a threshold operation. So far this process corresponds to the classical idea of

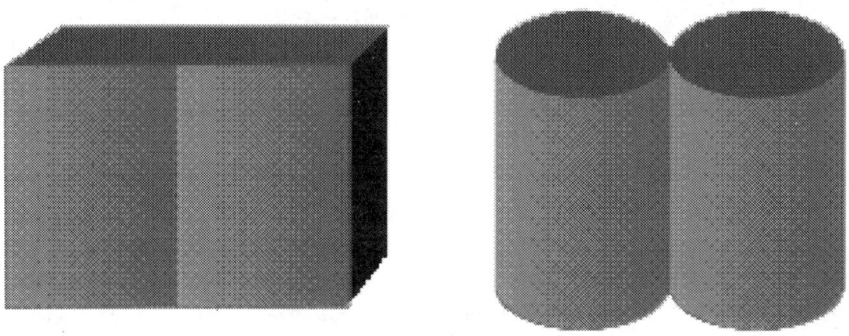

Figure 8. The left panel shows a stimulus similar to "two-squares-and-a-happening" of Land and McCann. Two identical gradients of intensity appear to have different lightnesses--the left rectangular patch appears darker than the right rectangular one. The right panel shows "two-cylinders-and-no-happening" in which the stimulus appears to be a pair of cylinders. The right panel has the same horizontal luminance distribution as the left panel. When the luminance gradients are seen as two cylinders, the lightness illusion goes away, and the two cylinders in the right panel appear to have identical lightnesses.

edge detection: smoothing, spatial differentiation, and then thresholding. If this signal is then re-integrated, one obtains an output in which the left region is uniform and a constant lower value than the right region. Part of the attraction of this filter algorithm is that the machinery (apart from the re-integration) is known to exist in the retina in the form of lateral inhibitory mechanisms.

The functional explanation, up to a point, is consistent and complementary to the mechanistic explanation. It says that the visual system is attempting to infer material reflectance because material reflectance is an object property and is invariant over illumination and viewpoint. The assumption is that reflectance changes are sharp and spatial changes in illumination are gradual--"shadows are soft". A mechanism which filters out the slow changes should leave a visual signal which represents reflectance. One of us, together with David Knill, pointed out that there is an alternative visual condition for which the functional explanation would suggest no illusion, whereas the classic mechanistic explanation would (Knill & Kersten, 1991a).

Knill and Kersten's idea was that gradual changes in intensity can also be due to gradual changes in shape, as well as in illumination. If the visual system is clever enough to decipher which is which, then the lightness illusion might go away when there is auxiliary information to indicate gradual shape as the "explanation" to the visual system of the two identical gradients. This is indeed the case, as can be seen in Figure 8. The right pair appears smoothly curved and does not show a strong lightness difference. The left pair of patches is seen as flat and the two halves appear to have different lightnesses. Figure 9 schematizes the two perceptual "explanations" for the image data. But what kind of visual mechanism could instantiate this function? One possibility is to try to compute three representations in Figure 9. This is closely related to the idea of "intrinsic images" (Barrow & Tenenbaum, 1978). It would require extending the multi-layer approach to include additional layers for each surface to estimate the reflectance and illumination. The data term of the energy potential could be expanded to include multiple causes of the data, d, and independent priors on reflectance (R), shape (S), and illumination (I):

$$p(R, S, I \mid d) \propto p(d - f(R, S, I))p(R)p(S)p(I).$$

The image formation function (e.g. a computer graphics geometric projection and rendering function) is $f(R,S,I)$. The actual implementation of such a set of constraints is not a straightforward extension of the multi-layer algorithm described above. In general one needs some efficient way of evaluating the consistency between one's guess of R, S, and I and the image data. Then one modifies this guess to get closer to the true answer. When we have multiple causes of d, even the problem of checking for consistency takes computer time which would be comparable to the complexity of the rendering problem in computer graphics. Although there is progress on the problem of computing surface properties from images in limited contexts (Poggio, Gamble & Little, 1988; Sinha & Adelson, 1993), a general robust solution is still a challenge for the future.

4. Summary

In this paper, we adopted a broad view of the relationship between finding structure in data, and the problem of vision. The perceptual problem of occlusion has clear relations to classical problems of data clustering based on similarity measures. Here similarity measures are based on depth, texture, or intensity of its feature elements. However, even for occlusion, we noted how one cannot simply group image pixels based on the values of features, but one has to consider prior probabilities on the geometric regularities of the

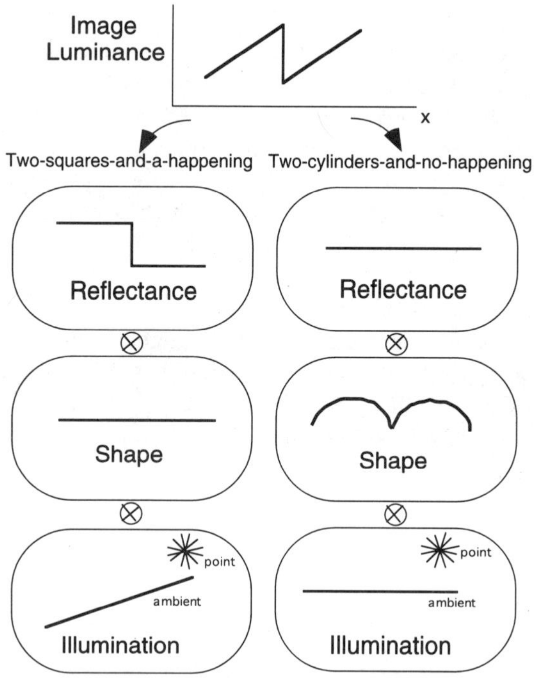

Figure 9. Functional explanations for the "two-squares-and-a-happening" and "Two-cylinders-and-no-happening" phenomena. Note that, as in Figure 8, the intensity gradients are not necessarily linear ramps, depending on the reflectance model and illumination.

underlying surfaces and the form of their bounding contours. The case of transparency, is even more interesting because, at least from a formal point of view, segmenting transparent surfaces involves one-to-many mappings--a consequence of the fact that a given image pixel intensity comes from more than one surface point in the world. A robust solution to the transparency problem would likely solve the occlusion problem too. Inferring occlusion is difficult for real images because of noise which makes local confidence in T junctions low. The final example of the two-cylinders lightness illusion carried the theme of inferring surfaces, their material and relations a step further. The two-cylinders lightness demonstration showed how the human perceptual system does something akin to mapping pixels in a one-to-many fashion. The outputs do not just correspond to different surface depth relations and material, but to the varied and general causes of image intensity changes including surface shape, surface material and lighting, for each surface. The use of intermediate-level representations and constraints to account for the perception of objects is a promising approach for a computational account of human vision. It may also provide stimulating problems for formal studies of pattern recognition and data organization.

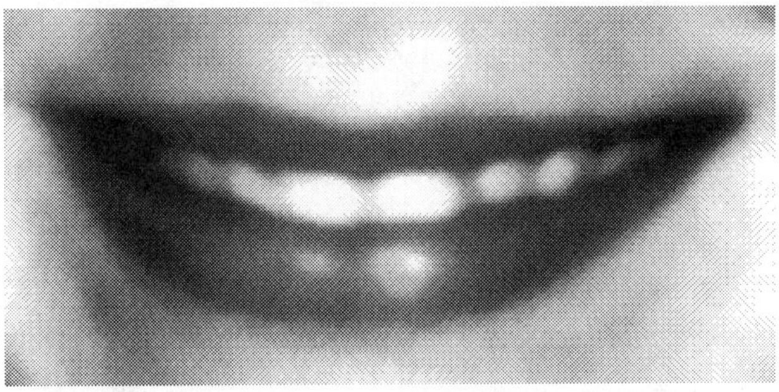

Figure 10. The data from Figure 1, plotted as gray-levels in x and y. The causes of the intensity changes due to shape and material are now immediately apparent. The image is of a smiling human mouth. The central ridge in Figure 1 is high because it derives from sparkling white teeth. The second smaller set of bumps below is not from teeth, however, and corresponds to specularities on the lips. The deep valley surrounding the two ridges is deep because of the darker pigmentation in the lips than the teeth or surrounding skin, and its real geometric position is closer to the camera than any of the other regions.

References

Barlow, H. B. *Why have multiple cortical areas*? Vision Research, **26** (1986), 81-90.

Barrow, H. G. & Tenenbaum, J. M. *Recovering Intrinsic Scene Characteristics from Images.* A. R. Hanson & E. M. Riseman. (Ed.) Academic Press, New York, (1978). (pp. 3-26).

Beck, J. *Surface Color Perception.* Cornell University Press, Ithaca, NY, (1972).

Blake, A. & Zisserman, A. *Visual Reconstruction* The MIT Press, Cambridge, Massachusetts, (1987).

Clark, J. J. & Yuille, A. L. *Data Fusion for Sensory Information Processing* Kluwer Academic Publishers, Boston, (1990).

Cox, I. J., Rehg, J. M. & Hingorani, S. *A Bayesian multiple hypothesis approach to edge grouping and contour segmentation.* International Journal of Computer Vision, **11** (1993), 5-24.

Darrell, T. & Pentland, A. *Discontinuity models and multi-layer description networks.* M.I.T. Media Lab Vision and Modeling Group, **Tech. No. 162** (1991).

Duda, R. O. & Hart, P. E. *Pattern classification and scene analysis* John Wiley & Sons., New York., (1973).

Finkel, L. & Sajda, P. *Object discrimination based on depth-from-occlusion.* Neural Computation, **4** (1992), 901-921.

Geiger, D. & Yuille, A. *A common framework for image segmentation.* International Journal of Computer Vision, **6** (1991), 227-243.

Geman, S. & Geman, D. *Stochastic relaxation, Gibbs distributions, and the Bayesian restoration of images*. I.E.E.E. Transactions Pattern Analysis and Machine Intelligence, **PAMI-6** (1984), 721-741.

Gibson, J. J. *The Perception of the Visual World.* Houghton Mifflin, Boston, MA, (1950).

Gilchrist, A. L. *Perceived Lightness Depends on Perceived Spatial Arrangement*. Science, **195** (1977), 185-187.

Gregory, R. L. *The Intelligent Eye* McGraw-Hill Paperbacks, New York, (1970).

Hurlbert, A. *Formal connections between lightness algorithms*. J. Opt. Soc. Am. A, **3** (1986), 1684-1693.

Julesz, B. *Experiments in the Visual Perception of Texture*. Scientific American, **232** (1975), 34-43.

Kersten, D. *Statistical limits to image understanding* C. Blakemore. (Ed.) Cambridge University Press, Cambridge, (1990).

Kersten, D., Bülthoff, H. H., Schwartz, B. & Kurtz, K. *Interaction between transparency and structure from motion*. Neural Computation, **4** (1992), 573-589.

Kersten, D. J. *Transparency and the Cooperative Computation of Scene Attributes* M. Landy & A. Movshon. (Ed.) M.I.T. Press, Cambridge, Massachusetts, (1991). (pp. 209-228).

Knill, D. C. & Kersten, D. *Apparent surface curvature affects lightness perception*. Nature, **351** (1991a), 228-230.

Knill, D. C. & Kersten, D. K. *Ideal Perceptual Observers for Computation, Psychophysics, and Neural Networks* R. J. Watt. (Ed.) MacMillan Press, (1991b).

Land, E. H. & McCann, J. *Lightness and retinex theory*. J. Opt. Soc. Am., **61** (1971), 1-11.

Madarasmi, S., Kersten, D. & Pong, T.-C. *The computation of stereo disparity for transparent and for opaque surfaces* C. L. Giles, S. J. Hanson & J. D. Cowan. (Ed.) Morgan Kaufmann Publishers, San Mateo, CA, (1993).

Madarasmi, S., Pong, T. C. & Kersten, D. *An energy minimization approach to surface segmentation using a multi-layer representation*. submitted to I.E.E.E. Transactions on Pattern Analysis and Machine Intelligence, (1994).

Malik, J. & Perona, P. *Preattentive texture discrimination with early vision mechanisms*. Journal of the Optical Society of America A, **7** (1990), 923-932.

Marr, D. *Vision: A Computational Investigation into the Human Representation and Processing of Visual Information.* W.H. Freeman and Company, San Francisco, CA, (1982).

Metelli, F. *The perception of transparency*. Scientific American, **230** (1974), 91-98.

Nakayama, K., Shimojo, S. & Ramachandran, V. S. *Transparency: relation to depth, subjective contours, luminance, and neon color spreading*. Perception, **19** (1990), 497-513.

O'Brien, V. *Contour Perception, Illusion and Reality*. Journal of the Optical Society of America, **48** (1958), 112-119.

Poggio, T., Gamble, E. B. & Little, J. J. *Parallel integration of vision modules*. Science, **242** (1988), 436-440.

Rock, I. & Palmer, S. *The legacy of Gestalt Psychology*. Scientific American, **263** (1990), 84-90.

Sinha, P. & Adelson, E. *Recovering reflectance and illumination in a world of painted polyhedra*. Proceedings of Fourth International Conference on Computer Vision. Berlin. (1993) 156-163.

Terzopoulos, D. & Waters, K. *Analysis and synthesis of facial image sequences using physical and anatomical models*. I.E.E.E. Transactions on Pattern Analysis and Machine Intelligence, **15** (1993), 569-579.

Trueswell, J. C. & Hayhoe, M. M. *Surface segmentation mechanisms and motion perception*. Vision Research, **3** (1993), 313-328.

Yuille, A. L., Cohen, D. & Hallinan, P. *Facial feature extraction by deformable templates*. Harvard Robotics Laboratory, **88-2** (1988).

First author address:
PSYCHOLOGY DEPARTMENT, UNIVERSITY OF MINNESOTA, MINNEAPOLIS, MN 55455
Current address: Max-Planck-Institut für biologische Kybernetik, 72076 Tübingen, Germany
Email: kersten@eye.psych.umn.edu

Second author address:
DEPARTMENT OF COMPUTER AND INFORMATION SCIENCES, UNIVERSITY OF MINNESOTA, MINNEAPOLIS, MN 55455
Current address: NECTEC, Rama VI Road, Rajthevi, Bangkok 10400, Thailand
Email: madarasm@nwg.nectec.or.th

DIMACS Series in Discrete Mathematics
and Theoretical Computer Science
Volume 19, 1995

VISUAL COMPUTATIONS AND DOT CLUSTER

STEVEN W. ZUCKER

ABSTRACT. We propose that the visual perception of dots underlies clustering operations when more domain specific information is not available. An early approach to labeling dot clusters from this perspective is reviewed, and some more modern experiments in curvilinear (curve) grouping are then illustrated. The process takes place in two stages. First is the inference of a discrete tangent field, i.e., an estimated representation of quantized tangents and curvatures at each possible trace point along the curve. We then present a new algorithm for inferring global curves through this tangent field, by inferring a covering of the global curve. The elements of the covering are short splines, each of which moves according to a potential distribution computed from the tangent field. The algorithm bears some analogy to biological mechanisms for curve detection, and hence supports the opening proposition.

1. INTRODUCTION

As a first step toward discovering structure in (two-dimensional) data sets, it is common to simply look at them [19]. While this process does not apply any domain-specific models to the data, it does suggest that the properties of the human visual system are relevant; in fact, one characterization of vision is the extraction of salient structure from displays such as dots. We take this position very seriously in this paper, and propose that the principles of organization within visual systems are applicable to the inference of structure for cluster analysis in dot patterns.

We concentrate on dot patterns in the plane, and note that there is an intuitive agreement with visual patterns. The dots could either organize into clusters, in which a group of dots would form a (2-dimensional) region, they could organize into (1-dimensional) curves, or they could be dispersed as a cloud of (0-dimensional) points. A more traditional image could be obtained by interpolating structure between the dots, the simplest version of which would be to place the point set on an image coordinate system and then blur it with, e.g., a Gaussian kernel. Two-dimensional clusters could then be defined as regions in the resulting image, 1-D clusters as curves, and clouds as noise.

We first began to explore the connection between visual algorithms and dot patterns about 15 years ago, and concentrated on 2-dimensional clusters. We proposed a heuristic system for the labeling of dot patterns ([21]), which is reviewed

1991 *Mathematics Subject Classification.* Primary 68T10.

Fellow, Canadian Institute for Advanced Research. Research sponsored by AFOSR, NSERC, and SERC. I thank my collaborators in various stages of this project: Allan Dobbins, Robert Hummel, Lee Iverson, and Chantal David. Demetri Terzopoulos assisted greatly with numerical issues.

1052-1798/95 $1.00 + $.25 per page

in the next section. Following this we shift the concentration to curves, and review a more recent attempt to model the biological processes of curve detection using computational and geometric techniques. To the extent that this is successful, it supports the proposition that, lacking more detailed models, cluster detection (in low dimensions) can benefit greatly from visual computations. Finally, however, we conclude with the caveat that visual perception implies certain constraints on embeddings and metrics, and that these will affect the clusters perceived in certain arrangements. Thus any attempt to explore the relationship between visual computations and arbitrary data clustering must proceed with the peculiarities of the visual system, and the visualworld in which it evolved, clearly in mind.

2. Labeling Dot Clusters

Imagine a scene consisting of a dark book lying on a light table. An image of this scene would contain a dark region within a light surround, and the outline of the book would be delimited by a set of "edge" pixels, or pixels right at the transition between light and dark. If the edge pixels had an orientation associated with them, then, locally, these could be interpreted as tangents, and an integral curve could be defined through them that would separate the figure (the book) from the ground (the table). (Dark) pixels interior to this curve would define the (image of the) book.

This leads to the model we proposed for labeling dot clusters ([21]). The goal was to label each dot with one of the following possibilities: as an $\mathcal{EDGE}$, with a certain orientation, as a point $\mathcal{INTERIOR}$ to the cluster, or as an isolated $\mathcal{NOISE}$ point. In notation: let $\{d_i, i = 1, 2, ..., n\}$ denote the n dots at coordinates (x_i, y_i), and let $\lambda \in \Lambda$ denote the possible labels:

$$\Lambda = \{\mathcal{INTERIOR}, \mathcal{EDGE}_{\theta_i}, i = 1, 2, \ldots, \Theta, \mathcal{NOISE}\}$$

where the orientation quantization was chosen with $\Theta = 8$.

Associated with each label λ was a measure of certainty, $p(\lambda)$, such that

$$0 \leq p(\lambda) \leq 1, \; \lambda \in \Lambda$$

and

$$\sum_{\lambda \in \Lambda} p(\lambda) = 1$$

for each dot.

2.1. A Relaxation Process for Dot Clusters. The goal is to select the proper label for each dot by maximizing a measure of certainty over all of them. This measure is based on the "compatibility" between the interpretations of pairs of dots. Letting $r_{i,j}(\lambda, \lambda')$ denote the compatibility between label λ on dot d_i and label λ' on dot d_j, they were set according to the following intuitions:

- $\mathcal{INTERIOR} - \mathcal{INTERIOR}$. The dominating term in this compatibility is that if the two dots are both interior to a cluster, then they should be close to each other; so the support function should be strong for nearby dots, and should drop off with distance. Denoting euclidean planar distance with d, the particular form chosen is:

$$r(\mathcal{INTERIOR}, \mathcal{INTERIOR}) = W_{II}(\exp^{-\alpha_{II}d} - C_{II})$$

The constants were: $W_{II} = 1.0$, $\alpha_{II} = 0.5$, so that dots located within two normalized distance units would exercise the stongest influence, and $C_{II} = -0.1$ for a small negative shift.

- $\mathcal{INTERIOR} - \mathcal{EDGE}$. The dominating feature of this compatibility is that INTERIOR points have to be "inside" $\mathcal{EDGE}$ points. We thus take the angle γ between the vector connecting the two points and the orientation of the $\mathcal{EDGE}$, resulting in:

$$r(\mathcal{INTERIOR}, \mathcal{EDGE}) = cos\gamma \exp^{-\alpha_{IE}d}.$$

Since $\mathcal{EDGE}$ points need not be as dense as INTERIORs, we take $\alpha_{IE} = 0.4$.

- $\mathcal{INTERIOR} - \mathcal{NOISE}$. $\mathcal{INTERIOR}$ points have the opposite relationship to $\mathcal{NOISE}$ points as they held to other $\mathcal{INTERIOR}$s; that is, the distance between $\mathcal{INTERIOR}$s and $\mathcal{NOISE}$ should be large:

$$r(\mathcal{INTERIOR}, \mathcal{NOISE}) = W_{IN}(\exp^{-\alpha_{IN}d} - C_{IN})$$

where $W_{IN} = -1.0, \alpha_{IN} = 0.4$, and $C_{IN} = 0.3$.

- $\mathcal{EDGE} - \mathcal{EDGE}$. This compatibility represents a Gestalt law of good continuation for the bounding contour surrounding the cluster; in early implementations it was set as in $\mathcal{INTERIOR} - \mathcal{EDGE}$ compatibilities; i.e., as the $\cos(\beta)$, where β was an appropriate angular form. Much more interesting methods for setting such compatibilities are discussed in the next section of this paper.

2.2. A Preliminary Relaxation Updating Process. Given an initial estimate of the measures $p_i^0(\lambda)$, indicating how strongly label λ is associated with dot (d_i), the p-measures were updated according to the following formulas:

$$p_i^{k+1}(\lambda) = \frac{p_i^k(\lambda)[1 + q_i^k(\lambda)]}{\sum_{\lambda \in \Lambda}[p_i^k(\lambda)[1 + q_i^k(\lambda)]]} \tag{1}$$

where

$$q_i^k(\lambda) = \sum_{j=1}^{n} \sum_{\lambda' \in \Lambda} r_{ij}(\lambda, \lambda') p_j^k(\lambda') \tag{2}$$

The numerator in eq. 1 basically combines information from a k-nearest neighborhood of dots, and attempts to maximize the consistency between them as a function of the support (q) from neighboring dots. The denominator is a renormalization term. We now understand the notions of support, consistency, and normalization in variational terms, and these are described in the next part of the paper (Sec. 3.2.4). Nevertheless, the above formulas approximate the correct ones throughout much of the space of possible labelings – see Hummel and Zucker, [12].

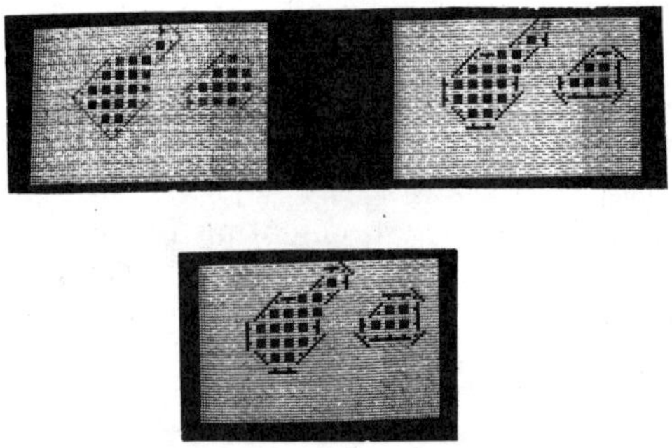

FIGURE 1. An illustration of the relaxation process for labeling dot clusters. INTERIOR labels are indicated as squares, and EDGE labels as short line segments with an indicator pointing toward the inside. The brightness of the labels indicates the value of p. Notice how the two clusters separate, and how a bounding contour is indicated around each, as the iterations proceed.

2.3. Results. An illustration of this dot labeling process is shown in Figure 1; additional examples can be found in [21] that illustrate its behaviour on non-linearly-separable clusters (intersecting "C"-shapes) and figure-"8" shaped clusters as well.

3. Labeling Curves in Images

In this section we focus on image curves, or 1-dimensional clusters of points. We partition the problem of curve detection into two distinct stages, because of the different nature of the computation for each stage.

The goal of our first stage is to infer a coarse distributed representation of the differential geometric structure of the curve at each point. For dot patterns these points might be separated somewhat, while for images the points should form a discretization of the trace of the curve. We call this representation a *discrete tangent field*, and it contains a quantized estimate of the tangent and curvature at each image point through which the curve passes. Orientation discontinuities are represented by multiple tangents at the same point, with an undefined value for curvature. The global curve is then inferred in our second stage by a process of dynamically fitting splines according to a potential distribution constructed from the discrete tangent field. This stage bears some relationship to "snakes", although the problems of initialization, length, and internal parameters are solved automatically as a function of the tangent field. The final result is a representation of each global curve by a smooth covering of second-order limited-length splines. For dot clustering, this second stage of interpolating a smooth curve through the trace points may not be necessary, although, in some cases, it could facilitate the matching to model predictions.

We explore the analogy between our model for curve detection and the neurobiology of primate visual systems. This complements the mathematical structure of the problem and is in concert with our theme that visual perception can provide a natural approach to dot clustering. Each aspect of the curve detection model reflects this. Both stages are parallel, and both model neural networks at the level of receptive fields and interactions between them. The first stage is metaphorically aimed at V1/V2, and, more speculatively, the second stage at V4/IT. But the flow of ideas is not only from biology to computation; it has already led to new possibilities in the other direction as well. The need to represent tangents and curvatures within the model has led to a new physiological hypothesis (that endstopped neurons can code orientation and curvature simultaneously) which in turn has led back to a new algorithm for estimating curvature from images. And the second stage of our model suggests how very large receptive fields (as are found in area IT) can be constructed in a way that still permits relatively fine sensitivities to orientation. The constructive potential field underlying this has fed back, in turn, to a computational understanding of how global entities can be inferred in parallel. The result, in computational terms, is an algorithm for inferring curves from images to subpixel accuracy.

We next present an overview of the full approach, followed by a more detailed development of the two stages in order. The relevant analysis applied to a fingerprint image is carried throughout.

3.1. Overview of the Curve Detection Algorithm. The algorithm consists of two distinct stages, the first leading to a coarse description of the local structure of curves, and the second to a much finer, global one. The various steps corresponding to the two stages are illustrated in Fig. 2.

3.1.1. *Stage 1: Inferring the tangent field.* Orientation selection is the inference of a local description of the curve everywhere along it. Formally this amounts to inferring the trace of the curve, or the set of points (in the image) through which the curve passes, its (approximate) tangent and curvature at those points, and their discontinuities [22]. We refer to such information as the tangent field, and note that, since the initial measurements are discrete, this will impose constraints on the (inferred) tangents, curvatures, and discontinuities.

This first stage of orientation selection is in turn modeled as a two step process:

- Stage 1.1. *Initial Measurement* of the local fit at each point to a model of orientation and curvature. A derivative of Gaussian approximation is used. This also models one of the structures involved in the neurobiology of edge and line detection—what are referred to as simple cell receptive fields in the visual cortex. Operators of different sizes, modeling receptive fields at multiple scales, are used. This array of information can be considered as analogous to the distribution of activity in a cortical hypercolumn. Implicit is considerable information about the local distribution of light contrast. The rate of change of tangent direction (or curvature) is made explicit by taking differences of the responses of smaller and larger receptive fields. This model of endstopped simple cells thereby encodes both the orientation and curvature estimate ([4]). This initial measurement information is illustrated in Fig. 2b. However, since the local measurements are inherently inaccurate, we require the next sub-stage:
- Stage 1.2. *Interpretation* into an explicit distributed representation of tangent and curvature at each point by establishing a consistent global interpretation of information gathered from the local measurements. Relaxation

labelling [12] provides the formal framework for this, and a discretization of the limiting process in differential geometry leads to a natural functional to guide the relaxation. This relaxation process places the earlier, more heuristic one on a more sound mathematical foundation. The result is a network which, contains two classes of structures: a range of smaller receptive fields for performing the initial measurements, and a range of larger projective fields which indicate the interaction terms between the smaller ones. We refer to the latter as curvature classes; they are illustrated in [23]. The result of the network on the fingerprint image is shown in Fig. 2c.

3.1.2. *Stage 2: Inferring a Covering of the Curve.* Since the tangent is the first derivative of a curve (with respect to arc length), the global curve can be recovered as an integral through the tangent field. Such a view typically leads to sequential recovery (numerical integration) algorithms. But such algorithms require starting points and some amount of topological structure (i.e., which tangent point follows which), and they violate our parallelism requirement. Instead we propose a different approach that is parallel.

The key idea is to recover the global curve by recovering a *covering* of it. Relevant global properties of the curve can then be obtained from the covering. The elements of the covering are unit-length splines, initially born at each (discrete) tangent location, but which then evolve according to a potential distribution constructed from the tangent field. The evolution takes two forms: (i) a migration in position to achieve smooth coverings at subpixel accuracy; and (ii) a "growth" to triple their initial length. Moreover, since our local splines are all born independently of one another, initially it is not known which go together into the covering of each distinct global curve. We model this by creating each one with a different "color", and include a second process by which splines covering the same position become the same color. In the end, then, the cover is given by a collection of overlapping splines, or short "snakes", each of which is the same color.

These dynamic spline fields are thus a generalization of"snakes", or fitting splines according to a minimization principal [13] (for an earlier attempt, see [10]), and we utilize the snake equations directly. There are key differences in their application, however: (i) our scheme is parallel, so questions about initial length, starting points, etc., never arise; and (ii) our scheme is constructive, with the potential field computed directly from the tangent field, so questions about features, degree of image blurring, etc., never arise. Imposing the tangent field as a structure intermediate between the image and the "snakes" was the key; previous approaches attempted to match snakes against the image directly, thereby attempting to fill too large a descriptive "gulf" in a single step. They ([13]) were forced to have a user solve these problems interactively.

Again, there are two conceptually distinct steps to Stage 2 of the algorithm:

- Stage 2.1. *Constructing the Potential Field* from the discrete tangent field. Recall that, in the network that computes the tangent field, the curvature classes define the spatial structure for interaction. They specify, for each tangent hypothesis, which other tangent hypotheses support it (under a particular curvature hypothesis). Under the straight curvature assumption, for example, the supporting tangents would be roughly co-linear with the orientation of the tangent being supported. Combining these spatial interactions with the individual tangent receptive fields, the result is that curvature classes can also be viewed as large projective fields; in the straight case under consideration, the curvature-class projective field would essentially look like (the positive part of) a large simple cell. The stage 2 potential field is constructed from precisely these curvature-class projective fields: simply take the field corresponding to each entry in the discrete tangent field, and add them together point-wise in space. The result is the global potential field illustrated in Fig. 2d for the fingerprint image.

- Stage 2.2. *Colored Spline Dynamics*
  Given the global potential field, the unit splines are born in the valleys corresponding to the discrete tangent positions. They evolve according to a variational scheme that depends on spline properties (tension and rigidity) as well as the global potential field. Finally, there is an inter-snake potential, which draws nearby snakes even closer together (and adjusts their color). The final positions of the snakes are shown in Fig. 2e.

3.2. Orientation Selection: Inferring the tangent field. The goal of this first stage is the discrete tangent field, or a coarsely quantized representation of the trace, tangents, curvatures, and (trace and orientation) discontinuities. The key issues for specifying the orientation selection network are (i) the representation, (ii) the nature of the initial measurements, and (iii) the structure of the relaxation networks which maximize consistency. We discuss each in turn.

3.2.1. *Representation.* Consider an approximation of an arbitrary image into a (2 x 2) binary array: only 3 distinct orientations are possible for a curve that passes through the south-west pixel. The lesson is that image quantization imposes a quantization on tangents, which in turn imposes a quantization on curvatures, etc.

Our representation is organized retinotopically, with a matrix of hypotheses about orientation and curvature associated with each pixel. Quantizing orientation into n discrete values and curvature into m, we thus have an $n \times m$ matrix of labels about the existence of a curve with the given orientation and curvature passing through the point. A copy of this matrix is associated with each image point.

With the proposed discretization, there is a set of n orientation labels represented by
$\lambda \in \{0, \ldots, n-1\}$, and Θ_λ, the orientation of the label λ, is given by $\Theta_\lambda = \epsilon\lambda$, where $\epsilon = \pi/n$. (Symmetry dictates only a half plane is necessarily represented.) The natural range for curvatures is limited to a given minimum radius r_{min} ($\kappa_{max} = 1/r_{min}$) which is related to the discretization of the plane. Adopting the convention that negative curvatures are confined to the left half-plane (associated with a particular orientation), a similar discretization for curvature is as follows. The label is $\kappa \in \{0, \ldots, m-1\}$, and K_κ, the curvature of the label κ, is given by $K_\kappa = -\kappa_{max} + \kappa\Delta\kappa$, where $\Delta\kappa = 2\kappa_{max}/m + 1$.

Thus building on the notation developed in Sec. 1 of this paper, we have $p_i(\lambda_i, \kappa_i)$ representing the certainty of a particular label pair at the pixel (x_i, y_i), or the confidence in the hypothesis about the existence of a curve with orientation Θ_{λ_i} and curvature K_{κ_i} passing through point i. The $p_i(\lambda_i, \kappa_i)$ are initially estimated from the image and, as in Sec. 1, they are updated based on their consistency with nearby estimates.

3.2.2. *Initial Measurements of Position, Orientation, and Curvature.* The initial values for $p_i(\lambda_i, \kappa_i)$ are obtained from measurements on the image. Such measurements are usually formulated as convolutions of operators against the image, as when line detectors, or operators with simple-cell-like organizations, are taken as image templates for unit tangents. But these often give spurious values.

The model of simple cell RFs that we use to obtain initial tangent estimates is a derivative of Gaussian (e.g. D_xG - odd-symmetric or $D_{xx}G$ - even-symmetric) which is split into a small number of subregions (2 to 5) along the direction of preferred orientation (normal to the derivative direction); and similarly in the orthogonal direction. We have developed a calculus for such operators, called Logical/Linear operators, and, for a full discussion, see [8]. The contrast sign nonlinearity makes the response of the operator considerably more specific, although still not uniquely specific, to the tangent of a smooth curve. It also has fascinating physiological implications; see [6].

Measuring Curvature as a "Difference" of Tangents. Estimating curvature poses a separate problem, and again a solution was obtained by appealing to the physiology. A common property of visual cortical neurons is endstopping or selectivity for the length of an oriented stimulus. In Dobbins *et al.* [4] we develop a computational model for endstopping, and show how it leads to quantitative predictions (now verified) about the response of endstopped simple (ES) neurons to curved stimuli. It amounts to a "non-linear difference" between simple cell responses, and provides the (non-linear) operators for our initial curvature measurements.

The model is based on the observation that there are simple cells whose receptive field size differs as a function of cortical layer. In particular, Layer VI of cat primary visual cortex contains cells with receptive fields notably longer than those in the laminae above it [1]. Now, if we let R_S denote the response of a short simple cell, and R_L denote the response of a long simple cell, with receptive fields sharing orientation preference and centered at the same retinal location, then their "difference" models the endstop property (the response stops for stimuli exceeding some length). But more importantly, the response of such an operator varies systematically with curvature. More precisely, letting $\phi(\cdot)$ denote a rectifying function (equal to its argument when positive and zero otherwise), the response of an endstopped simple cell (R_{ES}) is modeled by:

$$R_{ES} = \phi(c_S \cdot \phi(R_S) - c_L \cdot \phi(R_L))$$

where c_S and c_L are positive constants that normalize the area difference between the receptive fields. By varying the model parameters one can vary the curvature response of model instances. For example the symmetry of the components determines the curvature sign selectivity, and the relative size and gain determine the preferred curvature and curvature tuning breadth. This is described at length elsewhere [5]

At each of eight orientations a small number of simple instances are combined to define 5 discrete curvature classes — two on either side of the zero curvature class. The four curved classes are obtained from endstopped instances and the zero curvature estimate from a nonendstopped simple instance. Typical results above noise are shown in Fig. 2b; although they convey a rough idea of what the curve structure is, there are both responses where there is no curve, and ambiguous (multi-valued) responses where there is a single curve. We contend that no local operator can solve these problems in general, and further that a spatially-interactive process can. In the next sub-section we sketch a computational framework in which such a process can be posed, and follow this with a derivation of a natural functional whose minimization can guide it.

3.2.3. *Orientation Selection Network.* Given the initial measurements, the relaxation system effectively imposes a coarse second-order model for the curve around each putative tangent, and then selects those that minimize a certain functional over all responses. For completeness, in the next sub-section we sketch the computational framework of relaxation labeling within which the process is posed, and follow this with a derivation of the functional through a discretization of the differential geometry. The result is a network that models something like the interactions between endstopped and nonendstopped simple cells in the visual cortex.

3.2.4. *Relaxation Labeling.* The abstract structure of a relaxation labeling process builds upon that introduced in Sec. 1. Let $i = (x_i, y_i) \in I$ denote discrete coordinate positions in the image I, and let $\lambda \in \Lambda_i$ denote the set of labels at position i. (In the model of orientation selection, think of the distinct labels as indicating distinct orientations and curvatures.) The labels at each position are ordered according to the measure $p_i(\lambda)$ such that $0 \leq p_i(\lambda) \leq 1$ and $\sum_{\lambda \in \Lambda_i} p_i(\lambda) = 1 \quad \forall i$. Compatibility functions $r_{i,j}(\lambda, \lambda')$ are defined between label λ at position i and label λ' at position j such that increasingly positive values represent stronger compatibility. The network structure derives from the support S that label λ obtains from the labeling on it's neighbors $Neigh(i)$; in symbols,

$$S_i(\lambda) = \sum_{j \in Neigh(i)} \sum_{\lambda' \in \Lambda_j} r_{i,j}(\lambda, \lambda') p_j(\lambda').$$

The final labeling is selected such that it maximizes the average local support (for symmetric compatibilities):

$$A(p) = \sum_{i \in I} S_i(\lambda) p_i(\lambda).$$

Such a labeling is said to be consistent [12] An iterative, gradient ascent algorithm for achieving consistent labelings is presented in [17]. We now simply remark that such "computational energy" forms have become common in neural networks, and observe that there is another interpretation in game theoretic terms (in which the consistent points become Nash equilibria); see [9]. We now specify the compatibilities $r_{i,j}(\lambda, \lambda')$—in short, the functional $A(p)$—for orientation selection.
Compatibilities and Co-circularity. Curvature can be viewed as a relationship between neighboring tangents along a curve, and hence provides the mathematical foundation for determining compatibility relationships between labels. The more common view of curvature is via the osculating circle, or the circle that just "kisses" the curve at 3 points. Both, of course, are equivalent in the limit as the 3 points approach one another; observe that the tangents have distinct position and orientation constraints for a given curvature. But evaluating such a limiting process on a

discrete grid is impossible, and Iverson, in his thesis, has developed a minimization approach. For early ideas, see [16], who first defined the notion of co-circularity, a generalization of co-linearity to circles.

The derivation starts with Fig. 3. First moving point i to the origin and rotating by $-\theta_{\lambda_i}$, the vector (x_j, y_j) is the relative displacement between the two tangents. Therefore

$$\begin{aligned} \theta_{ij} &= atan_2(y_j, x_j) \\ d_{ij} &= \sqrt{x_j^2 + y_j^2} \end{aligned}$$

are the relative angle and distance respectively. To derive the geometric constraints, we start with the notion of *co-circularity*, the relationship that holds between two nearby tangents of the same osculating circle. Given the relative angle and distance, we can find the center of curvature and thus the (identical) curvatures κ_i and κ_j. The curvature and center of curvature restrict the orientation allowed at point j. These observations lead to the three constraining equations:

$$\begin{aligned} \kappa_i &= 2\sin\theta_{ij}/d_{ij} \\ \theta_j &= atan_2(x_j, 1/\kappa_i - y_j) \\ \kappa_j &= \kappa_i \end{aligned}$$

which must all be satisfied for the tangents to be cocircular.

Another way to interpret these equations is as a model for parallel transport of a tangent along an osculating circle so that its value can be properly compared with tangents at the new position. These geometric properties hold exactly only when the variables are continuous, but the representation for orientation selection is based on discrete values (each label represents a quantized sample point). Thus the problem is to find the discrete counterparts to the continuous cocircularity constraints; these then become the compatibility relationships between labels. The geometric constraints define a *constraint* manifold $\mathcal{C}$ in the 6-dimensional continuous variable space $(x_j, y_j, \theta_i, \kappa_i, \theta_j, \kappa_j)$. Since cocircularity can be defined in terms of three independent constraining equations, constraint manifold of valid solutions is a three-dimensional submanifold in the six-dimensional space. Now, the discrete values for the variables (X_j, say) lie at regular intervals in the continuous variable space, which the constraint manifold $\mathcal{C}$ rarely intersects (since the codimension of the embedding is 3). We take the distance d from the fixed point to the consistency manifold $\mathcal{C}$ to be inversely related to the degree with which the associated label set conforms to the constraints implied by the defining equations. Therefore, determining the compatibility of a pair of nearby labels can be reduced to finding the minimum distance from the fixed point associated with the label set X to the constraining manifold $\mathcal{C}$. This minimal distance can be computed using standard numerical minimization techniques.

Relaxation Implementation. Given the compatibilities between labels, we now describe a particular implementation of the relaxation process in which each label λ competes only against its negation $\neg\lambda$, and not directly against each other. (The inter-label competition is actually implicit in the compatibility functions.) The advantages of this scheme are that (i) multiple labels can survive at a single position, given that they are both supported; and (ii) only the "positive support" for each label needs to be represented (since $p(\lambda) + p(\neg\lambda) = 1$). Hence this implementation results in a *relaxation multi-graph* [16], or a parallel collection of inter-related relaxation graphs.

More formally, let ρ_i denote a special label multi-index:

$$\rho_i = \{(i, \lambda_i, k_i | i \in I, \lambda_i \in \Lambda, k \in K\}$$

that selects each (tangent, orientation) pair at position (x_i, y_i). Then the support functional is given by:

$$S_i(\rho_i) = \sum_{j \in Neigh(i)} r_{i,j}(\rho_i, \rho_j) p_j(\rho_j)$$

and the relaxation process maximizes the average local potential:

$$A(p) = \sum_{\substack{i \in I \\ j \in Neigh(i)}} r_{i,j}(\rho_i, \rho_j) p_i(\rho_i) p_j(\rho_j).$$

A result of the network running is shown in Fig. 2c.

3.2.5. *Colored Snake Dynamics: Inferring global coverings.* The tangent field, or the first stable representation of differential structure, has now been computed. The next stage is to find the global curves through it. In this section we present a new approach to inferring such curves, which is based on a notion of energy-minimizing spline fitting. As such, it builds upon the work of Kass, Terzopoulos, and Witkin ([13]), and specifically follows the treatment in [2]. We shall use the following expressions in the development: our goal is to recover a covering of the "global curves" through the tangent field; we shall refer to the elements of this cover as "splines", each of which evolves as a "classical deformable curve". We next review the classical dynamics of deformable curves, and, following this, we specify how to obtain the various parameters and potential fields.

3.2.6. *Dynamics of the deformable Spline.* Let $v(s,t) = (x(s,t), y(s,t))$ be a deformable curve, with $s \in \Omega = [0,1]$. The kinetic energy functional $T(v)$ is defined as

$$T(v) = \frac{1}{2} \int_\Omega \mu |v_t|^2 ds, \tag{1}$$

where μ is the (constant) mass density, and the potential energy functional $U(v)$ is defined as

$$U(v) = \frac{1}{2} \int_\Omega \omega_1(s) |v_s|^2 + \omega_2(s) |v_{ss}|^2 + I(v) + S(v) ds. \tag{2}$$

where

- - $\omega_1(s) x_s$ controls the tension of the curve;
- - $\omega_2(s) x_{ss}$ controls the rigidity of the curve;
- - $I(v)$ is the potential coming from the tangent field information (position, orientation, and curvature)
- - $S(v)$ is a force between neighboring curves to attract them toward one another when they get very close.

According to the principle of least action, the motion of the deformable curve during the time interval $[t_0, t_1]$ is described by those functions $x(s,t), y(s,t)$ for which the integral

$$\int_{t_0}^{t_1} L(v)dt \tag{3}$$

of the Lagrangian $L(v) = T(v) - U(v)$ is a minimum. ¿From the calculus of variations it is known that any extrema of $\int_{t_0}^{t_1} L(v)dt$ must satisfy the generalized Euler equations (or, in this particular case, the Lagrange equations of motion). Setting $f(v) = \frac{1}{2}(\mu|v_t|^2 - \omega_1(s)|v_s|^2 - \omega_2(s)|v_{ss}|^2 - I(v) - S(v))$, the Euler-Lagrange equations of motion become, after adding a dissipation term and simplifying:

$$\begin{aligned}\mu x_{tt} + \gamma x_t - \frac{\partial}{\partial s}(\omega_1(s)x_s) + \frac{\partial^2}{\partial s^2}(\omega_2(s)x_{ss}) &= -\tfrac{1}{2}I_x(v) - \tfrac{1}{2}S_x(v)\\ \mu y_{tt} + \gamma y_t - \frac{\partial}{\partial s}(\omega_1(s)y_s) + \frac{\partial^2}{\partial s^2}(\omega_2(s)y_{ss}) &= -\tfrac{1}{2}I_y(v) - \tfrac{1}{2}S_y(v).\end{aligned}$$

Determining Potentials and Parameters. Using the information coming from the first part of the process, it is possible to determine the model's parameters, and then use the deformable curves in a fully automated system (i.e. without user interaction). Recall that the potential energy function of a deformable curve $v(s,t)$ is

$$U(v) = \int_{\Omega} \omega_1(s)x_s + \omega_2(s)x_{ss} + I(v) + S(v)ds$$

where $\omega_1(s)x_s$ controls the tension, $\omega_2(s)x_{ss}$ controls the rigidity, $I(v)$ is the potential coming from the tangent field, and $S(v)$ is the attractive force between overlapping curves. We deal with each term in turn.

Tangent Field Potential $I(v)$. The entries in the tangent field are given by those hypotheses whose certainty does not drive to zero; we shall denote them by the sparse array:

$$T(x,y) = \begin{cases}\Theta_\lambda, & \text{if } p_i(\lambda,\kappa) > 0;\\ 0, & \text{otherwise.}\end{cases}$$

where Θ_λ denotes the orientation of the λth tangent label. Values for the curvature are available as well. Now, given the tangent field $T(x,y)$, a reliable image potential can be constructed by assuming that each tangent (and curvature) represents a rough hypothesis about what the curve is doing in the neighborhood of (x,y). Since this hypothesis is built up through estimates of where the tangents are and how they interact with their neighbors, it is only natural to take the curvature-classes, or the domain of interactions, as the primitives. For each entry in the tangent field, then, we take an elongated Gaussian approximation to the tangent centered at the appropriate position and orientation; an example of which is shown [2]. The total field is simply the pointwise addition of each of these local component fields. Observe how the overlap between these components smoothes the "jaggies" imposed by the initial image sampling, and permits sub-pixel accuracy. These large constructed tangent field potential distributions may bear some resemblance to the large receptive fields observed in areas V4/IT [3], and perhaps they should be modifiable by "top-down" attentional influences as appears to be the case in IT (as shown by Moran and Desimone [15]).

Inter-Spline potential $S(v)$. If we were working with continuous and exact data, all curves should overlap exactly to give the exact curves. To compensate for all the inherent inaccuracies (numerical quantization, noise, errors in the first part of the process), and to accelerate convergence, an *inter-spline force* is added, such that nearby curves will tend to glue together. This force has shorter range, and smaller amplitude, than the tangent field potential. It depends on the splines, and varies when they vary, unlike the fixed tangent field potential. It also drives the initially different colors toward an equilibrium color when the splines overlap.

Initial Spline Position and Length. The initial estimated positions for the dynamic splines come from the tangent field. Although these are reliable, they are coarse and may not overlap because of noise, image quantization, or arc length problems. We capitalize on this reliability by creating, at each tangent $T(x, y)$, a small spline of length 1 oriented according to the tangent and curvature estimates. By construction, we know that this spline will be born in a valley of the tangent field potential distribution, so the ambiguities inherent in local minima problems are avoided by construction. The initial splines are then permitted to migrate, to smooth out the curve and find the true local minima. But the inference of a cover for the global curves requires that the splines overlap, so that each point on the curve is covered by at least one spline. We therefore let the splines extend in length while they migrate in position, until they each reach a prescribed length $L_p = 3$. The covering is thus composed of these *extensible splines* which have grown in the valleys of the tangent field potential.

Tension and Rigidity. It is a natural choice to relate the tension parameter $\omega_1(s)$ to length, and the rigidity parameter $\omega_2(s)$ to curvature. In our model, the unit length initial curves increase to a prescribed length L_p such that they can join together. This result can be achieved by setting $\omega_1(s)$ at each time t as a function of the difference between the actual curve length L_v^t and the prescribed length, i.e.

$$\omega_1^t = f(L_v^t - L_p)$$

where

$$L_v^t = \int_\Omega \sqrt{(x_s^t)^2 + (y_s^t)^2}\,ds.$$

The rigidity term, ω_2, is naturally related to curvature, and we set it according to the estimate of curvature available from the tangent field. Straighter estimates are more rigid, and the highest curved ones less so.

Dynamic Splines: Results. An example of the dynamic splines in motion is shown in [2]. Initially they are born at the positions dictated by the tangent field, with unit length and orientation identical to the coarse tangents. With iteration they seek the valleys of the potential distribution (subject to their internal continuity constraints), and grow in length (to $L_p = 3$) to fill in short gaps and arc-length variation within the quantized grid. The result is shown in Fig. 2(d).

4. Sparse Clusters

The previous sections dealt with a biologically motivated model for the clustering of points that fall along curve-like sets; i.e., which form one-dimensional clusters in the plane. The density of points within these sets was further assumed to be sufficently high that an initial stage of logical/linear "convolutions" could provide initial estimates of tangent and curvature. Finally, the tangent field produced by this initial stage was assumed sufficently dense to support the development of a potential distribution to drive the spline coverings. Clearly, such algorithms are restricted in application to rather special sets.

The more general problem of cluster detection involves grouping collections of points distributed over two dimensional sets. Again, density matters, because, if the set is sufficently dense, then a process of "edge detection" (as in Sec. 2) could be developed in the same detail as the curve detection process in Sec. 3. In particular, the initial logical/linear operators would need to be odd-symmetric, and distributed around 2π radians so that a "figure" side could be distinguished from a "background" side. Otherwise, the theory would be rather similar and has been developed [8]. The remaining important difference is that integral curves are formed through the tangent field keeping the sense of the local edge tangents consistent.

Sets of points distributed too sparsely to support this differential-geometric approach pose a more difficult problem. In the Introduction we suggesed that Gaussian convolutions provide a way to "blur" the dots into image-like regions. This was intended mainly as an introductory metaphor, and not a serious clustering algorithm. The Gestalt psychologists have studied the perception of grouping in dot patterns, and have suggested the following heuristic principles of perceptual grouping [20]. All other aspects of the simulus being equal, group those elements of a pattern together that are:

- spatially proximal;
- temporally proximal;
- geometrically similar;
- give rise to pattern continuity;
- result in closure;
- uniform density;
- common fate;
- symmetric;
- exhibit a common orientation
- maximize *pragnanz*: In summary, those elements that optimize an intuitive measure of "figural goodness".

The first, fifth, sixth, and eigth principles are most appropriate for sparse patterns; orientation and good continuation taken together are the principles that most concisely summarize the previous section on curve detection. Clearly, however, much remains to be done before algorithms are developed that successfully incorporate these principles into more general grouping algorithms.

We end this section with a cautionary remark. Although the bulk of this paper was aimed at supporting the argument that visual algorithms are applicable to determining clusters in planar point sets, we now caution the reader that implicit in this position are a number of assumptions about distances, metrics, and embeddings. As a result, specific advantages for vision, e.g., the proximity principle in Gestalt grouping, may or may not be relevant for clustering. As illustrated in Fig. 4, according to this principle slight changes in the embedding can lead to large changes in the apparent clusters. Whether such sensitivity questions arise in the general situation remains to be determined, as the example shown here has a very particular structure. How such situations should be characterized (for cluster ananlysis), and whether embeddings should be developed that avoid such sensitivity problems, is an open question.

5. Conclusions

In this paper we explored the hypothesis that dot clustering is naturally related to the visual perception of dot patterns. It follows that dot clustering algorithms should therefore be related to visual computations, which are also partitioned into different classes. Dot clusters may be sparse or dense, and they may align along one dimension or two. For dense patterns aligning along one dimension the visual computation is curve detection, and much of the paper described one such algorithm that is biologically motivated.

The algorithm for inferring curves from images was based on a particular intermediate representation—the discrete tangent field—which contains a representation of the trace of the curve, or the set of points through which it passes, together with a coarse estimate of the tangent and curvature at those points. Orientation discontinuities appear as multiple tangents at a position, with no curvature estimate. A relaxation algorithm demonstrated for computing the tangent field, based on minimizing a functional of position, tangent, and curvature. The second stage of the algorithm synthesizes a covering of the global curves through the tangent field. Seeking a covering rather than a single global curve fit leads us to a new parallel algorithm for dynamically moving splines, or "snakes". The splines are initially born with orientations at positions dictated by the tangent field, and migrate according to a potential function computed from the tangent field. As they migrate, the individual splines increase in length and, when they overlap, equilibrate to the same identifying "color". There is no danger of smoothing over discontinuities, because these are given by the tangent field, nor are there problems of convergence, because the initial splines are born in the valleys of the potential field. The final covering for each global curve is then given by the family of splines with the same "color".

Throughout the development, mathematical constraints have been supplemented by ideas from physiology. This was of interest computationally, in that physiological ideas suggested new ways in which curvature could be computed and resulted in an entirely parallel system, but also physiologically, in that computational ideas suggested novel interpretations of endstopped receptive fields and higher-order (e.g., V4/IT) visual function.

Sparse patterns are more difficult to treat from a biological perspective than curves. Little is known about their representation in the brain, although much experience has been gathered since the pioneering experiments of the Gestalt psychologists. Perhaps the informal principles of perceptual grouping that emerged from the Gestalt movement will provide further insight into formal methods for the grouping of sparse dot patterns.

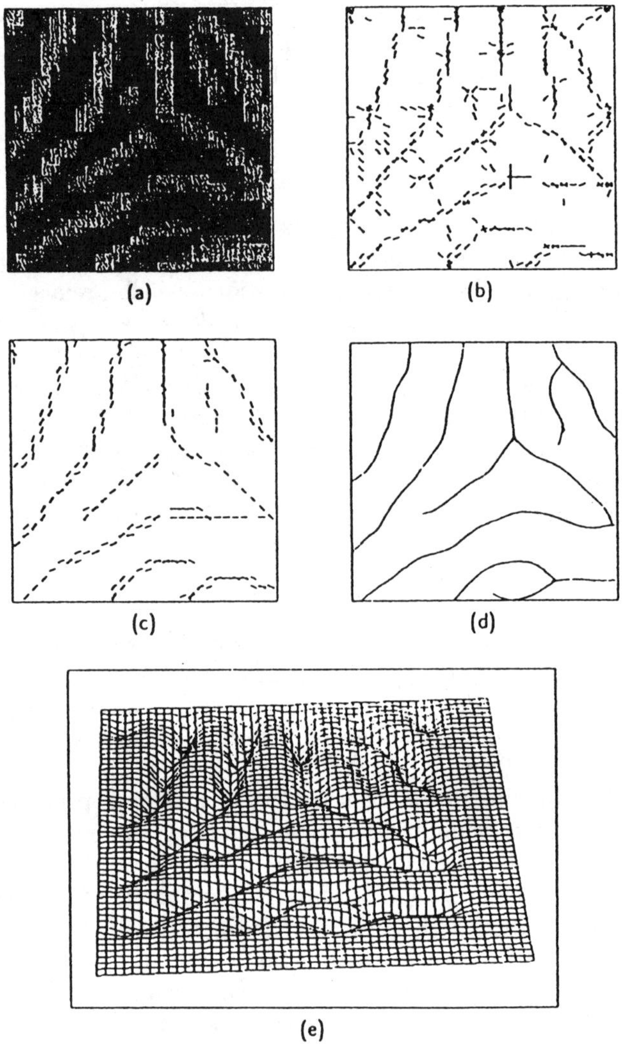

FIGURE 2. The different stages of the curve detection algorithm, illustrated on a portion of a fingerprint image (a). In (b) the initial estimate of the tangent field is shown, as obtained from Logical/Linear operators. In (c) the tangent field has been refined by the co-circularity-based relaxation process; (d) are the coverings of the global curves, while (e) is the potential distribution. Each of these stages is described more fully in the text. (After [23]).

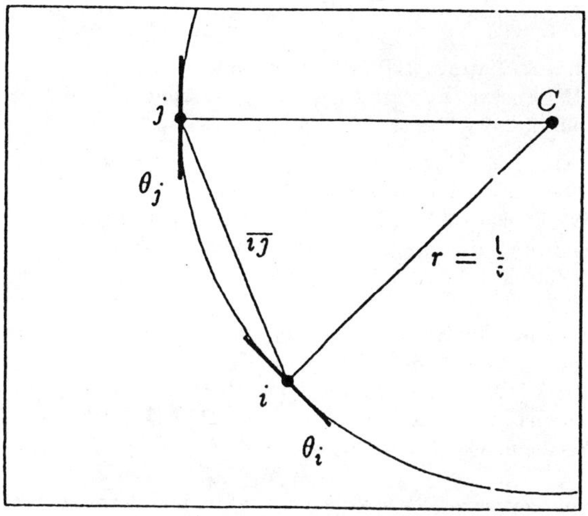

FIGURE 3. The geometric relationships necessary for defining the compatibilities between two label pairs at points i and j.

FIGURE 4. An illustration of the Gestalt grouping principle by which closer dots form clusters. If you imagine moving the even-numbered dots (counting from the left) toward the right, at a sudden point the clusters will shift. Visual displays in which the spacing is close to this shift point will be highly unstable, a property that clustering algorithms should perhaps avoid.

REFERENCES

1. J. Bolz, and C. Gilbert, Nature, **320** (1986), 362-365.
2. C. David, and S. W. Zucker, *Potentials, valleys, and dynamic global coverings*, Int. J. Computer Vision, **5** (1990), 219 - 238.
3. R. Desimone, S. Schein, J. Moran, and L. Ungerleider, *Contour, color, and shape analysis beyond the striate cortex*, Vision Research, **25** (1985), 441 - 452.
4. A. Dobbins, S. W. Zucker, S.W., M. S. Cynader, *Endstopping in the visual cortex as a substrate for calculating curvature.* Nature, **329** (1987), 438-441.
5. A. Dobbins, S. W. Zucker, S.W., M. S. Cynader, *Endstopping and curvature*, Vision Research, **29** 1989, 1371 - 1387.
6. P. Hammond, and D. M. MacKay, *Influence of luminance gradient reversal on simple cells in feline striate cortex*, J. Physiol. **337** (1983), 69-87.
7. R. A. Hummel, and S. W. Zucker, *On the foundations of relaxation labeling processes IEEE Trans. Pattern Analysis and Machine Intelligence*, **PAMI-5** 1983, 267-287; reprinted in Fischler, M., and Firschein, O., *Readings in Computer Vision*, Morgan Kaufman, Palo Alto, 1987.
8. L. Iverson, and S. W. Zucker, *Logical/linear operators for image curves*, IEEE Trans. Pattern Analysis and Machine Intelligence, in press.
9. D. Miller, and S. W. Zucker, *Efficent simplex-like methods for equilibria of nonsymmetric analog networks*, Neural Computation, **4** (1992), 167 - 190.
10. U. Montanari, *On the optimum detection of curves in noisy pictures*, CACM, **14** 1971, 335 - 345.
11. J. J. Nelson, B. J. Frost, *Intracortical facilitation among co-oriented, co-axially aligned simple cells in cat striate cortex* Exp. Br. Res., **61** (1985), 54-61.
12. R. Hummel, and S. W. Zucker, *On the foundations of relaxation labelling processes*, IEEE Trans. PAMI, **5** (1983), 267–287.
13. M. Kass, A. Witkin, and D. Terzopoulos, *SNAKES: active contour models*, Int. J. Computer Vision, **1** (1988), 321 - 332.
14. J. Maunsell,and W. Newsome, *Visual processing in monkey extrastriate cortex*, Ann. Rev. Neuroscience, **10** (1987), 363 - 401.
15. J. Moran, and R. Desimone, *Selective attention gates visual processing in the extriastriate cortex*, Science, **229** (1985), 782 - 784.
16. P. Parent, and S. W. Zucker, *Trace inference, curvature consistency, and curve detection*, IEEE Trans. Pattern Analysis and Machine Intelligence, **11** (1989), 823 - 839.
17. P. Parent, and S. W. Zucker, *Radial projection: An efficent update rule for relaxation labeling*, IEEE Trans. Pattern Analysis and Machine Intelligence, **11** (1989), 886 - 889.
18. F. Schmitt, and F. Worden, (eds.), *The Neurosciences: Fourth Study Program,* MIT Press, Cambridge, 1979.
19. J. Tukey, *Exploratory Data Analysis*, Addison Wesley, New York, 1972.
20. Wertheimer, M., *Laws of organization in perceptual forms,* Psych. Forsch.,4, 301-350 (1923); trans. in Ellis, W., A Source Book of Gestalt Psychology, Routledge and Kegan Paul, London, 71-88 (1938).
21. S. W. Zucker and R. A. Hummel, *Toward a low-level description of dot clusters: Labeling edge, interior and noise points*, Computer Graphics and Image Processing, **9** (1979), 213 - 233.
22. S. W. Zucker, C. David, A. Dobbins, and L. Iverson, *The organization of curve detection: Coarse tangent fields and fine spline coverings*, Proc. 2nd Int. Conf. on Computer Vision, Tarpon Springs, Fla, 1988.
23. S. W. Zucker, A. Dobbins, and L. Iverson, L., *Two stages of curve detection suggest two styles of visual computation*, Neural Computation, **1** (1989), 68 - 81.

COMPUTER VISION AND ROBOTICS LABORATORY, CENTRE FOR INTELLIGENT MACHINES, MCGILL UNIVERSITY, MONTRÉAL, QUÉBEC, CANADA H3A 2A7, AND, ISAAC NEWTON INSTITUTE FOR MATHEMATICAL SCIENCES, UNIVERSITY OF CAMBRIDGE, CAMBRIDGE, CB3 0EH, U.K.